# FOODS
# EXPERIMENTAL
# PERSPECTIVES

Seventh Edition

**Margaret McWilliams**
Ph.D., R.D., Professor Emerita
California State University, Los Angeles

**Prentice Hall**

Boston   Columbus   Indianapolis   New York   San Francisco   Upper Saddle River
Amsterdam   Cape Town   Dubai   London   Madrid   Milan   Munich   Paris   Montreal   Toronto
Delhi   Mexico City   Sao Paulo   Sydney   Hong Kong   Seoul   Singapore   Taipei   Tokyo

**Editorial Director:** Vernon R. Anthony
**Acquisitions Editor:** William Lawrensen
**Editorial Assistant:** Lara Dimmick
**Director of Marketing:** David Gesell
**Marketing Manager:** Thomas Hayward
**Senior Marketing Coordinator:** Alicia Wozniak
**Marketing Assistant:** Les Roberts
**Project Manager:** Holly Shufeldt
**Art Director:** Jayne Conte
**Cover Designer:** Bruce Kenselaar
**Cover Image:** Shutterstock
**Full-Service Project Management and Composition:** Integra Software Services Pvt. Ltd.
**Printer/Binder:** Edwards Brothers Malloy
**Cover Printer:** Lehigh-Phoenix Color
**Text Font:** Garamond

Credits and acknowledgments borrowed from other sources and reproduced, with permission, in this textbook appear on the appropriate page within the text. Unless otherwise stated, all artwork has been provided by the author.

**Library of Congress Cataloging-in-Publication Data**
McWilliams, Margaret.
  Foods : experimental perspectives/Margaret McWilliams.—7th ed.
    p. cm.
  Includes bibliographical references and index.
  ISBN-13: 978-0-13-707929-2 (alk. paper)
  ISBN-10: 0-13-707929-X (alk. paper)
  1. Food—Composition.  2. Food—Analysis.  3. Cooking.  I. Title.
TX531.M38 2012
664'.07—dc22

                        2010035636


10 9 8 7 6

**Prentice Hall**
is an imprint of

www.pearsonhighered.com

ISBN 10:    0-13-707929-X
ISBN 13: 978-0-13-707929-2

*To my food science and dietetic students
who made my teaching career such a
pleasure and intellectual challenge!*

# Brief Contents

# Contents

## Chapter 10    Vegetables and Fruits                          195

## PART VI—FOOD SUPPLY PERSPECTIVES                          455

### Chapter 19    Food Safety Concerns and Controls          457

*Note*: Every effort has been made to provide accurate and current Internet information in this book. However, the Internet and information on it are constantly changing, so it is inevitable that some of the Internet addresses listed in this textbook will change.

# Preface

Never before have so many people been focused on food as much as they are today. Television and the Internet have done much to create the current scene. What Julia Child started when she taught America how to cook the French way has now grown into multiple cooking shows; and an entire television network. News programs have fostered interest by keeping the public posted on food recalls and related issues, as well as commenting on the connection between obesity, health, and eating. The heightened interest has reached the point where shoppers can be seen reading labels in grocery stores and restaurants. The food industry is in the limelight much of the time, and the federal government keeps adding to these complex issues.

Careers in this realm require a depth of understanding far beyond what can be seen on television. This book is designed to help readers broaden and strengthen their scientific knowledge of food and its safe preparation. Part I surveys the consumer marketplace and also explores opportunities in food-related careers. Then, it examines research basics, including sensory and objective evaluation. Part II focuses on the physical aspects of food and its preparation. Parts III, IV, and V discuss carbohydrates (sugars, starch, vegetables, and fruits), lipids, and proteins (meat, eggs, milk, and baked products). Part VI includes food safety, preservation, and additives. Laboratory experiments to augment your study are presented in my *Experimental Foods Laboratory Manual*, eighth edition, also published by Prentice Hall.

"Food for Thought" and other boxes highlighting unique ingredients are features of this new edition that broaden this look at the world of food. Numerous pictures, objectives, margin notes and definitions, summary charts, recommended Web sites, and expanded study questions are designed to enhance learning.

This revision has been updated to include new information pertinent to each chapter, ranging from the food scene and marketplace to the spectrum of foods and the science that forms the basis for their handling and preparation. Particularly extensive revisions occur in Chapters 1, 10, 15, and 19.

Space limitations make it impossible to include background information that usually is included in an introductory class. You will find this material in my book *Food Fundamentals*, also published by Prentice Hall.

You have chosen an important field. I hope that you will enjoy your study of food and that it will serve you well as you proceed in your career.

# Acknowledgments

Once again three special people have given generously of their time and expertise to help me with this edition, and it is a special privilege to thank my local "support team": Three cheers to Dr. Roger McWilliams for his timely and accurate oversight of the physics in this book; to Dr. Antoinette Empringham for her probing questions and suggestions that do so much to add both dimension and focus to the content and also for her accurate proofreading; and to Pat Chavez, for not only sharing gourmet meals and photo props but also using her sharp "printer's" eyes for typographical errors to help avoid offending your eyes while reading this book. I also want to thank Bill Lawrensen, my editor, and Lara Dimick for their assistance throughout this revision. To Maren Miller go kudos for helping me through production and its snarls.

What a pleasure it is to be able to thank Barbara Boyer, R.D., a former neighbor and now professional colleague and food safety specialist, for her thorough review of Chapter 19.

I also want to thank Shiny Rajesh and her colleagues for the unforgettable day I spent with them at Integra, the company in Pondicherry, India, doing the production on this edition.

My deep gratitude goes to my reviewers who gave such useful comments and encouragement during development of this edition. The reviewers were Mia M. Barker, Ph.D., Indiana University of Pennsylvania; Kimberli Pike, Ball State University; Lisa Ritchie, Harding University; and Janelle M. Walter, Baylor University.

Margaret McWilliams

# Other Prentice Hall books by Margaret McWilliams

Experimental Foods Laboratory Manual (8th edition)
Foods Fundamentals (10th edition)
Illustrated Guide to Food Preparation (11th edition)
Fundamentals of Meal Management (6th edition)
Food Around the World (3rd edition)

# 1

# Research Perspectives

This dinner featuring a slice of rib roast, baked potato, steamed broccoli, and a glass of milk provides a variety of flavors, textures, and color for dining pleasure.

# CHAPTER 1

# Dimensions of Food Studies

## Chapter Outline

## OBJECTIVES

After studying this chapter, you will be able to:

1. Appreciate the interrelationships between consumers and their food, including scientific and safety factors influencing agriculture, production, and marketing.

2. Describe various career opportunities that are centered on food.

3. Outline how to conduct basic classroom laboratory experiments.

## INTRODUCTION

Food is a vital part of life for everybody, and for many people it is the central focus of their careers. All of us eat, but what we choose varies widely, depending on availability, affordability, and preferences. Consumers play a major role in defining the marketplace and its inventory. Farmers and fishermen produce the basic commodities of our food supply, often with the assistance of research scientists who focus on studies to improve yield and safety. On the way to market, these basic ingredients may be transformed into convenient, varied, and tempting products with extended shelf life.

Food transportation systems are so effective today that fresh produce of high quality is available in most markets throughout the year, and the variety includes seasonal items grown locally, as well as imports from around the world. Vast arrays of packaged and convenience items are also displayed in abundance on grocery shelves and in the freezer section. Food manufacturers have been highly creative in developing items that meet the needs of busy consumers with little time to prepare meals at home from basic ingredients.

Unfortunately, the plentiful and tempting food supply, in combination with lifestyle choices, has created a nation with an increasing number of overweight and obese people

3

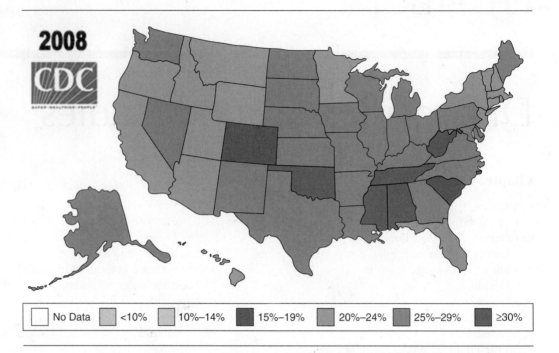

**Figure 1.1**   Incidence of obesity in the United States, 2008.

(Figure 1.1). Now, the food industry is challenged with helping consumers alter their practices and choices to bring about weight loss and to improve health.

Among key issues confronting the food industry today are safety, biotechnology, and health promotion (functional foods):

- Safety has always been a concern for food manufacturers and handlers, but the specter of possible terrorist actions has accentuated the need for constant vigilance in safeguarding products.

- Agricultural researchers have been developing genetically modified organisms that will have increased resistance to insects, a better amino acid profile, or other useful qualities.

- Nutrition researchers are constantly expanding scientific understanding of the body's requirements. Consequently, many food products are being modified by increasing the levels of some key nutrients and other substances that help protect the body. Food companies develop and market new products, conduct basic research on food-related topics, and monitor and shape consumer trends in food preferences and dietary concerns. Clearly, the combined expertise of dietitians, food scientists, and chefs is being used today to meet the challenge of bringing a healthy diet and dining pleasure to consumers.

All of these aspects converge in the marketplace and afford a wide variety of career opportunities centered on food. Academic preparation for these positions combines a strong foundation in science with the pleasures of eating. Knowledge of the principles of physics and chemistry that govern the changes occurring during food preparation is essential for professionals involved in the food industry. However, these aspects of food science need to be blended with an understanding of the qualities in food that create consumer satisfaction and promote good health. Consumers rely on the food industry to provide a bounty of products that are convenient and pleasant to eat, while also providing the nutrients their bodies require.

This chapter examines the roles of consumers, producers, and industry in shaping the food marketplace; it describes career opportunities and the academic path toward professional positions that focus on providing food for people.

## CONSUMERS: WHO ARE THEY?

### Demographics

Defining consumers is a daunting task, one that is constantly shifting because of multiple factors and the time frame being considered. In 2000, the U.S. Census Bureau determined that the nation's population was 281,421,906, but the estimated population in 2008 increased dramatically to 304,059,724, a rise of 8 percent. The percentage changes were distributed quite unevenly: Northeast (2.5), Midwest (3.4), South (11.5), and West (12.1).

The median age in 2008 was 36.8 years. Infants and children to age 19 comprised a little over a quarter (27 percent) of the population; just over a third (34.3 percent) of the people were between ages 19 and 34; about a quarter (25.7 percent) were between ages 35 and 64; and 12.8 percent were 65 or older.

The group below the age of 35 is expected to comprise only about 45 percent of the population by 2030 (Sloan, 1998) compared with more than 61 percent in 2008, but an increase in the percentage of the population ages 65 and older is predicted by 2030. Such age shifts can be expected to alter consumer demands for foods during the upcoming decades.

The ratio of women and girls to men and boys is another demographic of interest to the food industry because of its influence on food patterns. In 2008, there were 154,135,120 females and 149,924,604 males (50.6 and 49.3 percent respectively). However, the ratio of women to men increased with the age-group. The percentage of females and males age 18 and older was 51.3 and 48.7 respectively, and rose to 57.6 and 42.4 in the group 65 and older. Boys outnumbered girls below age 18, but there were somewhat more adult women than men; among people in their nineties, there were only 37 men per 100 women.

Even greater changes in food expectations will occur as cultural diversity continues to change in the United States as a consequence of both immigration and differences in birthrates between the various groups comprising the American population in the upcoming decades. According to the U.S. Census Bureau, the Hispanic population is increasing five times faster than the total population. Projections of racial distribution of the U.S. population (Table 1.1) anticipate a reduction in the percentage of whites alone (not Hispanic), a slow increase in blacks, and a slightly larger increase in Asians. In contrast, the percentage of Hispanics of any race is predicted to almost double between 2000 and 2050.

### Lifestyles

In addition to the cultural heritage of consumers, their preferred lifestyles definitely influence the food marketplace. Frequency of meals or even the definition of a meal can evoke very different expectations and behaviors among consumers. Some people, particularly the elderly, still expect to eat three times a day, but for many, eating seems to be appropriate any time of day and any number of times. Perhaps the most noticeable behavior is the lack of agreement on when and how often to eat during the day. Ready availability of prepared foods and snacks for consumption at work, school, home, or anyplace in between has prompted people to eat frequently, if not always wisely.

**Table 1.1**   Projected Percentage of U.S. Population by Race and Hispanic Origin

| Race | 2000 | 2010 | 2020 | 2030 | 2040 | 2050 |
|---|---|---|---|---|---|---|
| White alone, not Hispanic | 69.4% | 65.1% | 61.3% | 57.5% | 53.7% | 50.1% |
| Hispanic (of any race) | 12.6 | 15.5 | 17.8 | 20.1 | 22.3 | 24.4 |
| Black alone | 12.7 | 13.1 | 13.5 | 13.9 | 14.3 | 14.6 |
| Asian alone | 3.8 | 4.6 | 5.4 | 6.2 | 7.1 | 8.0 |
| All other races[a] | 2.5 | 3.0 | 3.5 | 4.1 | 4.7 | 5.3 |

[a]Includes American Indian and Alaska Native alone, Native Hawaiian and other Pacific Islander alone, and two or more races.
Adapted from U.S. Census Bureau, 2004, "U.S. Interim Projections by Age, Sex, Race, and Hispanic Origin."

Take-out food eaten in a car or at home is a popular solution to meals in today's society. Prepared items from the deli counters at grocery stores are frequently purchased on the way home for the evening meal (Sloan, 2006). These options are significant reasons why only about a third of the evening meals in the United States is prepared at home, and convenience foods from the shelf or the freezer often play a significant role in these meals. Busy people are selecting time-savers such as fruit, vegetable, and salad packages that are already washed, sliced, cut up, and ready. Increasingly, preparation responsibilities are being relegated to the food industry, and consumers are becoming managers rather than cooks in many homes.

## Health

Health and the impact that food has on it are among the top interests of Americans. In response, the food industry has developed many products with added nutrients and others with higher fiber, lower salt, or other formulations to provide healthier food choices. The relationship between weight and health has been the subject of much research striving to understand and manage weight.

The nationwide emphasis on the "epidemic of obesity" is focusing attention on helping consumers make wiser food choices for reducing weight and achieving better health. The Center for Disease Control (CDC) found that 11 percent of children ages 2–5, 15 percent ages 6–11, 18 percent ages 12–19, and 67 percent of adults over 20 were overweight in 2005–2006. Blame for this national health crisis frequently has been aimed at the food industry. Now, food manufacturers and commercial food establishments are directing considerable effort to create products that contain fewer calories; smaller portions are also being promoted.

# PRODUCERS

## Fishermen

Fish of many types are important sources of nutrients in diets today. These may be caught in the sea by independent fishermen or by commercial fishermen who often operate from large vessels sailing on extended missions far from shore. The increased demand for seafood has resulted in seriously depleted quantities of some of the more popular types, such as abalone, some types of tuna, and orange roughy. Quotas have been implemented in some regions in an attempt to re-establish the quantities that used to be found in the commercial fishing grounds.

Fish farming is a growing industry in freshwater, as well as in the ocean. Catfish, shrimp, abalone, mussels, and salmon are some of the types that have been farmed successfully (Figure 1.2). As this industry has grown and matured, problems of safety have emerged, and research is being conducted to study ways of preventing contamination to sustain viable production from ponds and enclosures over an extended period. Despite some concerns, fish farming appears to be a source of food that will play an increasingly important role in feeding consumers.

## Farmers

Crops for consumers and food companies are grown by farmers across the nation and also in other countries. Whether they are grown on farms operated as a commercial enterprise or as a family business, the food that is produced is an essential, but sometimes precarious source of income. Although irrigation is utilized in many areas when needed, even this source of water is being reduced in California and other drought-affected states as rain and snow levels have declined (possibly an effect of global warming). In 2009, many farmers in California suffered serious losses because their water rations were inadequate to support their crops. In some cases, the impact was so serious that some lost their farms.

**Figure 1.2**   Researchers and fish farmers sample catfish to check the efficiency of a floating platform grader near Belzoni, Mississippi. (Courtesy of Agricultural Research Service.)

The recent erratic weather patterns have resulted in flooding in some regions and drought in others. Either condition can have a very significant impact on crop yields (Figure 1.3). Added to this picture may be abnormal temperatures that alter the usual growth patterns: For example, an unusually late spring delays planting; an early killing frost may end the growing season before the crop is mature and ready for harvest. In 2009, planting of corn was late in the Midwest due to bad spring weather, and harvesting was delayed because rain continued so long in the summer that the field corn was slow in drying to the 15-percent moisture level required to prevent mold growth during storage. Some farmers had to pay to have their harvested corn dried, a factor that further reduced income. By Halloween, 20 percent of the Midwestern corn crop had been harvested in 2009 compared to an average of 58 percent by that date in the preceding five years; comparable figures for the Midwestern soybean harvest were 44 and 88 percent. These are some of the factors that influence price and availability of foods in the marketplace.

Consumers influence what crops farmers decide to raise and what technologies to consider. The pressure for organic produce has resulted in considerably more crops that meet the standard for this designation. A related issue is biotechnology, also referred to as genetically modified organisms (GMO or GM).

In Sweden, the pressure to eat a diet that minimizes the carbon footprint from producing various foods became a national issue in 2009. This concern involves not only raising crops, but also using livestock practices that reduce carbon dioxide emissions and minimize the amount of energy required to produce each food. By publicizing this information on labels, Sweden hopes that people there will alter their diets to select foods that have a relatively low impact on the environment and that this example will encourage other nations to follow.

## Organic Concerns

As environmental concerns have moved to the forefront on the social agenda, there has been a strong trend toward buying "natural" foods. Despite the evidence of improved yields

**Figure 1.3**    Weather plays a significant role in determining crop yields from farms. (Courtesy of Agricultural Research Service.)

and blemish-free fruits and vegetables when farmers apply chemical fertilizers and pesticides to their crops, many consumers insist upon organically grown food. Their assumption that the food grown using only manure is higher in nutrients than that with chemical fertilizers often is not true.

---

## FOOD FOR THOUGHT: World Farming Challenges and Nonprofits

Concerns about the environment and the many people who are involved in agriculture have prompted a variety of efforts directed at improving the situation. Approaches toward helping vary, but two nonprofit organizations that are working to help meet these challenges are Rainforest Alliance and World Cocoa Foundation.

Rainforest Alliance, formed in 1987, is an organization that is trying to improve soil and water conservation, integrated pest management and integrated management of waste on farms, ecosystem conservation, wildlife conservation, and fair treatment and good conditions for workers (including access to healthcare, education, and equitable wages) in rainforest regions around the world. Its program includes a certification seal on crops from fields that have been certified as meeting the standards specified by the Rainforest Alliance.

World Cocoa Foundation was founded in 2009 and is funded by the Bill and Melinda Gates Foundation and 12 chocolate companies. Its programs are designed to help cocoa farmers in 15 countries around the world to implement farming practices that will improve productivity and quality of their crops, learn business skills, promote diversification of income, and enhance access to support services.

**Figure 1.4**   Produce that meets the legal requirements can be marketed with the organic label.

**Natural** and **organic** foods are not necessarily synonymous. When used in reference to foods, *natural* means that original food ingredients have been used and that artificial or chemical additives have not been included. Using beet pigments rather than a red chemical dye to color a food product is one example of using a natural coloring agent. The legal definition to label a food *organic* is that the plant or animal food has been produced without using growth hormones, antibiotics, or petroleum-based or sewage sludge-based fertilizers. Designation as organic does not mean that the food is higher in nutrients than the same type of food that has not been produced according to organic requirements. Nevertheless, many consumers often are willing to pay more for organic produce because they think they are buying more nourishing food.

The first federal legislation regarding producing and marketing organic foods was the **Organic Food Production Act of 1990.** However, the impact of this act was not evident until passage of legislation in 2002 that implemented the earlier act. The **National Organic Program**, administered by the Agricultural Marketing Service of the U.S. Department of Agriculture (USDA), went into effect in late 2002 (Figure 1.4).

Plant and animal foods marketed as organic may carry the USDA Organic seal if they meet the following requirements:

- To be eligible for the designation of organic or 95 percent organic, at least 95 percent of produce (by weight) must not have been treated with sewage sludge-based or petroleum-based fertilizers, conventional pesticides, ionizing radiation, or bioengineering.

- Food mixtures in packages may be labeled "made with organic ingredients" if 70 percent of the ingredients (by weight) meet the requirements for being designated organic.

## Biotechnology

Demand for natural foods and more and improved food to feed the world is spurring the efforts of **biotechnology** to develop plants and even animals that have specific desired characteristics or traits (Figure 1.5). Researchers using bioengineering techniques in animals

**Natural**
A food product made without chemical or artificial additives.

**Organic**
Legally defined as plant or animal food produced without using growth hormones, antibiotics, or petroleum-based or sewage sludge-based fertilizers.

**Organic Food Production Act of 1990**
Federal legislation that regulates production and marketing of organic foods.

**National Organic Program**
Federal legislation passed in 2002 to implement the Organic Food Production Act of 1990.

**Biotechnology**
Development of new products by making a genetic modification in a living organism.

---

## FOOD FOR THOUGHT: Ultragrain®

The conflict that some people face when choosing bread that is good for them (whole grains) and the one they prefer (soft, white) has been addressed by ConAgra. This food company spent more than eight years developing a new strain of wheat and a modified milling process to produce a flour for people who want whole grain nutrition benefits and soft white bread. The new strain of wheat is grown in the Mid-west, the part of the country where the hard wheat traditionally used for bread making is grown. New milling equipment was developed to mill this new wheat into flour with a uniform texture and small specks of bran. These efforts on the farm and in the mill resulted in a flour that is being marketed as Ultragrain®. Compared with other hard wheat flours, Ultragrain® is sweeter and milder in flavor and has a lighter color. Commercial bread bakers and other food manufacturers are seeing whether this new flour will win acceptance among those seeking the characteristics of white bread and the nutrients of whole grains.

---

**Figure 1.5**   Biotechnology is the avenue for developing crops with such advantages as increased resistance to disease. (Courtesy of Agricultural Research Service.)

have met with limited success, but this is an active field of scientific studies. They are working toward such objectives as animals with faster growth rates, more lean muscle mass, strengthened resistance to disease, or improved use of dietary phosphorus to lessen the environmental impacts of animal manure. In the plant world, research is aimed at developing healthier plants (resistant to insects and diseases and capable of producing a high yield) that can be raised with less fertilizer or insecticide than the traditional crop would require. Other research may have the objective of increasing the nutritive value of the food.

**Genetic engineering** is another term for biotechnology. A gene is a segment of DNA that encodes enough information to synthesize a protein. By identifying specific genes that provide the codes for making proteins that impart desirable traits, researchers have then been able to transfer the desired genes to other organisms to develop plants or animals that continue to replicate the desired gene(s) in succeeding generations.

Research in biotechnology to improve specific qualities of a plant results in food crops that are designated as **genetically modified organisms (GMO)**, sometimes simply referred to as GM. Corn and soybeans are major crops that have been the object of efforts to create plants that effectively can resist infections from insects, viruses, and/or fungi. Roundup-Ready® is one brand of genetically modified crops (e.g., soybeans) already being grown in the United States. Biotechnology also has led to the development of sunflower, peanut, and other oilseed plants with reduced levels of *trans* fatty acids and higher smoke points (see Chapter 11).

Continuing research projects to develop a range of crops with modified nutrient value and health benefits include:

- altering oil composition of oilseed rape and soybeans
- soybean protein for use as meat substitutes
- potatoes to reduce discoloration from bruising for commercial storage and also to lower moisture content (to reduce oil absorption during cooking)
- vitamin A levels (increased) in rice (Golden Rice™) to aid in preventing blindness in Southeast Asia
- increasing the antioxidant content in some vegetables

Other research in biotechnology is aimed toward improving flavor, color, and/or texture of tomatoes (e.g., FlavrSavr® tomato), corn, and squash (Figure 1.6). Additional research is being directed toward creating plants that are drought resistant, heat resistant, and/or able to survive increased salinity.

In many countries in various parts of the world, particularly in Europe, concerns initially were raised regarding the growing and marketing of food resulting from biotechnology. Nevertheless, farmers in many parts of Asia, South America, and the United States have been raising GMO crops for more than a decade. In 2005, one billion acres around the world were planted in GMO crops, and these crops were raised by 8.5 million farmers in 21 countries. European nations were slower to accept the new strains of crops. However, farmers in the European Union (EU) planted 107,719 **hectares** of GMO crops in 2008, which was an increase of more than 20 percent over 2007. Spain is the leading producer of these crops among the EU. In 2009, more than 200 different GMO crops were grown or being developed in 46 countries around the world.

Resistance to insect attack on corn and a few other crops is being incorporated by inserting a gene from *Bacillus thuringiensis* (a soil bacterium), resulting ultimately in the formation of a toxin that serves as an insecticide in the plant. Crops that incorporate this gene are designated as **Bt**. Bt maize is being grown in Europe despite the fact that France has banned it. Ongoing research and monitoring are being done to detect possible negative consequences of Bt crops. On the positive side is a significant decrease in worldwide fertilizer use (estimated to be almost 30 percent during the decade from 1996 to 2006) because of Bt maize and cotton.

Governmental regulatory agencies are involved in approving and regulating the products of biotechnology. In the United States, the FDA is responsible for food products of plant

**Genetic engineering**
Biotechnology in which a genetic modification is achieved by removing, adding, or modifying genes.

**Genetically modified organism (GMO)**
Plant or animal foods developed by genetic manipulation to alter nutrient levels or other characteristics; also designated as GM.

**Hectares**
Area equivalent to 10,000 square meters or 2.471 acres.

**Bt**
Designation that a seed has been modified by splicing a gene from *Bacillus thuringiensis* to promote resistance to insects.

**Figure 1.6**  Endless Summer tomatoes are an example of bioengineering to create produce for today's markets. (Courtesy of Agricultural Research Service.)

biotechnology. Agricultural products are under the purview of the USDA, primarily its Animal and Plant Health Inspection Service (APHIS). Herbicides and pesticides are under the aegis of the U.S. Environmental Protection Agency (EPA). These agencies are all involved in the regulation and oversight of some products of biotechnology.

The possibilities for improved food products as the result of biotechnology (specifically by techniques of genetic engineering) are diverse and multiplying. Examples include increased essential amino acid content in corn and soybeans, naturally decaffeinated coffee, plant oils with modified fatty acid content, potatoes and tomatoes with higher solids content (to decrease energy needed to remove water during processing), and controlled ripening of fruits and vegetables that are difficult to ship to markets in satisfactory condition.

Other plant crops have also been genetically modified and are now being farmed successfully. For example, rape (*Brassica napus*) has been genetically modified to increase its tolerance of herbicides and improve health benefits by achieving high levels of lauric and oleic acids and little erucic acid in the **canola oil** produced from it. Canola oil was named to indicate that it is an oil developed in Canada and also to distinguish this desirable oil from unpleasant images that the name rape seed might imply. Golden Rice™ was developed to help increase the levels of vitamin A in diets in Asia, where this vitamin is often deficient. This new rice variety was created by inserting a gene from *Erwinia uredovora* (a bacterium) and two genes from daffodils, which causes a yellowish color due to production of

**Canola oil**
Oil from rape (*Brassica napus*) seeds, a genetically modified variety of rape.

β-carotene in the endosperm. Summer squash modifications have been directed toward improving crop yields by increasing resistance to viruses, such as zucchini yellow mosaic virus, that attack the plants.

## INDUSTRY

The food industry provides consumers with a wide range of products that incorporate ingredients produced by fishermen and farmers. In some cases minimal processing is done, while in others, numerous ingredients and processing steps may be required. The challenge for these companies is to create and market products that meet consumer expectations and concerns. To do this, they are constantly monitoring consumer issues, conducting internal research and development, and checking quality and safety. Consumer concerns about the relationships between diet and health have fostered considerable research and product development in the areas of functional foods, nutraceuticals, and nanotechnology. The National Center for Food Safety and Technology (NCFST) is a unique food research consortium of the U.S. Food and Drug Administration's (FDA) Center for Food Safety and Applied Nutrition (CFSAN), Illinois Institute of Technology (IIT), and the food industry. It was established to bridge the gap between the food industry, government agencies with responsibilities for food safety, and academics in this field.

## Functional Foods

**Functional food** refers to food and food products that provide possible health benefits beyond just the nutrients they contain. **Nutraceutical** is sometimes used to describe a functional food, but it also may be used more broadly to cover supplements and medicinal herbs. Some of the pigments and certain other compounds naturally occurring in plants, animals, and fish may contribute to human health when they are included in the diet.

> **Functional food**
> Food that may provide health-promoting qualities beyond just the nutrients it provides.

> **Nutraceutical**
> Sometimes used to describe not only functional foods, but also supplements and medicinal herbs.

Considerable publicity has been given to the importance of such foods (e.g., the "5-a-day" campaign by the USDA to increase consumption of fruits and vegetables). Soy in various forms is an ingredient in many different types of products from beverages to energy bars because of the potential health benefits associated with its isoflavone content. Some eggs with enhanced omega-3 polyunsaturated fatty acid levels now are available for people striving to lower their cholesterol levels.

***Phytochemicals.***    **Phytochemicals** are compounds in foods that have biological activity that may promote health benefits beyond the effects of such nutrients as vitamins and minerals. Some examples of phytochemicals are catechins in tea, lycopene in tomatoes and red peppers, and beta glucan in oats (Table 1.2).

> **Phytochemicals**
> Chemical compounds in plants that are important to promote healthful reactions in the body but are not classified as nutrients required for life and growth.

Considerable research is underway to explore the relationship between diet and health problems, particularly cancer and heart disease. These two diseases concern many consumers

**Table 1.2**    Sources and Potential Benefits of Some Phytochemicals

| Phytochemical | Source | Potential Health Benefit |
|---|---|---|
| Indoles | Brussels sprouts, cabbage, cauliflower, broccoli | Help block cancer growth |
| Lignan | Flax seed | May interfere with estrogen and help block cancer growth |
| Beta glucan | Oats | Reduce risk of cardiovascular disease |
| Carotenoids | Orange, dark green, and red fruits and vegetables | Antioxidant action promotes health of eyes, other tissues |
| Lutein | Spinach, kale, collard greens | May help protect against eye diseases in the elderly |
| Lycopene | Tomatoes | Reduce risk of prostate and cervical cancers, degeneration |
| Flavonoids | Tea, wine, celery, cauliflower, apples | Raise HDL cholesterol; antioxidant action protects cells |
| Catechins | Green tea | Inhibit carcinogenic processes |
| Genistein, daidzein | Soybeans, dried legumes | May reduce cancer risk; block estradiol at receptor sites |
| Resveratrol | Skin of red grapes, red wine | Lower cholesterol; may reduce cancer risk |

**Figure 1.7**    Fruits and vegetables are essential in the diet for their phytochemicals, vitamins, and minerals.

because of the toll they take each year. Researchers are trying to determine whether various phytochemicals may aid in fighting or protecting against some diseases. Results sometimes have been contradictory; early findings of effectiveness have been countered by subsequent studies that failed to find similar results. Nevertheless, studies have not shown any harmful effects from eating foods high in various phytochemicals.

The fact that some researchers have found positive ties between diet and the incidence of these problems has added impetus to the importance of eating plenty of fruits, vegetables, and cereal products, all of which are rich in various minerals, vitamins, and phytochemicals. In fact, phytochemicals (Figure 1.7) such as lycopene, genistein, catechins, and tocotrienols, are becoming buzz words.

**Prebiotic**

Healthful bacterial culture added during manufacturing to enhance and/or modify a dairy product that does not survive in the digestive tract; or a carbohydrate readily digestible by bacteria in the intestine, but not by humans.

**Probiotic**

Bacterial culture added to a dairy product because of its health-promoting capability and viability in the intestines.

***Prebiotics and probiotics.***    **Prebiotics** and **probiotics** are other illustrations of functional food components. Various bacterial cultures such as *Streptococcus thermophilus* and *L. delbrueckii ssp. bulgaricus* are added as prebiotics to create yogurt and other milk products resulting from the action of the culture during manufacturing. Prebiotic cultures may be killed by the acidity encountered in the digestive tract after the food is ingested.

Other prebiotics are incorporated into some designer foods (see next section) to ultimately feed bacteria in the intestinal tract and promote the desired bacterial flora there. Fructooligosaccharide, inulin, and even honey are examples of ingredients that may serve as prebiotics to enhance the growth of desirable intestinal microflora and possibly to reduce diarrhea, urogenital infections, and colon cancer.

Probiotics (cultures of useful health-promoting bacteria that are viable in the intestines) now are being added to some dairy products to create functional foods. The cultures used as probiotics are selected for their ability to survive the very low pH levels (1.5–2) of the stomach and bile acids; they also may be microencapsulated in alginate gels to protect their viability as they travel through the gastrointestinal tract into the intestines. Among the probiotics used are certain strains of such bacteria as *Lactobacillus acidophilus, L. bulgaricus, L. helveticus, Bifidobacterium adolescentis, B. lactis,* and *B. subtilis.*

**Table 1.3** Some Food and Drug Administration-Approved Heart-Health Claims

| Nutrient/ Health Relationship | Section in Title 21 CFR[a] |
|---|---|
| Sodium/ hypertension | 101.74 |
| Dietary saturated fat and cholesterol/ risk of coronary heart disease | 101.75 |
| Fruits, vegetables, and grain products that contain fiber, particularly soluble fiber/ risk of coronary heart disease | 101.77 |
| Soluble fiber from certain foods (oat and psyllium products)/ risk of coronary heart disease | 101.81 |
| Soy protein/ risk of coronary heart disease | 101.82 |
| Plant sterol or stanol esters/ risk of coronary heart disease | 101.83 |
| Whole-grain foods/ risk of heart disease | Docket No. 99P–2209 |
| Potassium/ risk of high blood pressure and stroke | Docket No. 00Q–1582 |

[a]Code of Federal Regulations.

Health concerns have motivated many consumers to seek food products containing ingredients that help in avoiding or improving outcomes of cancer, strokes, heart attacks, and other physical threats. As an aid to consumers, the FDA identified some apparent relationships between diet and health and approved some heart-health claims (Table 1.3).

## Designer Foods

The increasing recognition of the relationship between diet and health has prompted food companies to develop a wide variety of products with added nutrients and/or ingredients to enhance the health benefits of these items to meet consumer demand. Foods that are developed specifically to meet the demand for products that promote improved health often are referred to as **designer foods**. These are also considered to be functional foods.

The FDA established definitions for some of the wording that can be used on food labels (Table 1.4). The intent is to aid consumers in understanding the nutritive content of the various products they are buying so they can make healthful choices. Food manufacturers can only use these defined words on their packages if the product conforms to the definition.

**Designer food**
Manufactured food that has been created to meet consumer demand for a food that may be effective in promoting health and avoiding or minimizing the risk of certain physical problems.

**Table 1.4** FDA Food-Labeling Definitions

| Label Wording | Definition |
|---|---|
| Calorie free | <5 calories/g |
| Low calorie | 40 calories or less/serving |
| Low sodium | 140 mg or less/serving |
| Very low sodium | 35 mg or less/serving |
| Low cholesterol | 20 mg or less and 2 g or less saturated fat/serving |
| Lean (meat, poultry, seafood) | 10 or less g fat, $4\frac{1}{2}$ g saturated fat, <95 mg cholesterol/$3\frac{1}{2}$ -oz serving |
| Fat free or sugar free | <1/2 g fat or sugar/serving |
| Low fat | 3 g or less/serving |
| Low saturated fat | 1 g or less/serving |
| High fiber | 5 or more g fiber/serving |
| Reduced | 25% less of specified nutrient or calories than the usual product |
| Light | $\frac{1}{3}$ fewer calories or $\frac{1}{2}$ the fat of the usual food |
| Good source of | At least 10% of Daily Value of specified vitamin or mineral/serving |
| High in | 20% or more of Daily Value of specified nutrient/serving |
| Healthy | Decreased fat, saturated fat, sodium, cholesterol, and at least 10% of the Daily Value of vitamins A and C, iron, calcium, protein, and fiber |
| Fresh | Raw or unprocessed (not even frozen or heated), no preservatives added |

Another option that food manufacturers may consider in package labeling and marketing is to make a health claim. The FDA has developed precise rules for making claims about each of the following:

- Calcium/osteoporosis
- Fat/cancer
- Saturated fat and cholesterol/coronary heart disease (CHD)
- Fiber-containing grain products, fruits, and vegetables/cancer
- Sodium/hypertension (high blood pressure)
- Fruits and vegetables/cancer
- Folic acid/neural tube defects
- Dietary sugar and alcohols/dental caries (cavities)
- Soluble fiber from certain foods, such as whole oats and psyllium seed husk/heart disease.

## Nanotechnology

**Nanotechnology**
Development, creation, and application of atomic, molecular, or macromolecular particles between 1 and 100 nanometers in size.

**Nanotechnology** is a rather new area of research and development, but the potential applications in the food chain from the field to consumers are numerous and likely will play significant roles in improving safety, quality, and nutrient content of our food. This technology develops, creates, and uses particles at the atomic, molecular, or macromolecular range of approximately 1–100 nanometers (1 nanometer equals 1 billionth of a meter); these nanostructures have unique properties.

Nanotechnology may be of significance in various ways on farms in the future. Nanosensors may be of use in monitoring moisture levels and water flow, which may lead to reduced pollution from runoff in areas where livestock are being raised. Other nanoparticles might be created that can neutralize pathogens in crops and animals. These are some of the avenues being explored in this emerging field. A cloth that is embedded with nanoparticles of insecticide releases the chemical very slowly, which reduces the need for fresh applications.

Applications of nanotechnology in food technology and science are being explored in a variety of directions. A particularly active area is food packaging materials. Edible films that include oil (from cinnamon or oregano) or nanoparticles of such elements as silver, zinc, or calcium are being developed as a type of antimicrobial packaging. The problem of gas permeability of plastic containers is being addressed by integrating nanoclays to impede entry of oxygen and the resulting oxidation that occurs over time when plastic containers are used. Another food packaging material being developed uses chitin, a polysaccharide that contributes rigidity to the shell-like coverings of crabs, lobsters, and other crustaceans. The fiber that can be spun from a solution of chitin by electrospinning is of nano dimensions and can be incorporated in packaging materials to impart antimicrobial properties.

By creating nano-size lipids, it is possible to form multiple emulsions that provide improved spreadability and stability of such foods as a low-calorie mayonnaise without adding extra thickening agents. Flavors can also benefit from nanotechnology. Nanoscale assays of how taste buds interpret various flavors create opportunities for modifying a flavor by introducing bitter blockers or enhancers of sweet or salt. The sweet and salt enhancers could be of particular value in developing foods with reduced levels of sugar or salt.

Nanotechnology may also be of help in detecting *E. coli* and other harmful microorganisms. A silicon chip with a protein from the microorganism can scatter light when a laser encounters cell mitochondria in a contaminated sample of food, and a digital camera records the scattering. This method will provide immediate detection so that the food can be removed from the system immediately, thus providing an accelerated response to such risks. Rapid *Salmonella* detection is becoming possible by coating antibody particles with nano-sized particles that fluoresce.

## Security in the Food Supply

Food safety is not a new issue triggered as a result of terrorism. Food-borne illnesses were recognized many, many years ago, and numerous governmental regulations and considerable personnel have been committed to protecting people from such risks (see Chapter 19).

A recent example of legislation to protect the health of consumers with severe food allergies is the **Food Allergen Consumer Protection Act of 2004 (FALCPA)**. This act requires that milk, eggs, fish, crustacean shellfish, tree nuts, peanuts, wheat, and soybeans be named as ingredients, followed by the source of the allergen so that sensitive people can avoid the specific allergens that might even be fatal to them.

**Food Allergen Consumer Protection Act of 2004 (FALCPA)**
Legislation requiring the listing of specific food allergens and their sources.

The gradual growth of agencies responsible for food safety has resulted in widely scattered governmental agencies, leaving gaps in some oversight areas and overlaps in others. The potential threat of terrorism has added to the challenge. Homeland security efforts encompass virtually all aspects of life in the United States today, including safeguarding the food supply from possible terrorist efforts. Tamper-proof packaging has been a fact of life for many years, but security plans triggered by the events of September 11, 2001, now go far beyond that protection.

In 2005, the federal government announced the **Strategic Partnership Program— Agroterrorism (SPPA)**. This program is a partnership of USDA, FDA (from HHS), DHS, and the FBI, all of which will collaborate with states and private industry to protect U.S. food and commodities from terrorism. These government agencies are developing and implementing the strategies deemed necessary to safeguard the nation's food and water supplies.

**Strategic Partnership Program—Agroterrorism (SPPA)**
Partnership of federal agencies (USDA, FDA, DHS, and FBI) with states and private industry charged with safeguarding food supplies and commodities.

Among some of the key units now involved with increasing protection of the food supply are the USDA and the Centers for Disease Control and Prevention (CDC). The FDA is providing guidance on such diverse challenges as (1) improved inspection of facilities and sampling products, (2) more careful examination of imported foods, (3) strengthened communication and collaboration with appropriate federal and state agencies, and (4) improved security at food plants and along the marketing and distribution chain. Cooperation and communication between and within agencies are essential to safeguarding the nation's food and water.

---

### FOOD FOR THOUGHT: Safety with Pressure

Concerns about food safety and consumer desire for the convenience of shelf-stable foods have led to the establishment of a research program to develop sterilized foods using high pressures and moderate heat. The special equipment needed for this innovative approach was made by Flow International, Inc., and installed at the National Center for Food Safety and Technology in Chicago. This center provides an opportunity for academic researchers (including those from the Illinois Institute of Technology and the University of Illinois) to work cooperatively with specialists from both industry (Procter & Gamble, ConAgra Grocery Products, and Kraft Foods) and government (FDA and the U.S. Army). Experiments are still being conducted on low-acid products that need to be shelf-stable. Safety of low-acid foods that will be marketed as needing no refrigeration requires sterilizing the food itself and also destroying any spores (see Chapter 19) that may be present. Research to determine processing parameters that are needed to assure sterilization of shelf-stable foods is being conducted at the National Center for Food Safety and Technology.

High-pressure processing begins with preheating prepackaged food in a water bath before subjecting the packages to pulses of pressure for a controlled period of time. Rapid cooling occurs before packages are removed from the processing cylinder. Some low-acid foods such as guacamole already are in markets, but require refrigeration.

## CAREER OPPORTUNITIES

Because food is absolutely essential to survival, careers centered on any aspect of food will always exist even though the focus and products may undergo considerable evolution over time. Advances in science and technology, food preferences, lifestyles, economics, and environmental factors combine to alter the food scene and to create the dynamic opportunities for careers based on food. Some careers are oriented toward interrelationships between food and health, some are based on feeding people in settings away from their homes, and still others are centered on basic food science and the development of marketable products for consumers (and the innumerable steps in bringing food from farms to consumers).

Careers in the food industry exist in such specialties as quality management, marketing, research and development, packaging, labeling, and compliance. The settings might include particular areas such as ingredients, food products, or food service. Work in governmental agencies and laboratories that administer and monitor various aspects of the food industry is yet another dimension for careers with food.

The great emphasis on food and health has spurred the building of bridges between the domains of professionals in the food industry, dietetics, nutrition, governmental agencies with responsibilities involving food and consumers, and food service (including restaurants and institutions). Chefs are among the food professionals who have played a significant role in shaping the public's food expectations and preferences, sometimes locally in restaurants and even nationally through television shows.

National attention on the growing problem of obesity in the United States and the health problems that often are compounded by excess weight are generating considerable interest among food professionals and consumers. Awareness of the need for healthful dietary habits has created public demand for products that are consistent with good nutrition, but also are appealing to eat. The dialogue that is developing in the various food-related professions is beginning to blur the divisions between the different segments. This change is opening opportunities for individuals to create unique positions that utilize the particular strengths they bring to the food industries.

Sound academic preparation is essential for persons planning to become food professionals, and that curriculum must include strong courses in the basic sciences that underlie food-chemistry, microbiology, biology, and physics. This science foundation needs to be incorporated into the study of food and its preparation and evaluation. Appreciation of the role that food plays in influencing the quality of life, because of its wonderful sensory and nutritional contributions, should be developed so that food professionals will always remember the needs of consumers as real people. Oral and written communication skills are essential for all professionals. Knowledge of research techniques and computer literacy are part of the tools of the food professional. Basic understanding of business is another requisite for most careers in this field.

Students wishing to focus on a health-related food career will pursue a degree in nutrition and/or dietetics. For many of these positions, applicants are required to be Registered Dietitians (R.D.). Positions are quite varied in this field and include clinical dietetics, food service administration, community nutrition, sports nutrition, consulting, and nutrition counseling, and are found in industries based on nutrition-related products.

The rapidly expanding realm of hospitality, hotels, and restaurants is the source of many positions requiring extensive food knowledge and culinary skills. Preparation for careers in this arena may be obtained through a dietetics curriculum or from programs tailored specifically toward this field. Emphasis is placed on the applied aspects of food and its service to groups of people. Entrepreneurial-minded students may use this curriculum to prepare for eventually owning and operating their own restaurants. Others may focus on managerial positions in one of the large corporations that dominate this arena. Still others may wish to enter the field of catering, either on their own or as employees.

Food businesses provide career opportunities that are quite diverse. Product development, quality assurance, food analysis, processing, packaging, microbiology and food safety, sensory evaluation, physical testing, labeling and governmental regulation, and marketing are some of the niches available in the food industry. Depending on the particular career objective, a student might prepare to enter the food business by obtaining a degree in nutrition and dietetics, food science, food technology, food service, hotel and restaurant (hospitality or culinary) management, or business.

## EXPERIMENTING WITH FOOD

A scientific attitude and research orientation are needed to enter the world of the food professional. Basic chemical and physical principles are the foundation of the food science that undergirds the nation's food supply. A thorough understanding of these principles allows the food professional to apply them to achieve the best possible results with the resources available.

The experimental approach to food study integrates theory and professional research studies with laboratory work. Valuable knowledge of the influence of ingredients and preparation procedures can be gained by performing experiments designed to illustrate key scientific principles involved in food preparation. The characteristics of foods can be identified and measured using subjective evaluation (vision, olfaction, taste, and feel) and objective (mechanical) tests. Frequently, experiments in a class in experimental food science are presented to the class as a group, and individuals conduct a portion of the work so that a suitably broad array of samples will be available to illustrate selected principles. Additional insights into food research are gained when an experiment is planned, conducted, evaluated, and reported individually.

Students of experimental food science quickly find that the emphasis is primarily on the theoretical "why" more than on the "how" approach to food and its preparation. The combination of scientific theory with the laboratory-based illustrations presented in food science courses provides deep understanding of foods: their structure, composition, and behavior. This knowledge provides the cornerstone on which professionals in food science, dietetics, and food service management function.

In an experimental foods class, attention can focus on the effects of modifying ratios and types of ingredients in food items and of altering methods. One product prepared correctly can serve as the control, and several variations made at the same time can illustrate the effects of varying ingredients and/or methods. Careful examination and testing of these samples are of great value in developing a clear understanding of the scientific principles undergirding the field of food science.

At the conclusion of this study of experimental food science, students will be able to evaluate a broad range of foods accurately, to identify possible errors in their preparation or formulation, and to plan appropriate corrective measures for subsequent preparation of the items. This knowledge is essential for anyone involved in food production, whether it be in supervising production in an institutional food service setting, developing new products, handling production or quality control in a food plant, or working with food in an educational or home context.

The dynamic nature of food and its susceptibility to changes during handling, storage, and preparation provide constant challenges to the professional working in the field of food. Solutions for controlling quality can be effected appropriately when the underlying principles are understood and their practical illustrations have been experienced.

### Science Adventures in a Seattle Kitchen

Dr. Nathan Myhrvold is not the usual cook in a Seattle kitchen; in fact, he is so unique that the *New York Times* featured him in an article on November 17, 2009. This former Microsoft technologist has left the world of computers and Bill Gates and now explores his interests in food and other branches of science at Intellectual Ventures, his company that pursues such diverse challenges as creating the perfect Peking duck (using dry ice and a dog brush with stiff bristles to create pinholes in the crust to drain the fat) and freeze-dried lobster tail. Eventually, some of his food experiments will be featured in a cookbook that he and his staff are developing. Chefs, a photographer, writers, and editors are part of the team to augment this scientist in his bustling kitchen research laboratory.

## Metrics

Laboratory work in food science is done using the **metric system**. Classroom experiments as well as research laboratories use metric units. However, consumer information is presented in household measures (teaspoons, tablespoons, cups for recipes, and inches to describe the sizes of baking dishes). Professionals working in a developmental laboratory with the possibility of applying the research to the consumer market need to be able not only to work in the metric system but also to convert between the language of the consumer kitchen and the research laboratory.

The metric system is a method of expressing length (distance), area, volume, and weight in an orderly fashion in basic units that are quantified by expressing values in decimals, the system of tens. Hence, length is expressed in meters, area in square meters, volume in cubic meters or liters, and weight in grams. Food experimentation utilizes primarily volume and weight.

To achieve reasonable numbers when working in the metric system, prefixes are appended to the unit of measure. A large array of prefixes can be used within this system, as shown in Table 1.5, but food experimentation usually is based on the following: kilo (k), centi (c), and milli (m). By use of a prefix, 1,000 g can be expressed simply as 1 kilogram (Kg); similarly, a hundredth of a meter is 1 centimeter, and a thousandth of a liter is 1 milliliter (ml).

Because household recipes frequently express the quantities of solids in volumetric measures, conversion to weight or volume (if a liquid) in the metric system is necessary if these recipes are to serve as the basis for experimental work. Conversely, metric experimental amounts must be converted to common household measures if recipes are to be made available to the public. These conversions can be done by determining the weight of the required household measure for the various ingredients. Liquids conversions are based on a household measuring cup equaling 236.6 ml or cubic centimeters (cc). Table 1.6 presents some of the equivalent measures and conversion factors that may be useful when converting recipes.

A cup of flour does not weigh the same as a cup of butter or a cup of chopped nuts. The difference in mass of various ingredients that would be measured volumetrically in household recipes requires that the weight of this volume be known if a household recipe is to be used as the basis of an experimental formula. It sometimes is necessary to determine this weight by weighing the desired volume of the ingredient. Fortunately, tables have been developed by experiments that provide uniform weights for a cup of many of the ingredients commonly used in food preparation. Table 1.7 includes the weights of a few selected ingredients to illustrate the need for determining the weight of various ingredients when recipes are converted for experimental use or back to household measures.

Just as metrics is the language of quantities in the realm of scientific research, so is the Celsius scale used for measuring temperature in the laboratory of the food scientist. This choice is not surprising in view of the fact that the Celsius scale is related to the decimal

**Table 1.5**    Prefixes in the Metric System

| Prefix | Symbol | Numerical Definition |
|--------|--------|----------------------|
| Tera | T | $1{,}000{,}000{,}000{,}000 = 10^{12}$ |
| Giga | G | $1{,}000{,}000{,}000 = 10^{9}$ |
| Mega | M | $1{,}000{,}000 = 10^{6}$ |
| Kilo | k | $1{,}000 = 10^{3}$ |
| Hecto | h | $100 = 10^{2}$ |
| Deka | da | $10 = 10^{1}$ |
| Deci | d | $0.1 = 10^{-1}$ |
| Centi | c | $0.01 = 10^{-2}$ |
| Milli | m | $0.001 = 10^{-3}$ |
| Micro | μ | $0.000{,}001 = 10^{-6}$ |
| Nano | n | $0.000{,}000{,}001 = 10^{-9}$ |
| Pico | p | $0.000{,}000{,}000{,}001 = 10^{-12}$ |

**Table 1.6**   Equivalent Measures and Conversion Factors Commonly Used to Convert Between Household and Metric Measures

**Equivalent**

*Weights*[a]
1 kg = 2.2 lb
454 g = 1 lb
28.35 g = 1 oz (avdp)
1 g = 0.035 oz (avdp)

*Measures*[b]
1 l = 1.06 qt
1 gal = 3.79 l
1 qt = 946.4 ml
1 c = 236.6 ml
1 fl oz = 29.6 ml
1 tbsp = 14.8 ml

*Length*[c]
1 in = 2.54 cm
1 m = 39.37 in

[a]kg = kilograms, lb = pounds, g = grams, oz = ounces, avdp = avoirdupois (weight).
[b]l = liters; qt = quarts; gal = gallons; ml = milliliters; c = cups; fl oz = fluid ounces (volume); $m^3$ = cubic meters; tbsp = tablespoons
[c]in = inches, cm = centimeters, m = meters.

**Table 1.7**   Average Weight of a Measured Cup of Selected Foods

| Food | Form | Weight of 1 Cup (g) |
|------|------|---------------------|
| Almonds | Blanched | |
| | Whole | 157 |
| | Chopped | 117 |
| Baking powder | Double-acting | 207 |
| Cheese, Cheddar | Shredded | 98 |
| Cornmeal | | |
| White | Uncooked | 140 |
| Yellow | Uncooked | 151 |
| Eggs | Whites | 255 |
| | Whole | 251 |
| | Yolks | 240 |
| Flour | Rice, white unsifted | 149 |
| | Rye, dark stirred | 127 |
| | Soy, full-fat, unsifted | 96 |
| | Wheat, all purpose, unsifted | |
| | Spooned | 126 |
| | Dipped | 143 |
| | Wheat, all purpose, sifted, spooned | 116 |
| | Wheat, cake, sifted, spooned | 99 |
| | Gluten, sifted, spooned | 136 |
| | Self-rising, sifted, spooned | 106 |
| Gelatin | Flavored | 187 |
| Margarine | Regular | 225 |
| | Soft | 208 |
| Rice | White, raw | |
| | Long grain | 192 |
| | Short grain | 200 |
| | Parboiled | 181 |
| Sugar | Brown, packed | 211 |
| | Confectioner's sifted | 95 |
| | Granulated | 196 |
| Yeast | Active dry | 142 |

Adapted from Fulton, L., Matthews, E., and Davis, C. Average Weight of a Measured Cup of Various Foods, *Home Economics Research Report No. 41*, Agricultural Research Service, U.S. Department of Agriculture: Washington, DC, 1977.

system, with the freezing point of water being designated as 0 and the boiling point at sea level as 100. To relate Celsius temperatures to the Fahrenheit temperatures commonly used on oven indicators and household thermometers, food professionals need to be able to convert from one of these common temperature scales to the other. By simply knowing that the boiling point of water is 212°F or 100°C and remembering that the number 32 (the temperature of freezing in Fahrenheit) and either 5/9 or 9/5 must be used, you can derive the formulas for conversions quickly (Figure 1.8). For example, to convert from Celsius to Fahrenheit, derive the formula by converting from 100°C to 212°F:

$$100°C \times 9/5 = 20 \times 9 = 180$$

$$180 + 32 = 212°F$$

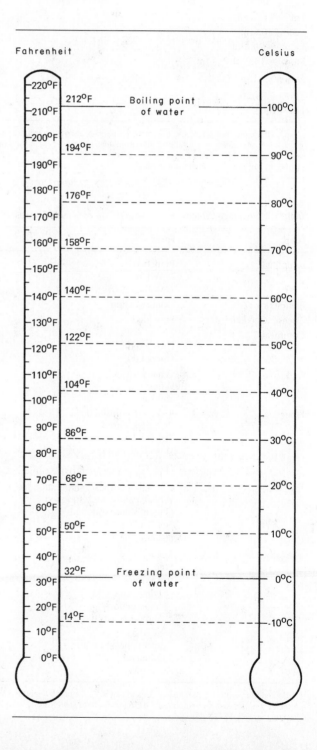

**Figure 1.8**    Comparison of the Fahrenheit and Celsius temperature scales.

Thus, the formula is

$$\left(\underline{\phantom{xxx}}°C \times \frac{9}{5}\right) + 32 = \underline{\phantom{xxx}}°F$$

The formula for converting from 212°F to 100°C is

$$212°F - 32 = 180$$
$$180 \times 5/9 = 20 \times 5 = 100°C$$

This formula then is

$$(\underline{\phantom{xxx}}°F - 32) \times \frac{5}{9} = \underline{\phantom{xxx}}°C$$

Table 1.8 provides some corresponding temperatures in Fahrenheit and Celsius.

## Taking Control

Experiments provide meaningful information only when controls are established and maintained to eliminate (as much as possible) unintentional variables. All aspects of the experiments, from the ingredients and their preparation to the evaluation process, need to be considered carefully to identify potential experimental errors that can lead to invalid results.

Researchers conducting experiments intended ultimately for publication spend considerable time developing their methods to eliminate uncontrolled variables and then replicating their experiments numerous times to ensure accurate results. Carefully documented techniques are repeated each time by the same laboratory personnel to eliminate the variability that would occur if others were to do the same task. Such measures enable researchers to obtain results that can be replicated again and again.

Unfortunately, it is not practical for individual students to prepare in class all of the variations needed to illustrate the impact of altering ingredients and/or methods used in food preparation. However, a class laboratory situation can provide considerable information if various class members prepare some of the samples required to demonstrate the scientific principles underlying the day's experiment. This broad involvement of class members in sample preparation obviously introduces the possibility of considerable experimental error. To keep this error to an absolute minimum, individual members of the class need to become personally responsible for thoughtfully and carefully preparing the variations assigned to them.

Control of the quantity of ingredients used in preparing samples is essential if results from objective and subjective testing are to be meaningful. In the laboratory, solids are weighed carefully, and liquids are measured volumetrically in graduated cylinders, pipettes, or burettes. Far greater accuracy is possible with these procedures than with standard household techniques. The diameter of household measuring cups is so great that precise measurements cannot be obtained, whereas the very narrow diameter of a graduated cylinder or other laboratory volumetric equipment reduces the potential for imprecise measurements of liquids. Other errors in household measurements occur as the result of packing

**Table 1.8**    Selected Examples of Comparable Temperatures in the Fahrenheit and Celsius Scales

| Celsius (°) | Fahrenheit (°) | Celsius (°) | Fahrenheit (°) |
|---|---|---|---|
| − 10 | 14 | 150 | 300 |
| 0 | 32 | 163 | 325 |
| 10 | 50 | 176 | 350 |
| 20 | 68 | 191 | 375 |
| 30 | 86 | 204 | 400 |
| 40 | 104 | 218 | 425 |
| 80 | 176 | | |

of ingredients when they are transferred into the measuring cup. By weighing solid ingredients, the researcher can eliminate this source of potential errors.

Liquid ingredients should be measured in the smallest volumetric device capable of holding the entire measurement at one time. This practice minimizes the use of devices with a larger diameter than is absolutely necessary. For example, 80 ml of oil can be measured more accurately in a graduated cylinder with a capacity of 100 ml than in one that holds 500 ml. On the other hand, if this same 80-ml measurement were made by filling a 10-ml graduated cylinder eight times, error would be introduced because some oil would cling to the graduated cylinder each time it was emptied. Each liquid measurement should be made at eye level by reading the bottom of the **meniscus**, the usually concave surface of the liquid. Pipettes and burettes provide greater accuracy than do graduated cylinders.

Dry ingredients and solids such as fats are weighed on balances, using careful and accurate laboratory techniques. **Trip balances** are available in some instructional laboratories. Ingredients can be weighed to an accuracy of 0.1 percent if 100 g of food is being weighed. This degree of accuracy requires that the container holding the food on the left pan be counterbalanced precisely with a similar, lighter container on the right pan to which shot or water has been added to zero the balance. The trip balance is an inexpensive, suitable choice because of its durability and satisfactory sensitivity for most laboratory experiments conducted by a class. It is preferred over the spring-type **dietetic balance** because of its greater sensitivity. The error of a dietetic balance is approximately 0.5 g when 100 g of food is weighed, whereas the trip balance error for this quantity is about 0.1 g.

Some **torsion balances** are used for weighing quantities greater than 2 kg, for this quantity cannot be weighed satisfactorily on a trip balance. They also are needed for weighing very small amounts, as might be the case for spices. The sensitivity of torsion balances makes them the preferred balance for all weighing, but cost and possible need for repairs limit their availability in instructional laboratories.

Direct-reading, **top-loading electronic balances** with a sensitivity of 0.01 g are yet another, but definitely costly, alternative for weighing spices or other foods needed in very small quantities. Top-loading electronic balances are desirable because of their convenience and accuracy. Unfortunately, their cost limits their availability in student laboratories. These balances save time by making it possible to weigh the container for the food and then use the tare mechanism to offset that weight so that only the weight of the actual food is indicated. Top-loading electronic balances are of particular merit when small amounts of ingredients need to be weighed because they are extremely sensitive and accurate, even when tiny quantities are required.

## Conducting and Evaluating Classroom Experiments

When the entire class conducts experiments, more samples can be prepared to demonstrate the effects of selected variables than can be produced by an individual researcher. This advantage is countered by the fact that individual techniques of the class members introduce uncontrolled variables into the experiment. Using electric mixers operating at a specified speed for a defined period of time can obtain some control of mixing techniques. In preparations requiring other mixing techniques, the class members should be sure to use the same design of beater, spoon, or other mixing tool; the total amount of mixing and the rate also need to be defined and heeded.

In some experiments, it is possible to set up an assembly line to prepare the samples for the class. When this is done, the same part of the preparation of each sample is done by the same person. The mixture passes in sequence along the assembly line. The obvious advantage of preparing samples for the class in this manner is that variations in preparation are kept to a minimum.

The completed products should be evaluated carefully according to the standards of quality appropriate for the type of food being tested. To distinguish the unique qualities of each of the variations available, each sample needs to be examined very carefully under good light

**Meniscus**
Curved upper surface of a liquid column that is concave when the containing walls are wetted by the liquid and convex when they are not.

**Trip balance**
Balance with two pans; the one on the left is used to hold the food being weighed, and the one on the right is used to hold the weights needed to counterbalance the left pan. Riders also are available for counterbalancing.

**Dietetic balance**
Single-pan, spring balance suitable for portion control, but not sufficiently accurate for food experimentation.

**Torsion balance**
Very sensitive (within 0.02 g) laboratory balance particularly useful for weighing very small quantities or quantities greater than 2 kg.

**Top-loading electronic balance**
Very accurate, electrically operated balance.

and in a quiet place. To avoid biases from other class members, all should refrain from any comments while evaluation is being done.

Comments should be recorded clearly as each sample is evaluated. A chart listing the variations and the characteristics to be evaluated will make the process smooth and accurate. If the samples are arranged on the display table in the same order as the chart and each sample is clearly labeled, the possibility of confusing the samples when recording results will be kept to a minimum.

## SUMMARY

Consumers, farmers and fishermen, and the food industry are constantly reshaping the food marketplace in response to increasing cultural diversity, environmental, economic, and lifestyles concerns. Among current key issues are food safety (including terrorism and environmental and microbiological risks) and health. Functional foods, designer foods, organic foods, phytochemicals, prebiotics, probiotics, genetically modified organisms (Bt), biotechnology, and nanotechnology are now in the vocabularies of consumers.

Careers related to food may involve the role of food and health or the feeding of people away from home or may be in the broad food industry (from basic food research to all aspects of bringing food products to the consumer, including governmental roles).

Food research, which had its origins primarily in the search for answers to problems observed in the home, is now shifting its emphasis increasingly toward the technological challenges presented as a considerable portion of the food consumed is prepared in factories and processing plants. Whether the food under study is intended for preparation in the home or on a vast commercial basis, the scientific principles underlying the properties and behavior of the various components are the same. A course in experimental foods considers the chemical and physical principles underlying the preparation of the diverse foods commonly served in the United States. Laboratory experiments can be conducted to learn evaluation techniques and to correlate cause and effect in food preparation. The language of experimental foods, like that of other sciences, is metrics. Food professionals need to be able to use the metric system and also to convert between the conventional system and metrics when necessary. Conversions between the Celsius and Fahrenheit scales frequently are required in research.

Accuracy of measurements is a key aspect of control in food experimentation. Volumetric devices (pipettes, burettes, and graduated cylinders) are used for measuring liquids, with the bottom of the meniscus being the location for determining the exact volume. Trip balances usually are the main type of balance used to weigh solid ingredients in classroom laboratories. Direct-reading, top-loading electronic balances are desirable for weighing very small quantities, but less expensive and slightly less precise torsion balances may be used satisfactorily for weighing ingredients needed in very small amounts or in quantities exceeding 2 kg.

For best results in classroom experiments, all aspects of sample preparation should be controlled as much as possible. Electric mixers are recommended whenever feasible to regulate the speed and duration of mixing. Evaluating class samples needs to be done in an accurate and organized manner.

## STUDY QUESTIONS

1. Identify and describe four characteristics of consumers where you live. How has each of these influenced the food markets in your town? Would the markets in a city in another state be the same as your market? If not, explain how they might differ.

2. Define "organic" and "natural" (in the context of food) and clarify any distinction between the two terms. Is there a difference in price and quality of organic produce versus the same type of regular produce in your market?

3. Define (a) phytochemical, (b) functional food, (c) prebiotic, (d) probiotic, (e) biotechnology, (f) GMO, (g) Bt, (h) genetic engineering, (i) nanotechnology.

4. What are two reasons why some people are opposed to GMO foods? What are two reasons GMO foods are appropriate to grow?

5. Identify the career that you wish to enter and describe the academic preparation you will need.

6. Why is it important for food researchers to weigh all ingredients carefully or measure them in a calibrated volumetric pipette, burette, or graduated cylinder?

7. Explain how to weigh 225 g of flour on your laboratory balance.

8. Convert the following measures:
   a. 40°F = ____°C
   b. 150°C = ____°F
   c. 375°F = ____°C
   d. 55°C = ____°F
   e. 14 tbsp = ____ml
   f. 472 ml = ____c
   g. $1\frac{1}{3}$ c = ____ml
   h. 7 fl oz = ____ml
   i. 1236 g = ____kg
   j. 236 ml = ____l
   k. 1 tsp = ____ml
   l. 4 tsp margarine = ____g
   m. 3 in. = ____cm

## BIBLIOGRAPHY

Anonymous. 2003. Operation liberty shield: New food security guidance. *FDA Consumer 37* (3): 18–19.

Becker, C. C. and Kyle, D. J. 1998. Developing functional foods containing algal docosahexaenoic acid. *Food Technol. 52* (7): 68.

Bidlack, W. R. 1998. Phytochemicals: Potential new health paradigm. *Food Technol. 52* (9): 168.

Bledsoe, G. E. and Rasco, B. A. 2002. Addressing the risk of bioterrorism in food production. *Food Technol. 56* (2): 43–47.

Boyce, B. 2010. Pandemics aren't just for people: How diseases can affect crops. *J. Amer. Dietet. Assoc. 110* (1): 18.

Bruemmer, B. 2003. Food biosecurity. *J. Am. Dietet. Assoc. 103* (6): 667–691.

Bugusu, B. 2009. Nanoscale science creates novel food systems. *Food Technol. 63* (9): 36.

Clark, J. P. 2006. High pressure processing research continues. *Food Technol. 60* (2): 63.

Clark, P. 2006. Evaluating nonthermal processes. *Food Technol. 59* (12): 79.

Clark, J. P. 2007. Pulsed electric field processing. *Food Technol. 60* (1): 66.

Committee on the Impact of Biotechnology on Farm-Level Economics and Sustainability; National Research Council. 2010. *Impact of Genetically Engineered Crops on Farm Sustainability in the United States*. National Academies Press. Washington, DC

Decker, K. J. 2010. Feeding healthy boomers. *Food Product Design 20* (3): 66.

DeHeij, W. B. C., et al. 2003. High-pressure sterilization: Maximizing the benefits of adiabatic heating. *Food Technol. 57* (3): 37–41.

Deis, R. C. 2003. Facts on functional foods. *Food Product Design 13* (4): 41.

Deis, R. C. 2003. What's in a claim: Labeling functional foods. *Food Product Design 13* (November Supplement): 15.

Dunford, N. T. 2001. Health benefits and processing of lipid-based nutritionals. *Food Technol. 55* (11): 38–44.

Frederick, K. 2009. Booster shots for energy and health. *Food Technol. 63* (9): 26.

Gerdes, S. 2003. You gotta have heart-healthy ingredients. *Food Product Design 13* (2): 113.

Hasler, C. M. 2009. Exploring the health benefits of wine. *Food Technol. 63* (9): 21.

Hicks, K. B. and Moreau, R. A. 2001. Phytosterols and phytostanols: Functional food cholesterol busters. *Food Technol. 55* (1): 63.

Hollingsworth, P. 2001. Close watch on America's food supply. Business/Marketing. *Food Technol. 55* (11): 20.

Katz, P. 2002. Fine line between functional foods and supplements. *Food Product Design 12* (3): 101.

Klahorst, S. J. 2004. Nutrigenomics: Window to the future of functional foods. *Food Product Design: Functional Foods Annual Supplement 14*: 5.

Lesney, M. S. 2003. A time to sow? *Today's Chemist at Work 12* (3): 22–25.

Liu, K. 1999. Biotech crops: Products, properties, and prospects. *Food Technol. 53* (5): 42.

Menkir, A., et al. 2010. Relationship of genetic diversity of inbred lines with different reactions to *Striga hermonthica* (Del.) benth and the performance of their crosses. *Crop Science Soc. America 50*: 602.

Meyer, R. S., et al. 2000. High-pressure sterilization of foods. *Food Technol. 54* (11): 67–72.

Ohr, L. M. 2003. Getting more from vitamins and minerals: Nutraceuticals and functional foods. *Food Technol. 57* (4): 87.

Ohr, L. M. 2009. Functional fat fighters. *Food Technol. 63* (10): 59.

Paeschke, T. 2003. Dropping calories, maintaining taste and functionality. *Food Product Design 13* (1): 33.

Pszczola, D. E. 2006. Reaping a new crop of ingredients. *Food Technol. 60* (7): 51.

Ravishankar, S., et al. 2009. Edible apple film wraps containing plant antimicrobials inactivate foodborne pathogens on meat and poultry products. *J. Food Sci. 74* (10): M440.

Shah, N. P. 2001. Functional foods from probiotics and prebiotics. *Food Technol.* 55 (11): 46–53.

Sloan, A. E. 1998. Food industry forecast: Consumer trends to 2020 and beyond. *Food Technol.* 52 (1): 37

Sloan, A. E. 2002. The natural and organic foods marketplace. *Food Technol.* 56 (1): 27–37.

Sloan, A. E. 2006. What, when, and where America eats. *Food Technol.* 60 (1): 18.

Spinelli, N., Jr. 2003. Research chefs and the development process. *Food Technol.* 57 (5): 18.

Tabashnik, B. E. 2008. Insect resistance to *Bt* crops: Evidence versus theory. *Nature Biotechnol.* 26: 199.

Wallace, T.C., et al. 2009. Unlocking the benefits of cocoa flavanols. *Food Technol.* 63 (10): 34.

Witwer, R. S. 1999. Marketing bioactive ingredients in food products. *Food Technol.* 53 (4): 50.

## INTO THE WEB

*http://www.cdc.gov/obesity/data/trends.html*—Trends in obesity rates in the United States.

*http://www.rainforest-alliance.org/*—Web site for Rainforest Alliance.

*http://www.worldcocoafoundation.org/*—Web site for World Cocoa Foundation.

*http://irps.ucsd.edu/assets/021/8422.pdf*—A critique of Rainforest Alliance certification.

*http://www.ams.usda.gov/AMSv1.0/NOP*—Web site for the National Organic Program.

*http://www.ams.usda.gov/*—Site for the U.S. Department of Agriculture.

*http://www.nifa.usda.gov/nea/biotech/in_focus/ biotechnology_if_animal.htm*—Overview of some animal biotechnology research possibilities and potential benefits.

*http://www.csrees.usda.gov/plantbreedinggeneticsgenomics .cfm*—Overview of some plant biotechnology research.

*http://www.monsanto.com/biotech-gmo/asp/default.asp*— Overview of plant biotechnology and some references.

*http://www.soyconnection.com/pdf/usbs_position/English/ 8007_USB_BioTechBro_v1.pdf*—Discussion of importance of biotechnology.

*http://www.agbios.com/dbase.php?action=Synopsis*— Database on global status of approved genetically modified plants.

*http://www.isaaa.org/*—International Service for Acquisition of Agri-biotech Applications; information on biotech programs around the world.

*http://www.bt.ucsd.edu/*—Information on Bt crops.

*http://www.ars.usda.gov/research/publications/publicatio ns.htm?SEQ_NO_115=235716.* Impact of Bt crops on nontarget organisms and insecticide use patterns.

*http://www.ncfst.iit.edu/platforms/research_ncfst.html*— Site for the National Center for Food Safety and Technology (NCFST).

*http://micro.magnet.fsu.edu/phytochemicals/index.html*— Images of phytochemicals and some discussion of them.

*http://lpi.oregonstate.edu/infocenter/phytochemicals.html*— Information on phytochemicals.

*http://www.fda.gov/Food/GuidanceComplianceRegulatory Information/GuidanceDocuments/FoodLabeling Nutrition/FoodLabelingGuide/ucm064919.htm*—FDA guidelines for health comments on food labels.

*http://www.fda.gov/food/labelingnutrition/labelclaims/health claimsmeetingsignificantscientificagreementssa/default .htm*—Information on health claims that may be on labels.

*http://www.nano.gov/*—Home page for the National Nanotechnology Initiative.

*http://www.ars.usda.gov/is/pr/2008/081215.htm*—Information on detection of *Salmonella* using nanotechnology.

*http://www.nanotechproject.org/process/assets/files/2706/94_ pen4_agfood.pdf*—Woodrow Wilson International 2006 report on nanotechnology in agriculture and food production and potential applications.

*http://pubs.acs.org/doi/abs/10.1021/jf903170b*—Article in Agriculture and Food Chemistry on phytoglycogen octenyl succinate, a carbohydrate nanoparticle to improve lipid oxidative stability of emulsions.

*http://www.senomyx.com/flavor_programs/receptorTech .htm*—Flavor receptor technology.

*http://www.fda.gov/Food/LabelingNutrition/FoodAllergens Labeling/GuidanceComplianceRegulatoryInformation/ ucm106187.htm*—Information about the Food Allergen Consumer Protection Act of 2004.

*http://www.nytimes.com/2009/11/17/science/17prof.html? pagewanted=all*—*New York Times* story on a science kitchen in Seattle.

Already noted for developing Oatrim and Z-trim, ARS chemist George Inglett has come up with another healthful food ingredient-Nutrin (bowl in foreground).

# CHAPTER 2

# The Research Process

## Chapter Outline

## OBJECTIVES

After studying this chapter, you will be able to:

1. Develop a clear statement of the purpose of a prospective research project.
2. Identify bibliographic references pertinent to the project.
3. Propose the research design, including method, evaluation, and data recording.
4. Describe how the research is to be conducted.
5. Explain how to interpret and report results of an experiment, including use of some statistics.
6. Outline how to write the report of a research project.

## INTRODUCTION

Classroom laboratory experiments often are planned to demonstrate pertinent scientific principles by having samples prepared and evaluated by members of the class. This approach is valuable to help illustrate the significance of the principles and the effects of variations on food products. There also is much that can be learned about research by planning, conducting, and reporting an individual experiment in which you define the purpose and justification. By identifying the problem and actually proceeding through the several steps involved in conducting your research project, you can begin to appreciate the realm of food research and the many factors involved in conducting, evaluating, and reporting such research. The following sections provide an overview of the process and serve as a guide for doing independent food research.

## DEFINING THE PURPOSE

Drafting a clearly stated purpose is the first step in designing a research study. Initially, it may be helpful to write down the general subject to be studied. This could be as vaguely defined as "roasting meat with aluminum foil." From this beginning, a specific statement

29

can be developed. The following statement of purpose for an individual project is an example:

> The purpose of this experiment is to determine the effect on muffins of substituting fructose for sucrose at two levels.

This statement indicates that two levels of substitution will be used, that the control will be sucrose and the substitute will be fructose, and that the product being tested is muffins. The specific characteristics to be evaluated are not indicated in this particular statement. Instead, the general term *effect* is used, which leaves the evaluation methods to be identified when the experimental method is developed. If it is appropriate to define *effect* more specifically, the statement of purpose might read as follows:

> The purpose of this experiment is to determine the effect on volume, texture, flavor, tenderness, and moisture when fructose is substituted for sucrose at two levels in muffins.

Although somewhat cumbersome, this type of statement clearly stipulates the scope of the experiment and aids in the development of the methods needed, particularly the method of evaluation.

Statements of the problem are appropriate for descriptive research studies. However, researchers wishing to apply statistical analysis to their results need to develop a *hypothesis* and a *null hypothesis*. This permits statistical testing to determine the probability that the variable being tested caused the results obtained in the study.

A **hypothesis** is a stated assumption of a consequence that is the outcome of the application of a variable in a research project. Researchers also state a **null hypothesis**, which is a statement that there will be no significant difference resulting from the application of the variable in the experiment.`

An example of a hypothesis is:

- There will be a significant difference in the volume of a shortened cake made with sucrose and one made with fructose.

An example of a null hypothesis is:

- There will be no significant difference in the volume of a shortened cake made with sucrose and one made with fructose.

Since research is often costly in both time and money, experimentalists should consider the justification for conducting the intended experiment. This justification should be developed concomitantly with the statement of purpose. In the case of the research topic just suggested, the justification may be based on the fact that fructose in solution is known to be sweeter than sucrose and that its use in cakes might be useful in reducing the caloric content. If this experiment demonstrated that fructose could replace sucrose satisfactorily at lower levels in cakes, reduced-calorie cakes could be developed and marketed to meet consumer demand for desserts with fewer calories.

Justification frequently is based on consumer needs. In some instances, special diet requirements for physical conditions such as a high serum cholesterol level may provide the justification for an experiment using egg substitutes or various other special ingredients related to the condition. Sometimes the cost of similar ingredients may be the basis of a study determining the feasibility of substituting the less costly ingredient for the more costly one, or even of reducing the amount of the costly item. Other experiments may be based on the need for a longer shelf life for products to maintain their acceptability during the marketing process. These are only a few of the factors that may provide justification for conducting food research.

**Hypothesis**
Positive assumption to test logical or empirical consequences of applying a variable in a research project.

**Null hypothesis**
Statement that applying a research variable will not make a significant difference in a research project.

## REVIEWING THE LITERATURE

Research projects are most meaningful when they are planned after a thorough search of the literature on the topic being studied. Previous work published on the research topic can provide insight into anticipated problems, appropriate methods, theories and facts related to the topic, and evaluation techniques. Considerable time can be saved and the quality of the research can be enhanced.

A computer search of the literature can be done using certain key words that are pertinent to the topic. For the preceding example, appropriate key words for the preliminary computer search would include *fructose, cakes, shortened cakes, sucrose,* and *sugar.* References regarding test methods can also be searched via computer.

Many journals have appropriate articles relating to specific research topics. Table 2.1 presents a list of some of the pertinent journals that may be helpful.

The table of contents in the most recent issues of the journals that may be most appropriate for your research topic may be useful in locating a pertinent article that has just been published. The bibliography in the article also will prove helpful. Many of the journals provide an annual index of articles. A quick examination of the index provides a review of articles in that publication that may be useful.

If a computer search is not a possibility, or the results of the search are delayed, alternative approaches to finding appropriate research articles may prove helpful. *Biological and Agricultural Index,* an index issued monthly, has a quarterly cumulative issue. Another index is the *Applied Science and Technology Index.* As the name suggests, indexed articles focus on applications and technology, rather than food science. Consequently, this index does not include articles from the *Journal of Food Science,* which is an especially important technical journal in the field. *Current Contents: Agriculture, Biology, and Environmental Sciences,* which publishes weekly the tables of contents from journals pertinent to food research, affords yet another entry to the literature in food research. All of these publications provide the bibliographic information needed to locate articles to help gather the research information available related to the research topic. You will need to obtain the articles themselves to gain the information needed in your literature review.

Abstracts also can be of help in gathering background information for your research topic. Whenever possible, read the actual article, rather than relying solely on the abstract. However, articles written in foreign languages or appearing in foreign journals may be available only in the form of the abstract. Several sources of abstracts may be available in the library and on the Internet. These include *Food Science and Technology Abstracts, Chemical Abstracts (Section 17),* and *Biological Abstracts.* A survey of their subheadings will help to identify the sections of the abstracts that might contain appropriate entries for a given topic.

**Table 2.1**   Some Research Journals Pertinent to Food Research Problems

| Journal | Types of Topics Covered |
| --- | --- |
| *Agricultural and Biological Chemistry* | Chemistry of basic products |
| *American Journal of Potato Research* | All aspects of potatoes |
| *Cereal Chemistry* | Applied research in the baking industry |
| | Some food science applications and scientific research on topics such as starch |
| *Cereal Foods World* | Various cereals |
| *Food Technology* | Food industry and problems related to product evaluation, control, and development |
| *International Journal of Poultry Science* | Many aspects of poultry |
| *Journal of Agricultural and Food Chemistry* | Chemical research on food |
| *Journal of the American Dietetic Association* | A few articles on nutrient content and food ingredients |
| *Journal of Dairy Science* | Milk and dairy products |
| *Journal of Food Protection* | Food microbiology |
| *Journal of Food Quality* | Some review articles on quality |
| *Journal of Food Science* | Very wide range of applied and basic food research articles |
| *Journal of Texture Studies* | Textural properties and characteristics theory and applications |

---

### FOOD FOR THOUGHT: AGRICOLA

AGRICOLA, although a word from Latin, is very much a product of the computer age. It is an acronym for AGRICultural OnLine Access. The U.S. Department of Agriculture established this online database in 1970 and continues to maintain extensive bibliographic information on a very broad base of agriculture-related disciplines, including food and human nutrition, animal and plant sciences, aquaculture and fisheries, farming, agricultural economics, extension, and education (Figure 2.1).

Users of AGRICOLA may search two data sets—Online Public Access Catalog (for books and other related types of publications) and Journal Article Citation Index (for journal articles). Bibliographic information is available using each of these search approaches. In some cases, the actual original text can be accessed via available links. Library materials can be obtained from the National Agricultural Library via http://agricola.nal.usda.gov/.

---

Computers can provide access to still other sources of information regarding pertinent articles. Two databases available by computer, but not in print, are AGRICOLA and Foods Adlibra™. The World Wide Web is a useful computer index.

After gathering the articles, careful reading and thought are necessary to relate the findings to the current project. Ideas for the statement of hypotheses and for the development of method often can be gleaned from the literature. Previous research can help validate the results obtained in the new experiment. Accepted analytical methods can be learned from the literature, thus enhancing comparison of the new results with previous research.

Be sure to record the complete citation for each article studied in an accepted bibliographic style. This style should be the one required either at the university or in the specific class; usually the style selected is based on the format used in an appropriate professional journal. In all forms, the basic information needed will include the authors' names (usually with initials), the title of the article, the volume (and issue, if necessary), the page number (either initial page or inclusive pages), and the year. Methodical notation of these pertinent data for each article will avoid the need to relocate the article later to complete the citation. Bibliographic styles differ from journal to journal, as can be seen from the following citations (*Food Technology* and *Journal of the American Dietetic Association,* respectively):

> Pillai, S.D. and Jesudhasan, P.R. 2006. Quorum sensing: How bacteria communicate. *Food Technol. 60* (4): 42–49.
>
> Stein, K. Contemporary comfort foods. *J. Am. Diet. Assoc.* 2008; 108:412.

## DESIGNING THE EXPERIMENT

When the purpose of the experiment has been stated clearly, a design can be developed to achieve the desired goal. Individual student experiments in an experimental foods course can be conducted meaningfully if only one variable is tested. In the example used in the preceding section, the variable is the substitution of fructose for sucrose; the levels of substitution need to be determined. The design of this experiment might be based on preparation of three cakes for each run: a control containing 100 percent sucrose and 0 percent fructose, an experimental cake containing 50 percent sucrose and 50 percent fructose, and a second experimental cake containing 0 percent sucrose and 100 percent fructose.

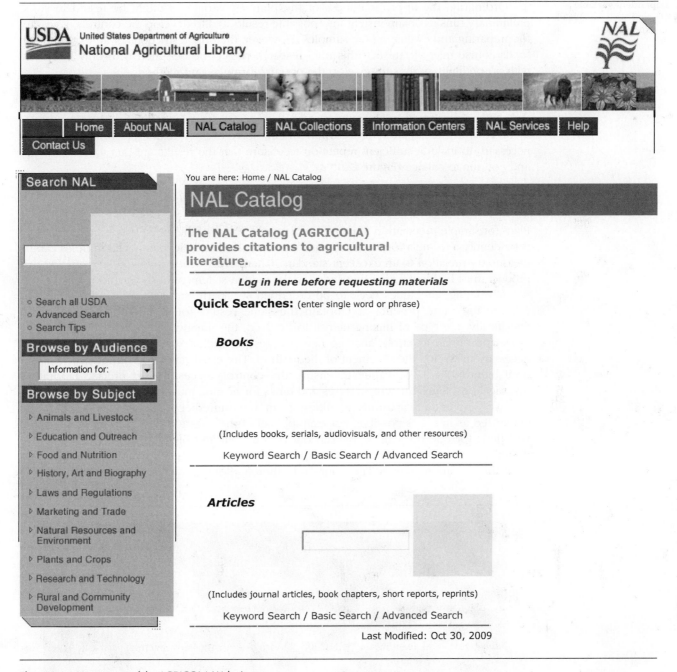

**Figure 2.1**   Home page of the AGRICOLA Web site.

A **variable** is any quantity or symbol that has no fixed value. In designing a research project, it is important to identify all aspects of the research that might vary. **Extraneous variables** are those that might add variations that are not truly a part of the experiment and that are not useful—for example, using two different brands of baking powder when that is not the focus of the experiment. Recognition of these undesirable extraneous variables is important so that they can be eliminated prior to conducting the experiment.

Two types of variables, independent and dependent, are of particular interest in designing an experiment. An **independent variable** (sometimes referred to as the *manipulated variable*) is defined by the researcher and is not measured. The **dependent variable** will have measured results or data as outcomes of the experiment. In the research comparing cakes made with sucrose and fructose, the independent variable is the type of sugar and the dependent variable is the volume of the cakes.

**Variable**
Quantity or symbol that has no fixed value.

**Extraneous variable**
Variable that is not intended to be part of the experiment and needs to be eliminated from or controlled prior to conducting the experiment.

**Independent variable**
Manipulated variable defined by the researcher.

**Dependent variable**
The measured variable of
an experiment.

Ordinarily, the appropriateness of the planned variations might be tested in some preliminary runs to obtain insight into possible results and to develop the controls necessary for preparing and evaluating the samples. However, the time restrictions of an experimental foods course may dictate that this mini research project be designed with only two runs, with no preliminary runs to develop the optimal formulas and method.

Publishable research requires extensive testing during the planning stage to establish the controls needed to eliminate errors resulting from variations in sample preparation. Actual data collection should begin only after repeatable results are obtained and the problems involved in the evaluation process have been solved. Once data collection begins, it is necessary to include sufficient repetition to ensure that the results are due to the variables and not due to chance (Figure 2.2).

## Method

Plans for sample preparation require careful consideration of the entire process. The first step is to identify a formula for the ideal control product so that the experimental samples can be measured in relation to an excellent standard. Then, every action involved in preparing that product must be identified and included in the written statement of the method.

All mixing techniques need to be described so clearly that anyone would be able to prepare the same product and obtain the same results following the stated method. Specifically, the type of mixing utensil to be used, the number of strokes (revolutions, or other appropriate control), and the rate (strokes per minute) are examples of the details necessary in writing the statement of the method. The usual statements in recipes (e.g., stir until blended) are too vague to ensure the controls necessary for precise laboratory investigations. Similarly, temperatures and times for heating must be specific.

Even potential variations resulting from the ingredients themselves need to be eliminated as much as possible. For example, all of the eggs to be used in the control and variations for one run can be broken out of the shell and blended together gently for a designated amount of mixing, and then the amount needed for each sample can be weighed from this common source. The flour and other staple ingredients needed for the whole

**Figure 2.2**   Dr. Pamela White, Dean of the College of Human Sciences at Iowa State University, is loading vials into the automatic injector of a gas chromatograph as she conducts an experiment on oils.

experiment should come from the same packages. If more than one package of an ingredient is needed for the entire experiment, the total amount should be mixed together thoroughly before starting and then stored appropriately for use throughout all of the runs. Of course, perishable ingredients will need to be procured throughout the experiment. Even though there will be some variation in ingredients of this type from one run to another, variation between the samples within a single run should be eliminated.

## Evaluation

Evaluation devices are a key part of the planning process. Careful thought should be given to ensure that all pertinent characteristics are tested appropriately. This means that the necessary objective measurements—for example, volume, tenderness, viscosity, pH, and chemical and physical attributes—are identified, and that plans for conducting each measurement are specified.

Sensory evaluation requires that a taste panel be designated and that a suitable scorecard be developed. If the panel is to be trained in the use of the scorecard, the training process also needs to be planned. Evaluations by objective and subjective methods are discussed in the next two chapters.

Plans for both objective and subjective evaluations need to be formulated to determine the quantities of sample required. Preparation of the samples for the evaluation also requires a thoughtfully developed written plan. For such objective testing as tenderness, the thickness of the samples must be controlled, a control usually achieved for a dough by rolling it on a board specially equipped with parallel guides that dictate the thickness of the dough prior to baking (Figure 2.3). In other cases, it may be necessary to develop a **template** to guide the cutting of specific samples for judges and for certain objective tests (Figure 2.4).

**Template**
Pattern guide to ensure accurate cutting of samples from a large sample, such as a cake.

Not only must the sample size be determined, but also the timing and temperature of samples for testing must be considered. Significant changes in textural characteristics occur in many foods as they cool after being heated. The notable increase in viscosity of a white sauce as it cools illustrates the need to plan testing circumstances. Baked products become firmer as they cool, making it necessary to specify the time for determining tenderness after removal from the oven. The gel structure of a starch-thickened pudding gradually tightens and becomes less fragile as it is allowed to cool without any agitation. These are but a few examples of the changes that necessitate careful planning of the conditions for evaluation.

## Data Recording

All information regarding a research project should be maintained safely and accurately. This record should include the statement of purpose, justification for the project, and the design. The formula and its variations, along with specific directions for preparing the samples, instructions for each phase of the evaluation, and samples of any forms to be used in the evaluation should all be recorded.

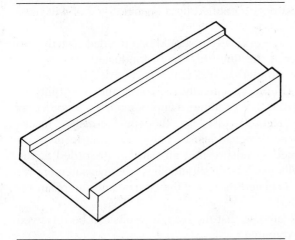

**Figure 2.3** Board for controlling thickness of pastry dough when preparing samples for testing in the shortometer. The rolling pin is rolled over the dough placed in the center until the dough is pressed down so that the rolling pin rolls evenly along the raised edges of the board.

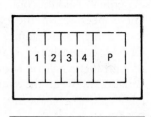

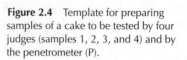

**Figure 2.4** Template for preparing samples of a cake to be tested by four judges (samples 1, 2, 3, and 4) and by the penetrometer (P).

If a computer is used for this purpose, frequent and careful proofreading of data entries is extremely important to avoid typographical errors in numerical values. Unless this proofreading occurs at the time data are being entered, errors can quickly take on an aura of truth that can taint the findings. Data should be backed up regularly at least once a week on a disk or portable data-management device.

A bound notebook should be maintained in the laboratory; daily entries need to include the date, all pertinent laboratory observations and notes, and all data collected as the experimental work proceeds for the day. This notebook also is the place for diagrams or drawings that are difficult to execute on the computer. The researcher's name, address, and phone number in the front of the notebook provide important identification to avoid loss or accidental use by another researcher in the laboratory.

## CONDUCTING THE EXPERIMENT

Each day the experiment is conducted, a new page should be started in the bound laboratory notebook. The date and any pertinent notes on unusual environmental conditions or ingredients should be recorded. Then the samples should be prepared, being certain that the exact procedures specified in the method are followed. This requires careful attention on the part of the researcher, for it is all too easy to do portions of the preparation by habit rather than by experimental design. Continuous attention to possible experimental errors is essential to a controlled experiment.

Thorough labeling of the variations throughout the preparation period and during evaluation will eliminate that potential for errors. The specific random numbers or symbols (e.g., □, △, ○) used in labeling samples each day must be recorded in the laboratory notebook for convenient reference. The same symbols can be used throughout the experiment, but the sample designated by each symbol should be changed each day according to the plan devised prior to the first run. In this manner, each sample bearing a particular symbol can be presented to the judges an equal number of times. This daily shifting of labels prevents the judges from expecting that the sample identified by a specific symbol will have certain characteristics.

The results of both objective and subjective evaluations should be recorded directly into the laboratory notebook and/or computer at the time the run is conducted. Accuracy is aided by always reporting the results of the various samples in the same order. By recording the results with the designation of the sample specifically in words, the possibility of erroneous interpretation of symbols is reduced. For example, samples might be recorded as "control," "50 percent fructose," and "25 percent fructose," or other suitable names; results then can be posted following each of these designations. Usually, recording results in a tabular form provides the structure needed to ensure that all data are recorded at the completion of each run (Figure 2.5). Initially, each table should show the measurement for each factor for each run, along with the calculated mean for the control and the mean for each of the variables prepared and evaluated.

These daily tables can then be consolidated so that the overall results can be reviewed and interpreted readily at the end of the experiment.

| Characteristic | Judge 1 | Judge 2 | Judge 3 | Mean |
|---|---|---|---|---|
| Aroma | | | | |
| 100% sucrose | | | | |
| 50% sucrose, 50% fructose | | | | |
| 100% fructose | | | | |
| Flavor | | | | |
| 100% sucrose | | | | |
| 50% sucrose, 50% fructose | | | | |
| 100% fructose | | | | |
| Tenderness | | | | |
| 100% sucrose | | | | |
| 50% sucrose, 50% fructose | | | | |
| 100% fructose | | | | |

**Figure 2.5**   Example of a chart for recording subjective data for 1 day.

## INTERPRETING AND REPORTING RESULTS

Tables for the evaluation of each subjective factor should show the rating given by each judge on each run, the mean of the judges' scores for each run, and the overall mean for that factor for the total experiment (Figure 2.6). This type of display permits easy tracking of the consistency of each of the judges for each specific factor. It also permits comparison between judges. If isolated examples of a variance from the usual evaluation are noted, the researcher can then look in the laboratory notebook for information that might aid in explaining this variance—for example, judge 1 had a bad cold and considerable nasal congestion that day.

When inconsistency from run to run is the pattern noted, there is a real reason to question the validity of the results. By probing for an explanation, the researcher may be able to identify flaws in the research method; perhaps the research can be conducted in the future to obtain reliable data for analysis. Obviously, researchers prefer to identify experimental problems early in the process so that the loss of valuable research time and resources is minimized.

### A Look at Statistics

The data collected during the research can be analyzed statistically by the use of **descriptive statistics** or **inferential statistics**.

**Descriptive statistics** Analysis of data by describing results in terms of calculations such as frequency, measures of central tendency, percentages, or various measures of dispersion.

| Run/date | Judge/mean | 100% sucrose | 50%/50% sucrose/fructose | 100% fructose |
|---|---|---|---|---|
| 1/Jan. 30 | 1 | 5 | 4 | 3 |
| | 2 | 4 | 4 | 4 |
| | 3 | 4 | 3 | 5 |
| | Mean | **4.3** | **3.7** | **4** |
| 2/Feb. 3 | 1 | 4 | 4 | 4 |
| | 2 | 4 | 3 | 3 |
| | 3 | 5 | 3 | 4 |
| | Mean | **4.3** | **3.3** | **3.7** |
| 3/Feb. 5 | 1 | 4 | 3 | 3 |
| | 2 | 5 | 3 | 3 |
| | 3 | 4 | 4 | 3 |
| | Mean | **4.3** | **3.3** | **3** |
| | Overall mean | **4.3** | **3.4** | **3.6** |

**Figure 2.6**   Scores for flavor.

***Descriptive statistics.*** For class experiments with few runs, descriptive statistical measures might include one or more of the following: frequency counts and distributions, **measures of central tendency** (mode, median, and mean), and **measures of dispersion** (range, mean deviation, standard deviation, variance, and standard error of the mean or difference between means).

Helpful insights into research results are obtained from simple calculations that yield a descriptive look at the collective data. Frequency counts and distributions determine the number of responses that fall into each of the various categories. These totals can then be calculated as percentages of the entire population being studied. To illustrate, if 50 subjects were interviewed and 5 of them fell in a specific group, this can be expressed as 10 percent of the subjects belonging to a specific group. When this calculation is done for all of the groups, a clear picture of the population in the study can be seen (Table 2.2).

When numerous data points are collected, measures of central tendency are often used to permit some comparisons. Determination of the mode is one possible measure of central tendency. **Mode** is the most common score or group. It is determined by plotting every data point on a chart and then finding which score or group contained the largest number of data points. In the example presented by the data in Table 2.2, the mode is 22 (Nature's Premium).

The **median** is the score in the middle position in an organized array of the data collected. In other words, there are as many scores on one side of the median as on the other side. This is simple to identify with an odd number of data. When the number of data is even, the value calculated to be halfway between the two middle points is the median score. Note that the median value does not necessarily define a numerical value halfway between the largest and smallest possible scores.

An example of determination of the median in an array with an odd number of data is

32
68
95 (median)
101
110

With an even number of data points, the midpoint between the two middle scores is the median:

11
12
16 (median = 17.5)
19
29
32

To calculate the **mean**, the arithmetic average of scores, total the data and divide by the number of scores. Despite the fact that the mean is influenced greatly when some data are far from most of the other scores, it is used frequently because it is the most reliable

Table 2.2    Brand of Orange Juice Preferred by Consumers in Newtonville

| Brand | Responses | |
| --- | --- | --- |
| | Number | Percentage |
| Esmerelda | 5 | 10 |
| Valley Pride | 15 | 30 |
| Nature's Premium | 22 | 44 |
| Superior | 3 | 6 |
| Naringenita | 5 | 10 |
| | 50 | |

measure of central tendency. The mean is used in some statistical calculations. The calculation of the mean for the preceding examples is

$$\frac{32 + 68 + 95 + 101 + 110}{5} = \frac{406}{5} = 81.2 = \text{mean}$$

$$\frac{11 + 12 + 16 + 19 + 29 + 32}{6} = \frac{119}{6} = 19.83 = \text{mean}$$

Each of these three measures of central tendency may be used in interpreting data. The median is used when scores are skewed markedly. The mode is useful in making a quick, rough estimation of the midpoint of data, but it is not suitable for use alone in analyzing data. The mean is well suited to represent the values obtained in the research study when statistical analysis is done.

**Percentage** sometimes is calculated to characterize the results of a study. An example might be to indicate the percentage of people who preferred one sample over two others in a preference test. **Percentile ranks** require that all scores be arranged in sequence so that the position of a specific score can be characterized—for example, the fifth score in an array of 10 is in the fiftieth percentile.

**Percentage**
Portion of a hundred.

**Percentile rank**
Relative position of a score within the total array of scores expressed on the basis of hundredths.

Measures of dispersion are used to indicate the conformity of the data points. **Variance** and **standard deviation** are two useful ways of assessing dispersion of values. These two measures are closely related, for standard deviation is the square root of the variance. Either of these may be used in determining tests of significance of the data obtained in a study.

**Variance**
Measure of the dispersion of data; the sum of the squares of the deviation of each value from the mean.

**Standard deviation**
Square root of the variance.

*Inferential statistics.* Statistics may be applied to make predictions about populations that meet the same criteria as those used for establishing the population of a study. These types of calculations are designated as inferential statistics. Studies using inferential statistics are based on carefully stated hypotheses—a working or alternate hypothesis and a null (negatively stated) hypothesis. The design of the study determines the specific statistics used for analysis.

Statistical analyses are conducted to determine the probability that the observed results are caused by the variable applied and not simply by chance. Calculations are appropriate to the type of statistical analysis selected for the experiment and its design.

The results of these calculations are compared with the appropriate statistical table(s) to determine whether or not the null hypothesis that was developed for the experiment should be accepted or rejected. These tables are designed with two levels of significance: .05 and .01. The **level of significance** is a proportion or decimal value at or below which the results of the research will be considered significant, and the null hypothesis can be rejected. At .05, there are only 5 chances in 100 that the result will occur by chance, and at .01 only 1 chance in 100 is the probability of the result occurring by chance.

**Level of significance**
The decimal value below which the results of the research will be considered significant, and the null hypothesis can be rejected.

If a variable is determined to have a significant effect at the .05 level but not at the .01 level, this is significant. However, this level clearly is not quite as significant as is a result at the .01 level of significance.

Another way of presenting statistical significance is by **level of confidence**, which is the percent of certainty that the variable caused the result. A statement of a 95-percent level of confidence is equivalent to a level of significance of .05. Similarly, a 99-percent level of confidence is comparable to a level of significance of .01.

**Level of confidence**
Percentage expression of certainty that the results caused by a variable are statistically significant.

Chi-square is an appropriate inferential test for a research study with one variable and two categories. It also can be used if there are two or more independent variables, but the calculations become increasingly complex as the number of variables increases.

When data are ordered sequentially or in rank order, other inferential tests can be used to determine the probability of occurrence. Such tests include the sign test, Wilcoxon matched pairs test, the Mann–Whitney $U$ test, the Kruskal–Wallis test, and the rank sums test. Selection of the appropriate statistics to be used is based on the design of the research project.

Interrelationships or the effect of a variable on various components often are of interest in food research. The goal is to determine the correlation between the variable and its effect

**Correlational research**
Research that determines interrelationships between variables.

**Regression**
Statistical methods applicable to correlational research.

**Student's *t*-test**
Statistical test to determine the significance of the mean of the experimental group versus the mean of the control group.

**Analysis of Variance (ANOVA)**
Statistical approach to determining differences between many sets of data.

on the various characteristics of the system. This is called **correlational research**; statistical analyses of correlational research are called **regression**.

Several tests can be used for regression statistical calculations. Among these are Pearson's Product Moment Correlation Coefficient (fortunately shortened to Pearson's r), Spearman's rho, C statistic, and Kendall's tau. The appropriate one to use depends on the form of the data.

Statistical significance can be established for the means of two groups (the experimental and the control groups) by using the **Student's *t*- test**. In the event that there is more than one experimental group in the research study, the mean of each of the experimental groups can be compared separately against the mean of the control using the Student's *t*-test to determine the significance of each of the group treatments.

**Analysis of variance (ANOVA)** is widely used to determine the statistical significance between many sets of data. To use ANOVA, the sample should be selected randomly or on the basis of probability. However, ANOVA sometimes is used even when the sampling was not done in these ways. A table of F values is needed to determine the significance. If the difference between treatments is to be considered significant, the F value must be equal to or greater than the value of the criterion F. There are several different ways by which ANOVA can be calculated. However, the results require comparison with the F table to determine significance.

## OVERVIEW OF THE REPORT

Preparing the research report is the final step. The format will vary, depending on the intended use. When reports are written as articles for a selected scientific journal, the style guide for the journal dictates the entire format. Similarly, research reports prepared as part of an experimental foods course usually need to conform to the guidelines established by the professor. Ordinarily, the sections included are:

- Introduction (stating the purpose and the justification for conducting the research)
- Review of literature
- Method (sample preparation and evaluation techniques, both subjective and objective)
- Results and discussion
- Conclusions (or Summary and Conclusions)
- Bibliography
- Appendices (optional, depending on the need for presentation of supporting information that cannot be placed appropriately in the text of the report)

Tables and other illustrative materials should be included in the report at the appropriate points to clarify the research findings to the reader. All illustrative materials should be referenced in the text and should have clearly-stated number designations and captions or titles.

In tables, each column needs to have a meaningful heading, and units can be indicated if the same designation is appropriate for all entries in a column. Solid lines above and below the area for the headings and also at the conclusion of the table are valuable guides to enable readers to grasp the content readily. If two or more subheads appear below a major column heading, the relationship can be shown by using a line under the major heading that extends over all of the pertinent subheads, as shown in Table 2.3.

Graphs provide a clear picture of data from experiments if they are constructed carefully. Both axes need to be labeled, and the units of measure must be identified. Accurate plotting of the data points will create an effective presentation of results from an experiment. A graph format suitable for reporting data is shown in Figure 2.7. Figure 2.8 presents data in the form of a bar graph. A pie chart (Figure 2.9) can be used when percentages or portions of the whole are to be depicted.

**Table 2.3** Tenderness and Volume of Angel Food Cakes Prepared with Three Different Sweeteners (Sucrose, Fructose, and Aspartame)[a]

| | Tenderness | | |
| Sweetener | Shear (lb) | Number of Chews | Volume (ml) |
| --- | --- | --- | --- |
| Sucrose | .002 | 5 | 175 |
| Fructose | .001 | 3 | 190 |
| Aspartame | .007 | 16 | 105 |

[a]Information in this table is fictitious and is presented simply to show the format of a table.

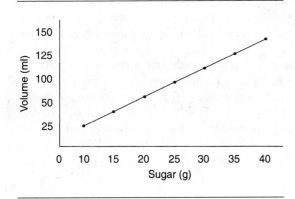

**Figure 2.7** Effect of sugar content on volume of plain cake.

## SUMMARY

Although much of the laboratory work in an introductory experimental food science class is done cooperatively by the entire group, a great deal can be learned through small, individual experiments conducted during at least a couple of laboratory periods. Ideally, this individual experiment will include identification of the problem, statement of the problem, review of related scientific literature, and design of the experiment. This design should include detailed procedures for preparing the sample and objective and subjective means of evaluating the products.

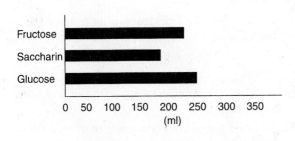

**Figure 2.8** Volumes of chocolate cakes made with different sweeteners.

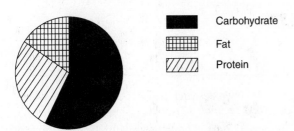

**Figure 2.9** Composition of rye bread.

By keeping a careful log using a computer and a bound laboratory notebook, the researcher can efficiently work with the data obtained. Preparation of tables showing all of the results will facilitate analysis. Much food research requires descriptive or inferential statistical analyses to determine the comparative significance of the findings.

Measures of central tendency that may be used in analyzing data include the mode, median, and mean. Percentages and percentile rank are other ways of expressing information gained in the study. Variance and standard deviation (square root of the variance) describe the dispersion of data.

Several types of inferential statistical tests are available to be applied in specific research designs. Chi-square is an inferential statistical test that is suited to a study with one variable and two categories or with two or more independent variables. Regression can be used to determine statistical significance in correlational research. Statistical testing of the significance of differences of the means of the experimental and control groups can be done using the Student's $t$-test.

ANOVA (analysis of variance) is of great use when analyzing many sets of data for statistical significance in a study. Levels of significance commonly identified are .05 and .01 (a confidence level of 95 percent and 99 percent, respectively) that the results are not due to chance.

Upon completion of the laboratory work and analysis, the entire project can then be written, either in the form required by a specific research journal to which it will be submitted or in the form specified by the professor. Appropriate tables and graphs should be included to support the findings.

## STUDY QUESTIONS

1. Write a statement of purpose that could be used for conducting an individual research project.

2. Outline a design for conducting the experiment identified in Question 1.

3. Select a research article on food in a professional journal; write a one-sentence statement of the purpose of the research.

4. Write the bibliographic citation for the article selected for study in Question 3.

5. Compare the types of articles and other information that appear in three different professional research journals.

6. Find examples of a table, a bar graph, and a pie chart in professional journals. Present the data from each one in a different format (e.g., plot a graph of the data provided in the table).

7. True or false. An independent variable can also be called the manipulated variable.

8. True or false. Percentages are an example of inferential statistics.

9. True or false. ANOVA is an example of inferential statistics.

10. True or false. Measures of dispersion are considered to be descriptive statistics.

11. True or false. Student's $t$-test is an example of inferential statistics.

12. True or false. Standard deviation is an example of a measure of central tendency.

13. True or false. The median and the mean are the same thing.

14. True or false. The mode and standard deviation are the same thing.

15. True or false. A level of confidence of 99 percent is the equivalent of a level of significance of .01.

## BIBLIOGRAPHY

Burns, R. B. 2000. *Understanding Research Methods.* 4th ed. Sage Publications. Thousand Oaks, CA.

Freedman, D., et al. 1997. *Statistics.* 3rd ed. W.W. Norton. New York.

Jenkins, M. L. Y. 1993. Research issues in evaluating "functional foods." *Food Technol.* 47 (4): 76.

King, J. W., et al. 2001. Thinking visually: Graphic tips for technical presentations. *Food Technol.* 55: 49–56.

Kurzer, M. S. 1993. Planning and interpreting "designer food" feeding studies. *Food Technol.* 47 (4): 80.

Labuza, T. P. 1994. Shifting food research paradigms for the 21st century. *Food Technol.* 48 (12): 50.

McDonald, P. 1995. Science libraries of the future: Research in the electronic age. *Food Technol.* 49 (4): 92.

Montgomery, D. C. 2004. *Design and Analysis of Experiments.* 6th ed. Wiley. New York.

Rupnow, J. and King, J. W. 1995. Primer on preparing posters for technical presentations. *Food Technol.* 49 (11): 93.

Rupnow, J. H., King, J. W., and Johnson, L. K. 2001. Thinking verbally: Communication tips for technical presentations. *Food Technol.* 55 (1): 46–48.

Turabian, K., et al. 2007. *Manual for Writers of Research Papers, Theses, and Dissertations.* 7th ed. U. of Chicago Press. Chicago, IL.

## INTO THE WEB

*http://www.crlsresearchguide.org/08_Focusing_A_Topic.asp* —Suggestions for focusing a statement of purpose for a research project.

*www.eatright.org/Public*—American Dietetic Association Web site.

*www.ift.org/publications*—Institute of Food Technologists, Food Technology, Journal of Food Science.

*www.foodproductdesign.com*—Food Product Design publication.

*www.adajournal.org*—Journal of the American Dietetic Association.

*www.ars.usda.gov*—Agricultural Research Service of the U.S. Department of Agriculture.

*www.usda.gov*—U.S. Department of Agriculture.

*www.fda.gov*—Food and Drug Administration of the U.S. Department of Health & Human Services.

*http://www.cpsd.us/crls/library/PDFs/Citation_Guides/cite_general_websites.pdf*—Citation guide for Web sites.

*http://www.experiment-resources.com/research-designs.html*—Information on types of research designs.

*http://mste.illinois.edu/hill/dstat/dstat.html*—Brief introduction to descriptive statistics.

*http://psychology.ucdavis.edu/Sommerb/sommerdemo/stat_inf/intro.htm*—Brief introduction to inferential statistics.

Fresh-cut apple slices like this one quickly turn brown and mushy when exposed to air. ARS chemist Dominic Wong was part of a team that discouered that certain calcium salts protect apple slice from changes in color, taste, or texture.

# CHAPTER 3

# Sensory Evaluation

## Chapter Outline

## OBJECTIVES

After studying this chapter, you will be able to:

1. Describe the sensory organs and their actions.
2. Identify the characteristics of food that can be evaluated by using human senses.
3. Explain how to select and train sensory panelists.
4. Decide the appropriate sensory testing to incorporate in a research study.
5. Design scorecards for sensory evaluation.

## THE IMPORTANCE OF EVALUATION

All people consciously or unconsciously evaluate food quality. Food choices in the marketplace are influenced by previous experiences with specific brands and various foods. Evaluations at home probably are not conducted scientifically, but they nevertheless determine whether or not a particular item will be bought again. Sales serve as an endorsement or proof of consumer acceptance, which then tells the food manufacturer that this quality level is preferred over that of similar products in the marketplace. Thus, the individual consumer's choice, combined with the decisions of countless other individual consumers, dictates the quality of food being produced in this nation.

Testing of food quality in the marketplace is too costly for food manufacturers to undertake on a broad basis without considerable preliminary research. **In-house testing** and evaluation are done on a scientific basis, with food scientists planning and supervising experiments. Careful and thorough tests are conducted to develop product formulations or processing techniques that are anticipated to be successful in the marketplace.

**In-house testing**
Evaluations conducted within a food company prior to field testing and test marketing.

45

**Sensory evaluation**
A synonym for subjective evaluation; measurements determined by using the senses of sight, smell, taste, and sometimes touch.

**Subjective evaluation**
Evaluation by a panel of individuals using a scoring system based on various characteristics that can be judged by using the senses.

**Objective evaluation**
Use of mechanical devices to measure physical properties of a food.

Trained sensory panelists evaluate the samples and provide guidance for improvement of the product. This type of testing is termed **sensory evaluation** or **subjective evaluation** because individual decisions based on the use of the senses determine the scores, not mechanical devices. However, mechanical testing usually is done as well to provide additional information about the food. This type of testing is categorized as **objective evaluation**.

Sensory testing done in-house measures the acceptability of the product to the staff, but it cannot be assumed that the general public will have the same assessment. Therefore, products often are tested further by focus groups of 4–12 typical consumers who fit the specific demographic characteristics of interest to the food company. Larger scale (200–500 consumers) testing can be conducted at a central location or by in-home testing of the product. Consumer tests provide valuable sensory information related to potential consumer acceptance of the product.

Subsequently, products may be test-marketed in a small area to obtain information from a selected sampling of the potential market. The results of this test determine whether the product is released into the general market, discontinued, or modified prior to general marketing.

Clearly, the food industry relies very heavily on evaluation in developing new products and in maintaining quality control in existing food items. These companies utilize evaluation techniques as basic tools. Both sensory evaluation and objective evaluation are vital sources of information to researchers and to people who are responsible for quality control.

Those who supervise institutional food service also need to evaluate quality and identify changes to enhance customer satisfaction. A professional can evaluate food precisely and relate this evaluation to the preferences of consumers. Studies of plate waste also provide valuable information regarding acceptability.

Professional evaluation requires careful analysis of foods and knowledge of the techniques for measuring their properties and characteristics. Practice in evaluation also is essential. This chapter presents the background needed for conducting meaningful sensory evaluation of food products.

## PHYSIOLOGICAL BASES OF SENSORY EVALUATION

### Olfactory Receptors

**Olfactory receptors**
Nasal organs capable of detecting aromas.

Odors are important preliminary cues to the acceptability of a food before that food enters the mouth. These odors are volatile chemical compounds that interact with the approximately 10 million **olfactory receptors** in the nose. Additional information is contributed by the olfactory receptors when food is in the mouth, because volatile compounds travel upward from the mouth to the nasal cavity where they are detected (Figure 3.1). The receptor sites for odors are organs housed in the upper area of the nasal cavity in the olfactory epithelium, which is a yellow, mucus-coated area.

---

### FOOD FOR THOUGHT:  Designing for Flavor

Anheuser-Busch is a name that causes consumers simultaneously to think of beer and the television image of the Clydesdale team pulling the beer wagon. The flavor of this company's Budweiser beer has been carefully monitored and gradually modified to meet the flavor expectations of consumers. The amount of hops was cut by more than half over the years to reduce the bitterness and aroma imparted by the cone-shaped hops flower used in the brewing process.

However, microbreweries and small beer companies have invaded the beer market in the last few years. The result is an array of beer flavors to appeal to the preferences of consumers, and niche markets have been created for a variety of beers with varying flavor and color. Small breweries are finding shelf space in local specialty markets and are cutting into the sales of the leaders in the national marketplace. To counter this threat, Budweiser is beginning to increase the quantity of hops to enhance the bitter taste and somewhat darker color that today's beer drinkers appear to be seeking.

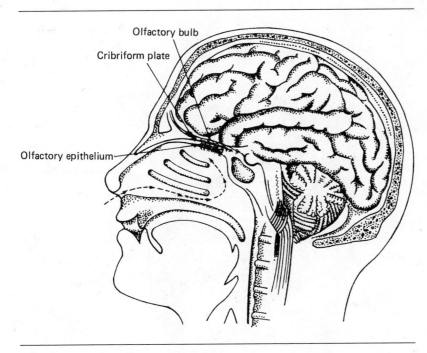

Olfactory bulb

Cribriform plate

Olfactory epithelium

**Figure 3.1**    Diagram of the olfactory receptors.

Odor perceptions occur when aromatic compounds migrate to the region in the nose where the olfactory receptors are located. Volatile compounds can be brought into contact with the oral receptors by (1) sniffing a food or its headspace, (2) swallowing the food (which creates a partial vacuum that draws the volatiles into the nose from the oral cavity), and (3) deliberately exhaling sharply after swallowing the food. The third technique of exhaling vigorously through the nose immediately after swallowing is particularly important in determination of **aftertaste**, the flavor that lingers in the mouth after the food is gone.

The olfactory sense is so sensitive that the nose can detect the presence of some odorous compounds at a concentration as low as $10^{-19}$M (molar). Individual variation in sensitivity exists as a consequence of possible nasal obstructions or sinus complications. Thus, some individuals are physiologically better suited than others to be sensory panel members even though they might be less well informed in the technical aspects of food science.

## Taste Receptors

Approximately 10,000 **taste buds**, most of which are found in the tongue, are the avenue by which the taste of food is perceived (Figure 3.2). Taste buds are found in the **papillae** of the tongue. Two types of papillae contain taste buds. The mushroom-like **fungiform papillae** on the sides and tip of the tongue generally contain taste buds, and the **circumvallate papillae** (elevated, large papillae in the form of a "V" toward the back of the tongue) always contain taste buds.

The taste buds on the tongue are of primary importance to adults in perceiving taste, but children have increased perceptions because of the additional taste buds they have in their cheeks, as well as in the hard and soft palates. These taste buds atrophy as children grow older. The number of functioning taste buds in adults is reduced to about 4,000–6,000, which is about half the number operating effectively in most children. The elderly have only about half as many functioning taste buds as they had in early adulthood. This trend of diminishing taste perception over the life span is typical, but there is considerable individual variation in the ability to taste at all ages.

Examination of the tongue shows that much of the surface is smoother than the regions containing the fungiform and circumvallate papillae. Fine-structured filiform papillae predominate over much of the tongue, and this type of papilla does not contain taste buds. Lack of taste buds in the comparatively smooth area of the tongue makes it necessary to shift

**Aftertaste**
The aromatic message of the flavor impression that lingers after food has been swallowed.

**Taste buds**
Tight clusters of gustatory and supportive cells encircling a pore, usually in the upper surface of the tongue; organs capable of detecting sweet, sour, salt, and/or bitter.

**Papillae**
Rough bulges or protuberances in the surface of the tongue, some of which contain taste buds.

**Fungiform papillae**
Mushroom-like protuberances, often containing taste buds and located on the sides and tip of the tongue.

**Circumvallate papillae**
Large, obvious protuberances always containing taste buds and distinguished easily because they form a "V" near the back of the tongue.

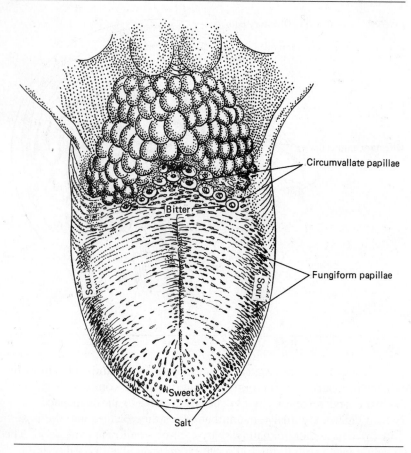

**Figure 3.2** Diagram of the taste receptor region of the tongue.

food around in the mouth so that all of the taste buds can be stimulated by the various compounds in the food. No stimulation can occur where the filiform papillae are found.

The taste message is initiated when a substance is dissolved on the tongue and comes into contact with the taste cell. The external shape of the receptor changes slightly with this contact. Potassium ions then escape from the cell, generating an electrical impulse to the brain. If many molecules are impinging on taste cells, the message will be relatively intense.

For many years four basic tastes were recognized: salty, bitter, sweet, and sour. In the 1980s, Dr. Kikunae Ikeda, who was researching flavor in seaweed broth, identified **umami**, which today generally is considered to be the fifth basic taste. L-glutamate and monosodium glutamate (derivatives of glutamic acid, an amino acid) are the substances that comprise umami to enhance the flavor of some foods.

These tastes are found in many foods in varying ratios and amounts. To perceive the basic tastes, the compounds imparting them need to be in aqueous solution. The ions that form transmit electrical messages along the nerves on the pathway from the taste buds to the brain.

Individual taste buds are capable of noting more than one basic taste, although the greatest ability to detect sweetness is in the taste buds on the tip of the tongue; saltiness also is noted particularly clearly by the taste buds on the tip of the tongue. Sour is detected most readily on the sides of the tongue and bitterness at the back of the tongue (see Figure 3.2).

The perception of a salty taste is caused by ionization of inorganic salts, such as sodium chloride. It is important to note that dry salt cannot be tasted, but when it is moistened by water or saliva, the taste is perceived. The other tastes—sweet, sour, bitter, and Umami— also require solution of the compounds for perception.

Hydroxyl ($OH^-$) groups are credited with contributing to the taste described as sweet. When dissolved in water, the various sugars have different apparent levels of sweetness. In pure solutions, fructose is the sweetest of the sugars, followed in descending order by sucrose, glucose, and finally lactose. Some amino acids, alcohols, and aldehydes also are perceived as sweet.

**Umami**
Taste sensation that enhances savory qualities of flavor but does not have a distinct taste by itself.

Sour is the taste impression created by the presence of the hydrogen ions ($H^+$). In the case of organic acids, the presence of the hydroxyl ($OH^-$) group may modify the apparent sourness of a compound.

The ability to detect phenylthiourea is found in approximately 75 percent of the population, but the remaining quarter cannot detect this bitter compound. This ability is genetic in origin, providing evidence that some people are limited as tasters. The bitter quality of phenylthiourea is attributed to the $-C = S$ moiety. This configuration is found in cabbage and turnips.

$$\underset{N}{\overset{|}{\phantom{x}}}$$

In addition to the basic tastes perceived by the taste buds, umami is perceived as a savory taste. Compounds such as **monosodium glutamate** and related amino acid-based compounds can serve as **flavor enhancers** and are sometimes added to savory product formulations.

Intensity of flavors can be enhanced or diminished under certain circumstances. The interactions of the pertinent compounds can react synergistically with **flavor potentiators** to provide a flavor experience that is greater than that of any of the components alone. These potentiators, unlike typical spices, enhance the quality of other substances already present rather than adding their own distinctive flavor. Monosodium glutamate is a familiar example of a flavor potentiator. Some 5′-nucleotides (such as monophosphates of inosinic and guanylic acids) are others that are used in meat and poultry products.

**Flavor inhibitors** appear to block the taste sites, thus preventing the normal taste response to a particular food. An example is the miracle fruit of Nigeria, which blocks the perception of sour. When sour perception is blocked in a fruit such as a lemon, the remaining taste is sweet.

Sometimes bitterness or other negative taste or flavor may need to be masked in a food that is modified to enhance nutritive value. Surprisingly, monoammonium glycerrhizinate (a derivative of licorice) is an effective flavor inhibitor. It not only masks bitterness, but is synergistic in promoting sweetness; it is as much as 100 times more effective as a sweetener than sucrose when the two are used together.

The level at which a taste can be noted is designated as the **threshold level**. This level varies somewhat from individual to individual. Below threshold levels, however, various taste compounds still can have an influence on the overall perception of taste. This effect of **subthreshold levels** enables salt to increase the apparent sweetness of a sugar solution or to reduce the apparent sourness of an acid; the effect on citric and acetic acids is less pronounced than that on tartaric and malic acids. Acid at subthreshold levels increases the apparent saltiness of sodium chloride. Sugar at subthreshold levels reduces saltiness, sourness, and bitterness (Table 3.1).

Tannins and various other polyphenols contribute to taste by creating a puckery sensation in the mouth, a characteristic that is easily detected when drinking tea that has been brewed for an extended period. This characteristic is called **astringency**.

Certain chemical compounds in foods can create feelings of coolness and heat (or even a bit of pain) in the mouth. These characteristics of coolness or pungency contribute to the flavor a food is perceived to have. Examples of compounds that feel cool in the mouth are menthol or peppermint and some sugar alcohols or polyols (e.g., xylitol and sorbitol).

**Monosodium glutamate**
Flavor potentiator; sodium salt of glutamic acid.

**Flavor enhancer**
Additive used to improve food flavor without contributing a specific identifiable taste.

**Flavor potentiator**
Compound that enhances the flavor of other compounds without adding its own unique flavor.

**Flavor inhibitor**
Substance that blocks taste perception.

**Threshold level**
Concentration of a taste compound at a barely detectable level.

**Subthreshold level**
Concentration of a taste compound at a level that is not detectable but can influence other taste perceptions.

**Astringency**
Puckery feeling in the mouth created by some compounds, such as tannins.

**Table 3.1**   Effect of Subthreshold Levels

| Ingredient Increased | Effect |
|---|---|
| Salt | Increases sweetness |
|  | Decreases sourness |
| Acid | Increases saltiness |
| Sugar | Reduces saltiness |
|  | Reduces bitterness |

**Capsaicinoids**
Capsaicin and related compounds responsible for the fiery quality of chili and other peppers.

**Glucosinolates**
Sulfur-containing irritants in mustard and horseradish.

**Flavor**
The sensory message blending taste and smell perceptions when food is in the mouth.

**Trigeminal cavity**
Olfactory receptors, taste buds, and oral cavity, the three parts of the body required for perceiving flavor.

Various compounds classified as **capsaicinoids** are the very pungent substances in chili peppers, black pepper, and ginger that give a burning sensation in the mouth and even on contact with the hands and eyes. **Glucosinolates** in mustard and horseradish contribute pungency and cause flavor excitement. Unlike the fiery substances in the capsicum or pepper family, these irritating sulfur-containing compounds impact the nasal passages, sometimes causing sneezing.

Ultimately, odor signals are blended with the taste messages from the tongue to provide the overall impression termed **flavor**. Flavor in food is extremely important in determining acceptability and quality. All of the factors discussed earlier contribute to what is collectively considered the flavor of a food. The word is simple, but its dimensions are complex. Perception of flavor in a specific food is determined by the combined action of the taste buds, the olfactory receptors, and the mouth cavity of the diner. Together, these three are sometimes referred to as the **trigeminal cavity**. The importance of the use of the nose as well as the mouth in detecting flavors is evident when nasal passages are congested by a cold.

## Visual Receptors

Some of the first cues a person receives about a food are its shape, texture, and color, messages that are received through the eyes. These pieces of information are possible because of the remarkable design of the eyes (Figure 3.3). In order to see, light must be refracted and the rays focused to give a sharp image. Actually, the image formed is an inverted one, but the brain enables the visual message to be inverted to the true physical orientation of the object. This is possible because of memory of objects gained from experience so that the inversion seems automatic.

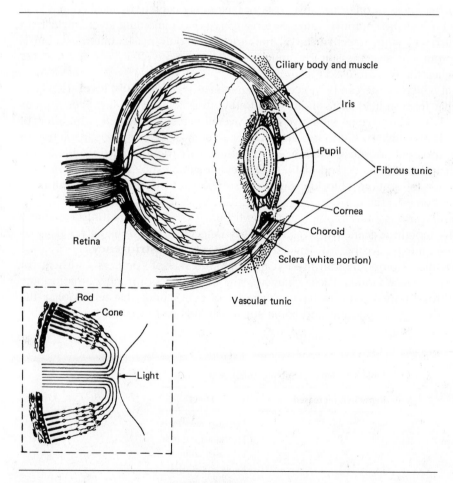

**Figure 3.3**    Diagram of the eye; inset shows rods and cones.

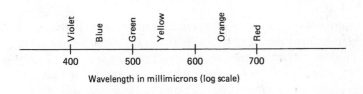

**Figure 3.4**   Wavelengths of colors in the visible spectrum.

The **rods** contribute to the formation of visual messages by breaking down rhodopsin into its two components (retinene and scotopsin) when light impacts them. This breakdown sends an electrical message that is conducted by the rods to the bipolar neurons, then to the ganglion neurons, and finally to the optic nerve.

A similar situation is found in the **cones**, but the red, green, and blue pigments in the three types of cones require a much stronger light for the generation of an electrical charge than is necessary for the rods. Because of this difference, color can be seen only when the light is considerably brighter than moonlight. Fortunately, the three colors that are available in the three types of cones provide messages that become mixed to give the rainbow of color that is seen in foods (Figure 3.4).

**Rods**
Elongated dendrites of photoreceptor neurons that transmit visual images in dim light, revealing movement and varying intensities of black and white.

**Cones**
Cone-shaped dendrites of photoreceptor neurons that enhance the sharpness of visual images and add the dimension of color to vision.

## SENSORY CHARACTERISTICS OF FOOD

Foods have several characteristics that require evaluation by sensory methods in order to gain knowledge of the human perception of them. These attributes include appearance, aroma, flavor, and texture. The following sections introduce the various characteristics and probe the qualities to consider when judging them as a part of a sensory evaluation.

### Toby, the Tea Taster

Lipton, a name that has long been a leader in producing and marketing tea around the world, gained its popularity because of its high quality and consistent color and flavor. Much of the credit for its tea was due to Toby, the man in his laboratory in Englewood Cliffs, New Jersey, who tasted teas brewed from leaves grown in Ceylon (now Sri Lanka), India, and other tea-growing regions in the world. His sensory evaluation determined whether or not he recommended that Lipton would buy the leaves from the various growers.

It was truly amazing to watch this diminutive, very proper British gentleman, dressed in a three-piece suit, conduct his tests. In preparation for Toby, an assistant brewed a cup from each type of tea being tested that day (perhaps as many as 20 different cups). The tea leaves from each cup were displayed on a lid covering the hot beverage. When Toby walked in, the assistant was ready to take careful notes as testing began.

First, Toby picked the damp leaves off the lid, rolled and rubbed them around in his hand as he looked at them carefully, and finally inhaled deeply while holding them near his nose. Then he picked up the cup and, with a gigantic slurp, transferred a sample of the beverage into his mouth. Emily Post or Miss Manners would never have approved, but this was a most effective way of engaging his olfactory receptors with his taste buds. To add even more information to his assessment, he vigorously swished the beverage all around his mouth to assure maximum contact. When he had made his decision regarding the sample, that test was ended by very accurately spitting the whole mouthful into the spittoon standing nearby on the floor. That was the grand finale for the first sample. Instructions were given to the assistant regarding the purchase decision, and Toby was ready for the next sample, and the next, and the next, until all had been evaluated. His three-piece suit still looked like he was dressed to enter a Board room, and Lipton was ready to buy the right leaves from distant ports to maintain its consistent, high-quality tea.

## Appearance

The appearance of a food is evaluated on the basis of several subcategories. Perhaps the aspect of appearance that is often deemed the most critical is color. Extensive use of food-coloring agents in commercial products attests to the value placed on color appeal in foods.

In fact, color often triggers the mind to expect particular flavors. Commonly, people will expect a red-colored food to have a flavor of strawberries or other red fruit and will identify such a food as being strawberry even when it is something quite different in flavor. Because the color of food establishes expectations of the actual product, evaluation and control of color are important aspects of product development and production.

Expectations of richness in vanilla puddings are generated by a creamy color. Uniform, golden-brown crusts generally are desired in baked products. Browning units have been added in some microwave ovens to produce the expected crust color in baked products. Browning also is an important aspect of cooked meats.

Although color can be measured objectively, this attribute is so important from the human standpoint that it should be included on scorecards for sensory evaluation. Color is evaluated by sensory methods to obtain information on the desirability or acceptability of the food color in human terms—that is, the psychological importance of the color of the food.

Surface characteristics of food products also contribute to their appearance. A baked custard with a very wrinkled surface does not meet accepted standards for the product. Scrambled eggs with a very dry surface also would be rated as less acceptable than those with a suggestion of moistness. Fudge with a glossy surface is rated high, whereas a batch with a coarse, gritty appearance would be scored considerably lower. These are but a few examples of surface characteristics.

Each food can be evaluated on the basis of the desirable surface features commonly understood to be signs of quality in the particular food item. In some products, the volume of the item will influence the evaluation of appearance. This clearly is true in the rating of a soufflé or cake.

In addition to exterior appearance, many products need to be evaluated on the basis of their interior appearance. Lumps in a pudding or gravy are visible to the eye, as well as obvious on the tongue. Cell size, uniformity of cells, and thickness of cell walls are all of interest in assessment of cake quality, and these are judged by appearance of the interior of the cake. The presence of layering in foam cakes and soufflés also is noted easily by checking the interior appearance. Often, interior appearance is judged most conveniently and accurately by cutting a clean slice with a sharp knife from the top to the bottom of the product.

## Aroma

The odor or aroma of foods frequently is of considerable importance, particularly if the food ordinarily is served hot or warm. The acceptability of the aroma is important to the overall acceptability of the food. A pleasing aroma beckons people to sample the food, whereas a strong, irritating aroma discourages diners.

Aroma can penetrate from a distance when comparatively volatile compounds are abundant, as is true in boiling cabbage, for example. The aroma of boiling cabbage often is quite identifiable in another room, even when the cabbage cannot be seen.

The volatility (and, therefore, the detection) of aromas is related to the temperature of the food. High temperatures tend to volatilize aromatic compounds, making them quite apparent for judging; cool or cold temperatures inhibit volatilization. This latter observation is illustrated by considering the evaluation of the aroma of ice cream, a test that clearly would provide little information. Because aroma is influenced so much by temperature, it is important that aroma of foods is judged when they are at the temperature at which they ordinarily would be served and consumed.

Aroma can be evaluated by sniffing the food. It may be helpful to the judge to fan the air above the sample with the hand to direct the aromatic compounds toward the nose. In planning experiments, it is important to avoid competing aromas from different samples. The nose quickly becomes saturated with odors.

Tasting booths or other areas where aroma is to be evaluated should be free of extraneous, competing aromas. Ordinarily, when evaluating aroma, judges concentrate on acceptability, but in some experiments the relative strengths of aroma of the various samples may need to be determined.

## Flavor

Flavor represents the composite assessment of taste and the blend of odor in the mouth. This is a very important attribute of a food and yet is difficult to communicate. Often, the mechanism for evaluating flavor subjectively is simply the level of acceptability of the total flavor.

Occasionally, the presence of an aftertaste may be of concern, as may be the case with saccharin-sweetened items. This aspect of flavor should be assessed as a separate entry on the scorecard rather than being encompassed within the single rating for flavor.

The apparent hotness or burning sensation from a highly seasoned food may be another characteristic related to flavor that is best assessed as a separate category on the scorecard. Some other products may require assessment of a particular component of the flavor, such as the comparative sweetness of samples.

The term *hot* is used in two ways in food evaluation. One definition refers to the physical temperature; the other refers to the burning sensation in the mouth after a spicy food, such as hot peppers, is eaten. Hot (spicy) foods may effectively mask subtle taste and odor evaluations.

The temperature at which a food is served may have an important influence on the ability to detect taste and to evaluate flavor. The extremes, whether very hot or very cold, limit the ability of people to judge food accurately. The best temperature range for flavor evaluation is 20–30°C (68–86°F). However, this range may be inappropriate for evaluation if the food is served either above or below this temperature range. Ice cream provides a clear example of the importance of evaluating a food at its serving temperature rather than at the temperature range ideal for detecting taste and flavor.

## Texture

*Texture* is an expansive term requiring careful definition for persons serving on a sensory panel, as well as thoughtful inclusion in the scorecard. The textural qualities of a food have a relationship to the appearance of a product, as described previously, and to its evaluation in the mouth, which relies on the **mouthfeel** of the food.

**Mouthfeel**
Textural qualities of a food perceived in the mouth.

The specific aspects of texture that are to be evaluated sometimes need to be listed in separate categories on the scorecard, but often the item is listed simply as mouthfeel. This is quite a general term and may lead to confusion unless each judge is informed of the specific aspect of mouthfeel that is to be evaluated. For instance, for a corn chip, one judge might be evaluating crispness to obtain the score for mouthfeel while another judge may be reporting on tenderness.

The various aspects of mouthfeel include grittiness, slickness, stickiness, hardness, crispness, toughness, brittleness, pastiness, lightness, crunchiness, smoothness, viscosity, moistness, burning, cooling, astringency, spiciness, and tingling. Not all of these are appropriate for any single food, but they suggest characteristics to be considered when evaluating texture. Acceptable mouthfeel is vital to repeated consumption of food products and must be developed optimally for a product to be successful in today's competitive marketplace.

The tenderness of a number of items can be evaluated meaningfully by querying the judges regarding this textural feature. For baked products, for example, tenderness may range from samples so tender that they readily become nothing but crumbs to products so tough that they are extremely difficult to bite or chew. The researcher ordinarily seeks a product with optimal tenderness, neither too tender nor too tough. Judges can convey this information by using a guide provided on the scorecard.

---

**FOOD FOR THOUGHT: Crispy or Crunchy?**

Each person biting into an apple probably has a different expectation of what it will feel like to take that bite and then chew it. Will the apple give a sharp snap when bitten, or will the flesh be rather soft, even slightly mealy? Apple growers want their crop to have the texture consumers want.

An interesting way of studying food textures from a sensory approach is being used at the Institute of Food Research in Norwich, England. As one example, four sensors were attached to measure the action of the main jaw muscles, and subjects also indicated when they swallowed, as well as how intense they perceived the flavor to be. The purpose was to determine the patterns subjects defined when they chewed and swallowed a bite of apples of different varieties. Using a supercomputer to analyze the enormous amount of data gathered in testing many volunteers, researchers determined that each person has a unique pattern, which allows identification of the person who made the specific pattern.

Differences in apple textures also were identified using this method. Crisp apples that were considered crunchy required about 20 seconds more chewing time than did mature, softer apples. Researchers hope to be able to develop this type of sensory evaluation to the point where it can aid in bringing apples that meet consumer expectations to market with less waste.

---

**Number of chews**
Subjective test in which a judge chews similar bites of food to the same endpoint and records the actual number of chews required to reach that point for each sample.

An alternative means of reporting tenderness is to ask the judges to report the **number of chews**. Individual judges control their technique for this test. The judge is instructed to use a bite of controlled size for the number of chews test and to chew the sample in the same location in the mouth to exactly the same endpoint for each sample tested. This number is recorded as the number of chews. Although different judges will have different numbers of chews because of differences in their tooth surface area for chewing, the relative scores of the various judges for the same samples should be consistent in their rank order. Since judges perform this test, it is considered a sensory evaluation device, but the mechanical nature of the chewing also makes it possible to view it as a somewhat objective testing method.

## SENSORY PANELS

Classes in experimental foods usually will employ two different formats for sensory evaluation. The laboratory experiments conducted by the entire class offer each student the opportunity to evaluate all of the products prepared by the class so that the study of scientific principles in food is reinforced by the laboratory examples. The second type of evaluation is that done by a sensory panel to evaluate the products prepared in individual research projects, which often are conducted as part of the course. This latter type of evaluation is similar in nature to that done by the food industry in research and product development.

### Selecting Panel Members

Panel member selection should be based on the factors identified as important for the specific study. Laboratory research frequently requires judges to be able to detect very subtle differences in aromas and flavors of products. In experiments where a specific capability is needed, potential panel members can be screened for their ability to discriminate on key aspects. This permits selection of a panel that is physically qualified to serve.

Laboratory testing often extends over a period of weeks or months, which dictates that the personnel ordinarily will be regular employees of the company so that they will be available for the tests when needed. From this potential pool, the actual panelists are chosen. Interest in participation is necessary to ensure that panelists meet their obligations as panel members.

The health of the potential panelist is important because of the need to be available for tests on a regular and continuing basis. Clarity of the nasal passages also is necessary for flavor and aroma assessment. Persons with chronic sinus conditions or colds may be poor panelists because of their physical inability to perceive flavors. These physical qualifications are more important for serving as a panelist than is depth of education in food science, although such knowledge is an asset.

Quite the opposite situation exists when a food company wishes to obtain information about consumer acceptance of a product that is ready to enter test marketing. Consumers who happen to be available at the testing site and who are willing to answer a few questions about the samples on the day of the testing usually constitute the **consumer panel** for this type of testing. Panelists can be screened based on a company's criteria (e.g., demographics or potential use of product). Typically, consumer panels may range from 200 to 500 people, which is far larger than the panels used in research laboratories. The questions asked of consumer panels (acceptability or preference) should be answerable by untrained panelists.

**Consumer panel**
Sensory evaluation panel selected from people who happen to be available at a test site and are willing to participate.

Panels may be of two types: untrained and trained. The **untrained panel** is composed of participants with no preparation regarding evaluation of the product. Consumer panels often are made up of untrained panelists.

**Untrained panel**
Sensory evaluation panel that has not been trained regarding the use of the scorecard and the evaluation of the various product characteristics.

## Training Panelists

The training of panelists for laboratory testing varies with the complexity of the testing. As an absolute minimum, members of a **trained panel** should be briefed prior to the actual collection of data; the researcher should review the scorecard with the judge, clarify any questions, and ask the judge to explain orally to the researcher what each score represents. This verbalization regarding the meanings of the scores enables the researcher to identify differences in the interpretation of terms among the various judges. When differences exist, the scorecard may require modification, but sometimes an explanation to the judges may be sufficient.

**Trained panel**
Sensory evaluation panel that has been thoroughly trained regarding the use of the scorecard and the evaluation of various product characteristics.

Another technique that sometimes is effective is to have all panel members discuss the product and the scorecard together the first time they evaluate the samples. The goal of this aspect of training is to ensure that all judges use the scorecard according to the same word definitions. Unless adequate work is done with the panel at this stage, experimental error often is introduced because of the differences in interpreting the terms on the scorecard. Such differences sometimes can be caused by the judges or by the researcher, whose familiarity with the project may cause interpretations of the scorecard that were not intended by the judges.

Another aspect of panel training is teaching any special sampling techniques. A common technique is for panel members to rinse their mouths with room-temperature water before they taste the samples. After evaluation, the residue can be spat into a disposable cup rather than swallowed. Samples should be tested in the sequence in which they are presented to the panelist.

Ordinarily, judges are not informed of the variable(s) being tested. Every attempt should be made to avoid biasing the judges by giving them information regarding the purpose of the experiment. Such an explanation is not appropriately part of the panel-training process.

Some food companies form a **descriptive flavor analysis panel (DFAP)**, which is trained to analyze flavor in extremely great detail (Figure 3.5). This type of panel evolved from the original work, the **flavor profile method**, developed by the Arthur D. Little Company. Members of these panels need to be selected on the basis of their communication skills, as well as their knowledge of food and food processing.

**Descriptive flavor analysis panel (DFAP)**
Thoroughly trained panel that works as a team to describe precisely in words the flavor of a sample.

The great amount of time involved in training flavor profile panelists and in their evaluation sessions underlines the importance of having committed and permanent employees as panelists. Training of panelists for a DFAP requires at least nine weeks of intensive work to develop the consistency of ratings required to score products accurately and consistently. As much as $4\frac{1}{2}$ months may be involved in training for certain panels.

**Flavor profile method**
Sensory evaluation method; highly trained panelists talk together to develop a detailed description of the flavor of a product.

**Figure 3.5**    Descriptive flavor analysis panel tastes a variety of commercial and formulated spaghetti sauces. Before conducting a sensory analysis on sauce enhanced with Provesta flavor enhancers, the panel must be fully oriented to the flavors found in typical spaghetti sauces. (Courtesy of Provesta Corporation.)

## ENVIRONMENT FOR SENSORY EVALUATION

With the exception of the profile panels, trained judges ideally will not have an opportunity to interact with one another while evaluating the samples. Individual tasting booths are recommended to avoid potential interactions between panelists. These are fairly compact compartments with a space for the samples to be placed and for the scoring to be done. Suitable side partitions prevent eye contact or exchanges between judges. The booths may have small sinks for spitting out samples, and the light will be appropriate for the samples being tested.

The temperature should be comfortable and the air free of odors other than those from the samples. Smoking should not be permitted in the area. Appropriate light may include a colored light, usually red, to obscure food color if the color is anticipated to bias the judges—for example, the evaluation of meats cooked to varying degrees of doneness. The varying colors of the samples will bias the judges' decisions unless red light is used to mask the color differences between samples.

When individual booths are not available, judges can be seated in quiet surroundings and preferably in different parts of the room to prevent interaction. If at all possible, the judges should conduct their evaluations away from all other people. Concentration is necessary for a quality performance in food evaluation.

The American Society for Testing and Materials has issued guidelines for the physical requirements of a laboratory designed for sensory evaluation of food. Positive air pressure, filtered air, and controlled temperature and lighting are some of the key aspects of the guidelines. Samples for evaluation should be prepared far enough away from the testing area to avoid any possible cooking aromas in the tasting area.

## SAMPLE PREPARATION AND PRESENTATION

An important aspect of planning an experiment is developing the evaluation devices. Included in this planning is the need to know how the sampling of the products will be done, both for sensory and for objective testing. Careful thought needs to be directed toward anticipating all factors that could modify judgment of the samples. For instance, it is necessary not only to decide on the size of the piece of cake to be given to a judge, but also to identify the specific location from which each individual judge's piece will be removed. Reliable data cannot be collected for sensory evaluation if the edge of a cake is used one time and the center is used another time.

Consistency from day to day is vital to the successful conduct of experiments. For many products, a template should be developed to serve as a guide in the preparation and distribution of the samples for the judges and also for the objective tests. Using a template reduces the likelihood of variability in the samples used for evaluating each run. Sample size need not be large. A liquid sample of 15 milliliters or a solid of about 30 g often is enough for panelists.

Plans for sampling must include directions for heating or chilling samples that require this type of control. In a busy laboratory, chilling may prove to be a particular problem because of the frequent opening of refrigerators.

Judges should be given samples that are carefully marked with a symbol that does not connote ratings or quality; for example, a random number lacking meaning (e.g., 667), or letters in the middle or end of the alphabet. Using *A, B, C* or *1, 2, 3* is not recommended because of their possible influence on judges. Symbols are rotated among samples at each session to eliminate bias. A wax pencil can be used to mark the sample plates with the appropriate identifying symbol.

The necessary silverware, a glass of water at room temperature, and a scorecard with a pencil should be placed, along with the appropriately labeled samples, in the scoring area.

## TYPES OF TESTS

Overall, types of sensory evaluation include (1) descriptive, (2) affective, and (3) difference testing. **Descriptive testing** enables researchers to characterize their products through selective, critical scoring of specific attributes of each sample and thus requires a panel that is well trained. Profile methods are a form of descriptive testing. Judges can do descriptive testing by acting individually to evaluate samples using the detailed scorecard developed by the researcher.

**Descriptive testing**
Using descriptive words in sensory evaluation to characterize food samples.

**Affective testing** is valuable in developing new food products and in evaluating quality. This type of testing requires judges to evaluate the acceptability of the samples. Preference for one product over another may be determined by affective testing. Ranking samples in order of preference is another way of using affective testing.

**Affective testing**
Sensory testing to determine acceptability or preference between products.

Hedonic (pleasure) scales in affective testing range from one extreme to the other—for example, like very much to dislike very much. Consumer panels often are used for this type of affective testing.

**Difference testing** (often referred to as discrimination) can be used to test the sensitivity of judges, as well as to perform such a practical function as determining whether or not a food company should buy an inexpensive ingredient to replace a more expensive one in formulating a food product. Researchers need to define the information desired before deciding on the type of test(s) to be conducted.

**Difference testing**
Sensory testing designed to determine whether detectable differences exist between products.

### Single Sample

Particularly in the early stages of product development, presenting a **single sample** can be useful in defining the direction in which the project should move. This type of test can be designed to test acceptability. The advantage of this method is the ease of conducting the test.

**Single sample**
Presenting one sample early in an experimental project to determine acceptability and to aid in the decision on future development of the product.

### Difference Testing

***Paired comparison.*** The **paired comparison** is a test of difference in which a specific characteristic is designated. The judge is asked to test the two samples presented to identify the sample with the greater amount of the characteristic being measured. The judge has a 50 percent chance of being right by chance alone in paired-comparison testing (Figure 3.6).

**Paired comparison**
Difference test in which a specific characteristic is to be evaluated in two samples, and the sample with the greater level of that characteristic is to be identified.

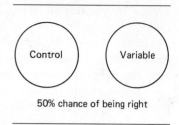

**Figure 3.6**   Diagram for presentation of a paired-comparison test.

**Duo–trio test**
Difference test in which two samples are judged against a control to determine which of the two samples is different from the control.

**Triangle test**
Difference test in which three samples (two of which are the same) are presented, and the odd sample is to be identified.

**Rank-order test**
Preference or difference test in which all samples are ranked in order of intensity of a specific characteristic.

***Duo–trio test.***   The **duo–trio test** is another test of difference. In this test, the control sample is presented first; it is followed by two other samples, one of which is the same as the control. The judge is requested to identify which of the last two samples is different from the control. Again, there is a 50 percent chance of being right by chance alone in a duo–trio test (Figure 3.7).

***Triangle test.***   As in the duo–trio test, three samples are given in the **triangle test**, but all three samples are presented simultaneously. The judge must identify the odd sample. Note that in the triangle test also, two samples are alike; however, the difference in the method of presentation reduces the chance of guessing the right answer to 33.3 percent. The triangle test is designed to determine difference (Figure 3.8).

***Rank order.***   The **rank-order test** is valuable when several samples need to be evaluated for a single characteristic. In this type of difference or preference testing, the samples are simply ranked in the order of intensity of the characteristic being measured.

## Descriptive Testing

Difference testing detects deviations between samples, and preference testing provides information about acceptability of samples. However, **descriptive testing** is a dimension of food sampling that is not available with the other two forms of testing. Words describing the various sensory attributes of food samples are essential to the success of this type of evaluation, and yet the precise vocabulary to convey sensory perceptions is difficult to identify. For example, even as familiar a food as an onion is frustratingly hard to characterize in words that accurately convey such sensory characteristics as aroma, flavor, and texture.

Descriptive testing usually is conducted using a scorecard containing precise word descriptions that structure the form of responses by judges. Each of the characteristics of a sample to be evaluated by the judges is described over a range, and the judge selects the specific description matching the sample for each item on the scorecard. The responsibility for selecting the appropriate vocabulary to elicit an accurate picture of the samples rests

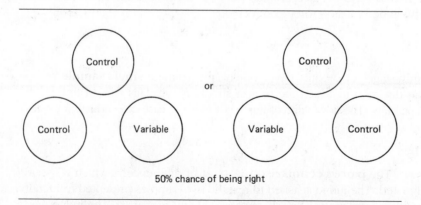

**Figure 3.7**   Diagrams for two possible presentations of a duo–trio test.

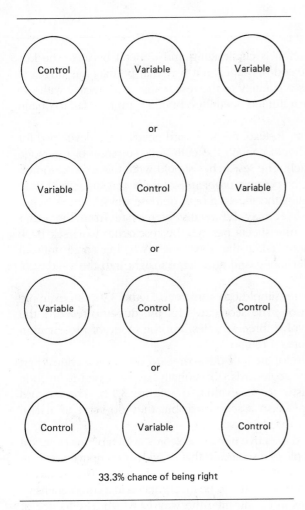

or

or

or

33.3% chance of being right

**Figure 3.8**   Diagrams for possible presentations of a triangle test.

with the researcher who developed the scorecard, and a well-constructed scorecard will give the desired information.

In another approach to descriptive testing, a group of highly trained panelists (DFAP) works together to develop the vocabulary needed to provide specific descriptions of food samples. Unlike the first method in which judges evaluate individually, the whole panel arrives at a single description in this second method, called **profiling**.

Several food researchers have developed techniques and language for evaluating texture by profiling. The great diversity in textual properties makes this aspect of sensory evaluation quite complex. Among the qualities that a panel doing texture profiling might consider are chewiness, cohesiveness, adhesiveness (to the palate, hands, and lips), moistness or wetness, denseness, gumminess, fracturability, hardness, mouth coating, flakiness, fibrousness, and viscosity. Texture profiling can be done in considerable detail by a trained panel, just as is true for flavor profiling.

**Profiling**
Detailed word description (usually of flavor) developed by a highly trained panel against which subsequent production is evaluated to maintain quality of production.

## Affective Testing

The preferences and attitudes of consumers toward food products are vital to success in the marketplace. Therefore, consumer panels often are used to indicate preference of one sample over another. A specific quality (e.g., sweetness) can be identified on the scorecard, and the judge then rates preference for one of the samples. Hedonic rating scales can be used to measure the degree of pleasure experienced with each sample. Sometimes, the motivation regarding the frequency that a judge might desire to eat the sample is measured as an approach to determining the acceptability of the various samples.

## SCORECARDS

Designing scorecards for sensory evaluation is challenging and difficult because the key qualities of the product need to be evaluated on paper in a way that permits the judges to transmit their assessments of the samples accurately to the researcher. A scorecard with too much detail and clutter may discourage careful judgment; too brief a form may fail to obtain some important information.

No single scorecard fits all experiments. Instead, the scorecard needs to be developed for the specific experiment. All scorecards should contain the date and the name of the judge (unless it is an anonymous consumer panel). The researcher should write this information on the sheets before the judging. This avoids the possible problem of a judge forgetting to identify the sheet. It is important to be able to identify the sheets in both of these ways.

Sometimes a data point from a sheet may be relatively remote from the other comparable data points. Identification of the sheets permits the researcher to check back into the laboratory notebook and the notes taken that day to see whether some unusual circumstance might explain that particular rating and add some insight into the validity of the data point in question.

Scorecards for difference testing are the simplest to construct. Figure 3.9 is an example appropriate for use in a paired comparison. The scorecards for the duo–trio test and the triangle test are similar, because both involve three samples, but the mode of presentation leads to slightly different scorecards (Figures 3.10 and 3.11).

It is in the **descriptive scale** that development of the scorecard becomes a critical part of the planning for an experiment. These scorecards occasionally are designed to include careful descriptions of each characteristic to be evaluated (Figure 3.12). These word descriptions are quite cumbersome to place in a scorecard, but they do serve as useful guides to the scoring of the product.

Descriptive scorecards can be structured to derive numerical scores, which permit statistical analysis of the data. Figure 3.13 is an example of a scorecard that provides descriptive words in a logical progression for some characteristic that is to be evaluated by the judge. The characteristics included are only some of those that probably should be incorporated in a comprehensive evaluation of a muffin. However, the ones listed illustrate the way in which a sequence of descriptive words can be developed. Note that the researcher could easily assign numbers in sequence to each of the scales (e.g., very pale = 1; slightly pale = 2; golden brown = 3; slightly too dark = 4; and burned = 5). Thus, the judges' scores would be recorded as the appropriate numbers, rather than words. The numbers can then be used for analysis; ultimately the words on the scorecard that appropriately describe the numerical results can be added to the discussion of results.

Rating scales are ordinarily designed with an odd number of points, usually 5 or 9. A 5-point scale is convenient, but some judges think that such a limited scale restricts their

**Descriptive scale**
Array of words describing a range of intensity of a single characteristic, with each step on the scale representing a subtle degree of intensity.

Figure 3.9    Scorecard for paired-comparison test.

Judge _____
Date _____
The more tender sample is _____ .

Figure 3.10    Scorecard for duo–trio test.

Judge _____
Date _____
Compare each of the two samples with the control sample. The sample that is different from the control is _____ .

```
                    Judge _____
                    Date _____
Two of these samples are the same, and
one is different. The different sample is
_____ .
What was different about this sample?
```

**Figure 3.11**   Scorecard for triangle test.

ability to communicate fine levels of difference to the researcher. In contrast, a 9-point scale permits finer distinctions to be noted because the range represented by each score is smaller.

Some characteristics can range from one extreme to the other. An example might be the crust color of a white cake, which might range from very pale to too dark. The points on the scale should be balanced so that the most desirable score is at the midpoint of the scale.

A common problem in the construction of a scorecard is illustrated in Figure 3.14. Note that in the categories crust color, interior color, contour of surface, thickness of cell walls, cell size, and flavor the optimal score is 3; however, in the scoring of aftertaste, the optimal score is 1. This arrangement can be used, but judges should be cautioned that the scoring for the last category deviates from the established pattern. The researcher, in handling the data and writing the project report, will need to keep the difference in the meaning of the numbers in this category firmly in mind and will need to reinforce this for the reader of the report.

When the goal of an experiment is to identify significant changes resulting from variations such as the use of different ratios of an ingredient or different ingredients, specific information may be needed from the judges. These scorecards require considerable thought regarding the characteristics that may be important in the testing and the way in which these characteristics will be recorded. Many variations of such scorecards can be made. The scorecard presented in Figure 3.15 is one suggested approach to this evaluation problem.

Name _____
Date _____

| Characteristic | Descriptive rating (circle the term that best applies to each characteristic) | | | |
|---|---|---|---|---|
| Color of surface | Pale | Slightly brown | Pleasing golden brown | Very dark brown |
| Tenderness | Extremely crumbly | Somewhat crumbly | Easily broken, slightly crumbly | Tough, little tendency to crumble |
| Texture | Heavy, thick cell walls | Slightly heavy, medium-thick cell walls | Somewhat coarse cell size, but relatively uniform | Tunneling toward top, fine cells between tunnels |
| Flavor | Aftertaste | Very slight aftertaste | Pleasing, no aftertaste | Burned |

**Figure 3.12**   Descriptive scorecard for rating muffins.

| Characteristics of muffins (Circle your choice for each) | | Judge _____ Date _____ Sample _____ | | |
|---|---|---|---|---|
| *Crust color* | | | | |
| Very pale | Slightly pale | Golden brown | Slightly too dark | Burned |
| *Contour of surface* | | | | |
| Sunken | Flat | Gently rounded | Somewhat pointed | Pointed |
| *Interior color* | | | | |
| Stark white | White | Creamy | Somewhat yellow | Yellow |

**Figure 3.13**    Descriptive scorecard illustrating logical progressions suitable for numerical interpretation of data.

Preferably, the items included for evaluation on a scorecard should be arranged in the same sequence in which they logically will be evaluated. Noninvasive tests (e.g., aroma, exterior color, and other external characteristics that may be included in the scoring) should be placed at the beginning of the list, as in Figure 3.14. This ensures that the judge will have the sample intact to evaluate these characteristics.

| Scorecard | Judge _____ Date _____ | | |
|---|---|---|---|
| | **Sample** | | |
| *Characteristic* | 517 | 727 | 464 |
| Crust color    1 = pale; 3 = golden brown; 5 = burned | | | |
| Contour of surface    1 = flat; 3 = rounded; 5 = pointed | | | |
| Interior color    1 = white; 3 = creamy; 5 = yellow | | | |
| Thickness of cell walls    1 = very thick; 3 = normal; 5 = too thin | | | |
| Cell size    1 = small; 3 = moderate; 5 = large | | | |
| Flavor    1 = unsweet; 3 = pleasing; 5 = too sweet | | | |
| Aftertaste    1 = none; 3 = slight; 5 = distinct | | | |

**Figure 3.14**    Detailed scorecard for evaluating muffins with various sweeteners.

| | Sample | | |
|---|:---:|:---:|:---:|
| *Characteristic* | 517 | 727 | 464 |
| Crust color<br>  1 = much too pale; 3 = somewhat too pale;<br>    5 = pleasing golden brown; 7 = somewhat<br>    too brown; 9 = much too brown | | | |
| Contour of surface<br>  1 = absolutely flat; 3 = somewhat rounded; 5 =<br>    pleasingly rounded; 7 = somewhat pointed;<br>    9 = very pointed | | | |
| Interior color<br>  1 = much too white; 3 = somewhat white; 5 =<br>    pleasingly creamy; 7 = somewhat too yellow;<br>    9 = much too yellow | | | |
| Thickness of cell walls<br>  1= extremely thick; 3 = somewhat too thick;<br>    5 = normal thickness; 7 = somewhat too thin;<br>    9 = much too thin | | | |
| Cell size<br>  1 = much too small; 3 = somewhat too small;<br>    5 = moderate; 7 = somewhat too large; 9 =<br>    much too large | | | |
| Flavor<br>  1 = absolutely not sweet enough; 3 = not<br>    nearly sweet enough; 5 = pleasingly sweet;<br>    7 = somewhat too sweet; 9 = much too<br>    sweet | | | |
| Aftertaste<br>  1 = extremely distinct; 3 = somewhat distinct;<br>    5 = none | | | |

Muffin scorecard[a]

Judge _____
Date _____

[a]You may also use numbers 2, 4, 6, and 8. These values are considered to be midway between the preceding and subsequent descriptions.

**Figure 3.15**   Modified form of Figure 3.14.

On the interior of the sample, physical characteristics that can be assessed visually (color of crumb and other characteristics that can be evaluated without eating) are the next items listed. The last items are characteristics such as mouthfeel, which can be evaluated only by placing a bite of the sample in the mouth. This arrangement permits an orderly evaluation of the sample as the judge simply works straight down the scorecard. Without this organization of the card, judges might skip around the card to test the various items and then discover that they do not have an appropriate sample left to score certain items.

Each item on a scorecard needs to be worded specifically so that only one quality is scored. For instance, cell size and uniformity could not be listed as a single entry on a scorecard. Two separate entries would be needed if both cell size and cell uniformity were to be evaluated.

The design of the scorecard in Figure 3.13 makes it necessary to give separate scorecards for each sample; each card must have the random number or symbol of the sample clearly marked, and the appropriate scorecard should be placed with its corresponding sample. Figure 3.14 provides information similar to that obtained using the format of Figure 3.13, but judges can mark their scores for the three samples in the columns provided. In this case, judges will need to be instructed that a score of 2 may be assigned for a product that falls between 1 and 3 (or 4 for one between 3 and 5).

The scorecard in Figure 3.14 can be strengthened considerably by modifying the scale to move the optimum score to 5 on a 9-point scale, as shown in Figure 3.15. The expansion of the scale provides the opportunity for finer distinctions between items such as crust color in which the optimum score is 3 when the scale is restricted to only five possible scores, as is the case in Figure 3.14.

Note that the scale for *aftertaste* in Figure 3.15 ends at 5 rather than continuing to 9. The scale cannot be extended beyond the ultimate, which needs to be the optimum value of 5. Other categories may need to be handled in this same fashion on scorecards dealing with various products.

When constructing the rating scale for a consumer scorecard, it is important to decide whether (1) a specific description of the characteristics or (2) acceptability is the information needed. For instance, judges might be asked to rate the sweetness of samples as not sweet, barely sweet, sweet, quite sweet, extremely sweet. Even if they all agree that the same sample is the sweetest, the researcher has no way of knowing whether the sweetest sample pleased the judges the most. Since satisfaction with a food product is important to maintaining repeat sales over time, researchers developing new products must be concerned with the hedonic ratings, not just the descriptive scores.

**Hedonic scale**
Pleasure scale for rating food characteristics.

A table utilizing the **hedonic scale** ranging from unacceptable to very acceptable is relatively easy to construct and is effective when the desirable and undesirable characteristics of a few samples are sought in an experiment. Figure 3.16 is an example of this type of table; note the use of numerical ratings.

In some testing situations, a picture on the scorecard may prove to be worth quite a few, if not a thousand, words. If the researcher is working with young children, with people who cannot read well, or with people who have limited use of the English language, a picture scale can prove invaluable in communicating the level of pleasure the food brings to the panelist. Such a scorecard will have simple drawings of a face, with the expression altered slightly from one picture to the next to create a picture rating scale. The pictures on such a scorecard range from a very smiling face to a very deeply frowning face, preferably with a total of nine drawings. The center part of the scale has a face with the mouth in a

**Name** _____
**Date** _____

| Characteristic | Sample | | |
|---|---|---|---|
| | 517 | 727 | 464 |
| Color of surface | | | |
| Tenderness | | | |
| Texture | | | |
| Flavor | | | |

Scale: 9 = Like extremely
        8 = Like very much
        7 = Like moderately
        6 = Like slightly
        5 = Neither like nor dislike
        4 = Dislike slightly
        3 = Dislike moderately
        2 = Dislike very much
        1 = Dislike extremely

**Figure 3.16**    Scorecard for muffins utilizing hedonic ratings.

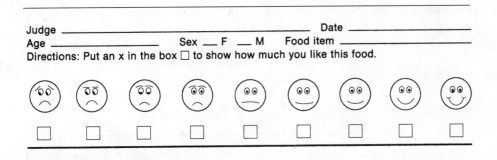

Judge _____  Date _____

Age _____  Sex __ F __ M  Food item _____

Directions: Put an x in the box ☐ to show how much you like this food.

**Figure 3.17** Scorecard using the "smiley" scale.

straight line, which depicts neither pleasure nor displeasure. Because of the smiling faces, this type of scale is dubbed the **smiley scale** (Figure 3.17).

## SUMMARY

Sensory evaluation is a critical part of food experimentation because it is the means of determining how people, the consumers, will react to a food. Such information is needed in basic research and in the food industry. Sensory evaluation encompasses use of all of the senses as they come into contact with the food being evaluated:

- Visual evaluation includes judgments on color, as well as on contour and texture.
- Olfactory sense is utilized in evaluating the aroma of the food, and it also contributes to the overall perception of flavor.
- Taste buds are a significant aspect of flavor evaluation because of their ability to identify sour, sweet, salt, bitter, and umami taste components of flavor.
- Tactile evaluation is important and generally is identified as mouthfeel.
- Auditory cues may be a part of food evaluation. Crispness is a textural characteristic with both tactile and auditory stimuli.

Sensory evaluation may be conducted to determine differences among food items. Familiar examples of difference testing are duo–trio and triangle tests. Often, it is desirable to determine preference and also to obtain specific information about food samples. Affective and descriptive testing can be performed using a prepared scorecard. Examples of scorecards are presented in this chapter, but each experiment should be evaluated by developing a scorecard specifically designed to procure the evaluation information needed.

Ranking of food samples is yet another type of evaluation that may be of considerable help in the food industry. The most sophisticated type of sensory evaluation is the profiling method, a technique that is used in flavor identification and that relies on the services of a highly trained panel.

**Smiley scale**
Sequential series of very happy and continuing through to very unhappy faces used in evaluating food products when respondents are unable to use the language easily.

## STUDY QUESTIONS

1. Why is sensory evaluation of importance in the food industry?

2. Describe the structures in the body involved in (a) color perception, (b) taste perception, and (c) odor detection.

3. Explain the interaction between taste and odor in the perception of flavor.

4. What is the difference between a duo–trio test and a triangle test? How is each conducted?

5. Describe the ideal environment for conducting sensory evaluation in the laboratory.

6. Outline the process by which you would select and train a taste panel.

7. Design a scorecard for an experiment you might conduct in the laboratory.

8. True or false. The smiley scale is suitable for testing using children as panelists.

9. True or false. Preference can be determined using a paired-comparison test.

10. True or false. Difference can be measured using a numeric scale.

11. True or false. A triangle test can determine preference.

**12.** True or false. Descriptive scorecards are useful when doing product development.

**13.** True or false. Flavor profiling is done individually using a detailed scorecard.

**14.** True or false. Taste is determined only in the mouth.

**15.** True or false. The olfactory receptors are needed to determine both aroma and taste.

## BIBLIOGRAPHY

Allison, A. M. and Work, T. 2004. Fiery and frosty foods pose challenges in sensory evaluation. *Food Technol. 58* (5): 32.

Civille, G. V. and Lyon, B.G. 1996. *Aroma and Flavor Lexicon for Sensory Evaluation.* ASTM. West Conshohocken, PA.

Decker, K. J. 2003. Where there's smoke, there's flavor. *Food Product Design 13* (4): 85.

Gacula, M. C., Jr. 2004. *Descriptive Analysis in Practice.* Wiley-Blackwell. Hoboken, NJ.

Gerdes, S. 2004. Perusing the food-color palette. *Food Product Design 14* (9): 94.

Griffiths, J. C. 2005. Coloring foods and beverages. *Food Technol. 59* (5): 38.

Hazen, C. 2003. Unveiled: Secrets of masking flavors. *Food Product Design 13* (November Supplement): 69.

Khatchadourian, R. 2009. The taste makers. *The New Yorker* (November 23): 84.

Lawless, H. T. and Heymann, H. 1998. *Sensory Evaluation of Food: Principles and Practices.* Chapman and Hall: New York.

Lioe, H. N., et al. 2010. Soy sauce and umami taste: A link from the past to current situation. *J. Food Sci. 75* (3): R71.

Marcus, J. B. 2005. Culinary applications of umami. *Food Technol. 59* (5): 24.

Mattes, R. D. 2002. The chemical senses and nutrition in aging: Challenging old assumptions. *J. Am. Dietet. Assoc. 02* (2): 192.

Meilgaard, M., et al. 2006. *Sensory Evaluation Techniques.* CRC Press. Boca Raton, Fl.

Papadakis, S. E., et al. 2000. Versatile and inexpensive technique for measuring color of foods. *Food Technol. 54* (12): 48.

Popper, R. and Kroll, J. J. 2003. Conducting sensory research with children. *Food Technol. 57* (5): 60.

Rodriguez, N.C. 2002. Sensory: Creating quality with sensory shelf-life studies. *Food Product Design 12* (4): 86.

Rodriguez, N. C. 2003. The power and potential of descriptive panels. *Food Product Design 13* (7): 101.

Rodriguez, N. C. 2005a. Communicating sensory information to R & D. *Food Product Design 14* (11): 86.

Rodriguez, N. C. 2005b. Sensory: Innovations in a cost-cutting mode? *Food Product Design 14* (11): 113.

## INTO THE WEB

*http://ucce.ucdavis.edu/files/datastore/234-416.pdf*—Overview of sensory evaluation techniques.

*http://www.tragon.com/who/our-facilities.php*—Description of test kitchen research facility.

*http://senselab.med.yale.edu/ORDB/default.asp*—Researchers' database for chemosensory receptors of some vertebrates or genes and proteins.

*http://www.leffingwell.com/olfaction.htm*—Scientific discussion of olfaction.

*http://www.britannica.com/EBchecked/topic-art/584034/3415/The-taste-buds-of-the-circumvallate-papillae-are-made-up*—Diagrams of the circumvallate papillae.

*http://users.rcn.com/jkimball.ma.ultranet/BiologyPages/T/Taste.html*—Brief discussion of taste buds.

*http://www.innerbody.com/image_nerv12/nerv124.html*—Interactive diagram of the tongue and taste receptors.

*http://www.sweetmarias.com/article.sensory-evaluation.html*—Overview of sensory evaluation.

*http://jds.fass.org/cgi/reprint/49/6/628.pdf*—Paper demonstrating varying ability of judges to discriminate flavor characteristics in milk.

*http://www.ag.auburn.edu/~kerthcr/671/Sensory%20Panel%20Methods.pdf*—Overview of taste panels, their selection, and training.

*http://www.probrewer.com/resources/library/siebel-tastepanel.pdf*—Paper regarding use of taste panels.

*http://smartstore.bpex.org.uk/articles/dodownload.asp?a=smartstore.bpex.org.uk.16.2.2009.14.22.58.pdf&i=297799*—Outline of process for evaluating pork using a trained taste panel.

*http://www.medallionlabs.com/Downloads/descriptive_sensory_Eval.pdf*—Description of Medallion Labs' trained sensory analysis team and its potential applications.

*http://www.astm.org/Standards/sensory-evaluation-standards.html*—American Society for Testing and Materials site for sensory evaluation of specific materials.

*http://www.olivecenter.ucdavis.edu/publications/Sliced%20Table%20Olives%20Consumer%20Testing.pdf*—Description of an experiment using descriptive testing.

*http://www.marketviewresearch.com/sc/psamples.asp*—Statistics on paired-comparison test.

*http://cat.inist.fr/?aModele=afficheN&cpsidt=19147830*—Article on duo–trio test order.

*http://www.stats.gla.ac.uk/steps/glossary/nonparametric.html*—Discussion of statistics in research using rank order.

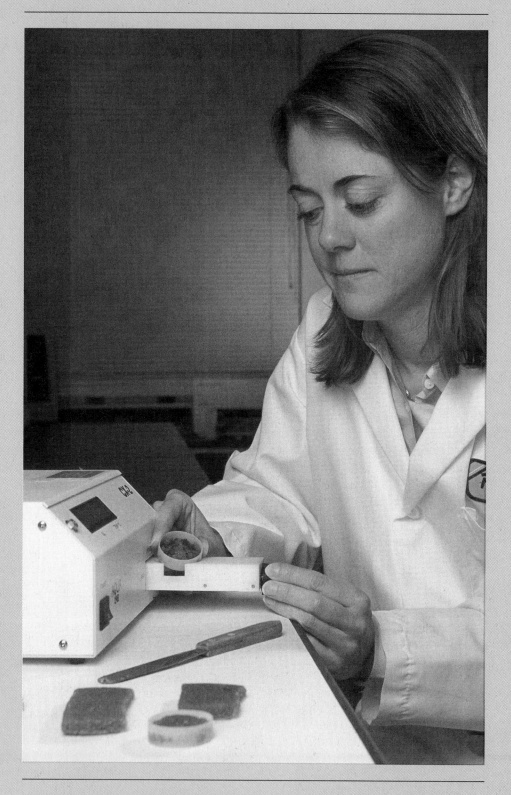

ARS food technologist Tara H. McHugh performs a test to evaluate the shelf life of the 100 percent natural pear bars that she helped develop.

# CHAPTER 4

# Objective Evaluation

## Chapter Outline

## OBJECTIVES

After studying this chapter, you will be able to:

1. Describe the physical methods for evaluating pertinent characteristics of various foods: for example, volume, specific gravity, moisture, texture, rheology, color, and cell structure.

2. Discuss the chemical methods that may be used in evaluating foods.

## INTRODUCTION

Because of its reliance on human panelists, sensory evaluation gives important (but sometimes variable) results during the course of an experiment. Variability of sensory evaluation is reduced by careful screening prior to training and use of large numbers of participants for consumer panels, both of which are costly. The objective data that can be obtained in the laboratory can be correlated with sensory data as a cost-effective and rapid means of obtaining the needed information and reliability important for research and development.

The goal of objective testing is to obtain highly reliable data on the food characteristics that are amenable to physical testing. Objective testing may utilize equipment of varying degrees of sophistication to test such attributes of a food as rheological properties, tenderness, and volume. Quantifiable results are of great significance in food research. These data obtained by objective tests supplement and/or reinforce the data obtained subjectively through sensory evaluation. By utilizing the information obtained from sensory and objective testing, maximum knowledge can be derived from the experiment.

## GENERAL GUIDELINES

Objective testing provides data that have the potential to give researchers a false feeling of security. When numerical values can be read from a testing device, these numbers assume an authenticity that may or may not be warranted. Once the figures are recorded in a laboratory notebook, reasons for questioning their validity may vanish. The good researcher is always on guard against the acquisition of false data. Constant vigilance regarding maintenance and operation of the machines and the preparation of samples for testing must be maintained.

Some of the basic guidelines for objective testing follow:

1.  *Conduct several objective tests appropriate to the experiment for which equipment is available.*

    Even when a sensory measurement is made on an attribute such as tenderness, an objective test is important to corroborate or challenge the panelists' ratings. Also, the pH of mixtures prior to and after heating may be pertinent. This information often is valuable in explaining results. Flow properties of batters or certain other mixtures prior to heat treatment can be measured, and that information is used in interpreting and explaining results.

2.  *Obtain necessary testing devices.*

    In the preliminary testing phase of an experiment, all steps in the preparation of the product need to be analyzed and the final product studied to determine whether there are specific characteristics that might be tested objectively if additional testing equipment could be procured or developed for the experiment. Other laboratories and departments within the university or local food companies may be able to provide access to the necessary equipment. Technicians or personnel in the campus machine shop or maintenance area often are able to construct useful devices when these are described and interpreted to them.

3.  *Be meticulous about maintenance of objective equipment.*

    In any laboratory, there always is the possibility that somebody may adjust, move, or use the equipment employed in a study. This increases the chance of malfunction of the equipment and the need for readjustments in calibration. Normal usage can contribute to mechanical failures, too. Routine lubrication and checking of equipment is needed. Before using any equipment, check to be certain the machine is operating correctly. Always be suspicious of equipment, and institute appropriate controls to be sure that the objective measurements will be made as accurately as possible.

4.  *Carefully define the samples to be used for objective testing.*

    A template of the item being tested is often an essential tool in obtaining comparable samples for objective tests. For example, strips of meat from bottom round must be exactly the same dimensions and from the precise location on the muscle if they are to give comparable data regarding tenderness, as measured by use of the shear apparatus. Similarly, the dimensions of pastry or cookie samples tested for tenderness on the shortometer must be identical. To obtain these samples, the thickness of the mixture prior to baking must be controlled precisely. Decisions need to be made regarding possible removal of surface skin on custard to be tested on the penetrometer.

5.  *Establish operating conditions for objective testing.*

    For consistency in obtaining results, there must be no variation in the collection of objective data. For example, the temperature of a starch paste being utilized for a line-spread test must be specified and controlled so that the effect of temperature on viscosity of starch pastes is not an uncontrolled variable in the measurement. Gelatin mixtures need to be tested at the temperature specified by the researcher and at the time a specified storage period has elapsed so that the effects of age and temperature are not uncontrolled variables in the experiment.

    These are but a few examples of the need to control size, storage, and temperature of samples. Each experiment needs to be designed to eliminate uncontrolled variables in objective testing.

## PHYSICAL METHODS

### Volume

**Seed displacement** Volume of firm food products can be determined by displacement, usually by seed displacement. Samples for determining volume of baked products are prepared by weighing comparable amounts of dough or batter of each product to be baked for the test. A **volumeter** is a convenient machine for measuring this. The volume of seeds (ordinarily rapeseeds) in a closed system is determined without the sample and then with the sample placed inside; the actual volume of the sample is the difference between the two measurements (Figure 4.1).

If a volumeter is not available, a box somewhat larger than the sample can be used to measure the seeds required to fill the box level with and without the sample. Similarly, cakes may be measured in their pans if the sides of the pan extend well above the cake itself.

**Index to Volume** Another approach to the comparison of volumes is the **index to volume**, a measurement made by first tracing a detailed outline of a cross section of the food. In this method, it is essential that the slice be taken from exactly the same location on each sample. This tracing can be done with a sharply pointed pencil or a pen, or by making a clear **inkblot** of the cross section by inking the cross section of the sample with a stamp pad before pressing the inked sample onto paper. Photocopying a cross section of each sample is another way of obtaining the precise outline. A **planimeter** is then used to trace the entire outline of the sample on the photocopy, carefully following all indentations and protrusions so that the final measure recorded on the planimeter represents the exact circumference of the slice (Figure 4.2).

### Specific Gravity

**Specific gravity** is a measure of the relative density of a substance (in this instance, a food) in relation to the density of water. The measurement is obtained by weighing a given volume of the sample; then that weight is divided by the weight of the same volume of water (1 milliliter of water weighs 1 gram at 4°C or 39°F). This technique is used for comparing the lightness of

**Volumeter**
Device for measuring volume of baked products; consists of a reservoir for storing the seeds, a transparent column for measuring volume, and a lower compartment in which the sample is placed.

**Index to volume**
Indirect means of comparing volume by measuring the circumference of a cross section of the product.

**Inkblot**
In food research, an inked impression that is made on paper by first pressing a cross section of the sample onto an inked stamp pad.

**Planimeter**
Engineering tool designed to measure distance as its pointer is traced around a pattern.

**Specific gravity**
Ratio of the density of a food (or other substance) to the density of water.

**Figure 4.1**   Volumeter for measuring volume of cakes, breads, and other baked products that are suited to seed-displacement measurement.

**Figure 4.2**    A planimeter can be used to trace the area encompassed by a sample (shown in this inkblot or in a photocopy).

products physically unsuited to the volume measurements previously described. For example, specific gravity can be used to compare egg white foams. Specific gravity has been used to distinguish between varieties of potatoes when studying their cooking properties; those with low specific gravity (waxy potatoes) had cooking characteristics different from those of the nonwaxy potatoes, which had a comparatively higher specific gravity (Figure 4.3).

**Figure 4.3**    Specific gravity of potatoes in a brine solution (1 part salt to 11 parts water by volume) is an objective test for differentiating potato characteristics. Nonwaxy potatoes for baking have a high specific gravity (1.115) and sink to the bottom. Waxy potatoes for boiling float because of their low specific gravity (1.070).

## Moisture

***Press Fluids.***   Although juiciness or moisture content can be judged subjectively, it sometimes is desirable to obtain an objective measure of this characteristic as well. The juiciness of meats, poultry, and fish can be measured by use of a succulometer, a machine that applies controlled pressure to a sample. The juices, called **press fluids**, are expressed from a weighed sample. After the appropriate pressure has been applied for a controlled length of time, the sample is again weighed. The difference between the two weights represents the amount of juice contained in the original sample. The greater the weight loss, the greater is the juiciness of the sample. A pressometer or Carver press also can be used to measure press fluids.

**Press fluids**
Juices forced from meat or other food under pressure.

***Wettability.***   Baked products can be tested for moisture level by conducting a test for **wettability**. For this test, the sample is weighed before being placed for 5 seconds in a dish of water. Immediately at the end of the elapsed time, the sample is removed from the water and weighed again to determine the weight gain. High moisture retention is synonymous with good wettability, a sign that a cake probably will be appropriately moist when judged subjectively.

**Wettability**
Ability of a cake or other food to absorb moisture during a controlled period of time; high moisture retention means a cake is sufficiently moist.

***Drying Oven.***   A slow method that sometimes is used to determine **moisture content** of a sample is a drying oven. A much faster drying method utilizing microwaves is the basis for microwave moisture analyzers. Regardless of the drying technique, the basic procedure is the same. The weight of the original sample is determined, and then the food is dried until the dried weight remains constant. The difference between initial and final weight is calculated, and that value is divided by the original weight and multiplied by 100 to calculate the percentage of moisture in the original sample.

**Moisture content**
Initial − dried weight/ initial weight) $\times$ 100 = % moisture.

***Karl Fischer Titrator.***   A much faster way of analyzing moisture content became available in 1990 when the Karl Fischer titrator analyzed water content of various food samples in 10 minutes or less, depending on the type of food. Food to be analyzed by this method is homogenized in a high-speed blender at speeds up to 7500 rpm to release the water. Following complete homogenization, the water is titrated with Karl Fischer reagent until all the water has reacted with the reagent. The calculation for water content is handled by a microprocessor, which is built into the machine. This costly approach to moisture analysis is well suited to production facilities where a quick response is important.

## Texture

*Texture* is a general term for qualities in food that are determined by touch. Machines can reveal textural properties by measuring force, time, distance, and/or energy. Qualities of texture in foods that can be measured by a variety of machines include fracturability, shear, crushing, tensile strength, puncture strength, torque, and snapping strength.

Innovative testing equipment is available to food researchers and manufacturers today to measure aspects of texture of concern in specific food products. Objective equipment may be used in basic research and product development in the laboratory, or it may be a key factor to maintain quality control in production facilities. Many of the testing devices utilize computers for recording and analyzing data.

***Warner–Bratzler Shear.***   A **Warner–Bratzler shear** is a device commonly used to measure the tenderness of meat. Meat samples of carefully controlled dimensions are placed through an opening in a thin metal plate, and the force required for the two parallel bars to shear the meat as they pass down opposite sides of the plate holding the sample is recorded (Figure 4.4). Core samples must be cut from exactly the same position from each meat sample using a coring tool. The standing time and temperature of the samples also must be the same at the time of coring and also at testing on the shear.

**Warner–Bratzler Shear**
Objective testing device for measuring the force required to shear a sample of meat or other food with measurable tensile strength.

## FOOD FOR THOUGHT: Accessories after the Facts

Unique problems in measuring various aspects of texture are being solved by the development of accessories to be used with texture analyzers. One example is the Volodkevich Bite Jaws, which was designed to imitate the way an incisor tooth functions when biting meat. The tenderness of meat and raw and cooked vegetables can be measured with this attachment to a texture analyzer.

The prominence of wheat flour tortillas in American diets today generated a need for a means of measuring their tensile strength. A unique attachment called the Burst Rig was developed, in which the test tortilla is clamped in circular plates so that a 2.5-cm spherical probe can be pushed through the center of the tortilla until it breaks through. A Chen–Hoseney Dough Stickiness Cell is another accessory that can be used with a texture analyzer. Measurement of dough stickiness is important in a tortilla factory to avoid problems in the production line.

A new accessory of merit in studying the texture characteristics of crisp and crunchy foods is the Video Playback Indicator. The rapidity with which crackers and other crisp or crunchy foods break or snap makes it difficult to observe every aspect of the action, but this accessory makes it possible to play back a test frame by frame. The timing that the video displays is coordinated with the texture analyzer as it tests the food.

Crispness of cereals, snack foods, crackers, and other food items is noted by sound as well as by tactile experience. The Acoustic Envelope Detector is an accessory that is used with a texture analyzer to acquire acoustic data. These data are the measures of the acoustic energy that is released when crisp foods are subjected to testing by a texture analyzer. The sounds made by biting into a potato chip or chewing a celery stick add to the eating experience, but such sounds have not previously been a part of objective evaluation. The Acoustic Envelope Detector is designed to fill this gap.

**Shear Press**
Objective testing machine that measures compressibility, extrusion, and shear of food samples.

**Compressimeter**
Objective equipment that measures the force required to compress a food sample to a predetermined amount.

**Penetrometer**
Machine that measures tenderness by determining the distance a cone or other device penetrates a food during a defined period of time and using only gravitational force.

**Bloom gelometer**
Modification of a penetrometer designed especially for measuring the tenderness of gels.

**Percent sag**
(Depth in container − depth on plate/ depth in container) × 100 = %.

***Shear Press.*** The **shear press**, a related device, is a machine that compresses, extrudes, and shears the sample at the same time. This is a suitable method for measuring textural characteristics of some fruits and vegetables.

***Compressimeter.*** The **compressimeter** (Figure 4.5) is related to the shear press, but it measures only compressibility, not shear strength. The firmness or compressibility of rather porous baked products can be determined by using this machine. The usual technique for operating the compressimeter is to apply pressure until the sample has been deformed a specific amount and then to measure the force that was required to accomplish this amount of deformation. The greater the force required, the firmer the product. The texture test system (Figure 4.6) is a similar device.

***Penetrometer.*** A **penetrometer** also can measure tenderness of some foods (Figure 4.7). This device consists of a plunger equipped with a needle or cone that penetrates the sample through gravitational force for a selected period of time. The distance the test device penetrates into the sample is measured to determine the comparative tenderness of samples. The larger the reading, the more tender is the product. Gels and many baked products are well suited to tenderness measurements using the penetrometer. The **Bloom gelometer** is a special type of penetrometer in which lead shot drops into a cup, which forces a plunger into the sample. When sufficient weight has been added to the cup to move the plunger a set distance, the test is completed, and the amount of shot required is determined as the measure of the test.

***Percent Sag.*** Another test to determine the comparative tenderness of a gel is the test for **percent sag**. For this test, the depth of a sample such as jelly is measured in its container

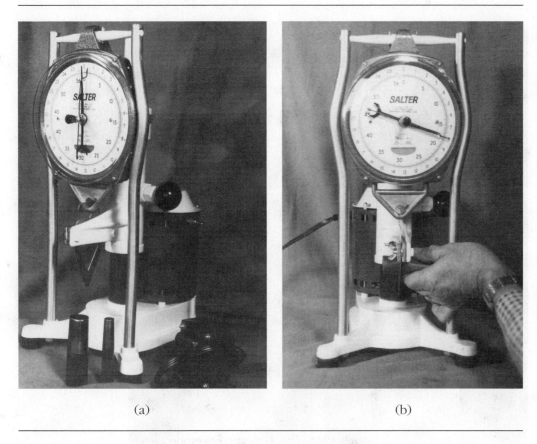

(a)                                          (b)

**Figure 4.4**    (a) Warner–Bratzler shear for measuring tenderness of meats and other foods with sufficient resistance for measuring the force required to shear the sample. (Note the coring tools for obtaining controlled sample size.) (b) Shear in operation. (Courtesy of G-R Electric Manufacturing Co.)

by using a probe. The product then is unmolded onto a flat plate. Once again the depth of the product is measured with the probe. This second measurement is subtracted from the original measurement; this figure is divided by the original depth of the sample and multiplied by 100 to obtain the percent sag of the sample. The greater the percent sag, the more tender the gel.

***Shortometer.***    Pastry, comparatively tender and crisp cookies, and crackers are well suited for tenderness measurements on the **shortometer**. This device consists of a platform containing two parallel, dull blades on which the sample rests. A motor actuates a third dull blade that presses down until the sample snaps. The force required to break it is the measure of the tenderness of the product (Figure 4.8).

**Shortometer**
Device designed to measure the tenderness of fairly tender, crisp foods.

***Universal Testing Machine.***    Instron's **universal testing machine** is a unit with several different testing devices that can be used to measure different aspects of texture in foods. Applications include assessment of food quality on production lines (Figure 4.9).

The Instron universal testing machine was so named because of its ability to test various facets of food textures. In some foods, hardness might be an important characteristic to measure, while cohesiveness might be of more interest in caramels or similar foods. The universal testing machine can provide a record showing seven aspects of texture from various food samples. These are cohesiveness, adhesiveness, hardness, springiness, gumminess, chewiness, and fracturability. The versatility of this machine to provide information regarding quality control during food processing has made it an important testing device in the food industry.

A convenient handheld device is available to test the texture of crops such as apples in the field rather than in the laboratory. This test provides information about ripeness, firmness, and texture. The testing mode can be "push-pull" or "tension-compression."

**Universal testing machine**
Multipurpose, complex machine capable of measuring various textural properties of food samples.

**Figure 4.5**    Compressimeter measuring compressibility of bread. (Courtesy of
C. W. Brabender Instruments, Inc., So. Hackensack, NJ, and Brabender® OHG,
Duisberg, Germany.)

***Farinograph.***    The farinograph (Figure 4.10) measures and depicts gluten development
during mixing of batters and doughs. This is done by dragging a device holding many metal
bristles through the dough ingredients to mix them and develop the gluten. A recorder
traces the increasing tension in the dough during mixing and then its weakening as the
gluten strands begin to break from the agitation and stretching.

***Extensograph.***    Brabender's extensograph can be used to determine stretching
characteristics of dough using various flours and/or added ingredients.

**Texturometer**
Simulation device that
measures physical textural
properties such as
hardness, cohesiveness,
and crushability of foods.

***Glutograph.***    The glutograph is another machine for measuring stretching of gluten and
recovery, which also reveals elasticity of the sample during testing.

***Texturometer.***    The General Foods **texturometer** tests textural properties similar to those
tested by the universal testing machine. Surprisingly, adjustments can be made in the

**Figure 4.6**    Texture test system. (Courtesy of General Kinetics, Inc.)

General Food texturometer to simulate various chewing actions in the mouth, including biting with the front teeth and chewing with molars.

***Masticometer.***    **Masticometers** of various designs approximate the chewing action of the jaw. The measurements obtained from the masticometer are comparative measures of tenderness of meat or other food.

**Masticometer**
Machine that measures comparative tenderness of meat and other foods by simulating chewing action.

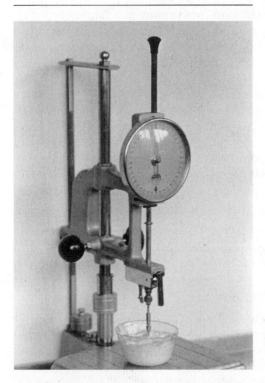

**Figure 4.7**    Penetrometer for measuring tenderness of gels and other foods that permit gradual passage of a cone or other testing device.

**Figure 4.8**    Shortometer for measuring tenderness of pastry and other crisp foods. (Photo courtesy of Magnuson Corporation.)

## Rheology

**Rheology**
Study of deformation and flow qualities of matter.

**Newtonian**
Classification of materials having a flow rate that is not affected by shear rate—for example, water and sugar syrups.

**Non-Newtonian**
Classification of materials having a flow rate that is influenced by shear rate—for example, chocolate and emulsions.

**Thixotropic**
Ability of a gel to become more fluid with increasing shear and then to regain previous viscosity after shearing stops.

**Consistometer**
Device for measuring the spread or flow of semisolid foods in a specified length of time.

**Line-spread test**
Measurement of flow of a viscous liquid or semisolid food by determining the spread of a measured amount of sample in a specified time at 90° intervals on the template of concentric rings.

**Rheology** is the study of the flow of matter and the deformations that result from flow. The *viscosity* of a fluid is its ability to develop and maintain a shearing stress and offer resistance to flow. This is a bit easier to visualize if you think of a liquid as having a top plate that is being forced sideways on the liquid (Figure 4.11) while the bottom plate of the liquid is moving sideways much less rapidly, and the fluid in the middle is moving at a rate between the top and bottom plates. This movement is the shear rate. Viscosity generally decreases with an increase in temperature of the fluid.

Fluids are categorized as either **Newtonian** or **non-Newtonian**, a distinction based on work by Sir Isaac Newton. Water, sugar syrups, and wine are examples of Newtonian liquids. Newtonian fluids have viscosities that are independent of the shear rate. This means that the viscosity of a Newtonian fluid will be the same even when a rotational viscometer is operated at different speeds.

Non-Newtonian fluids, such as emulsions and tomato paste, will flow at an altered rate when subjected to shear stress. Mayonnaise and catsup are examples of fluids that thin with increasing shear, but they gradually return to their original viscosity after shearing ceases. Chocolate also responds in this manner, a characteristic that is useful to commercial candy makers. Shear causes the chocolate to become soft enough to be spread as a coating over the center, and yet this soft chocolate will become firm rather quickly after the shearing action stops. This behavior characterizes **thixotropic** fluids.

*Consistometer.*    The Bostwick **consistometer** and the Adams consistometer measure the consistency (viscosity) or spread of semisolid foods. If a measured amount of semisolid food is placed on the slanting trough of the Bostwick consistometer, the distance of flow in a given period of time can then be measured to determine the relative consistency of foods that flow at atmospheric pressure. The Adams consistometer measures the spread of the sample (the area covered) in a specified length of time.

*Line-Spread Test.*    The **line-spread test** is a simple variation of the measurement of viscosity by consistometers. A measured amount of sample is placed in a column centered on measured, concentric rings and is allowed to flow for a measured length of time (up to 2 minutes), after which the spread at each 90° increment of the circles is read; the line-spread value is the mean of the four values obtained (Figure 4.12).

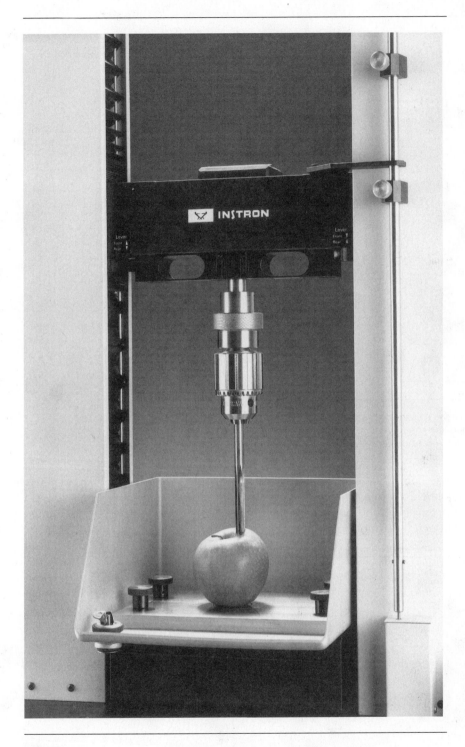

**Figure 4.9**    Puncture test of an apple being conducted on the Instron Model 1011 Test Instrument. (Courtesy of Instron Corporation.)

***Viscoangiograph.***    The Brabender **viscoangiograph** is another piece of objective equipment for measuring consistency. It can determine the viscosity of starch pastes at controlled, selective temperatures (Figure 4.13).

***Viscometer.***    **Viscometers** (also sometimes called viscosimeters) are functionally similar to the Brabender viscoangiograph. As their name implies, these objective testing devices measure viscosity of a variety of foods with flow properties. Measurement of viscosity by

**Viscoangiograph**
Device designed to control the temperature of a starch paste and to measure its viscosity.

**Viscometer**
Objective testing device for measuring viscosity of liquids that flow on the basis of rotational resistance or capillary action.

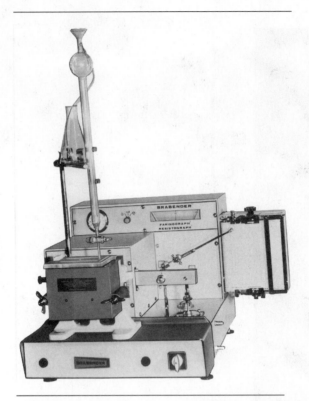

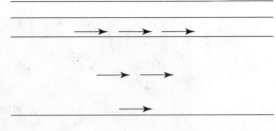

**Figure 4.11**    Diagram illustrating the shear rate of a liquid (difference between rate of movement in the top layer and the rate near the bottom).

**Figure 4.10**    Farinograph is used to measure and record the development of gluten in batters and doughs. (Courtesy of C. W. Brabender Instruments, Inc., So. Hackensack, NJ, and Brabender® OHG, Duisberg, Germany.)

viscometers is based on either rotational operation (a measure of torsional force) or capillary action. The Brookfield viscometer (Figure 4.14) illustrates measurement based on rotational operation, for it measures the drag the test sample places on a spindle that is rotated mechanically through the sample.

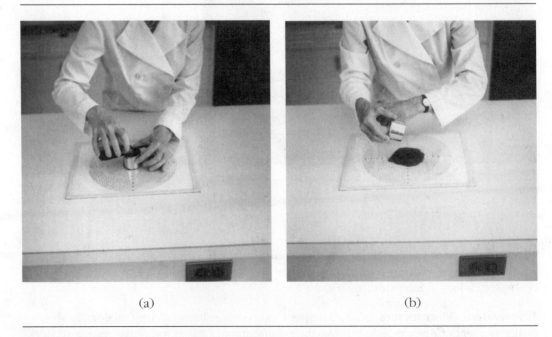

(a)                                              (b)

**Figure 4.12**    Line-spread test for measuring rheological properties of sauces and other foods that flow: (a) pouring the sample; (b) timing the test.

A **jelmeter** (a special viscometer similar to a pipette) often is used to test the viscosity of pectin-containing juices to determine the amount of added pectin needed for gelling fruit juices to make jams and jellies.

## Color

Color is measured according to either the **CIE** (Commission Internationale de L'Eclairage) or the **Munsell system**. These systems consider the spectral color, degree of saturation, and brightness. A spectrophotometer can be used to save considerable time in defining colors under the CIE system.

In the Munsell system, hue, value, and chroma are identified on a numerical scale. To measure hue, red is expressed numerically as +a, green as −a, yellow as +b, and blue as −b. Using these measured numerical values, hue is then calculated using the formula:

$$\text{hue angle } \theta_s = \tan^{-1} \frac{a}{b}.$$

Chroma is determined by using these values in the following formula:

$$(a^2 + b^2)^{\frac{1}{2}} = \text{chroma}.$$

Value is determined based on a range of 100, with 0 representing black and 100 representing white. Color samples are available to match with the color of the food being identified.

**Jelmeter**
Pipette-like viscometer designed to measure the adequacy of the pectin content of fruit juices to make jams and jellies.

**CIE**
Commission Internationale de L'Eclairage; group that established a system of measuring color based on spectral color, degree of saturation, and brightness.

**Munsell system**
System of identifying colors on the basis of hue, value, and chroma using a numerical scale.

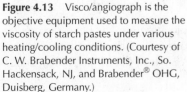

**Figure 4.13**   Visco/angiograph is the objective equipment used to measure the viscosity of starch pastes under various heating/cooling conditions. (Courtesy of C. W. Brabender Instruments, Inc., So. Hackensack, NJ, and Brabender® OHG, Duisberg, Germany.)

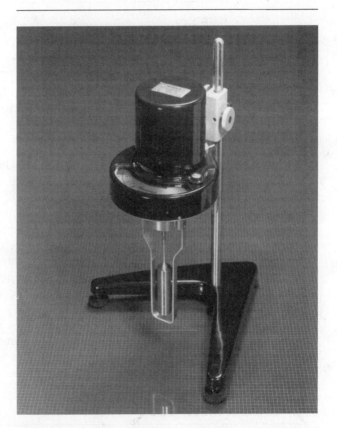

**Figure 4.14**   Brookfield viscometer (Model LVT Dial Reading Viscometer) rotates through sample to measure viscosity. (Courtesy of Brookfield Engineering Laboratories.)

**Figure 4.15**    Photocopying provides a convenient, detailed record of the appearance of bread (or other suitable baked products).

**Color-difference meter**
Objective machine, such as the Hunter color-difference meter or Gardner color-difference meter, capable of measuring color differences between samples by utilizing the CIE or Munsell color system.

The Hunter **color-difference meter** is commonly used to identify food colors. For an overview of color and its measurement, see Setser (1984).

## Cell Structure

Cell structure of baked products is an important characteristic to measure, and yet it is difficult to obtain meaningful results using objective testing equipment. Judges on sensory panels provide crucial information on cell structure when their scorecards are designed properly. However, an objective approach also is valuable. One key technique for compiling information about cell structure (including uniformity, size, and thickness of cell walls) is by making photocopies of cross-sectional slices (Figure 4.15). This simple technique successfully reveals the third-dimensional view into the cells on the cut surface of the sample. Another benefit of this technique is that the actual size is represented clearly, which permits use of the planimeter to determine the circumference, as noted previously in the discussion of index to volume.

A photograph is another valuable tool to depict information about cell texture. Digital cameras are particularly useful because of the convenience of transmission via the computer and Internet. A clearly marked ruler placed adjacent to the sample is a simple and effective way of adding some concept of size to photographs.

## CHEMICAL METHODS

### Nutrient Analysis

Particularly since the advent of nutrition labeling, considerable attention has been devoted to analysis of the nutrient content of food. The accepted methods for conducting these analyses are contained in a publication by the Association of Official Analytical Chemists (AOAC, 2002).

## pH

Measurement of the pH of food mixtures often is essential to determine acidity or alkalinity. Although pH papers are available, a pH meter is a more precise way of obtaining accurate information on this facet of food. The pH meter utilizes a glass-indicating electrode and a reference electrode to complete an electrical circuit and measure the effective hydrogen ion concentration (pH) of the food being tested (Figure 4.16).

Most foods have a pH of 7 or less, which means that they are acidic in reaction: The lower the pH number, the more acidic the food. Bicarbonate of soda is a food ingredient that is alkaline, and egg whites are quite alkaline; that is, their pH is high. Although the general pH value is known for many foods, the pH of a specific food sample may vary slightly from the expected value. In fact, the pH of foods varies not only with the type of food but also with its freshness and the environmental conditions (e.g., temperature, moisture level, and surrounding air or gases). Ingredients included in food products also influence the pH of food mixtures. For these reasons, measurement of the pH of food samples can provide important information in food research.

Various models of pH meters are available to meet specific needs in the laboratory or in the field. In some instances, a pH meter that adjusts to changes in temperature during the course of an experimental run can be useful. Basic models usually are appropriate for classroom use.

## Sugar Concentration

Refractometers can determine the concentration of a sugar solution; the concentration is reported on the **Brix scale**. Light is refracted as it passes through sugar solutions, with the specific values calibrated in degrees Brix, an indication of the percentage of sucrose in the solution. A portable refractometer (Quick-Brix™ refractometer) can be used to

**Brix scale**
Hydrometer scale designed to indicate the percentage concentration of sugar in sugar solutions.

**Figure 4.16**    pH meter (Courtesy of Beckman Instruments, Inc.)

determine sugar level in such produce as grapes. This application is well suited to the wine industry where the amount of sugar in the fruit impacts the finished product significantly.

## Saltiness

Flame photometry analyzes the sodium content of foods. Actually, the salts of interest include not only sodium but also potassium, magnesium, calcium, ammonium, and lithium.

## Aroma

**Electronic nose**
Testing machine that develops diagrams of the flavor components in a headspace sample.

**Chromatography**
Separation of discrete chemical compounds from a complex mixture by the use of solvents or gases; separation may be accomplished by use of a GLC or a HPLC or by other somewhat less sophisticated means.

**Gas–liquid chromatograph (GLC)**
Machine that separates individual compounds from a mixture by passing them with a carrier gas along a special column that adsorbs and releases individual compounds at different rates.

**High-pressure liquid chromatograph (HPLC)**
Machine that under pressure separates a sample dissolved in liquid into its individual components as they are adsorbed along a special column and finally eluted individually from the column.

**Mass spectrometry**
Identification technique in which a pure compound is bombarded by a high-energy electron beam to split into various ions that then are sorted and finally recorded as a mass spectrum that reveals the actual compound.

**Infrared spectroscopy**
Identification technique in which a pure compound is subjected to infrared (wavelengths from 2,500 to 16,000 nm) energy to vibrate the molecule and create a spectrum of peaks indicating its structural features (e.g., an aldehyde group or benzene ring).

Although aroma has been an aspect of sensory evaluation for years, an objective approach to aroma analysis was developed in the late 1990s to serve as an adjunct approach. The generic name for this type of testing device is the **electronic nose**. Various detectors may be used in an electronic nose, including a series of sensors (e.g., metal oxides, conducting polymers, or quartz microbalance).

Electronic nose measurements use the vapors from headspace of the sample. Data can be compiled as flavor profile analysis diagrams (Figure 4.17) or discriminant factor analysis plots. These findings can be compared with similar data obtained under identical testing conditions from other samples. Electronic noses are used effectively as a tool in analysis, quality control, and/or product matching. Applications include water analysis, identification of sources and quality of such crops as coffee and grains, freshness of fish, rancidity in oils, detection of bacterial growth in fresh foods, and quality control in processing such items as cheeses and beer.

## Flavor

Flavor analysis using **chromatography** is an active research area (Figure 4.18). The ultimate goal is to identify the numerous volatile and nonvolatile substances that combine to produce the overall flavor impression of a specific food. The key to this type of research is the use of the **gas–liquid chromatograph (GLC)** and the **high-pressure liquid chromatograph (HPLC)**. These instruments separate the complex mixtures of chemical compounds into the individual compounds in foods that constitute the aroma and flavor characteristics of the sample.

Both types of chromatographs contain a special column prepared with an active material that will adsorb and release different compounds at different rates. A detector where the individual compounds exit the column is linked to a recorder to make a permanent record of the peaks, indicating transit time and quantity of each compound.

The components that are separated by a chromatograph are then collected individually or directed into a mass spectrometer for subsequent analysis. Identification of these compounds may involve the use of **mass spectrometry** or **infrared spectroscopy**. *Mass spectrometry* reveals the molecular weight, various structural features, and the actual identity of a compound. *Infrared spectroscopy* shows the structural features of an unknown compound, but not its total size. The ultimate goal of this sophisticated basic research is to synthesize flavors in the laboratory and then to produce them for industrial use so that nature's flavors can be replicated accurately.

## Electronic Tongue

Electronic tongue devices are being developed to identify the presence and relative amounts of compounds that can be detected in a solution of food by sensors sensitive to basic tastes (sweet, salty, sour, bitter, and umami). Researchers at the University of Illinois, Urbana (Lim, et al. 2008), developed a device about the size of a credit card that can identify any of 12 sweeteners in a food simply by dipping the card into a solution of the food and waiting two minutes for the distinctive patterns of the sweeteners to appear.

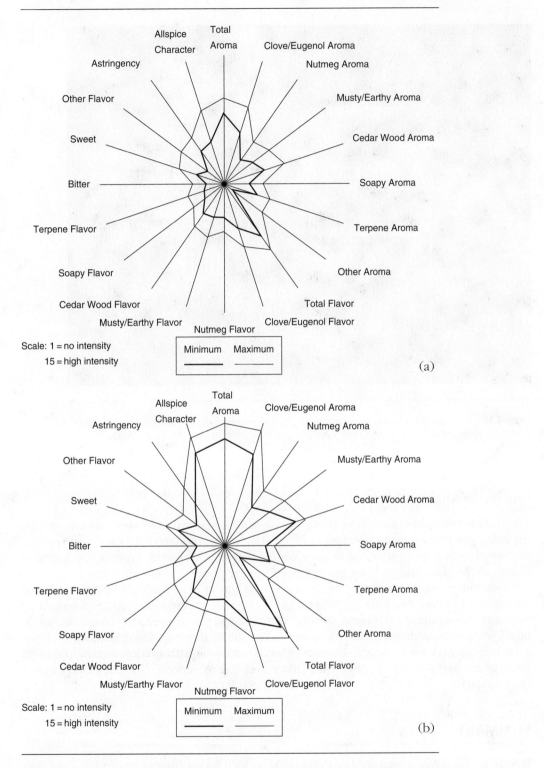

**Figure 4.17**   Flavor profile range for (a) Jamaican allspice and (b) Guatemalan allspice obtained using electronic nose measurements. (Courtesy of M. G. Madsen and R. D. Grypa, McCormick & Company. *Food Technology* 54 (3): 44–46. 2000.)

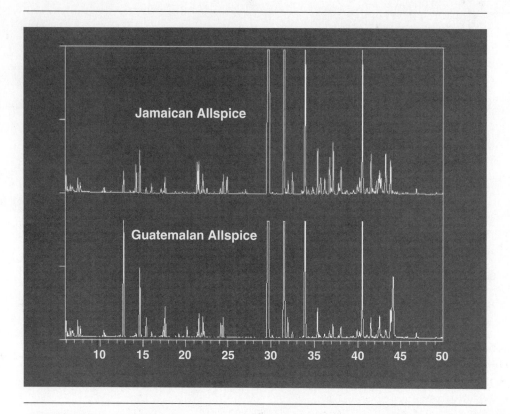

**Figure 4.18**    GLC chromatograms of headspace vapors from Jamaican and Guatemalan allspice. (Courtesy of M. G. Madsen and R. D. Grypa, McCormick & Company. *Food Technology* 54 (3): 44–46. 2000.)

## Proximate Analysis

The carbohydrate, lipid (fat), and protein content of a specific food is of interest, in part because these are the energy-yielding nutrients. *Proximate analysis* of a food determines the amount of each of these nutrients in a known quantity. Results can be expressed in grams of carbohydrates, fats, and proteins in a specified serving portion or in percentages.

Determination of proximate analysis can use the official methods specified by the Association of Official Analytical Chemists (AOAC). These methods are modified frequently and are now published online in the 18th edition by the AOAC International at http://www.eoma.aoac.org/. Analytical methods also are developed and published by the American Association of Cereal Chemists (AACC), and the 11th edition is now online at http://www.aaccnet.org/ApprovedMethods/, and the American Oil Chemists' Society (AOCS, 2003).

## SUMMARY

Research and development of food products, as well as quality control in food production, require careful and accurate evaluation techniques. Evaluation is of two types: (1) sensory or subjective and (2) objective. Objective testing can provide extremely valuable data if the sampling techniques are well controlled and if the equipment is functioning properly.

Physical measurements can be made by use of a wide range of devices varying in their level of engineering sophistication. Volume can be measured by use of a volumeter, which is a seed-displacement method. Index to volume is a related measure of volume. A planimeter traces the detailed circumference of the sample to obtain the index to volume.

Specific gravity, a comparison of the density of the sample relative to that of water, is a measurement useful in assessing light products such as foams, which are not suited to testing in a volumeter.

Moisture or juiciness in meats and related foods can be determined by placing the sample under pressure in a succulometer to express the press fluids. Wettability of baked products, determined by weighing a sample before and after a 5-second dip in water, is helpful in determining moisture levels in cakes. Use of a drying oven is another technique for determining moisture content.

Texture of foods has several parameters, and various instruments measure different aspects of texture. The Warner–Bratzler shear is used widely for measuring the tenderness of meats. A shear press measures compressibility, extrusion, and shear at the same time. The compressimeter is used to determine the firmness of bread and related baked products. Penetrometers of various designs provide another means of measuring tenderness. Percent sag is an unsophisticated but useful technique for measuring gel tenderness when a penetrometer is not available. Other items for testing texture include the masticometer, universal testing machine (Instron), and the texturometer (General Foods). Photocopying and photography afford effective means of providing permanent evidence of cell structure, including cell size, uniformity, and thickness of cell walls.

Rheology, the study of the flow of matter and deformation resulting from flow, is of interest in many Newtonian and non-Newtonian food products, including chocolate. The flow properties of semisolid foods can be tested by a simple line-spread test if the Bostwick or the Adams consistometer is not available. Viscosity of starch pastes can be measured by using the Brabender amylograph. Viscometers, operating by rotation or capillary action, measure flow properties.

Color can be measured and reported very precisely using either the CIE or the Munsell system for color identification. A spectrophotometer is a useful measuring device. The Hunter color-difference meter is used frequently in color measurements.

Chemical measurements used in food evaluation include nutrient analysis, a particularly timely subject of evaluation. The pH meter is an important piece of equipment for checking acidity or alkalinity at various points during the preparation and evaluation of products. Refractometers enable researchers to ascertain the concentration of sugar in solutions, and flame photometry is used to identify selected elements, particularly sodium.

The electronic nose is used to analyze aromas. Sophisticated flavor research utilizes the gas-liquid chromatograph (GLC) or high-pressure liquid chromatograph (HPLC) to separate volatile flavoring compounds from the total flavor sample. Purified compounds from the mixture are then identified using infrared spectrophotometry and sometimes mass spectrometry. The ultimate goal is to identify all key components of natural flavors and then to formulate synthetic flavors identical to the flavors found in the natural foods.

Proximate analysis of the carbohydrate, lipid (fat), and protein content of food samples can be done chemically, according to methods developed by recognized professional groups (AOAC, AACC, and AOCS).

## STUDY QUESTIONS

1. Outline the exact way in which you would prepare a sample of round steak for use in the shear. Be sure to identify all aspects requiring control to ensure uniform samples.

2. Does the volumeter result in a more accurate volume measurement than that obtained by the planimeter to determine the index to volume? Explain your answer.

3. Identify the technique you would select for measuring the tenderness of each of the following and explain your rationale for the choice:
   - baked custard
   - pastry
   - grape jelly
   - bread
   - round steak

4. Outline the method for conducting a line-spread test.

5. Explain the contribution of the following machines when used in flavor research:
   - GLC
   - infrared spectrophotometer
   - HPLC
   - mass spectrometer

6. True or false. Volume of a cake can be measured using a shortometer.

7. True or false. The thickness of pastry dough needs to be controlled precisely if a shortometer is to be used to measure tenderness.

8. True or false. A shortometer can be used to measure tenderness of a beef sample.

9. True or false. A penetrometer can be used to measure sag of a baked custard.

10. True or false. Temperature is a factor that needs to be controlled when measuring viscosity of starch pastes.

11. True or false. A penetrometer is an appropriate device for measuring viscosity of stirred custard.

12. True or false. A gas chromatograph is of little importance in flavor analysis because it does not separate individual compounds from the vapors.

13. True or false. A pH meter defines the sweetness of a food.

14. True or false. A planimeter can be used when comparing volume of cakes.

15. True or false. A penetrometer can be used to measure tenderness of jam.

## BIBLIOGRAPHY

AOCS. 2009. *Official Methods and Recommended Practices of the AOCS.* 6th ed. American Oil Chemists Society. Champaign, IL.

Barnes, K. W. 1995. Introduction to food analysis techniques. *Food Technol. 49* (6): 48.

Bartlett, P. N., et al. 1997. Electronic noses and their application in the food industry. *Food Technol. 51* (12): 44.

Bourne, M. C. 2002. *Food Texture and Viscosity.* Academic Press. San Diego, CA.

Clark, R. 1997. Evaluating syrups using extensional viscosity. *Food Technol. 51* (1): 49.

Duxbury, D. 2005a. Analyzing fats and oils. *Food Technol. 59* (4): 66.

Duxbury, D. 2005b. Hyphenated techniques offer greater specificity. *Food Technol. 59* (10): 68.

Duxbury, D. 2005c. Determining moisture content of foods. *Food Technol. 59* (12): 76.

Giese, J. 2000a. Color measurement in foods as a quality parameter. *Food Technol. 54* (2): 62.

Giese, J. 2000b. Online instruments for the food industry. *Food Technol. 54* (4): 98.

Giese, J. 2001. Pitt Con 2001 features new food analytical instruments. *Food Technol. 55* (4): 88.

Giese, J. 2002a. Lab developments featured at food expo. *Food Technol. 56* (8): 96.

Giese, J. 2002b. GMO testing. *Food Technol. 56* (11): 60.

Giese, J. 2003. Texture measurement in foods. *Food Technol. 57* (3): 63.

Leake, L. 2006. Electronic noses and tongues. *Food Technol. 60* (6): 96.

Lim, S. H., et al. 2008. A colorimetric sensor array for detection and identification of sugars. *Organic Letters 10* (20): 4405.

Mackey, A. 1967. *Recent research with potatoes.* Proceedings, 6th Annual Washington State Potato Conference.

Musto, C. J., et al. 2009. Colorimetric detection and identification of natural and artificial sweeteners. *Anal. Chem. 81*: 6526.

Peppard, T. L. 1999. How chemical analysis supports flavor creation. *Food Technol. 53* (3): 46.

Weaver, C. M. and Daniel, J. R. 2003. *The Food Chemistry Lab.* 2nd ed. CRC Press. Boca Raton, FL.

Zhang, C., et al. 2006. Colorimetric sensor arrays for the analysis of beers: A feasibility study. *J. Agric. Food Chem. 54*: 4925.

Zhang, C. and Suslick, K. S. 2007. Colorimetric sensor array for soft drink analysis. *J. Agric. Food Chem. 55*: 237.

## INTO THE WEB

*www.grscientific.com*—Catalog of company's laboratory equipment, including the Karl Fischer titrator.

*http://us.mt.com/us/en/home/industries/level_1_food_bev. html*—Access to several types of objective equipment for testing various foods.

*www.ohaus.com*—Information on balances and moisture analyzers.

*http://www.cwbrabender.com/MoistureTesterMTC.html*—Description of a moisture tester that incorporates weighing and drying.

*http://lab.mt.com/food/*—Video of operation of a moisture analyzer.

*http://us.mt.com*—Karl Fischer titrator information.

*www.brabender.com*—Information on various texture-testing equipment.

*www.instron.com*—Information on texture-testing equipment.

*www.stablemicrosystems.com*—Information on texture analyzers and accessories by Stable Microsystems, Inc.

*http://www.cwbrabender.com/FarinographE.html*—Manufacturer's information on using the farinograph.

*http://www.cwbrabender.com/ExtensographE.html*—Manufacturer's information on using the extensograph.

*http://www.cwbrabender.com/GlutographE.html*—Manufacturer's information on using the glutograph.

*http://www.cwbrabender.com/AmylographE.html*—Manufacturer's information on use of the amylograph for measuring starch pastes.

*http://www.cwbrabender.com/ViscographE.html*—Manufacturer's information on the viscograph.

*www.beckmancoulter.com*—Information on pH meters and other laboratory equipment.

*http://us.mt.com*—Information on refractometer types and applications.

*http://www.standardbase.com/tech/FinalHUTechFlame. pdf*—Description of flame photometry and how to perform a test.

*http://www.burkardscientific.co.uk/analytical/flame_ photometer.htm*—Example of a flame photometer.

*http://www.iit.edu/&sim;jrsteach/enose.html*—Description of electronic nose and its applications.

*http://www.alpha-mos.com/resources/pdf/instruments/ electronic_ nose*—Example of an electronic nose.

*http://www.files.chem.vt.edu/chem-ed/sep/lc/lc.html*—Brief description of chromatography.

*http://ull.chemistry.uakron.edu/chemsep/lc/*—Discussion of liquid chromatography.

*http://www.cem.msu.edu/&sim;reusch/VirtualText/Spectr py/MassSpec/masspec1.htm*—Discussion of mass spectrometry.

*http://www.electronictongue.com/*—Information about the electronic tongue.

*http://www.scs.uiuc.edu/suslick/pdf/orglett.08.4405.pdf*—Article on sweeteners and the electronic tongue.

*http://news.nationalgeographic.com/news/2009/09/0909 01-electronic-tongue_2.html*—Brief article on the electronic tongue.

*http://www.sciencedaily.com/releases/2008/08/08080410 0254.htm*—Article on electronic tongue for identifying grape varieties and vintage of wines.

*http://www.alpha-mos.com/resources/pdf/instruments/ electronic_tongue/Alpha-MOS_ASTREE_Electronic_ Tongue.pdf*—Example of an electronic tongue.

*http://www.sciencedaily.com/releases/2008/05/08050509 1820.htm*—Description of an electronic mouth designed to mimic actions in the mouth, including chewing, saliva release, and the beginning of digestion.

*http://www.eoma.aoac.org/*—Address for 18th edition of Official Methods published by AOAC.

*http://www.aaccnet.org/ApprovedMethods/*—Address for 11th edition of Approved Methods, published by AACC.

# 2

# Physical Perspectives

A cold mountain lake fed by snow from above sets the stage for studying about water and its importance in food and its preparation.

# CHAPTER 5

# Water

## Chapter Outline

## OBJECTIVES

After studying this chapter, you will be able to:

1. Describe the chemical character of water molecules and its influence on their behavior.
2. Interpret how the factors that determine the freezing and boiling points of water create their effects.
3. Describe bound water in foods.
4. Discuss the significance of water activity in foods.
5. Explain how the quality of water can influence food preparation.

## A CLOSE LOOK AT WATER

Water is a common constituent of foods and is used so frequently in food preparation that it is easy to take it for granted. However, the universality of its occurrence and its unique characteristics make water a subject that needs to be studied carefully.

The behavior of water molecules is influenced strongly by the polarity of the molecule. The two hydrogen atoms are joined to the oxygen atom by two bonds that form an angle of almost 105° (Figure 5.1). This orientation concentrates the electrons of the oxygen atom toward one side of the molecule (away from the hydrogen atoms). The electrons contribute a negative charge to this portion of the water molecule, whereas the hydrogen atoms provide a positive component, thus making water a **dipole**.

This situation promotes the formation of hydrogen bonds between water molecules as the oxygen of one molecule bonds with the hydrogen of another. This hydrogen bonding between water molecules results in innumerable clusters of water molecules when water is in the liquid state. The propensity of water to form hydrogen bonds enables it to be in liquid form over a temperature range that is physically somewhat surprising. Although hydrogen bonds are comparatively weak bonds (being secondary rather than primary bonds), they are sufficiently strong to hold water molecules in close proximity unless a considerable amount of energy is introduced into the system.

**Dipole**
Molecule that is electrically asymmetrical—that is, one portion is slightly negative and another part is slightly positive.

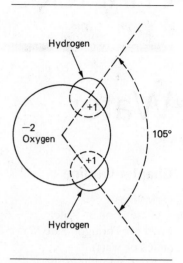

**Figure 5.1**    Diagram of a water molecule showing the bond angle of approximately 105° between the two hydrogen atoms.

## STATES OF WATER

Water can be found in solid, gaseous, or liquid form in foods. Consider water at atmospheric pressure. The state in which it is found reflects varying energy states: the crystalline, solid form (ice) represents the lowest energy state and steam the high-energy, gaseous state. As water cools, energy is lost from the system. Concomitantly, the molecules move slower and slower, until at 4°C (39°F) an organized bonding pattern between molecules begins to appear.

As the water continues cooling, its molecules separate slightly, causing a small increase in volume. At 0°C (32°F), freezing (the change from a liquid to a solid) occurs, but only after removing almost 80 calories of heat for each gram of water (Figure 5.2). The removal of this heat, called the **heat of fusion**, results in a change in state without a change in temperature. The bonding in ice leaves hexagonal openings in the immobilized, somewhat expanded lattice (Figure 5.3). As a result of the expansion (about 9 percent), ice is less dense than water and floats to the surface.

The reverse situation—that is, the change from a solid to a liquid—occurs when ice melts. For this transformation, just under 80 calories (actually, 79.7) need to be absorbed into the system for each gram of ice to provide the energy needed to break the bonds that formed when the water solidified. After ice has melted, the very cold water contracts in volume until it reaches 4°C (39°F), the point at which water is most dense. The warming of the water requires only one calorie per degree Celsius per gram of water, a sharp contrast with the high energy required to transform ice into water.

As water is heated, hydrogen bonds constantly break and form again, and the individual molecules move increasingly rapidly. Finally, some of the molecules move so rapidly that they overcome their attraction to other molecules of water and escape into the air as vapor or steam. For a gram of water to change into gas, 540 calories (actually, 539.4) are required (Figure 5.4). This energy requirement is called the **heat of vaporization**. The temperature of water ceases to rise when this point, the boiling point, is reached because any energy introduced thereafter is used to meet the high energy required for the change of physical state from liquid to gas. Table 5.1 and Figure 5.5 summarize the energy required for transforming ice into steam.

**Heat of fusion**
Heat released when a liquid is transformed into a solid (80 calories per gram of water); also called *heat of solidification*.

**Heat of vaporization**
Heat energy absorbed in the conversion of water into steam (540 calories per gram of water).

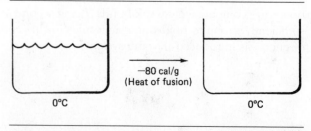

**Figure 5.2**    At the change of state from water to ice, 80 calories per gram are lost as the result of the heat of fusion, and the volume increases by 9 percent.

**Figure 5.3** Crystals are clearly visible as ice forms along the edge of a mountain stream. (Courtesy of Catherine Mulligan.)

## Factors Influencing the Freezing Point

If a soluble substance (such as sugar) or a substance that ionizes (such as salt) is added to water, the temperature at which freezing takes place is lowered. This variation is important in preparing, storing, and serving frozen foods, notably frozen desserts. At the freezing point, water in its liquid state is in equilibrium with the solid state, ice. For pure water, this equilibrium

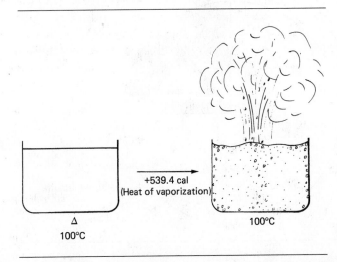

+539.4 cal
(Heat of vaporization)

$\Delta$
100°C

100°C

**Figure 5.4** When water is changed to steam, almost 540 calories are required as the result of the heat of vaporization.

**Table 5.1**    Energy Required to Transform 1 g of Ice to Steam

| Physical State | Change Occurring | Energy Involved |
|---|---|---|
| Ice (0°C, 32°F) | Water (0°C, 32°F) | 79.7 cal |
| Water | 0°C (32°F)–100°C (212°F) | 100.0 cal |
| Steam | 100°C (212°F) water to steam | 539.4 cal |
| Total energy required | | 719.1 cal |

is reached at 0°C (32°F), but the freezing point of solutions is reduced somewhat as a consequence of the reduction in vapor pressure effected by the solute.

Sugar is soluble in water, but the actual solubility is influenced by the temperature. Very low temperatures decrease the amount of sugar capable of dissolving in water, but even at freezing temperatures some sugar can dissolve. The dissolved sugar modifies the freezing point of the solution and reduces the vapor pressure.

If a gram molecular weight (342 g) of sucrose is dissolved in a liter of water, the freezing point of the solution drops 1.86C° (3.52F°). In practical terms, a solution of $1\frac{3}{4}$ cups of sugar in a liter of water freezes at about −1.9°C (28.5°F). The addition of more sugar (Table 5.2) would continue to depress the freezing point at this rate until no more sugar could dissolve at these colder temperatures. The effect of sugar on the freezing point increases the length of time required to freeze frozen desserts because of the need to remove more heat from the mixture. It also means that frozen desserts with high sugar content will melt more quickly than will those with little sugar.

Because salt ionizes, it has twice as great an effect per gram molecular weight as sugar. A gram molecular weight of salt (58 g or about $3\frac{1}{4}$ tablespoons) in a liter of water depresses the freezing point of the solution 3.72C° (7.04 F°). The solubility of salt, like that of sugar, is reduced as the temperature drops. The practical limit on the freezing point of a salt solution is about −21°C(−6°F) (about 328 g or $1\frac{1}{8}$ cups of salt per liter). This fact is used when freezing ice cream in crank-type ice cream freezers. Rock salt is added to the ice surrounding the container in these freezers because of the ability of salt solutions to produce cold temperatures (Chapter 14).

## Factors Influencing the Boiling Point

**Boiling point**
Temperature at which vapor pressure of a liquid just exceeds atmospheric pressure.

The **boiling point** is the temperature at which vapor pressure just exceeds atmospheric pressure. At sea level, water boils at a temperature of 100°C (212°F). However, both atmospheric pressure and vapor pressure can be modified by various means to change the temperature at which boiling occurs. Both commercially and in the home, examples of these techniques are available.

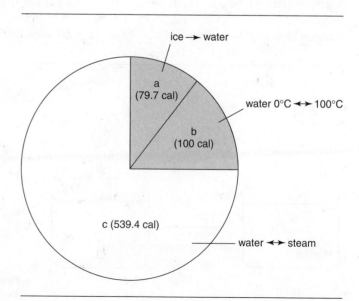

**Figure 5.5**    Calories required to (a) change 1 g of ice to water at 0°C, (b) heat 1 g of water from 0 to 100°C, and (c) convert 1 g of water at 100°C to steam (a total of 719.1 calories).

**Table 5.2**   Depression of Freezing Point of 1 Liter of Water with Varying Amounts of Sugar

| Sugar | | Depression of Freezing | Actual Freezing | |
|---|---|---|---|---|
| Measure (c) | Weight (g) | Point[a] (C°) | Temperature (C°) | (°F) |
| 1 | 200 | 1.09 | −1.09 | 30.04 |
| 2 | 400 | 2.18 | −2.18 | 28.08 |
| 3 | 600 | 3.26 | −3.26 | 26.13 |
| 4[b] | 800 | 4.35 | −4.35 | 24.17 |

[a] $\dfrac{\text{sugar wt (g)}}{342} \times 1.86 = $ depression of freezing point in C°.

[b] Typical concentration of sugar in ice cream.

Atmospheric pressure is one of the parameters influencing the temperature at which boiling occurs. Atmospheric pressure decreases with an increase in altitude because of the reduced amount of air above the surface of the earth at high elevations. At sea level, atmospheric pressure is approximately 14.7 pounds per square inch, compared with 12.3 pounds per square inch at 5,000 feet and only 10.2 pounds per square inch at 10,000 feet.

The vapor pressure that is sufficient to exceed atmospheric pressure decreases as altitude rises. This change in the vapor pressure parallels the drop in atmospheric pressure as altitude is increased. For every 500-foot rise in elevation, the temperature of boiling water drops 1F°; an increase of 960 feet in elevation decreases the boiling point of water by 1C°. Table 5.3 presents the effect of elevation on the temperature of boiling water. A longer cooking time is required for boiling foods in the mountains than is recommended at sea level because the temperature of boiling water is lower at very high elevations.

In commercial food-processing plants, it is feasible to employ a partial vacuum to simulate the effect of altitude on boiling points. By controlling the pressure appropriately within a partial vacuum, liquids can be brought to a boil at such low temperatures that cooked flavors and color changes do not develop. This technique has been used with considerable success in producing fruit juice concentrates.

The reduction in the temperature of boiling can be an inconvenience when food is prepared in the mountains. Fortunately, a pressure saucepan can be used to increase atmospheric pressure artificially. This device forms a tight seal, which traps vapors in the pan. As heat energy is conducted into the pan's interior, the pressure above the liquid increases, causing the temperature of the boiling water to be hotter. The pressure is increased to the desired level, and the heat source is controlled to maintain this pressure for the appropriate length of time. Careful cooling is required before the seal is released. This reduces the pressure in the system to the normal pressure of the surrounding atmosphere and prevents spraying food from the pan.

**Table 5.3**   Temperature of Boiling Water at Selected Elevations

| Elevation (ft) | Temperature of Boiling Water | |
|---|---|---|
| | °C | °F |
| Sea level | 100.0 | 212 |
| 1,000 | 98.9 | 210 |
| 2,000 | 97.8 | 208 |
| 3,000 | 96.7 | 206 |
| 4,000 | 95.6 | 204 |
| 5,000 | 94.4 | 202 |
| 6,000 | 93.3 | 200 |
| 7,000 | 92.2 | 198 |
| 8,000 | 91.1 | 196 |
| 9,000 | 90.0 | 194 |
| 10,000 | 88.9 | 192 |

Using a pressure saucepan saves cooking time because the food is cooked at a higher temperature than can be achieved in a regular saucepan, even when the lid is on it. A pressure saucepan or pressure canner is of particular merit in the canning of low-acid foods, such as vegetables and meats. These foods may harbor harmful microorganisms (notably *Clostridium botulinum*) that are extremely resistant to heat. Adequate temperatures for processing can be reached only by using pressure.

A pressure of 5 pounds at sea level provides a temperature in the pressurized pan of 109°C (228°F). By increasing the thermal energy in the pan, a pressure of 10 pounds can be produced, resulting in a temperature of 115.6°C (240°F); at 15 pounds pressure at sea level, the temperature reached is 121°C (250°F). Consistent with the previous description of the effect of altitude on temperatures of boiling water, the pressure must be increased by 5 pounds if the same temperature is desired at 10,000 feet as is sought at sea level. To illustrate, 20 pounds of pressure need to develop in the pressure cooker to reach a temperature of 121°C (250°F) at 10,000 feet, whereas 15 pounds is required at sea level to achieve 121°C (250°F) in the cooker.

Vapor pressure is the other factor influencing the temperature at which boiling of a liquid occurs. In true solutions, dissolving or ionizing substances in a liquid modify vapor pressure. Both sugar and salt are of potential interest in food systems, and sugar is particularly useful because of its palatability. These substances reduce the vapor pressure of a solution, which means that more heat energy must be supplied if the solution is to develop sufficient vapor pressure to overcome the pressure of the atmosphere and boil. In fact, the boiling temperature of a sugar solution increases 0.52C° (0.94F°) per gram molecular weight dissolved in the water. Table 5.4 illustrates this relationship.

Ions, like molecules in solution, reduce vapor pressure. The fact that salt ionizes into one ion of sodium and one ion of chloride explains why a gram molecular weight of salt elevates the boiling point of a salt solution 1.04C° (1.88F°). Although a gram molecular weight of salt has twice as great an effect on the boiling point of a salt solution as does a gram molecular weight of sugar (which does not ionize), salt is of little practical use to raise the boiling point. The taste becomes unpalatable, and the health risks are unacceptable at the levels of salt that would be needed if the temperature of the boiling solution were to be elevated sufficiently to reduce cooking times perceptibly.

Larger substances—that is, those forming colloidal dispersions or coarse suspensions—have so little effect on vapor pressure that their influence is imperceptible. This means that proteins and coarse, insoluble particles have no measurable influence on boiling points.

**Bound water**
Water that is bound to other substances and no longer exhibits the flow properties and solvent capability commonly associated with water.

## Bound Water

Typically, flow properties are associated with water. Water also is expected to be able to dissolve and dilute many substances. However, some water is not able to function in these typical ways because it is held tightly to other molecules and hence is called **bound water**. This bound water

**Table 5.4**    Boiling Temperatures of Varying Concentrations of Sugar Solutions

| Sucrose (%) | Boiling Point | |
|---|---|---|
| | °C | °F |
| 0 | 100.0 | 212.0 |
| 10 | 100.4 | 212.7 |
| 20 | 100.6 | 213.1 |
| 30 | 101.0 | 213.8 |
| 40 | 101.5 | 214.7 |
| 50 | 102.0 | 215.6 |
| 60 | 103.0 | 217.4 |
| 70 | 106.5 | 223.7 |
| 80 | 112.0 | 233.6 |
| 90 | 130.0 | 266.0 |
| 100 (molten sugar) | 160.0 | 320.0 |

fails to exhibit flow properties, and it does not serve as a solvent. Other changes in characteristics also are apparent, including high density, inability to be expressed from tissues, apparent lack of vapor pressure, and inability to enter the frozen state.

Water can be bound to other molecules in food to varying degrees. Pure water is not bound to any other types of molecules. However, water in tissue membranes is bound to some other molecules and loses its ability to flow freely. Some water in foods is held by hydrogen bonds in microcapillaries and shows behavior sharply different from that of free water. Water is adsorbed in monolayers on many proteins and other electrically attractive components of foods. No sharp divisions separate one type of water from another.

Because it lacks flow properties, bound water can contribute significantly to the textural characteristics of foods. A particularly graphic example of the change in flow properties that occurs when water is transformed from free water to tightly bound water is provided when gelatin is hydrated. A quarter of a cup of freely flowing water loses its ability to flow when a tablespoon of dry gelatin is sprinkled into it and allowed to stand briefly to allow time for water to be bound to the gelatin molecules. Bound water is not completely static, for there is some shifting between the bound and free water, but the water can no longer flow. The binding of water in gelatinizing starch mixtures and in denaturing proteins explains some of the textural changes observed when various food products are cooked.

## WATER ACTIVITY

**Water activity ($a_w$)** is a comparison of the vapor pressure of water in a food sample with the vapor pressure of pure water. Since vapor pressure is influenced by temperature, the food sample must be at the same temperature as the pure water that serves as the basis for comparison. This relationship can be seen in the following equation, assuming a common temperature:

**Water activity ($a_w$)**
Ratio of the vapor pressure of a food sample to the vapor pressure of pure water.

$$\text{Water activity} = \frac{\text{vapor pressure of water in sample}}{\text{vapor pressure of pure water}}$$

Water activity in all foods is always below a value of 1.0; this is due to the reduction in vapor pressure that is found to varying degrees in all foods.

Many familiar foods, including meats and produce, have a water activity of slightly less than 1.0—actually between 0.95 and 1.00 (Table 5.5). The levels of salt and sugar in these foods are comparatively limited, which means that the vapor pressure of water in the food sample will be very close to that of water itself. However, cheeses, salted hams, and dried foods have increased levels of salt and/or sugar, and the vapor pressure of water in the sample will decrease significantly. The result is that water activity in aged cheeses, jams, dried fruits, and similar items usually will be in the range of 0.80–0.90 (compared with a value of 1.0 for pure water).

**Table 5.5**    Water Activity ($a_w$) and Approximate Water Content of Selected Foods

| Food | $a_w$ | Approx. water content (%) |
|---|---|---|
| Beef chuck, raw | 0.95–0.99 | 50 |
| Watermelon | 0.95–0.99 | 93 |
| Butterhead lettuce | 0.95–0.99 | 96 |
| Pound cake | 0.90–0.95 | 18 |
| White bread | 0.90–0.95 | 38 |
| Aged Cheddar cheese | 0.80–0.90 | 41 |
| Corn syrup | 0.60–0.70 | 26 |

Adapted from Troller, J. A. and Christian, J. H. B. 1978. *Water Activity and Food*. Academic Press. New York; Pennington, J. A. and Douglass, J. S. 2004. *Bowes and Church's Food Values of Portions Commonly Used*, 18th ed. J.B. Lippincott. Philadelphia, PA.

**Table 5.6**    Approximate $a_w$ Range Favorable to Microorganisms

| $a_w$ Range of Food | Type of Microorganism |
| --- | --- |
| 0.85–1.0 | Bacteria |
| 0.87–0.91; 0.60–0.65 | Yeasts |
| 0.80–0.87; 0.60–0.75 | Molds |

Adapted from Beauchat, L. R. 1981. Microbial stability as affected by water activity. *Cereal Foods World* 26 (7): 345.

The data in Table 5.5 highlight the fact that total water content and water activity afford quite different perspectives regarding the behavior of water in relation to food. Butterhead lettuce and watermelon have a much greater proportion of water than does beef chuck, and yet lettuce and watermelon cannot be poured despite the abundance of water. The ways in which their water is held or distributed throughout their structures effectively prevent the water from flowing freely.

Conversely, corn syrup has such a comparatively low amount of water that you might predict that it could not be poured, but clearly it does flow, if sluggishly. This behavior can be traced to its low level of water activity, which is due to the very high level of glucose, a carbohydrate capable of forming extensive hydrogen bonding with the water molecules that are present.

Related to water activity is the uptake or loss of water by a food at varying levels of relative humidity at a constant temperature. The plotting of the uptake (termed *resorption*) or the loss of water (*desorption*) provides a record of $a_w$ (water activity) of a particular food at a particular temperature over varying levels of humidity in the environment. Such a plot is called an *isotherm*. Surprisingly, the plot for resorption is not identical with the plot for desorption. The $a_w$ values for resorption in the mid-moisture range tend to be higher than those for desorption—that is, food undergoing desorption will have a higher moisture content at $a_w$ 0.6 than will the same food if it is undergoing resorption.

One of the reasons that $a_w$ is so important in foods is that it is a significant factor in food spoilage and safety. Among the microorganisms inhibited by a level of $a_w$ at 0.95 or just slightly higher are *Pseudomonas, Escherichia,* and *Clostridium perfringens.* Many fresh foods, including meats, milk, and vegetables, have $a_w$ levels in the range of 0.95–1.0, which tends to make them unfortunately attractive to many bacteria (Table 5.6).

In the $a_w$ range of 0.91–0.95, other bacteria (*Salmonella, Vibrio parahaemolyticus,* and *Clostridium botulinum*) are quite capable of surviving and reproducing. Between $a_w$ 0.87 and 0.91 is a range where yeasts can flourish, and at the still lower $a_w$ range from 0.80 to 0.87 molds are active. A few microorganisms can live at $a_w$ levels below 0.80, but none can reproduce at $a_w$ 0.5 or lower. Dried foods that are at moisture levels of 12 percent or less are not susceptible to microbiological spoilage.

Water activity is influenced by temperature in some low-moisture foods, while certain other foods show little change in water activity with heating. As Table 5.7 shows, the water activity of a dry soup mix changed markedly when heated, while beef jerky changed only slightly. Food scientists designing new food products need to consider the effect of changes of ingredients on the water activity of the new product to help assure safe foods. For example, changes in product formulation when fat replacers are incorporated in a product will alter the usual water activity in a food such as cookies.

**Table 5.7**    Water Activity of Selected Foods at Varying Temperatures

| Food | Water activity at | | | |
| --- | --- | --- | --- | --- |
| | **10°C** | **20°C** | **30°C** | **40°C** |
| Distilled water | 1.000 | 1.000 | 1.000 | 1.000 |
| Dry soup mix | 0.191 | 0.239 | 0.292 | 0.302 |
| Beef jerky | 0.694 | 0.697 | 0.693 | 0.698 |
| Sausage | 0.942 | 0.943 | 0.944 | 0.938 |

---

### FOOD FOR THOUGHT: Safe Water for All

Although most Americans only need to turn a faucet to have safe water to drink, as many as 25 percent of the world's people have to carry water home from either wells or streams that may be the source of a variety of illnesses. Even in America, the task of assuring a safe water supply has been complicated significantly by the need to protect public water sources from possible terrorist actions. The Department of Homeland Security and the Environmental Protection Agency are the lead agencies in assuring our water supplies are safe to drink. Other agencies and committees assist in this effort and also in monitoring and forecasting adequacy of the water supplies throughout the nation. On the international level, the World Bank is involved in financing water supply and sanitation projects in regions where water supplies are unsafe and/or inadequate.

Bottled water is an important food product in developed countries around the world. Major producers belong to the International Bottled Water Association. In some countries, bottled water is essential because public water supplies are not safe, but sales of bottled water are very large in the United States despite the fact that healthy drinking water is available from the tap. Bottled water sold in the United States is under the jurisdiction of the Food and Drug Administration; tests occasionally lead to recalls.

Consumers need to be aware that bottled water often does not contain fluoride, which is an important nutrient usually available in public water supplies. Bottled waters containing fluoride will indicate its presence on the label.

Some environmentalists object to bottled water because of the recycling needs it creates. Another negative is the higher cost of bottled water over tap water. However, bottled water is convenient, particularly in circumstances where safe tap water is not available.

---

Sometimes it may be desirable to alter $a_w$ of a food. Different techniques are available. Drying is a method of preserving food (and concomitantly reducing $a_w$) that has been practiced for many centuries. Freezing is another preservation technique that reduces $a_w$, in this case by forming ice crystals and, therefore, removing water from solution in the food. The addition of salt and/or sugar introduces solutes into a food system and lowers $a_w$. Both of these solutes have been utilized for many years in preserved products such as ham and jams.

## WATER IN FOOD PREPARATION

For health reasons, water should be free of harmful chemicals, microorganisms, and any other materials that can be detrimental to health, and it also needs to have fluoride at a level of approximately 1 ppm to promote sound teeth. Usually, chlorine and filtration are used to assure that the water quality is appropriate for human consumption. Despite the excellent work of many communities in providing safe water, bottled water companies are enjoying huge sales of their products.

In food preparation, the importance of having a safe water supply is recognized, and the water is monitored constantly in cities and towns. However, people relying on their own wells or other private sources of water need to be sure to have the safety of the water checked regularly.

When cooking with water, its softness or hardness is important because **hard water** (water containing salts of calcium and magnesium) can influence the quality of products such as tea, coffee, and dried legumes. Boiling water containing calcium or magnesium bicarbonates will cause precipitation of at least some of these salts, leaving a powdery, metallic precipitate that builds up if a pan is used for this purpose frequently. The fact that these salts can be precipitated from the water means that it is classified as temporarily hard water.

**Hard water**
Water containing salts of calcium and magnesium.

Unlike the bicarbonate salts, the sulfate salts of calcium and magnesium do not precipitate out of water while it is boiling. Water containing calcium and magnesium sulfates is classified as permanently hard water. Hard water of either type causes cloudiness in tea (especially in iced tea) and coffee because of interactions between these salts and the tannins in the beverages. **Soft water** (either naturally soft or chemically softened by treating with lime or ion exchange resins of complex sodium salts) will yield more sparkling versions of these beverages than can be obtained when hard water is used.

Softening of dried legumes is impeded significantly when hard water is used for rehydrating and cooking them. The problem occurs because of interactions between the pectic substances present in the dried legumes and the salts contained in the hard water. The hardness of the water is a significant factor, with extremely hard water making it virtually impossible to soften legumes to the desired extent.

**Soft water**
Water treated with lime or ion exchange resins (complex sodium salts) to remove the metallic cations.

## SUMMARY

Water is a common, yet vital, ingredient in food and food preparation. The temperature of boiling water at sea level is 100°C (212°F) and that of freezing water is 0°C (32°F). By addition of a substance capable of forming a true solution or of ionizing, the vapor pressure of the solution is reduced, and the temperature of the boiling liquid is raised. An increase in altitude reduces the temperature of boiling because of the reduction in atmospheric pressure. A similar situation can be created commercially by using a partial vacuum. Conversely, elevated cooking temperatures can be obtained by using a pressure cooker or pressure saucepan to artificially increase atmospheric pressure.

Some water in foods is free water, which has flow properties and the ability to act as a solvent. However, water may be hydrogen-bonded within a food system and lose these important characteristics. This bound water influences the physical properties of food, but in different ways than free water.

Water activity ($a_w$) is the ratio of the vapor pressure of water in a food to the vapor pressure of pure water. In many foods, this value is between 0.95 and just less than 1.00. Highly salted foods and those with very high sugar content have values between 0.80 and 0.90.

Water activity values of a food vary depending on whether moisture is being resorbed or desorbed. Bacteria are viable when $a_w$ is between 0.85 and 1.0 in a food, with a specific bacteria flourishing only in a somewhat narrower range. Yeasts are active at lower $a_w$ (0.87–0.91 and 0.60–0.65), while molds of various types can live at $a_w$ ranges of 0.80–0.87 and 0.60–0.75. Water activity in a food can be influenced by drying, freezing, and adding solutes (e.g., salt or sugar).

Hard water may be temporarily hard because calcium and magnesium bicarbonates are present or permanently hard because calcium and magnesium sulfates are present. Coffee and tea will be cloudy when made with hard water. Softening of legumes is greatly delayed or even prevented by the salts in hard water when they interact with pectic substances during cooking.

## STUDY QUESTIONS

1. Explain why ice floats on water.

2. Compare the amounts of energy involved when water is changed to ice and when water is transformed to steam.

3. Explain why the boiling temperature of water can be altered by use of a (a) vacuum and (b) pressure saucepan.

4. What happens when cold water is added to gelatin and the mixture is allowed to stand briefly?

5. What is water activity? How can it be modified in a food?

6. Why is soft water important when making tea or coffee?

7. True or false. Rice will cook more quickly when boiled in the mountains than at the shore.

8. True or false. Increasing the amount of sugar in an ice cream mixture will extend the time needed to freeze the mixture.

9. True or false. Boiling temperature of water is not altered by adding brown sugar.

10. True or false. More energy is required to convert water to ice than to convert water to steam.

11. True or false. The temperature of boiling water rises if more heat is applied.

12. True or false. Bound water and free water exhibit the same flow properties.

13. True or false. The temperature of boiling water in a pressure saucepan at 5 pounds pressure is higher than at 10 pounds pressure.

14. True or false. If equal amounts of salt and sugar are in separate pans of boiling water, the temperature of the water will be the same in both pans.

15. True or false. Ice floats in water because the crystals are compacted.

## BIBLIOGRAPHY

Decker, K. J. 2003. Wonder waters. *Food Product Design 13* (5): 57–74.

Fennema, O. R. 1996. *Food Chemistry*. 3rd ed. Taylor & Frances. Boca Raton, FL.

Hollingsworth, P. 1995. Pouring it on. *Food Technol. 49* (6): 42.

Knorr, D., et al. 1998. Impact of hydrostatic pressure on phase transitions of food. *Food Technol. 52* (9): 42.

Miraglio, A. M. 2004. Drinking your way to better health. *Food Product Design 13* (10): 33.

Pszczola, D. E. 2005. Innovative chills ahead for frozen desserts. *Food Technol. 59* (3): 40.

Roos, Y. H. 1986. Phase transitions and unfreezable water content of carrots, reindeer meat, and white bread studied using differential scanning calorimetry. *J. Food Sci. 51* (3): 684.

Roos, Y. H., et al. 1996. Glass transitions in low moisture and frozen foods: Effect on shelf life and quality. *Food Technol. 50* (11): 95.

Singh, R. P. 1995. Heat and mass transfer in foods during frying. *Food Technol. 49* (4): 134.

## INTO THE WEB

*http://chemed.chem.purdue.edu/genchem/topicreview/bp/ch14/melting.php#top*—Discussion of freezing and boiling points.

*http://chestofbooks.com/food/science/Experimental-Cookery/Bound-And-Free-Water.html*—Discussion of bound water.

*http://www.decagon.com/water_activity/*—Information about water activity.

*http://members.ift.org/NR/rdonlyres/930A5554-0E9D-4D27-AEAD-BC4AAE7CA930/0/1106lab.pdf*—Paper on water activity and food quality.

*http://foodsafety.psu.edu/Foodpreservation/Water_activity_of_foods.htm*—Discussion of water activity levels in foods and growth of some harmful microorganisms.

*http://www.fda.gov/ICECI/Inspections/InspectionGuides/InspectionTechnicalGuides/ucm072916.htm*—FDA discussion of water activity.

*www.epa.gov/safewater/security*—Description of infrastructure for water supply security.

*www.bottledwater.org*—International Bottled Water Association Web site.

*http://www.hardwater.org/*—Information on hard water and ways of dealing with it.

*http://chemistry.about.com/cs/howthingswork/a/aa082403a.htm*—Chemistry of hard and soft water.

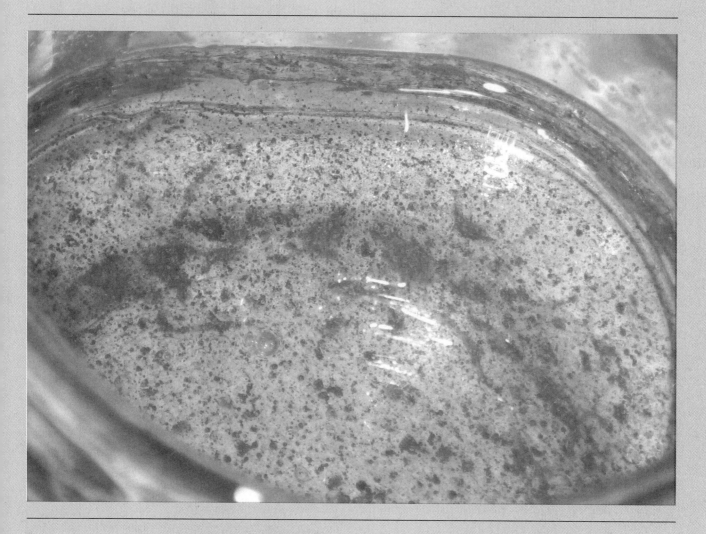

This water-in-oil emulsion is a colloidal dispersion consisting of droplets of red-tinted water dispersed in corn oil.

# CHAPTER 6

# Physical Aspects of Food Preparation

## Chapter Outline

## OBJECTIVES

After studying this chapter, you will be able to:

1. Explain the ways that different forms of energy interact with foods.
2. Discuss the various states of matter and the transitions that can occur.
3. Define types of dispersions and identify food examples.

## INTRODUCTION

Food and food products are influenced very significantly by physical principles that apply throughout harvesting, preparation, and storage. Awareness, appreciation, and understanding of the physical principles that affect the production of successful food products are keys to the education of food scientists. The following sections highlight some of the topics that illustrate the close relationship between physics and food.

## ENERGY AND FOOD

### Mechanical Energy

Energy transfer and/or conversion may be involved in the preparation of many food products, either to make the foods more palatable or to preserve them. The forms of energy used include mechanical, microwave, ionizing, and heat energies. **Mechanical energy** (available

**Mechanical energy**
Energy transferred to food through physical movements, such as beating.

from creaming, beating, and mixing) is transformed into a modest amount of heat energy and also can provide enough energy to accomplish a bit of denaturation of egg white and other proteins when they are beaten.

## Radiation

**Radiation**
Direct transfer of heat energy from its source to the surface of the food.

**Radiation** is a direct method of transmitting heat energy. Quanta of energy radiate from the material (heat source) to the food as electromagnetic waves or rays. The amount of radiation depends on the temperature of the material. As materials are heated, they begin to glow and even can glow white if hot enough. The glow is seen as visible light. At lower temperatures, the waves of electromagnetic radiation commonly are in the infrared range where they are not visible, but the heat can be felt and the wavelengths may be seen as a glow. **Radiant energy** is another term for energy traveling as electromagnetic waves.

**Radiant energy**
Energy traveling as electromagnetic waves.

Materials that produce a dark red glow are at about 550°C (1022°F); temperatures of about 800°C (1472°F) glow cherry red, and 1100–1200°C (2012–2192°F) will be yellow. These colors are just the visible part of the emitted radiation. The peak radiation is in the infrared range about 2–20 times the wavelengths of visible light. This means that the wavelengths of infrared radiation are mostly in the range of 1–10 microns. In this range of temperatures, the hotter the material, the more it will emit radiation in the visible range, and the peak radiation intensity will be closer to visible wavelengths.

**Angstrom (Å)—**
Unit of measure; for example,
1 cm = 100,000,000 Å.

Broiling is the familiar example of the use of radiation in food preparation. Most of the energy emitted when broiling is in the range of 1–10 microns wavelength. The light waves from a broiler are in the visible range at about 700 Å (red). Radiation heats only the surface of the food in a broiler; therefore, broiling uses a continuous energy source, with nothing intervening between the food surface and the energy source.

In many ranges, the oven door needs to be left ajar during broiling so that the energy source remains on continuously. Failure to do this causes heat to build up rapidly in the oven or broiler unit, and the thermostat will cycle the heat source off and on. When the heat source is off, the food is simply heated by conduction and not by radiation.

Restaurants provide another illustration of heating by radiation. Cooked food is kept warm under infrared heat lamps that emit their peak intensities in wavelengths longer than those emitted in broiling. In contrast to the visible glowing of an electric broiler rod, the glass of the infrared heat lamp does not glow.

**Irradiation**
Exposure to the gamma or beta rays emanating from a radioactive material or X-rays. Food does not become radioactive.

Food can be preserved commercially by **irradiation** using special equipment to produce electromagnetic waves of the desired length. Radioactive material (cobalt-60 or cesium-137) emits **gamma rays** (ionizing electromagnetic waves). An electron beam machine produces X-rays that also can irradiate food. As the waves and particles pass through food, they collide with molecules in the food and in microorganisms that may be present. These collisions result in chemical alterations of some of the molecules, producing some ion pairs and free radicals, which then undergo additional changes and alter the food by killing microorganisms, inactivating enzymes, and forming some different chemical compounds (Figure 6.1).

**Gamma ray**
Radiant energy of very short wavelength capable of penetrating food, but not lead.

Among the compounds that can form from water during irradiation of food is hydrogen peroxide, a very reactive compound that is toxic to microorganisms. These changes enable irradiated food to be stored for very long periods of time at room temperature without significant changes in other characteristics of the food. The food is not radioactive.

**Rad**
Ionizing energy equal to $10^{-5}$ joule per gram of absorbing material.

The energy involved in irradiation is expressed as **rads** or multiples of rads (Table 6.1). A rad is an amount of ionizing energy comparable to $10^{-5}$ joule per gram of absorbing material. The rads impacting a food are determined by the length of time of irradiation and the composition of the food. Gamma rays are used for irradiating food because these waves are able to penetrate throughout a food. Alpha particles are not used because they lack the ability to penetrate the interior of the food and, therefore, are effective only on the surface.

**Microwave**
Comparatively short (1–100 centimeters) electromagnetic wave.

**Microwaves** are yet another form of energy used in preparing foods. When microwaves impact food, they cause rapid vibration of water, fat, or sugar molecules in the food, which generates heat. Vibration occurs because these types of molecules are dipoles (see Chapter 5)—that is, they contain both positive and negative electrical charges that are separated by a short distance. When the microwaves hit the dipoles, the molecules begin to

**Figure 6.1** Interior of food irradiation chamber. Source ($^{60}$Co) is immersed in a 20-foot diameter and 20-foot deep pool of highly purified water directly below the floor and is raised by cables (center) when irradiation is being done.

vibrate against one another, creating friction that translates to heat within the food rather than on the surface, as is true of radiation in broiling.

Microwave ovens are common appliances in kitchens today. These ovens utilize radiation in a unique way to heat foods. A **magnetron tube** generates microwaves at either 915

**Magnetron tube**
Tube in a microwave oven that generates microwaves at a frequency of 915 or 2,450 megahertz.

**Table 6.1** Terms Designating Amounts of Radiant Energy

| Term | Definition |
| --- | --- |
| Rad | Basic unit; $10^{-5}$ joule/gram absorbing material |
| Gray | 100 rads |
| Kilorad | 1,000 rads |
| Kilogray | 1,000 grays or 100 kilorads or 100,000 rads |
| Megarad | 1,000 kilorad or 1,000,000 rads |

**Megahertz**
Measure of frequency
defined as 1 million cycles
per second.

or 2,450 **megahertz**. The electrical charge in these ovens cycles extremely rapidly, a fact readily apparent when it is recognized that one megahertz is defined as 1 million cycles per second. The frequency of these cycles causes excitation of water, fat, and sugar molecules as these energy waves penetrate as far as $1\frac{1}{2}$ inches into the food.

The heating that results from the microwaves is augmented by subsequent conduction of this heat energy through the food. Directions for microwave cookery recognize the potential for this subsequent distribution of heat by conduction. **Standing times** often are stipulated as a part of the total cookery procedure in the microwaving of various foods, particularly bulky and dense foods (meat, for example).

**Standing time**
Length of time that food is
allowed to stand in a
microwave oven without
operation of the magnetron
tube so the residual heat in
the food can be transferred
by conduction.

The various settings available in microwave ovens automatically cycle the magnetron tube on and off to regulate the duration of energy input in relation to standing time, thus eliminating the need for manual regulation of standing time. In other words, the reduced power settings that are available on microwave ovens actually are timed periods for cycling the microwave energy on and off.

One of the advantages of microwaves as a source of energy for heating foods is speed. When the magnetron is turned on, the microwaves are generated immediately, and the food starts to heat. The actual rate of heating is determined by the amount of energy entering the food to cause the molecules to vibrate. The amount of energy entering the cavity of the microwave oven is constant, regardless of the amount of food in the compartment.

When a small amount of food is present, the total energy is absorbed within the food. This same amount of energy is available if a large amount of food is placed in the microwave oven. However, this energy will be dispersed through all the food, with the result that there actually will be less energy available per unit of food when the oven load is large than when only a little food is present. Therefore, a large load requires a longer microwaving period than a small load to achieve the same temperature in the food. The speed advantage of a microwave oven for small families is apparent.

Not only does microwave cookery modify the way in which foods are heated, but it also alters the appearance of the crusts of baked products and of meats. The rapid rate of heating and the lack of intense heat on the exterior of foods in a microwave oven cause the foods to be pale on the surface. The usual browning reactions do not have an opportunity to occur.

The surface of microwaved foods is too cool to cause browning because the energy from the microwaves is transformed into heat within the food, and the heat then is conducted to the surface. At the surface, the heat quickly is lost to the surrounding air because the chamber of the microwave oven itself is not warmed by the microwaves and is considerably cooler than the food being microwaved.

## Conduction

In addition to microwave heating, food may be heated by conduction, convection, and radiation. The mode of energy transfer is different in each type, and the total impact on a food also is different. Consequently, the heating method may be dictated not only by the availability of a particular method but also by the desired end result.

**Conduction**
Transfer of energy from
one molecule to the
adjacent molecule in a
continuing and progressive
fashion so that heat can
pass from its source,
through a pan, and
ultimately throughout the
food being cooked.

**Conduction** is a particularly important energy transfer method. Thermal energy is transmitted directly from the heat source when it contacts the pan used to cook the food. The molecules of metal constituting the outer surface of the pan transfer the energy to the next metal molecules, which transfer the energy to the next molecules, and so on until the molecules of metal at the interior of the pan transfer the energy into the molecules of the food that are in contact with the pan. In turn, the molecules in the food pass the energy on to the other molecules that they contact. The final result is that the food in the pan is hot and so is the pan.

Conduction is a convenient means of transmitting heat energy, although it has some disadvantages. For example, time is required for the energy to pass from molecule to molecule, and some energy is wasted in heating the pan and the air near the pan of food. Because food is heated from the outside to the inside by conduction, the food in direct contact with the pan becomes hot much more rapidly than the food in the interior (Figure 6.2).

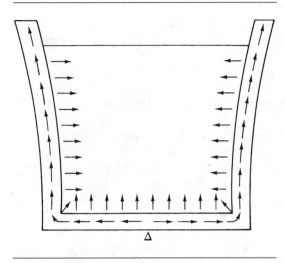

**Figure 6.2**   Conduction of heat occurs throughout the metal in the pan to the water.

Not all pans conduct heat uniformly well. Heavy cast iron and aluminum are popular pan materials with excellent heat-conducting properties. Thin metal pans tend to scorch some foods because some spots in the bottom of the pan may become too hot before the heat can be conducted uniformly across the bottom surface. Even inexperienced cooks can use heavy aluminum and cast-iron pans with a minimum risk of burning or scorching food. In contrast, stainless steel pans develop hot spots readily as the heat energy tends to concentrate in one or more small areas of the bottom of the pans (Figure 6.3). These hot spots often are responsible for burning or scorching, but the problem can be minimized by using a moderate or slow rate of heating and by stirring the food constantly. Pans made of stainless steel encasing a core of heavy aluminum heat quite evenly because of the aluminum.

**Figure 6.3**   Stainless steel develops hot spots due to uneven conduction of heat by the metal. (Courtesy of Plycon Press.)

If the food is heated in a large amount of water, the material of the pan is of limited significance. However, if milk and milk-containing foods are being heated and sauces are being thickened, using a heavy aluminum or cast-iron pan is important to avoid scorching and to assure uniform heating.

Often water is the medium in which a food is cooked. The temperature of the water rises as the heat is conducted to the water via the pan. In turn, water conducts the heat to the food cooking in the water and prevents burning even in stainless steel pans unless the all the water evaporates.

Fat also is an effective conductor and is, in fact, able to transfer far greater thermal energy than water can. The heat energy in water is limited by the boiling. However, fat can be heated to a much higher temperature without being limited by boiling. Air also can conduct heat, but it does so much less efficiently than either water or fat.

## Convection

**Convection**
Transfer of heat by the circulation of currents of hot air or liquid resulting from the change in density when heated.

When air or liquid is heated, currents of hot air or hot liquid develop and help spread heat energy through the container more quickly than conduction can. The circulation or **convection** of the heated material is the natural result of its reduced density. This lighter, heated air or liquid rises through the cooler food in a pan or through the air toward the upper portion of the oven. As the less dense, hot substance rises, it pushes aside the denser, cooler material. This causes the cooler portion to sink toward the bottom, where it comes into contact with the metal of the pan or the heating element of the oven. This material then is heated by conduction, which reduces its density and causes it to rise. As this hot material rises, the cooler portions sink toward the bottom and perpetuate the convection currents (Figures 6.4 and 6.5).

Convection also contributes to the heating of foods by conduction. The two actions complement each other and facilitate the preparation of quality products when used appropriately. The efficiency of convection heating saves time and energy. Commercial bakeries have used **convection ovens** for many years with excellent success. Accelerated movement of the air throughout the convection oven chamber shortens baking time by as much as half and promotes browning to achieve baked products with outstanding crust color.

**Convection oven**
Oven designed with enhanced circulation of heated air to increase heating by convection, reduce baking time, and promote optimal crust browning.

The success of commercial convection ovens stimulated appliance manufacturers to develop a unit for use in homes. Convection ovens for the home, like their commercial ancestors, augment the natural convection currents with the aid of a fan. Home convection ovens do save some baking time over the standard home oven in which conduction is the principal mechanism for heating.

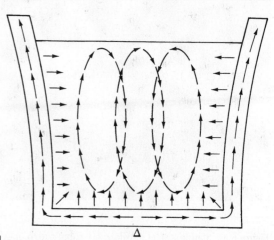

**Figure 6.4** Water is heated in a pan by a combination of conduction and convection energy transfer methods.

**Figure 6.5**    Whether in a standard oven or a convection oven, pans must be arranged so that heat can circulate by convection to bake products uniformly.

## FIGHTING POLLUTION: SWOSTHEE (Single-pan Wood Stoves of High Efficiency)

No matter where people live, they cook their food by some means. In developed countries, the heat source is gas or electricity, and sometimes it may be kerosene or propane. However, firewood and animal dung, euphemistically referred to as biomass, may be burned. Regardless of the type of fuel, convection is the basic way in which heat is transmitted from its source to the surface of the food.

Smog is a problem in many parts of the world, but it is an ever-growing health threat for people living in India, East Africa, and other areas of the developing world where cooking often is done over wood or dung fires. Since firewood and dung are low-cost and renewable energy sources, they are obvious choices for cooking food there. However, the efficiency of the fire and the smoke and pollution it creates can be influenced significantly by the design of the stove used, and considerable effort has been invested in creating and marketing low-cost, energy-efficient, wood-burning stoves for families and small businesses. The common designation for this type of cooking stove is **SWOSTHEE (Single-pan Wood Stoves of High Efficiency)**.

The need for improved wood-burning cooking stoves is extremely great; estimates today are that as many as 3 billion people are cooking indoors over fires fueled by firewood or dung. This translates to untold costs in terms of health care and productivity, not to mention the reduced quality of life when the accumulating smoke in the air not only blackens lungs but also obscures the horizon, transforming near-by buildings into blurry outlines. Approximately 18 percent of greenhouse gas emissions arise from the countless, seemingly insignificant cooking fires of people around the world. SWOSTHEE are an inexpensive and effective way of fighting global warming while improving health and well-being in areas most in need of help.

**SWOSTHEE (Single-pan Wood Stoves of High Efficiency)** Acronym for wood-burning stove designed for families to cook food efficiently using wood as the fuel.

## Mass Transfer

Movement of a food component into or out of a food that is being heated is referred to as **mass transfer**. For example, potatoes frying in deep fat lose some moisture from the outer surfaces into the cooking oil, and that moisture is replaced at the surface by water that has migrated from the interior of the French fry. The net result is that there is not only a reduction in the total amount of water in the center, but some of the water-soluble components also are transferred, along with the water. The entry of oil into the French fry during deep-fat frying is another example of mass transfer.

Water on the surface of a French fry (or other food being fried) is transformed into steam rather quickly. Formation of steam represents a heat loss from the surface of 540 kilocalories per gram of water vaporized. The result of this **endothermic reaction** is that the surface stays cool enough during frying to prevent rapid burning. The temperature for frying needs to be controlled because too high a temperature will cause mass transfer of a significant amount of water from the potato into the oil. The shrunken outline of the fries shows the occurrence of this problem.

## THE STATE OF MATTER

### Solids

*Solids* are defined as substances that do not flow under moderate stress. A solid is in a comparatively low energy state, its molecules moving slowly within a defined space. Close examination reveals that some solids are organized, crystalline structures, whereas others are amorphous—that is, lacking in organization.

Crystalline solids are organized aggregates of molecules, which are called *crystals*. Although they appear to be inert substances, considerable activity actually occurs within some of these systems. Crystalline candies, such as fudge, are familiar examples of crystalline solids. They contain water or other liquids in a dynamic state, one in which some crystals constantly are dissolving and new crystals are forming. The overall result is equilibrium; the relative proportion of liquid and crystals remains constant, but new crystals form as old ones dissolve in the liquid.

Because small sugar crystals dissolve more readily than large aggregates, the smaller crystals are replaced gradually by increasingly larger aggregates. The large aggregates form as some of the dissolved sugar recrystallizes to the solid state on existing crystals. This phenomenon occurs because crystal formation on an existing aggregate results in a lower energy state than would occur if numerous, very small crystals were formed.

Over a period of time, a crystalline candy will exhibit the tendency to become detectably grainy, the result of the formation of many large aggregates at the expense of the original small crystals. This phenomenon occurs despite the fact that the crystalline candy still is considered a solid throughout this ripening or aging process.

Solids that have a random or disorganized structure are classified as **amorphous**. Familiar examples of amorphous solids are hard candies, such as toffee. The sugar molecules in amorphous candies are virtually locked within the solid and lack the freedom of movement necessary for them to reorient themselves in an orderly fashion. The very high concentration of sugar molecules within amorphous candies serves as a serious deterrent to the freedom of movement that would be required for gross rearrangement of the molecules.

Amorphous solids actually may be fairly dynamic in nature, depending on their moisture content, temperature, and other storage conditions. Under appropriate conditions [usually very cold and low water activity ($a_w$) in the solid], the solid will be in a physical state termed a **glassy state**. When the temperature and/or the proportion of $a_w$ is increased, the material may begin to be somewhat more elastic or rubbery and exhibit reduced stiffness and viscosity. This change, termed **glass transition**, occurs over a range of low temperatures despite the fact that a **glass transition temperature ($T_g$)** exists for specific food solids.

The somewhat dynamic nature of some food solids (e.g., dried milk solids) can create quality and storage concerns for food companies. Clumping of dried milk solids is attributed to glass transition problems in which moisture is resorbed on the surface of the dry milk particles, with the result that the temperature for the glass transition can rise to the ambient temperature of the atmosphere in which the milk is held. This increases the plasticity of the dried milk and facilitates bonding between dried milk particles, which is observed as very hard clumps. If the problem of glass transition is to be avoided in the manufacturing of dried milk solids and related protein products, the dehydration process and packaging problems need to be geared toward achieving and maintaining a practical glass transition temperature. Related problems involving glass state transitions are found in the production of some high-starch products, breads, and frozen foods.

## Liquids

When energy is added to a food, the molecules within begin to move at an ever-increasing rate as the energy level increases. If sufficient energy (usually in the form of heat) is introduced, the molecules move so rapidly that some solid foods begin to change state and become liquids.

A liquid exhibits flow properties. Attractive forces operating between molecules in a liquid give some degree of cohesion despite their transitory status. This transitory bonding in liquids is of far shorter duration than the attractive forces that exist in solids. The energy in liquids is sufficient to maintain a high level of cleavage but subsequently to reform van der Waal's forces, hydrogen bonds, and other **secondary bonds**, thus resulting in the flow properties observed in various liquids.

The temperature of a liquid is important in determining the fluidity of the material. With an increase in temperature, molecules move more rapidly, which increases the tendency of molecules to escape from the liquid and become a vapor. This effort or pressure to be in the gaseous rather than the liquid state is termed **vapor pressure**. Vapor pressure increases as temperature rises.

A molecule in the interior of a liquid interacts equally with other molecules of the liquid in all directions. However, a different situation exists in the boundary layer of a liquid; here, at the interface between liquid and air (or other gas), interaction with other molecules occurs only on the portion of the molecule that touches other liquid molecules.

The outer or upper portion of the molecule has no other molecules of the liquid with which to cross-link. As a result, the molecule may have sufficient energy to break the attractive forces formed with other molecules of the liquid and move away from the liquid into the adjoining gaseous phase. This loss of molecules from a liquid that occurs at the surface as a result of vapor pressure is termed **evaporation** (Figure 6.6).

**Surface tension**, a phenomenon related to vapor pressure, is defined as the attraction between molecules at the surface of a liquid. This phenomenon is important because it controls the loss of molecules from a liquid. For evaporation to occur, vapor pressure needs to be greater than the forces represented by the surface tension.

The effect of surface tension can be observed by carefully examining a drop of water on a flat, clean surface or by placing a sewing needle absolutely flat on the surface of a bowl of

**Secondary bonds**
Attractive forces between atoms and functional groups that are less strong than the bonding that occurs when electrons are shared; examples are van der Waal's forces and hydrogen bonding.

**Vapor pressure**
Pressure exerted as molecules of a compound attempt to be in the gaseous rather than the liquid state.

**Evaporation**
Escape of liquid molecules into the surrounding atmosphere.

**Surface tension**
Attraction between molecules at the surface of a liquid.

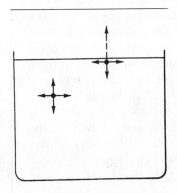

**Figure 6.6**    Interactions on molecules of water and in the boundary layer.

water. Note that the drop of water fails to spread infinitely to a uniform thickness but instead shows some tendency to round into a flattened, ball-like shape.

A needle will float on the surface of the water, causing a shallow deformation of the water's contour. Unless the needle is placed so that its entire length is in contact with the water at the same time, it will sink because its density is greater than that of water. However, the surface tension of water is strong enough to buoy the needle if the weight of the needle is distributed over the entire length of the metal shaft.

Mercury provides an even more dramatic illustration of surface tension. The surface tension of mercury is extremely high, which causes mercury to form sphere-like drops when the liquid is poured or spilled.

The surface tension of liquids varies with the liquid and with the temperature. When two disparate liquids are in contact, their surface phenomena are different. The tendency for molecules at the surface of a liquid to remain within this liquid rather than to intersperse with molecules in an adjoining liquid is referred to as **interfacial tension**. Interfacial tension of liquids is of particular interest in relation to food emulsions and is discussed in that context later in this chapter under "Emulsions."

**Interfacial tension**
The tendency for molecules at the surface of a liquid to remain with the liquid rather than intersperse with molecules of a second adjacent liquid.

## Gases

When a substance is in the gaseous state, the distance between molecules is extremely great when contrasted with the comparatively short distances in liquids and the even shorter distances in solids. Molecules dispersed in the gaseous state move constantly and relatively faster than molecules in liquids. Steam and carbon dioxide are two of the gases of considerable importance in some food products. Swiss cheese is a good example of a product in which the gas formed by microorganisms causes development of a porous texture. The texture characteristics and volume of baked products afford other illustrations of the importance of utilizing gases in making food products.

A particularly important property of gases is their ability to increase in volume and/or to develop pressure as heat is applied to them. This behavior is utilized in the preparation of

---

### Savoring the Sparkle

Champagne is generally considered to be the epitome of the sparkling wines, but not all champagnes are created equal. Their sparkle is created when yeast ferments the sugar in grape juice to ethanol and carbon dioxide, which is trapped in the wine. Pressure builds as more gas is generated in the bottles and can blow the lid off unless special wire collars are locked over the cork. It is rumored that Dom Pierre Perignon, a Benedictine monk, crafted the forerunner of this device as he toiled to develop his special sparkling champagne in the 1600s.

Production of champagne includes two periods of fermentation to generate 60–75 pounds per square inch of pressure from carbon dioxide in the beverage. This pressure is so strong that thicker bottles and the wire cover over the cork are necessary for safety and to prevent breakage. A considerable amount of gas escapes as a mist of very tiny bubbles when the bottle is opened, but the potential for forming millions more bubbles remains. The more important bubble formation occurs when champagne is poured into a glass (preferably a tall, thin flute). Even the tiniest bit of lint traps some carbon dioxide gas; then surface tension, pressure, and viscosity of the wine start to force countless tiny bubbles from the fiber at the rate of about 400 bubbles per second. The added bonus is that the bubbles are sufficiently flexible so they don't pop instantly. Instead, they can carry some flavors to the surface before the bubbles pop and release them into the air directly above. Bubbles of carbon dioxide breaking in the mouth stimulate sour receptors to add some tingle to the tastes and flavors conveyed by champagne.

---

### FOOD FOR THOUGHT: The State Matters

A dessert made in Vietnam provides a useful illustration of the importance of the physical state of ingredients. Preparation begins when sweet glutinous rice (a solid) and coconut milk (a liquid) are heated together until the starch in the rice is gelatinized completely. Thorough pounding or stirring of the rice results in formation of a highly viscous paste. A small ball of this gelatinized rice paste then is deep-fat fried in very hot fat until the exterior is golden brown and crisp.

Although this recipe sounds rather basic, it illustrates the impact that can result when the physical state of water is changed by heat. Much of the initial water in the coconut milk becomes bound with the starch during gelatinization. However, there is still some free water in the ball of paste that is dropped into the deep-fat fryer. The intense heat of the fat quickly generates steam from this water. This gas pushes very hard against the starch, causing rapid expansion of the ball before the crust has a chance to become rigid. The result is a dramatic ball of golden brown dessert with a crisp outer shell and a small amount of coconut-flavored starch nestled within. The huge increase in volume is possible because of the change in the state of water from a liquid to a gas.

---

baked goods, with temperature and time controlled to produce the desired textural characteristics (see Chapter 17).

Various leavening agents serve as sources of gases in batters and doughs to achieve the desired volume and somewhat open texture considered appropriate for specific products. The underlying principle in the action of a leavening agent is that heat causes gases to expand, and this expansion creates sufficient pressure to increase cell size in batters and doughs until their proteins coagulate. This action generally improves texture and increases tenderness by stretching the cell walls thinner and thinner during baking.

## DISPERSIONS

Foods represent mixtures or dispersions of two or more types of substances. These substances can be combined in several different ways. Of particular importance in determining the type of dispersion involved in a food mixture is the size of the molecules or particles to be dispersed. The three basic categories established or defined by particle or molecular size are true solutions, colloidal dispersions, and coarse suspensions.

## True Solutions

**True solutions** have the smallest particle size of the three types of dispersions, for molecules or ions of comparatively small dimensions are the dispersed units. Actually, only molecules less than one **millimicron** (1mμ) in diameter can be dispersed in a liquid to form a true solution, and this is true of simple molecules such as sugar. Starch and protein molecules are too large to form true solutions.

True solutions are characterized as the most stable of the three types of dispersions despite the fact that some solutions are made up of electrically charged ions distributed in the liquid. In food preparation, salt (as sodium and chloride ions) or sugar can be dispersed in water to form a true solution. The ions of salt or sugar molecules are designated as the **solute**, the water as the **solvent**.

True solutions contain varying amounts of ions or molecules of dissolved substances, depending on the solute and the temperature of the solvent. If a true solution contains less solute than the solvent can dissolve at that temperature, it is classified as an **unsaturated solution**. If the true solution contains as much solute as can be dissolved at that temperature, it is classified as a **saturated solution**. If a heated saturated solution is cooled a bit under carefully controlled conditions, more solute can be kept in solution than theoretically can be

**True solution**
Dispersion in which ions or molecules no larger than one millimicron are dissolved in a liquid (usually water).

**Millimicron**
Billionth of a meter.

**Solute**
Substance dissolved in a liquid to form a true solution.

**Solvent**
Liquid in which the solute is dissolved to form a true solution.

**Unsaturated solution**
True solution capable of dissolving additional solute at the temperature of the solution.

**Saturated solution**
True solution containing as much solute in solution as is possible to dissolve at that temperature.

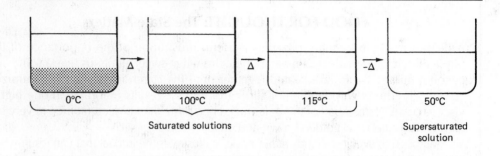

**Figure 6.7**    The solvent in saturated solutions can dissolve increasing proportions of the solute as the temperature rises, and careful cooling permits the dissolved solute to remain in solution, creating a supersaturated solution with excess solute.

**Supersaturated solution**
True solution containing more solute than theoretically can be dissolved at that temperature, a situation created by cooling a heated saturated solution carefully.

**Colloidal dispersion**
Two-phase system containing at least one colloid (substance measuring between 0.001 and 1 millimicron in diameter).

**Continuous phase**
Medium surrounding all parts of the dispersed phase so that it is possible to pass throughout the system in the continuous phase without traversing any portion of the dispersed phase.

**Discontinuous (dispersed) phase**
Phase distributed in a discontinuous fashion, making it necessary to pass through at least some of the continuous phase to reach another particle of the dispersed phase.

**Gel**
Colloidal dispersion of a liquid dispersed in a solid.

**Sol**
Colloidal dispersion of a solid dispersed in a liquid.

**Foam**
Colloidal dispersion of a gas dispersed in a liquid.

**Suspensoid**
Colloidal dispersion of a gas dispersed in a solid.

dissolved by the solvent at this cooler temperature. This seemingly impossible system is designated as a **supersaturated solution** (Figure 6.7).

As the temperature of a solvent increases, the amount of solute that can be dissolved in it to form a saturated solution increases. Table 6.2 illustrates this relationship.

Table 6.2 clarifies that a sucrose solution saturated at 20°C is an unsaturated solution at 40°C. Conversely, a sugar solution boiled to a temperature higher than 100°C has the potential to become supersaturated on cooling because the cooler solution cannot dissolve the amount of sugar already dissolved at the higher temperature. Sugar cookery is based on the interrelationships among concentration of sugar, temperatures, and solutions. The key concepts related to sugar cookery are discussed in Chapter 8.

## Colloidal Dispersions

**Colloidal dispersions** contain molecules intermediate in size between the tiny substances capable of forming true solutions and the gross particles found in coarse suspensions. Specifically, only substances between 0.001 and 1 millimicron can be dispersed in systems classified as colloids. All colloidal systems contain a substance of colloidal dimensions dispersed in a different material. The two components of a colloidal system are found in any of the three states of matter—solid, liquid, or gas—and the system is categorized on the basis of the states involved.

All colloidal systems have two phases: a *continuous phase* and a *discontinuous (dispersed) phase*. The **continuous phase** extends throughout the system, surrounding all parts of the other phase of the system. The **discontinuous (dispersed) phase** is, as the name implies, distributed in isolated or disconnected fashion throughout the entire colloidal system.

Technically, colloidal systems may be any combination of solid, liquid, or gas as the continuous or discontinuous phase, as long as the particle size of one phase is within the dimensions of a colloid. For example, a colloid might consist of a solid dispersed in a liquid.

Table 6.3 describes the colloidal systems of particular importance in foods. Colloidal systems often are designated on the basis of the state(s) of matter constituting the two phases. To illustrate, a **gel** can be categorized as a liquid in a solid. Similarly, a **sol** is a solid in a liquid, a **foam** is a gas in a liquid, and a **suspensoid** is a gas in a solid.

**Table 6.2**    Solubility of Sucrose in Water at Various Temperatures

| Temperature of Solution (°C) | Maximum Weight of Sucrose Soluble in 100 ml of Water (g) |
|---|---|
| 0 | 179.2 |
| 20 | 203.9 |
| 40 | 238.1 |
| 50 | 260.4 |
| 100 | 487.2 |
| 115 | 669.0 |

**Table 6.3**   Colloidal Systems in Foods

| Name of Colloidal System | Dispersed Phase | Continuous Phase | Example in Food |
|---|---|---|---|
| Emulsion | Liquid | Liquid | Salad dressing |
| Sol | Solid | Liquid | Gravy |
| Gel | Liquid | Solid | Baked custard |
| Foam | Gas | Liquid | Egg white foam |
| Suspensoid | Gas | Solid | Congealed whipped gelatin |

The behavior of colloids and the systems in which they are found is influenced by various physical characteristics. Some colloids are **hydrophilic** (water loving) and are hydrated readily. Other colloids exhibit the opposite behavior and are termed **hydrophobic** (water hating). Hydrophobic colloids do not readily attract water, which makes it possible for such substances to be precipitated with the addition of only a little salt.

**Hydrophilic**
Attracted to water.

**Hydrophobic**
Repelling water.

The chemical composition of a compound is significant in determining whether the substance is hydrophilic or hydrophobic. In practice, many substances in foods have some parts of their structures that are hydrophilic and other portions that are hydrophobic. The functional groups in molecules that are attracted to water are polar groups. These polar groups include the functional groups in the following organic compounds:

$$\text{organic acid } (-\overset{\displaystyle O}{\overset{\|}{C}}\diagdown_{OH}), \text{ aldehyde } (-\overset{\displaystyle O}{\overset{\|}{C}}\diagdown_{H}), \text{ ketone } (-\overset{\displaystyle O}{\overset{\|}{C}}-)$$

and others, as Table 6.4 shows. Nonpolar structures include cyclic structures and carbon chains. Organic compounds with polar and nonpolar groups tend to collect at the interface between two liquids and thus serve as stabilizing agents in emulsions because parts of their molecules are drawn toward the dispersed phase and parts toward the continuous phase.

**Table 6.4**   Polar and Nonpolar Groups

| Polar Group (Hydrophilic) | Nonpolar Group (Hydrophobic) |
|---|---|

**Emulsion**
Colloidal dispersion of a
liquid in another liquid
with which it is immiscible
(not able to be mixed).

***Emulsions.***    An **emulsion** is a colloidal system in which a liquid is dispersed in droplets in another liquid (the continuous phase) with which it is immiscible. The old familiar saying "oil and water don't mix" is a colloquial way of recognizing the instability of emulsions. An emulsion is formed by thoroughly shaking two immiscible liquids together. Shaking provides the energy needed to enable the liquid with the higher surface tension to form many small droplets or spheres surrounded by the other liquid. Formation of these numerous small spheres creates a far larger surface area for the dispersed liquid than occurred when the two liquids were in contact as two layers.

Formation of an emulsion requires energy to permit the continuous phase to stretch out and surround the forming droplets. The overall process is rather complex, for the continuous phase needs to be extended and more surface area needs to be available to surround the droplets that are to be the dispersed phase. The discontinuous or dispersed phase must be split into droplets, and the surfaces of these droplets must be coated with the emulsifying agent.

Beating, stirring, or shaking provides the energy needed to spread the continuous phase and form the emulsion. Some resting time can be useful in the formation of emulsions, for the rest provides time for the emulsifying agent to orient its molecules along the newly formed interfaces between the droplets and the continuous phase. Warming of the ingredients may ease the formation of an emulsion by increasing the fluidity of both phases, a situation that facilitates spreading of the continuous phase and splitting of the dispersed phase into the desired small droplets.

The droplets of the dispersed phase tend to coalesce when they bump into one another as they move through the emulsion because the one large droplet (created from two small droplets) represents a lower energy state than the two droplets. Because the one large droplet has less surface area than the two small droplets have cumulatively, the droplets in an emulsion continue to coalesce into larger droplets until ultimately the emulsion breaks or separates into two distinct phases.

**Oil-in-water emulsion
(o/w)**
Colloidal dispersion in
which droplets of oil are
dispersed in water, for
example, mayonnaise.

**Water-in-oil emulsion
(w/o)**
Colloidal dispersion in
which droplets of water are
dispersed in oil, for
example, butter.

Emulsions often are classified on the basis of the type of liquid constituting each of the phases. Thus, an emulsion may be classified as an **oil-in-water emulsion (o/w)** or a **water-in-oil emulsion (w/o)**. These categories express clearly that oil is dispersed as droplets suspended in water or that water is dispersed as droplets in oil, respectively (Figures 6.8 and 6.9).

The stability of emulsions is important in food products, for the behavioral characteristics of the two liquids are quite different when in an emulsion than they are in the non-emulsified state. One of the most obvious effects of an emulsion is that its viscosity is increased over the viscosity of either of the liquids. Mayonnaise illustrates this clearly. It is possible to add so much oil to a mayonnaise emulsion that the product actually can be sliced. This truly is remarkable when compared with the very fluid state of vinegar and the only slightly more viscous state of the oil used in preparing mayonnaise.

The stability of an emulsion is determined by

- viscosity of the continuous phase
- presence and concentration of the emulsifying agent
- size of droplets
- ratio of dispersed phase to continuous phase

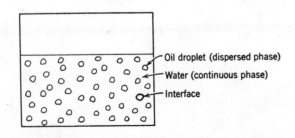

**Figure 6.8**   An oil-in-water emulsion.

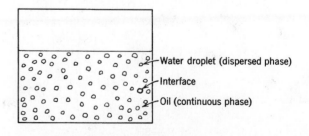

**Figure 6.9**   A water-in-oil emulsion.

A key factor in the formation and stability of emulsions is the presence of an **emulsifying agent**. An emulsifying agent consists of both polar and nonpolar groups and thus has some attraction toward both phases of the emulsion.

The polar groups orient the emulsifying agent toward the aqueous phase, and the nonpolar groups pull the molecule toward the oil. The net result of this tug-of-war is that the emulsifying agent is held at the interface of the two phases. The emulsifier forms a monomolecular layer that effectively coats the interfaces to impede the contact between oil droplets in an oil-in-water emulsion (or, conversely, water droplets in a water-in-oil emulsion). Because the emulsifying agent collects around the surface of the dispersed spheres, the droplets cannot touch one another directly and coalesce.

The nature of the emulsifying agent influences the type of emulsion that is formed. If the agent is attracted more strongly to water than to oil, the surface tension of the water is reduced more than that of the oil. The result is the formation of an oil-in-water emulsion.

The balance between hydrophilic and hydrophobic (or lipophilic) characteristics in various emulsifying agents is expressed on a 20-point scale referred to as the **hydrophilic/lipophilic balance (HLB)**. Emulsifying agents with an HLB between 8 and 18 are especially attracted to water and are considered suitable for forming oil-in-water emulsions, which are the type found in most food emulsions. Butter and margarine, both of which are water-in-oil emulsions, can be stabilized (if desired) with an emulsifying agent in the range of 3–6 HLB. Such agents are attracted more to oil than to water.

The best emulsifier available for forming emulsions in the home is egg yolk, which derives its unique abilities from its **lecithin** and protein content (see Chapter 16). Lecithin is a phospholipid that has two largely lipophilic fatty acid chains and a hydrophilic phosphoric acid component. The combination of these features promotes formation of an oil-in-water emulsion because of the way that lecithin is oriented toward both the oil and water. The proteins in the egg yolk also are attracted to the region of the interface between oil droplets and water, further stabilizing the emulsion.

The amount of emulsifying agent present significantly influences the stability of the emulsion. If enough agent is present to form a complete monomolecular layer around each droplet, the emulsion will be stable; any unprotected surfaces, however, will allow the droplets to coalesce when they come into contact. No benefit results from having more emulsifying agent than is necessary to provide total coverage of the interfaces. Up to the point of total coverage, an increase in the amount of emulsifying agent will increase stability. Thorough coverage with an emulsifying agent results in a permanent emulsion. Although it is possible to break permanent emulsions, usual conditions of handling and use do not damage these emulsions. Mayonnaise is a familiar example of a **permanent emulsion**.

Some stability in emulsions is imparted simply by the viscous nature of the emulsion. Sweet salad dressings for fruit may be stabilized by the use of a cooked, viscous sugar syrup or by honey. Although these ingredients are not very effective as emulsifying agents at the interface, their viscous nature produces a fairly thick dressing, one in which the dispersed droplets move rather slowly.

Collisions between droplets occur with decreasing frequency as the viscosity of an emulsion is increased. Breaking the emulsion into the two distinct phases is retarded by reduced contacts between droplets, thus promoting stability. Emulsions stabilized by this technique are termed **semi-permanent emulsions**.

**Temporary emulsions** are quite fluid systems with little emulsifier present. An oil-and-vinegar dressing is a familiar example of a temporary emulsion. The insoluble seasonings, such as ground pepper, provide only slight protection of the emulsion by orienting themselves at the interface when the oil and vinegar are shaken together. However, simply far too little of these substances is present to provide the protective coating needed to keep the oil droplets from coalescing, and they frequently bump into one another in the fluid emulsion. The finer the particle size of these powders, the more effective is their action as emulsifiers.

Emulsions, even permanent emulsions, separate under certain conditions. Temporary emulsions always separate quite rapidly because their droplets are able to move together freely in the fluid condition typical of these emulsions. However, more stable emulsions (such as a viscous honey-like dressing or even mayonnaise) also separate if their temperature is changed

**Emulsifying agent**
Compound containing both polar and nonpolar groups so that it is drawn to the interface between the two phases of an emulsion to coat the surface of the droplets.

**Hydrophilic/lipophilic balance (HLB)**
Twenty-point scale indicating the affinity of an emulsifying agent for oil versus water.

**Lecithin**
Phospholipid in egg yolk that is an effective emulsifying agent; its formula is

**Permanent emulsion**
Emulsion containing an amount of emulsifying agent sufficient to enable it to remain intact during ordinary handling and use.

**Semi-permanent emulsion**
Emulsion with rather good stability because of the viscous nature of the liquid constituting the continuous phase.

**Temporary emulsion**
Emulsion that has little emulsifying agent and is too fluid to restrict movement of droplets; such instability requires that the ingredients be shaken to form a temporary emulsion immediately before use.

drastically. Heating increases fluidity and encourages the emulsion to break. Freezing breaks emulsions because the water expands as ice crystals form. Jarring or shaking also can cause permanent emulsions to break, as a result of the different densities of the two phases.

***Sols.***    Sols are colloidal systems in which a solid of colloidal dimensions is dispersed throughout a liquid. This type of system has flow properties, which may range from rather fluid to extremely viscous, barely flowing. Regardless of the viscosity, the dispersed (discontinuous) solid is always distributed throughout the sol; it does not precipitate to the bottom (as can happen in coarse suspensions). Proteins in sols often are aided in remaining dispersed by the electrical charges on their external surfaces. Protein molecules of similar electrical charge repel each other, thereby making it improbable that they will join together and become too large a unit to remain in the colloidal sol.

Pectin, a complex carbohydrate, provides a useful example of a second way in which solids are able to disperse to form a sol. Pectin (see Chapter 7), like other carbohydrates and many proteins, is hydrophilic and attracts a layer of water that is bound tightly to the molecules by hydrogen bonds. Water thus forms an insulating shield for the pectin or other hydrophilic colloid, providing a layer that inhibits bonding between the molecules of the colloidal substances.

Sols are characterized as pourable, or at least able to flow slightly. This property is to be expected, because liquid is the continuous phase. Even when the concentration of solid colloidal material is high, some liquid still separates each of the solid particles, which ensures that at least some flow is possible. The actual ability of a sol to flow is determined by several factors, among which the temperature of the sol and the concentration of solid in the liquid are particularly important. Sols flow more readily at a high temperature than they do at a low one, a characteristic that must be considered in evaluating the viscosity of sols. The higher the concentration of the solid in a sol, the more viscous is the sol.

Several sols commonly are served. These include white sauces, gravy, stirred custard, and other thickened sauces. These sols need to be prepared with the proper ratio of ingredients to achieve the desired viscosity or flow properties at the temperature anticipated for serving. If they begin to thicken too much and to exhibit poor flow properties, they can be reheated to reduce viscosity, or more liquid can be added.

Other sols may be formed as a preliminary step in making a gel but will not actually be served as a sol. Gelatin is a familiar example in which a sol is formed first by hydrating and dispersing the gelatin molecules in hot liquid. Jams and jellies made with pectin and starch-thickened puddings are other illustrations of the need to form a sol prior to the desired gel structure.

***Gels.***    Frequently, a sol can be transformed into a second type of colloidal system—a gel (Figure 6.10). This transition occurs subtly as the energy level of the system drops during the cooling of a sol. The solids in the discontinuous phase move with increasing difficulty

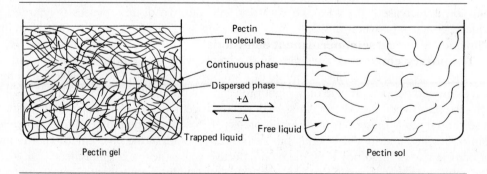

**Figure 6.10**    In a pectin gel, the pectin molecules are the continuous phase and the liquid is the dispersed phase; the sol is the reverse situation, that is, the pectin molecules are the dispersed phase and the liquid is the continuous phase.

through the continuous liquid phase and eventually start to associate with one another. A gel also is a colloidal system, but it is the reverse of a sol. A gel contains a solid matrix (the continuous phase) and a liquid (the discontinuous or dispersed phase). The solid in the gel is sufficiently concentrated to provide the structure needed to prevent the flow of the colloidal system. Some of the liquid in a gel is adsorbed to the molecules of the solid and is bound water. This binding of some of the liquid accounts, in part, for the loss of flow properties when a gel is formed. The remaining water is free but is trapped between the solid, interlocking molecules or strands of the solid component of the gel.

Some of the free liquid may be released if the gel structure is cut. Drainage of free liquid from a gel is termed **syneresis**. Syneresis is exhibited in varying degrees by various gels. The type of solid in the gel and its concentration are of particular importance in determining the amount of syneresis. A cranberry jelly provides a clear example of extensive syneresis when it is cut and allowed to stand for a period of time. Although syneresis is undesirable in cranberry jelly, this phenomenon is useful in cheese production. In fact, the production of cheese requires extensive cutting of the curd (a gel structure) to aid in removal of the liquid whey.

**Syneresis**
Weeping or drainage of liquid from a gel.

Because sols and gels are similar, but reverse types of colloidal dispersions, it is not surprising that many of them can be changed from one form to the other, depending on the conditions. Gelatin dispersions provide a familiar illustration of the reversibility of some sols and gels. When a gelatin dispersion is warm, it is a sol and exhibits obvious flow properties. When cooled adequately, the gelatin molecules form secondary bonds as they pass near one another. Gradually, so much energy is lost from the system as it cools that some relatively permanent bonds form among the gelatin molecules, establishing a solid network that is capable of entrapping the water in the system. The result is an apparent solid, the molded gelatin, which is capable of holding its shape when served.

*Foams.* *Foams* are dispersions of gases in a dispersing medium. In food foams, the dispersing medium usually is liquid, sometimes strengthened by a solid or modified into a solid by heating. The foams utilized most commonly in food preparation are those made with heavy cream, egg whites, egg yolks, gelatin, and concentrated milk products (Figure 6.11).

Formation of a foam is accomplished by providing energy to counteract the surface tension of the liquid and to stretch the fluid into thin films encompassing bubbles of gas. Liquids capable of forming foams have low surface tension and thus can spread or stretch easily and not coalesce readily. They also must have low vapor pressure.

If surface tension is high, the foam will be difficult to form because of the resistance to spreading and the strong tendency to expose the least possible surface area. High surface tension presents the further problem of limiting the stability of the foam. A foam formed with a liquid of high surface tension collapses very quickly as a result of the tendency to coalesce as the foam stands after beating ceases.

Liquids with high vapor pressure evaporate quickly. When the liquid surrounding a gas bubble evaporates, nothing is left to retain the gas that had been trapped in the liquid. Clearly, liquids with a low vapor pressure are necessary to form useful foams.

Additional stability can be provided to foams if some solid matter is incorporated into the films to increase the rigidity of the walls surrounding the gas. For example, particles of fat in chilled heavy whipped cream aid in stabilizing foams of this type.

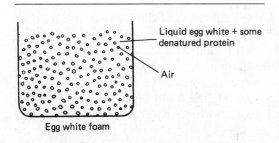

Egg white foam

**Figure 6.11** An egg white foam is a colloidal system in which air (gas) is dispersed in a liquid (egg white), and denatured proteins in the egg white add rigidity to the cell walls.

**Figure 6.12**    Poaching stabilizes islands of egg white foam by denaturing the proteins.

In protein foams, the denaturation of the protein provides stability to cell walls (Figure 6.12). This can be noted particularly clearly in the egg foams used so often in baked products, but other protein foams also benefit from the denaturation of the protein.

Foams are important because of their contribution to volume and texture of various food products. The bubbles form a porous texture, which may be quite variable in character, ranging from very fine to extremely coarse cells. These bubbles promote a feeling of lightness in foods containing foams.

The volume of foods is increased when foams are formed with some of the ingredients. In baked products, foams contribute significantly to the volume of the finished product in two ways. The air already trapped in the foam expands in the heat of the oven until the cell walls of the baking food lose their elasticity when the protein denatures. In addition, steam and the carbon dioxide formed from chemical and/or biological leavening agents during the period in the oven collect in and expand the existing cells of the foam.

## Coarse Suspensions

Many mixtures of substances with particles larger-than-colloidal dimensions (greater than 1 millimicron) are part of systems formed in food preparation. These systems are classified as *coarse suspensions*. They are influenced by gravity and tend to separate. An uncooked starch mixture and oatmeal flakes in water are illustrations of coarse suspensions. When starch or cereal is stirred with water, a coarse suspension results until the suspended food settles by gravity to the bottom of the container, leaving the water on top. When this separation occurs, the coarse suspension no longer exists; however, stirring once again creates a coarse suspension.

## SUMMARY

Heat in food preparation may be introduced by mechanical energy and/or electromagnetic waves (including microwave radiation); distribution occurs through conduction, convection, and mass transfer. Various modes of heating may be combined in the preparation of a single food item, with each method providing some specific effects on the character of the final product.

Mass transfer indicates the shifting of a food component into or out of a food during heating. It is illustrated by the loss of water from french fries during frying and the entry of fat.

The behavior of food during harvesting, preparation, and storage follows basic principles of physics. The three forms of matter—gas, liquid, and solid—are found in various foods and food products. Solids, whether crystalline or amorphous, are useful and stable. Amorphous solids can transform to a glassy state that can cause problems in some dried and frozen foods. Ice is an example of a crystalline solid that can be transformed into its liquid state if enough thermal energy is supplied. If still more thermal energy is applied, water can be transformed into its gaseous state.

Liquids exhibit vapor pressure and surface tension. Vapor pressure is the force exerted as the liquid molecules attempt to leave the liquid or evaporate. Surface tension is the energy or attraction between molecules at the surface of the liquid, a force tending to draw the liquid into its smallest possible volume. Vapor pressure and surface tension work in opposition. When two liquids are in contact with each other but are not miscible, their interaction is known as interfacial tension.

Gases are important in foods because of the change in volume as the temperature changes. The expansion of gases in batters and doughs during the baking of these products results in a vital expansion that enhances the volume, texture, and general palatability of baked products.

Food dispersions may be categorized as: (1) true solutions, (2) colloidal dispersions, or (3) coarse suspensions. True solutions may be unsaturated, saturated, or supersaturated, depending on the amount of solute in relation to solvent and on the temperature and treatment.

Molecules ranging in size from 0.001 to 1 millimicron can be dispersed in a liquid, gas, or solid to form colloidal dispersions of various types. Four types of colloidal dispersions frequently are used in food systems: (1) sols, (2) gels, (3) emulsions, and (4) foams. Sols are composed of a solid of colloidal dimensions dispersed in a liquid. Sols do not precipitate or separate; their common characteristic is that they exhibit flow properties.

Emulsions are colloidal systems in which one liquid forms the continuous phase and an immiscible liquid is dispersed in it in the form of small droplets. Salad dressings are a familiar example of oil-in-water emulsions, in which oil droplets are dispersed in vinegar (water). Stability of emulsions ranges from temporary to permanent and depends on the viscosity of the emulsion and the presence of emulsifying agents. The hydrophilic/lipophilic balance (HLB) determines whether an emulsifying agent promotes formation of an oil-in-water or a water-in-oil emulsion. Lecithin, a phospholipid in egg yolk, has both polar and nonpolar groups that draw it effectively to both the oil and vinegar components of the emulsion. This concentrates the lecithin at the interface (the surface of the droplets) and provides a protective coating. Ingredients that can coat the surface of droplets and act at the interface between oil and water are called emulsifying agents.

Sols can be converted into gels, which are colloidal dispersions in which the solid is the continuous phase, and the liquid is the discontinuous phase (just the opposite of sols). Protein and carbohydrate gels are common in the diet. The solid colloidal component cross-links to other solid particles to form a random, yet continuous solid network that traps the liquid to form a gel from a sol. In some cases, gels can be reversed to sols by heating. When gels are cut, some liquid may drain from them, a phenomenon termed syneresis.

Foams in foods ordinarily are air trapped in films of liquid. Frequently, some solid component helps strengthen the cell walls in foams to give some rigidity and stability to the foam. Proteins often are present to add strength to food foams. Egg white, egg yolk, gelatin, concentrated milk products, and heavy cream foams are familiar foams used in food products. Their formation and stability are influenced greatly by the surface tension of the liquid being whipped and its vapor pressure.

## STUDY QUESTIONS

1. Explain why it is possible to float a needle on water.

2. Theoretically, a supersaturated solution cannot exist, and yet such solutions can be made. Explain how this is achieved.

3. Define true solution, colloidal dispersion, and coarse suspension.

4. Why is the hydrophilic nature of the solid particles in a sol of importance to the behavior of the sol?

5. Why is the transition temperature of a glassy solid of importance in dried milk solids?

6. Explain the relationship of hydrophilic/lipophilic balance and the type of emulsion formed using various emulsifying agents.

7. Explain how an emulsifying agent is able to enhance the stability of an emulsion.

8. What is the relationship between sols and gels?

9. Why is a foam made with a liquid that has a high vapor pressure a comparatively unstable foam?

10. Compare the mechanisms of heat transfer involved when a meat loaf is baked in a conventional oven with those utilized when meat loaf is baked in a microwave oven equipped with a browning unit.

11. True or false. Viscosity influences stability of an emulsion.

12. True or false. It is possible to create a supersaturated solution by heating candy to boiling.

13. True or false. A microwave oven heats foods by conduction.

14. True or false. Dipoles in water and fat are important in heating food in a microwave oven.

15. True or false. High surface tension makes it difficult to form a foam.

## BIBLIOGRAPHY

Berry, D. 2004. To foam or not to foam. *Food Product Design* 14 (1): 69.

Berry, D. 2005. Through thick and thin. *Food Product Design* 15 (2): 34.

Chandrashekar, J., et al. 2009. Taste of carbonation. *Science 326* (5951): 443.

DeCareau, R. V. 1985. *Microwaves in Food Processing Industry.* Academic Press. Orlando, FL.

Duxbury, D. 2004. Phase transitions in foods: Basic science for the modern scientist. *Food Technol. 58* (8): 78.

Ehrenberg, R. 2009. Tongue's sour-sensing carbonation. *Science News 176* (10): 12.

Elias, P. S. and Cohen, A. J. 1983. *Recent Advances in Food Irradiation.* Elsevier: New York.

Friberg, S. E., et al. 1990. In *Food Emulsions.* ed. Larson, K. and Friberg, S. E. Marcel Dekker. New York, p. 21.

Gerling, J. E. 1986. Microwaves in food industry: Promise and reality. *Food Technol. 40* (6): 82.

Hazen, C. 2005. In the thick of it. *Food Product Design 15* (May Supplement): 24.

Holmes, Z. A. and Woodburn, M. 1981. Heat transfer and temperature of foods during processing. *CRC Crit. Rev. Food Sci. Nutr. 14* (3): 231.

Howard, D. 1991. A look at viscometry. *Food Technol. 45* (7): 82.

Knorr, D. 1993. Effects of high-hydrostatic-pressure processes on food safety and quality. *Food Technol. 47* (6): 156.

Liger-Belair, G., et al. 2006. Modeling the kinetics of bubble nucleation in champagne and carbonated beverages. *J. Phys. Chem. 110* (42): 21145.

Lydersen, A. L. 1983. *Mass Transfer in Engineering Practice.* Wiley. New York.

Morrison, I. D. and Ross, S. 2002. *Colloidal Dispersions: Suspensions, Emulsions, and Foams.* Wiley Europe. Chichester, England.

O'Hagan, P. 2004. New technologies: Why measure particle size? *Food Product Design 14* (11): 120.

Olson, D. G. 2004. Food irradiation future still bright. *Food Technol. 58* (7): 112.

Reitz, C. 1986. *Microwave Reference Guide 1986–1987.* International Microwave Power Institute. Clifton, VA.

Roos, Y. H., et al. 1996. Glass transitions in low moisture and frozen foods: Effect on shelf life and quality. *Food Technol. 50* (11): 95.

Russell, W. B., Saville, D. A., and Schowalter, W. R. 1992. *Colloidal Dispersions.* Cambridge University Press. Cambridge, England.

Sauter, E. A. and Montoure, J. E. 1972. Relation of lysozyme content of egg white to volume and stability of foam. *J. Food Sci. 37*: 918.

Schauwecker, A. 2004. New technologies: Under pressure. *Food Product Design 14* (4): 96.

Shoemaker, C. F., Nantz, J., Bonnans, S., and Noble, A. C. 1992. Rheological characterization of dairy products. *Food Technol. 46* (1): 98.

Singh, R. P. 1995. Heat and mass transfer in foods during frying. *Food Technol. 49* (4): 134.

## INTO THE WEB

*http://uw-food-irradiation.engr.wisc.edu/Facts.html*—Information on irradiation of food.

*http://www.fda.gov/Food/FoodIngredientsPackaging/IrradiatedFoodPackaging/ucm135143.htm*—FDA site discussing food irradiation.

*http://cgpl.iisc.ernet.in/site/Portals/0/Technologies/Biomass%20Stoves.pdf*—Discussion of some designs for SWOSTHEE.

*http://www.envirofit.org/?q=our-products/clean-cookstoves/IAP*—Discussion on indoor pollution caused by cooking fires.

*http://sify.com/news/now-biomass-cooking-stoves-in-india-news-national-jegrTCiicij.html*—Information on one-pan cooking stove.

*http://www.medicalnewstoday.com/articles/167671.php*—Article on the taste of carbonation.

# 3

# Carbohydrates

Biscuits and honey are a tempting way of eating simple and complex carbohydrates.

# CHAPTER 7

# Overview of Carbohydrates

## Chapter Outline

## OBJECTIVES

After studying this chapter, you will be able to:

1. Classify various carbohydrates on the basis of their chemical structures.

2. Discuss how chemical structure influences the characteristics of different carbohydrates.

3. Describe applications in food products for which various carbohydrates are well suited.

## INTRODUCTION

Carbohydrates are, as their name implies, hydrates of carbon. Whether the compound is the smallest (monosaccharide) or the largest (polysaccharide) carbohydrate, the ratio of hydrogen to oxygen in the molecule is essentially two to one, just as in water. Chemists categorize carbohydrates according to the number of basic units linked together: monosaccharide, disaccharide, oligosaccharide, and polysaccharide. Sometimes mono- and disaccharides are referred to as **simple sugars**, and polysaccharides are grouped together as **complex carbohydrates**.

The monosaccharides may contain three to six carbon atoms, which leads to their generic designations as **trioses**, **tetroses**, **pentoses**, and **hexoses**. Note that the common suffix is -*ose*, which designates that the compound is a carbohydrate. The various monosaccharides are combined to make the other carbohydrates; glucose is the monosaccharide that is found most frequently as the building block of more complex carbohydrates.

The sweet taste of the monosaccharides and disaccharides is of great importance in food products. Other characteristics of value are their ease of solubility, ability to contribute to mouthfeel in candies and syrups, browning at very high temperatures, and their contribution to volume in baked products.

Complex carbohydrates exhibit different characteristics, which vary with the specific type of compound. Starch is valued for its thickening ability. Cellulose and the hemicelluloses

**Simple sugars**
Monosaccharides and disaccharides.

**Complex carbohydrates**
Polysaccharides, such as starch and cellulose.

**Triose**
Saccharide with three carbon atoms.

**Tetrose**
Saccharide with four carbon atoms.

**Pentose**
Saccharide with five carbon atoms.

**Hexose**
Saccharide with six carbon atoms, the most common size unit.

129

modify textures when they are incorporated into food products, contributing a somewhat harsh mouthfeel. Gums and pectin serve as thickening agents that are valued for their limited calorie contribution.

## MONOSACCHARIDES

**Monosaccharide**
Carbohydrate containing only one saccharide unit.

Although **monosaccharides** occur in nature to only a limited extent, they nevertheless are of interest to food scientists. Formulas for several of the more common monosaccharides are presented in the sections that follow.

### Pentoses

Ribose, arabinose, and xylose are the key pentoses of interest in food products. Their chemical structures, presented as Fischer structures, are:

D-ribose    D-arabinose    D-xylose

Actually, these sugars often occur in nature with a linkage between the oxygen on the first carbon and one of the alcohol (—OH) groups, which results in a ring structure (Haworth structure), as shown:

α-D-ribose    α-D-arabinose    α-D-xylose

**Dextrose**
Synonym for glucose; so named because polarized light bends to the right in a glucose solution.

**Levulose**
Synonym for fructose; so named because polarized light bends to the left in a fructose solution.

**Aldose**
Hexose with one carbon atom external to the 6-membered ring.

**Ketose**
Hexose with two carbon atoms external to the 5-membered ring.

Each of these structures is an alpha ($\alpha$) form of the pentose sugar because of the orientation of the hydroxyl group on the first carbon ($C_1$). The orientation of the hydroxyl group is reversed in the beta ($\beta$) form. Note that the ring structure always involves the first carbon and one of the other carbon atoms (either carbon 4 or carbon 5).

### Hexoses

Glucose (also called **dextrose** because of its ability to bend polarized light to the right), fructose (sometimes referred to as **levulose** or left-bending), and galactose are especially important hexoses. Others found in foods include mannose, gulose, and sorbose. Hexoses with only one carbon atom external to a 6-membered ring are classified as **aldoses**; glucose, galactose, mannose, and gulose are examples of aldoses. When a hexose has two carbon atoms external to a 5-membered ring, the sugar is a **ketose**; fructose and sorbose are ketoses.

The Fischer structures of glucose, fructose, and galactose are of particular interest:

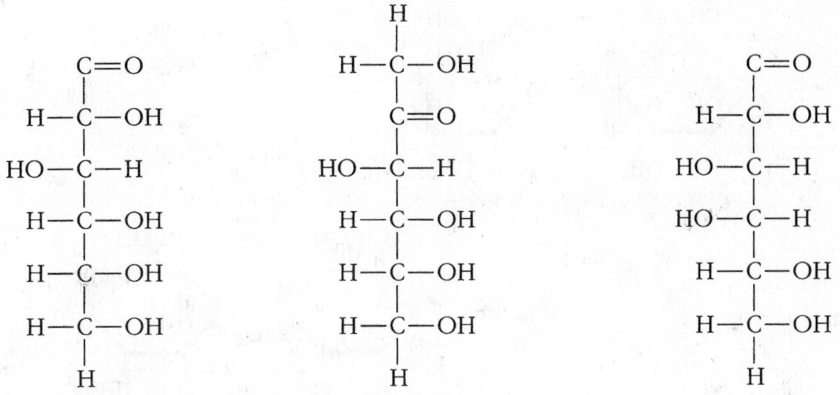

D-glucose D-fructose D-galactose

The Haworth or cyclic structures of these three common hexoses are:

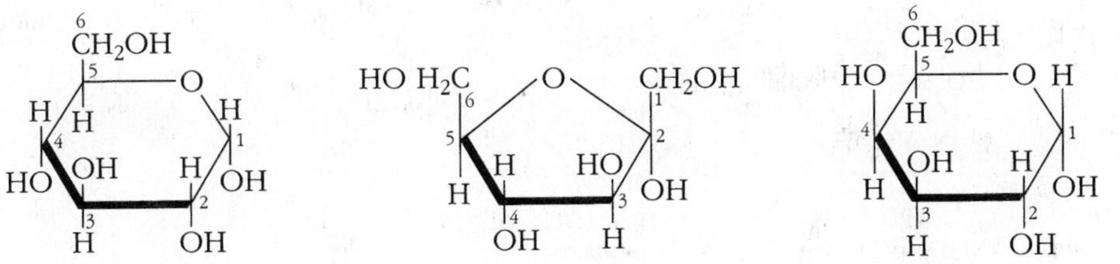

α-D-glucose α-D-fructose α-D-galactose

Glucose and galactose are aldoses and have the sixth carbon atom external to the ring. Note that fructose is a ketose with the first and sixth carbon atoms external to the ring.

## DISACCHARIDES

**Disaccharide** Carbohydrate formed by the union of two monosaccharides with the elimination of a molecule of water.

**Disaccharides** are structurally related to monosaccharides. Disaccharides form when two monosaccharides join together by a glycolytic linkage, eliminating a molecule of water. The three disaccharides common in foods (sucrose, maltose, and lactose) all contain glucose. Sucrose is formed from glucose and fructose:

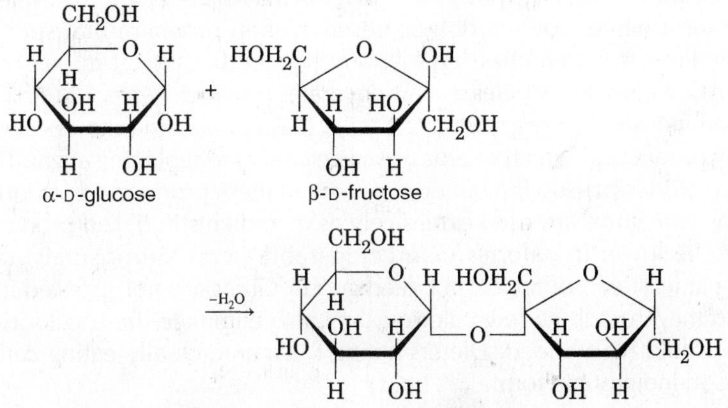


sucrose

The formation of maltose from two glucose molecules is as follows:

The formation of lactose from glucose and galactose is as follows:

---

## FOOD FOR THOUGHT: Carbohydrate Paradox

Innumerable weight-loss plans are featured in books, magazines, and other media. Fashions in diets come and go, but many of them focus on carbohydrates. Some restrict carbohydrate intake; breads, pasta, and sweets may be severely restricted or even eliminated for a while. Low-carbohydrate diets often proclaim that such foods are fattening, and meats are emphasized (despite the fact that the fat in meat contributes more than twice as many calories as carbohydrates provide).

The food industry has responded to the increased concern with obesity by developing many products in which alternative sweeteners are replacing sugar. This change reduces the calories per serving but does not mean these products are calorie free.

Carbohydrate gums are used extensively as ingredients in baked products that are designed to be lower in calories than comparable items. Surprisingly, these gums can replace at least part of the fat in baked goods. Gums are not digested in the small intestine, so they contribute few calories; they also eliminate the 9 calories per gram that the fat would have added. Dieters eating them are actually eating carbohydrate, although in an indigestible form.

## OLIGOSACCHARIDES

Many carbohydrates are still more complex than the disaccharides. **Oligosaccharides** are composed of between three and ten monosaccharide units, each joined to the next by the elimination of a molecule of water. These substances are less prominent in foods than either the mono- and disaccharides or the much larger polysaccharides. Oligosaccharides are formed during the transition of complex carbohydrates, such as starch, into simpler di- and monosaccharides.

Two trisaccharides in legumes are stachyose and raffinose. Other oligosaccharides identified in foods are maltotriose and manninotriose. The structure of raffinose is:

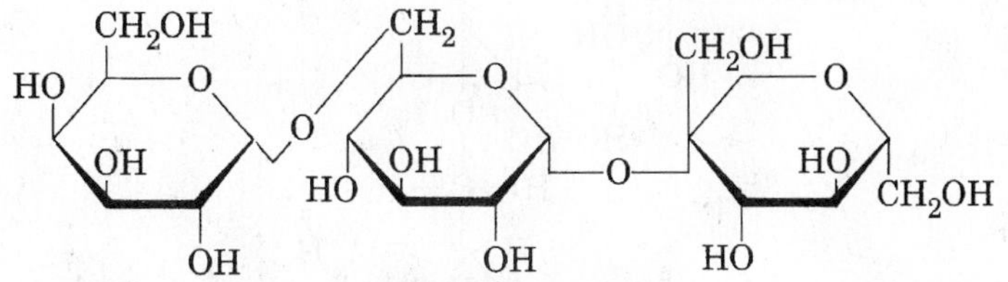

raffinose

Humans lack the enzyme needed to digest raffinose and stachyose, which causes flatulence. This can be overcome by taking the enzyme orally.

## POLYSACCHARIDES

### Dextrins

Technically, oligosaccharides are **polysaccharides**, but generally the term *polysaccharides* indicates much larger molecules. Some of the most important polysaccharides in food are composed only of glucose units linked together by α- or β-glucosidic linkages. The simplest of these substances are the **dextrins**. These molecules range widely in size but are distinctly shorter in chain length than starch, the related substance. Dextrin molecules are composed entirely of glucose, and these units are linked together by 1,4-α-glucosidic linkages, as shown previously for maltose.

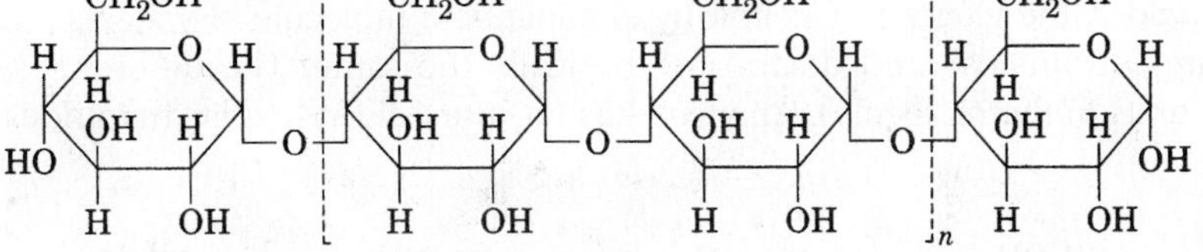

dextrins

Unlike the mono- and disaccharides, which are characterized as sweet in taste and quite soluble, dextrins exhibit rather different properties. Dextrins have slight solubility, a barely sweet taste, and limited thickening ability. They are formed when flour is being browned using dry heat.

### Dextrans

**Dextrans** are yet another type of polysaccharide and are found in bacteria and yeasts. Again, these are composed of glucose, but with a 1,6-α-glucosidic linkage. Branching occurs in dextrans, with the point of branching being unique to a particular species or strain. A portion of a dextran molecule follows. Note that dextrins and dextrans differ in their linkages, having 1,4- and 1,6-α-glucosidic linkages, respectively.

**Oligosaccharide**
Carbohydrate formed by the union of three to ten monosaccharides joined by the elimination of water.

**Polysaccharide**
Carbohydrate formed by the union of many saccharide units, with the elimination of a molecule of water at each point of linkage.

**Dextrins**
Polysaccharides composed entirely of glucose units linked together and distinguishable from starch by a shorter chain length.

**Dextrans**
Complex carbohydrates in bacteria and yeasts characterized by 1,6-α-glucosidic linkages.

$$\text{dextran}$$

## Starch

*Amylose.*   **Starch**, a glucose polymer of very large dimensions, is actually comprised of two fractions: *amylose* and *amylopectin*. The simpler of these is **amylose**, a large molecule consisting of considerably more than 200 glucose units linked by 1,4-α-glucosidic linkages. Amylose molecules are somewhat linear in their spatial configuration, enabling them to form some hydrogen bonds to each other under certain conditions. Amylose is slightly soluble but does not have a sweet taste. In the structure of amylose presented here, note that like the disaccharides, the glucose units link by eliminating a molecule of water. As can be seen, the structures of amylose and dextrin are basically the same. The difference is in $n$ (the number of units in the molecule). Amylose has far more glucose units than does dextrin.

$$\text{amylose}$$

*Amylopectin.*   **Amylopectin**, the other starch fraction, is more complicated structurally than is amylose, but it also contains only glucose units. There are two types of linkages in amylopectin: 1,4-α-glucosidic and 1,6-α-glucosidic. There are far more 1,4 linkages than there are 1,6 linkages. The usual configuration contains between 24 and 30 glucose units linked together consecutively between carbons 1 and 4, at which point a single 1,6 linkage occurs.

The 1,6 linkage results in disruption of the linear extension of the molecule wherever it occurs; the result is a branching of the molecule at this linkage. Other glucose units continue to be linked to the unit involved with the 1,6 linkage because the first and fourth carbons are still available for bonding to other units. In nature, this new segment will again have between 24 and 30 glucose units linked by 1,4 linkages before another 1,6 linkage causes additional branching of the molecule.

Amylopectin molecules are extremely large and may have a molecular weight of a million or more. The branching of amylopectin results in a molecule with little solubility; like amylose, it also does not contribute sweetness to food flavors. The structure of amylopectin is presented, in part, here.

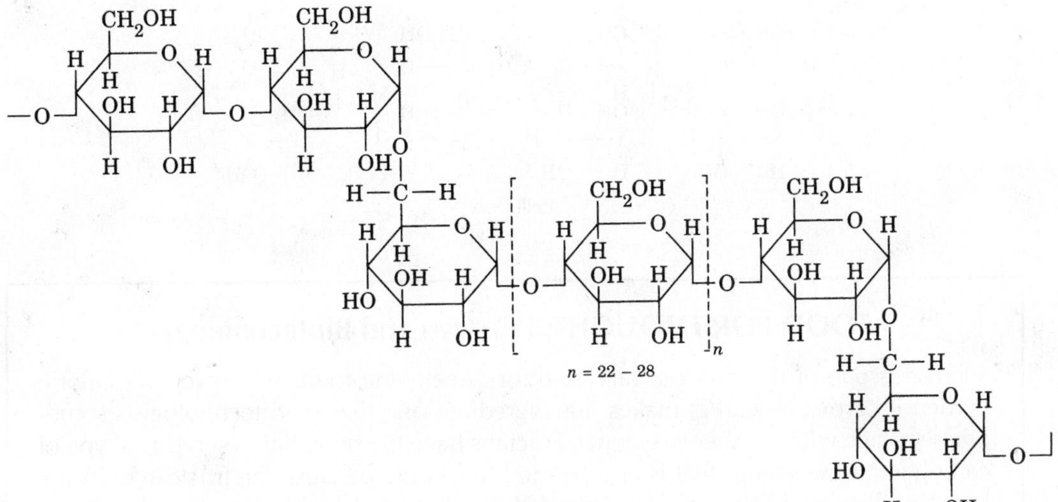

amylopectin

## Glycogen

**Glycogen**, which is the storage form of carbohydrate in animal tissues, is somewhat comparable to amylopectin in structure. The primary difference between amylopectin and glycogen is that linear segments of the glycogen molecule are generally between 8 and 12 glucose units long, rather than the 24–30 found in amylopectin. This gives glycogen a spatial arrangement even more bulky than that of amylopectin. The structure, in part, is presented next.

**Glycogen**
Complex carbohydrate that serves as the storage form of carbohydrate in animals; glucose polymer with 1,4-α-glucosidic linkages interrupted by 1,6 linkages about every 8–12 units, resulting in bulky branching.

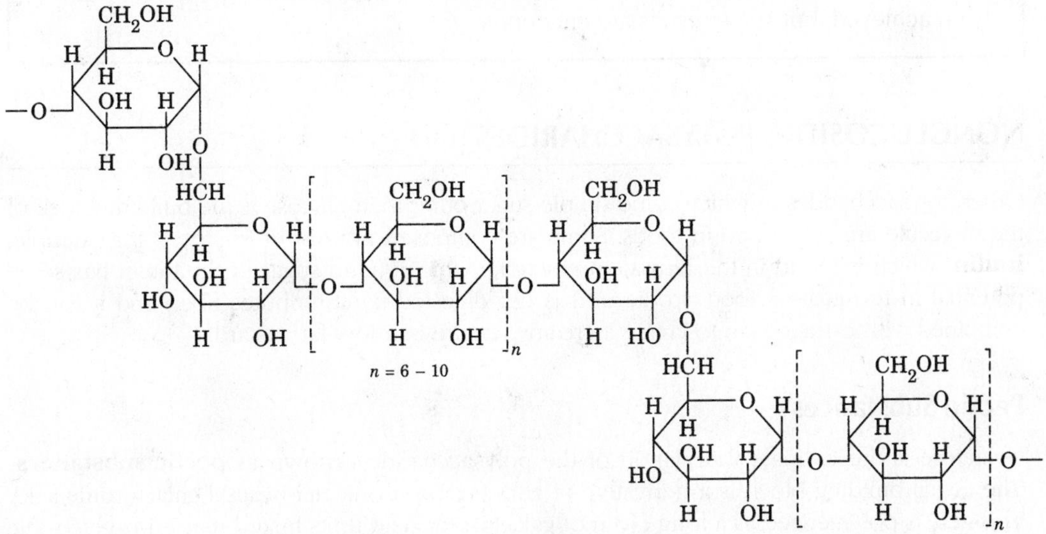

glycogen

## Cellulose

**Cellulose**
Complex carbohydrate composed of glucose units joined together by 1,4-β-glucosidic linkages.

The significance of the α versus the β linkage between glucose units is evident if amylose and **cellulose** are compared. Both have a basically linear configuration spatially, as a result of the 1,4 linkage. However, cellulose, which has a 1,4-β-glucosidic linkage throughout, is insoluble and is a key structural component of plants. In contrast, amylose is somewhat soluble. Humans can digest starch to provide energy to the body, whereas cellulose is not an energy source for people because of their inability to split its 1,4-β-glucosidic linkages.

A portion of the cellulose molecule is shown here. Note that the compound is a polysaccharide composed exclusively of glucose in the 1,4-β-glucosidic linkage.

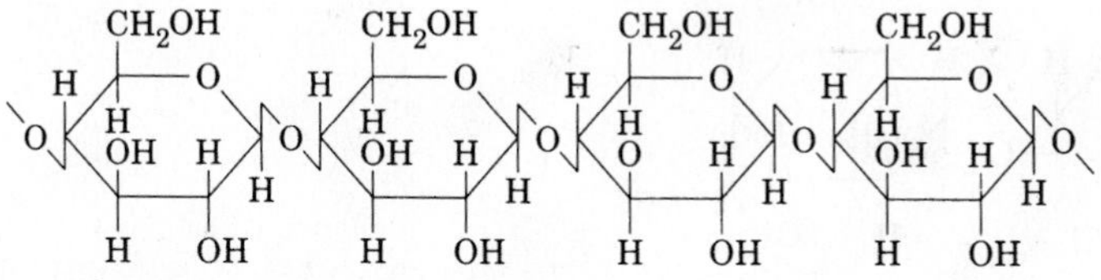

cellulose

---

### FOOD FOR THOUGHT: Fructans and Biotechnology

Sucrose is one of the major sugars used for sweetening foods. However, the caloric burden it carries (4 kcal/g) makes this ingredient one that food technologists sometimes replace with another sweetener. Fructans have the potential to serve as a type of carbohydrate sweetener that is not digested by people because the fructose units are combined through a β linkage. This carbohydrate group consists of polymers of fructose that can be formed in plants, where they can help promote drought resistance and also store energy in the vacuole of some cells.

Two enzymes appear to be particularly important in promoting the formation of fructans. Sucrose-sucrose-fructosyltransferase (SST) is the plant enzyme that catalyzes the formation of a trisaccharide (polymer of two fructose and one glucose units) and a glucose molecule from two molecules of sucrose. Fructan-fructan-fructosyltransferase (FFT) promotes polymerization of these trisaccharides to larger fructan molecules.

Research in biotechnology is underway to attempt to develop sugar beets with a high content of fructans in their roots. This has the dual goal of creating plants that are more resistant to drought and that also might serve as a commercially viable source of fructans for use as a noncaloric sweetener in the food industry. These goals have not yet been achieved, but some results are encouraging.

---

## NONGLUCOSIDIC POLYSACCHARIDES

**Inulin**
Complex carbohydrate that is a polymer of fructose.

Other polysaccharides in which some simple sugar other than glucose is the building block of the molecule are also found in foods. Some are composed of units of fructose—for example, **inulin**, which is found in the Jerusalem artichoke, chicory, and sugar beets. Inulin has some potential in formulating food products: it is not digested so contributes fiber, and it can be combined with carrageenan to create a creamy texture in a low-fat custard.

## Pectic Substances

**Pectic substances**
Group of complex carbohydrates in fruits; polymers of galacturonic acid linked by 1,4-α-glycosidic linkages with varying degrees of methylation.

Galactose is the ultimate foundation of the polysaccharides known as **pectic substances**. The actual building block is a derivative of galactose, a uronic acid called galacturonic acid. This acid is polymerized as a long chain of galacturonic acid units linked in a 1,4-α-glycosidic linkage. The acids in the molecule frequently are methylated to form methyl esters when the fruit is barely ripe. This methylated form of the pectic substances is termed *pectinic acid* or *pectin*. A portion of this structure is shown here. The significance of pectic substances in making jam and jellies is discussed in Chapter 20.

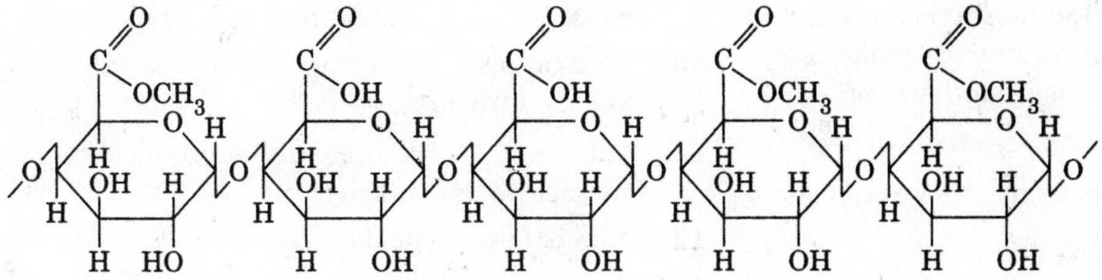

pectin

## Gums

Complex polysaccharides based on saccharides other than glucose are found in seeds, plant exudates, and seaweed (see Chapter 10). Depending on the source, these **gums** may contain a variety of sugars in their structures, although most have galactose as a common component. The seed gums also contain mannose. Plant exudates contain various other sugars including two pentoses (arabinose and xylose) and two unusual hexoses (rhamnose and fucose) in which the terminal carbon is a methyl group rather than an alcohol, as shown here.

**Gums**
Complex carbohydrates of plant origin, usually containing galactose and at least one other sugar or sugar derivative, but excluding glucose.

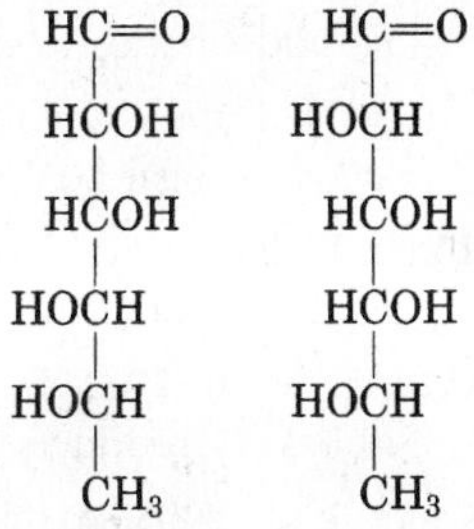

L-rhamnose    L-fucose

Gums are used widely as stabilizing agents and fat replacers.

## SUMMARY

Carbohydrates are composed of carbon, hydrogen, and oxygen; the hydrogen and oxygen are usually present in approximately the 2:1 ratio found in water. The simplest of the carbohydrates are the monosaccharides, which contain between three and six or more carbon atoms. Familiar pentoses are ribose, arabinose, and xylose; common hexoses are glucose, fructose, and galactose, although mannose, gulose, and sorbose also occur occasionally. The common disaccharides—two saccharide units joined together—include sucrose (composed of glucose and fructose), maltose (two units of glucose), and lactose (glucose and galactose).

Oligosaccharides (containing between three and ten glucose units) are not common components of foods but form polysaccharides and are hydrolyzed into basic components. Polysaccharides are very large polymers of saccharides joined together by 1,4 linkages (which usually are $\alpha$ linkages) and occasionally by 1,6-$\alpha$ linkages. Dextrins, cellulose (a polymer of glucose formed by 1,4-$\beta$ linkages), and the two fractions of starch (amylose and amylopectin) are common polysaccharides in foods. Pectic substances and gums are polysaccharides usually found as polymers of derivatives of galactose; gums sometimes contain other sugar derivatives, as well.

## STUDY QUESTIONS

1. Show how maltose is formed from two glucose units.

2. Which monosaccharide is the most common unit in the majority of the carbohydrates? In which complex carbohydrates is this compound absent?

3. How are dextrins distinguished from amylose?

4. Identify the two starch fractions and explain how they differ.

5. Describe the chemistry of pectic substances and compare them with starch.

6. How do gums compare chemically with the pectic substances?

7. True or false. The linkages in dextrins are β-glucosidic.

8. True or false. The linkage in sucrose is a β linkage.

9. True or false. Sucrose is a compound that forms when starch is broken down.

10. True or false. The components of lactose are galactose and fructose.

11. True or false. The breakdown products of galactose are lactose and fructose.

12. True or false. Cellulose contains β-glucosidic linkages.

13. True or false. Glycogen contains 1,4- and 1,6-α-glucosidic linkages.

14. True or false. Glucose is a component of all gums.

15. True or false. Amylose molecules are larger than amylopectin.

## BIBLIOGRAPHY

Backas, N. 2005. How sweet . . . and how hot it is! *Food Product Design 15* (2): 56.

Banasiak, K. 2004. Carbohydrates: To count or not to count. *Food Technol. 58* (5): 28.

Blankers, I. 1995. Properties and applications of lactitol. *Food Technol. 49* (1): 66.

Campbell, L. A., et al. 1994. Formulating oatmeal cookies with calorie-sparing ingredients. *Food Technol. 48* (5): 98.

Chinachoti, P. 1993. Water mobility and its relation to functionality of sucrose-containing food systems. *Food Technol. 47* (1): 134.

Crosby, G.A. 2005. Lignans in food and nutrition. *Food Technol. 59* (5): 32.

Darling, K. 2009. Starch on the side. *Food Product Design 19* (11): 46.

Deis, R.C. 2005. How sweet it is—using polyols and high-potency sweeteners. *Food Product Design 15* (7): 57.

Giese, J. 1993. Alternative sweeteners and bulking agents. *Food Technol. 47* (1): 113.

Grenus, K. 2005. Focus on fiber. *Food Product Design 15* (3): 92.

Hazen, C. 2006. New fiber options for baked goods. *Food Product Design 15* (10): 80.

Hubrich, B. 2004. Low-carb diets—does science support them? *Food Product Design 13* (11): 59.

Izzo, M., et al. 1995. Using cellulose gel and carrageenan to lower fat and calories in confections. *Food Technol. 49* (7): 1995.

Jeffery, M. S. 1993. Key functional properties of sucrose in chocolate and sugar confectionery. *Food Technol. 47* (1): 141.

Knehr, E. 2005. Carbohydrate sweeteners provide sweet results. *Food Product Design 15* (May Supplement): 38.

Kulp, K., et al. 1991. Functionality of carbohydrate ingredients in bakery products. *Food Technol. 45* (3): 136.

Kuntz, L. A. 2005. A starch that's hard to resist. *Food Product Design 15* (6): 19.

Ohr, L. M. 2006. Functional sweets. *Food Technol. 60* (11): 57.

Palmer, S. 2004. Sweetening power of polyols. *Food Product Design 14* (2): 30.

Pszczola, D. E. 2006. Fiber gets a new image. *Food Technol. 60* (2): 43.

Turner, J. 2003. Honey of an option. *Food Product Design 13* (4): 14.

Turner, J. 2005. Everybody loves carbs. *Food Product Design 15* (May Supplement): 12.

Verdi, R. J. and Hood, L. L. 1993. Advantages of alternative sweetener blends. *Food Technol. 47* (6): 94.

Wilshire, G. 2004. Low-carb going mainstream. *Food Product Design 13* (11): 38.

Zhao, J. and Whistler, R. L. 1994. Spherical aggregates of starch granules as flavor carriers. *Food Technol. 48* (7): 104.

## INTO THE WEB

*http://food.oregonstate.edu/carbohydrate*—References on carbohydrates.

*http://www.sciencemag.org/feature/data/carbohydrates.dtl*—Provides access to several aspects of carbohydrate chemistry.

*http://micro.magnet.fsu.edu/micro/gallery/sugars/sugar.html*—Photomicrographs of selected sugars.

*http://www.nature.com/nbt/journal/v16/n9/abs/nbt0998-843.htm*—Biotechnology and development of fructans (fructose polymers) for drought resistance and potential as a non-nutritive sweetener.

*http://www.actahort.org/books/744/744_45.htm*—Prebiotic effect of fructans from agave, dasylirion, and nopal.

*http://www.nutraingredients.com/Research/Inulin-gum-mix-creams-up-low-fat-custard*—Paper on inulin in custard product.

*http://www.gmo-safety.eu/en/glossary/#I*—Brief description of inulin.

*http://www.sensus.nl/applications.html*—Some applications of inulin.

Sap drips into bags from the taps that have been drilled into this sugar maple tree in anticipation of boiling the sap to make maple syrup. (Courtesy of Debra McRae.)

## CHAPTER 8

# Monosaccharides, Disaccharides, and Sweeteners

## Chapter Outline

## OBJECTIVES

After studying this chapter, you will be able to:

1. Compare the physical properties of some sugars occurring in foods.
2. Explain chemical changes that can occur during heating of sugars.
3. Discuss factors that influence the quality of crystalline candies.
4. Describe the factors that determine the characteristics of various amorphous candies.
5. Discuss the roles of sugars in jams and related products and in various baked products.
6. State reasons for selecting various sugars or sweeteners for specific food products.
7. Describe alternative sweeteners and their applications.

## INTRODUCTION

Sweet, one of the four basic tastes, occurs in foods through a variety of compounds. Although the various mono- and disaccharides numbered among the carbohydrates in nature are the common sweeteners used (Figure 8.1), non-carbohydrate compounds such as saccharin and aspartame are also incorporated into manufactured foods to enhance sweetness. The characteristics of each type of sweetener are unique to the specific sugar or sugar substitute, making it important for food scientists to be familiar with the behavior of possible sweeteners and the performance required of them in various food products. This

**Figure 8.1**    A drop of sap emerges from the tap in this sugar maple. (Courtesy of Debra McRae.)

chapter explores the chemistry and the applications of the common sweeteners available to the food industry today. Particular attention is directed toward the crystalline nature of many candies and the factors that influence crystallinity.

## PHYSICAL PROPERTIES OF SUGARS

### Sweetness

When dissolved, all sugars are sweet to the tongue, but some are sweeter than others. The relative ability to sweeten a food product is of interest because sugars contribute 4 kilocalories per gram. Theoretically, a sugar that is very sweet can be used in smaller quantities than a sugar that is less sweet, resulting in a reduction in calories without sacrificing sweetness. Table 8.1 shows the relative sweetness of some monosaccharides, disaccharides, and other sweeteners.

The temperature of the solution containing the sugar influences the relative sweetness values for sugars. For example, fructose is about 1.4 times sweeter than sucrose at 5°C (41°F), comparable in sweetness when the solution is at 40°C (104°F), but only 0.8 times as sweet at 60°C (140°F). However, maltose sweetness ratings are essentially independent of temperature.

Clearly, most of the non-sugar sweeteners provide far more sweetening than a comparable weight of any of the sugars can contribute. Even though these intense sweetening agents are of merit from the perspectives of taste and calories, sweetness is not the only important property in using sugars in foods, as is evident in the later section on "Functional Properties of Sugars."

### Hygroscopicity

**Hygroscopicity**
Ability to attract and hold water, which is a characteristic of sugars to varying degrees.

Sugars are able to attract and hold water to varying degrees. This capability, known as **hygroscopicity**, can be useful in maintaining the freshness of some baked products, but it can be a source of potential problems in texture when the relative humidity is high. An elevation in temperature also increases the absorption of moisture from the atmosphere. Table 8.2 presents a comparison of the hygroscopicity of selected sugars under varying temperature and relative humidity.

**Table 8.1**  Relative Sweetness of Selected Sugar Solutions (5%) and Other Sweeteners

| Sweetener | Relative Sweetness |
|---|---|
| Thaumatin[a] (Talin®) | 2,000–3,000 |
| Monellin[a] | 1,500–2,000 |
| Sucralose[a] (Splenda®) | 600 |
| Stevioside[a] | 300 |
| Saccharin[a] | 200–300 |
| Acesulfame-K[a] (Sunett®) | 130–200 |
| Aspartame[a] (NutraSweet®, Equal®) | 100–200 |
| Cyclamates[a] | 30–80 |
| Fructose | 1.3[b] |
| Xylitol[a] | 1.01 |
| Sucrose | 1.0[c] |
| Tagatose (Naturlose™) | 0.92 |
| Invert sugar | 0.85–1 |
| Xylose | 0.59 |
| Glucose | 0.56 |
| Galactose | 0.4–0.6 |
| Maltose | 0.3–0.5 |
| Lactose | 0.2–0.3 |

Figures compiled from multiple sources including Godshall, M. A. 1997. How carbohydrates influence food flavor. *Food Technol. 51* (1): 63.
[a]Non-sugar sweetener.
[b]Highly variable, depending on temperature. This is a representative value, but measurements may range from 0.8 to 1.7.
[c]Value of sucrose arbitrarily set at 1.0 for reference purposes.

## Solubility

The amount of sugar that will go into solution in water varies with the type of sugar and also with the temperature of the water. As the temperature of water rises, the amount of sugar able to dissolve in a given amount of water also increases. This fact is illustrated by the figures for sucrose solubility, shown in Table 5.4, p. 98. Table 8.3 presents the comparative solubility of various sugars.

Solubility is important because of its relationship to food texture. Candies containing fructose are softer than those containing other sugars because of the greater solubility of fructose. The very low solubility of lactose, the sugar in milk, is a particular textural problem in the manufacture of ice cream. The low temperature required in ice cream storage

**Table 8.2**  Hygroscopicity of Selected Sugars under Varying Conditions

| Sugar | Percentage of Water Absorbed | | |
|---|---|---|---|
| | 20°C (68°F) | | 25°C (77°F) |
| | RH[a] = 60%, 1 hour | RH = 100%, 25 days | RH 90%, equilibrium |
| β-maltose | 5.05 | | |
| α-lactose hydrate, pure | 5.05 | | |
| D-fructose | 0.28 | 73.4 | 41–43 |
| D-glucose | 0.07 | 14.5 | 17–18 |
| Invert sugar | 0.16 | 74.0 | |
| Sucrose | 0.04 | 18.4 | 50–56 |

[a]RH = relative humidity.

**Table 8.3**    Solubility of Selected Sugars at 50°C (122°F)

| Sugar | Grams of Sugar Dissolved in 100 ml Water |
|---|---|
| Fructose | 86.9 |
| Sucrose | 72.2 |
| Glucose | 65.0 |
| Maltose | 58.3 |
| Lactose | 29.8 |

promotes the formation of lactose crystals (detected by the tongue as a somewhat gritty texture), due to the low solubility of this particular sugar.

## CHEMICAL REACTIONS

### Hydrolysis

Disaccharides undergo hydrolysis when heated. An acidic medium favors this degradative reaction, as does the presence of water. However, even if seemingly dry sugars are heated alone, the uptake of a molecule of water and the resultant splitting into the two component monosaccharides occur. The reaction for the hydrolysis of sucrose, the disaccharide particularly susceptible to hydrolysis, is shown here.

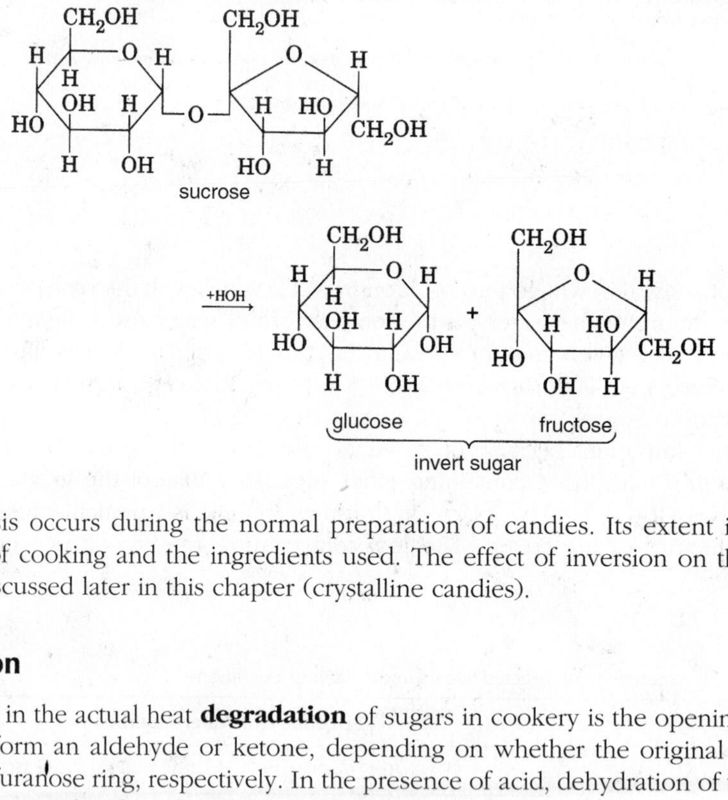

Hydrolysis occurs during the normal preparation of candies. Its extent is influenced by the rate of cooking and the ingredients used. The effect of inversion on the texture of candies is discussed later in this chapter (crystalline candies).

### Degradation

**Degradation**
Opening of the ring structure as the prelude to the breakdown of sugars.

The first step in the actual heat **degradation** of sugars in cookery is the opening of the ring structure to form an aldehyde or ketone, depending on whether the original sugar was a pyranose or furanose ring, respectively. In the presence of acid, dehydration of the molecule occurs as three molecules of water are eliminated. Organic acids and aldehydes are the result. These reactions can occur in an acidic medium, but they take place even more readily in an alkaline medium.

### Caramelization

When sugars are heated to such intense temperatures that they melt [170°C (338°F) for sucrose], a series of chemical reactions begins, which ultimately can lead to a charred or burned product. However, some caramelization of sugar creates pleasing color and flavor changes, with the color ranging from a pale golden brown to a gradually deepening brown before burning actually occurs. Similarly, the flavor begins to assume new and distinctive overtones as the mixture of sugar derivatives undergoes change.

The overall process of caramelization involves a number of steps, beginning with the inversion of sucrose (conversion to invert sugar). After the ring structures in the components of the invert sugar are broken, some condensation of the compounds occurs, which creates some polymers ranging in size from trisaccharides to oligosaccharides (as many as 10 subunits polymerized). Severe chemical changes at the very high temperatures involved also lead to dehydration reactions and the formation of organic acids and some cyclic compounds, as well as many other substances.

Caramelizing can be halted abruptly by adding boiling water to cool the extremely hot sugar mixture rapidly. Surprisingly, boiling water is much cooler than the caramelizing sugar. Of course, the addition of cool water also will halt the caramelization process. However, this practice is not recommended because of the extreme splattering and potential for burning one's skin that result when the two liquids come into contact and equalize their extreme differences in energy.

Evidence of the creation of acids during caramelization can be seen by stirring some baking soda into the caramelizing sugar, as is done in preparing peanut brittle. The carbon dioxide that forms when the soda neutralizes the acids creates a porous product as the gas expands in the hot, viscous candy solution.

## The Maillard Reaction

An extremely important browning reaction in the preparation of foods is the **Maillard reaction**. This reaction, like the series involved in caramelization, is classified as **non-enzymatic browning**. Actually, the Maillard reaction is a series of reactions involving the condensation of a **reducing sugar** and an amine.

Glucose, fructose, and galactose are reducing monosaccharides; similarly, lactose and maltose are reducing disaccharides. Lactose undergoes non-enzymatic browning most readily of the reducing sugars, followed in descending order by ribose, fructose, and glucose. These reducing sugars can combine with amines in milk and other protein-containing foods to cause non-enzymatic browning. Sucrose, however, is not a reducing sugar and does not participate in the Maillard reaction. It must undergo inversion to glucose and fructose before it can enter into this type of non-enzymatic browning.

The color changes during the steps of the Maillard reaction occur rather slowly and with less energy input than is required for caramelization. The progression is from an essentially colorless substance to a golden color, on to a somewhat reddish brown and then a dark brown. This range of colors can be followed as caramels are boiled to their final temperature or during the baking of a plain or white cake as the crust color develops. Similarly, the reactions can be traced by watching the color development in sweetened condensed milk when it is heated in a water bath. A pH of 6 or higher accelerates the Maillard reaction.

The Maillard reaction proceeds rather quickly at elevated temperatures, but it also can occur at room temperature during extended periods of storage. In fact, one of the early problems in developing packaged cake mixes was prevention of the Maillard reaction, which sometimes occurred during prolonged marketing operations. The series of reactions appears to consist of many steps, with the first step probably being as follows:

**Maillard reaction**
Non-enzymatic browning that occurs when a protein and a sugar are heated or stored together for sometime.

**Non-enzymatic browning**
Browning resulting from chemical changes that may be facilitated by heat.

**Reducing sugar**
Sugar having a free carbonyl that can combine with an amine, leading to non-enzymatic browning via the Maillard reaction.

$$
\begin{array}{ccc}
 & & H-N-R \\
 & & | \\
H-C=O \qquad H & & H-C-OH \\
| \qquad\quad\ \ \backslash & & | \\
(H-C-OH)_n + \quad N-R \rightleftharpoons (H-C-OH)_n \\
| \qquad\quad\ \ / & & | \\
CH_2OH \quad H & & CH_2OH \\
\text{aldehyde form} & \text{amino acid} & \text{glycosylamine} \\
\text{of an aldose} & \text{or protein} & \text{(addition product)}
\end{array}
$$

**Enolization**
Reversible reaction between an alkene and a ketone.

Through **enolization** and dehydration, colored pigments (melanoidins) are formed. The reactions outlined here are possible steps for this transformation.

$$
\begin{array}{c}
\text{H} - \text{N} - \text{R} \\
|\\
\text{H} - \text{C} - \text{OH} \\
|\\
(\text{H} - \text{C} - \text{OH})_n \\
|\\
\text{CH}_2\text{OH}
\end{array}
\xrightleftharpoons[+\text{H}_2\text{O}]{-\text{H}_2\text{O}}
\begin{array}{c}
\text{H} - \text{N} - \text{R} \\
|\\
\text{C} - \text{H} \\
||\\
\text{C} - \text{OH} \\
|\\
(\text{H} - \text{C} - \text{OH})_{n-1} \\
|\\
\text{CH}_2\text{OH}
\end{array}
\rightleftharpoons
\begin{array}{c}
\text{H} - \text{N} - \text{R} \\
|\\
\text{H} - \text{C} - \text{H} \\
|\\
\text{C} = \text{O} \\
|\\
(\text{H} - \text{C} - \text{OH})_{n-1} \\
|\\
\text{CH}_2\text{OH}
\end{array}
\longrightarrow
\begin{array}{c}
\text{H} - \text{N} - \text{R} \\
|\\
\text{H} - \text{C} - \text{H} \\
|\\
\text{C} - \text{OH} \\
||\\
\text{C} - \text{OH} \\
|\\
(\text{H} - \text{C} - \text{OH})_{n-2} \\
|\\
\text{CH}_2\text{OH}
\end{array}
\longrightarrow
$$

$$
\xrightarrow{-\text{Amine}}
\begin{array}{c}
\text{CH}_2 \\
||\\
\text{C} - \text{OH} \\
|\\
\text{C} = \text{O} \\
|\\
(\text{H} - \text{C} - \text{OH})_{n-2} \\
|\\
\text{CH}_2\text{OH}
\end{array}
\rightleftharpoons
\begin{array}{c}
\text{CH}_3 \\
|\\
\text{C} = \text{O} \\
|\\
\text{C} = \text{O} \\
|\\
(\text{H} - \text{C} - \text{OH})_{n-2} \\
|\\
\text{CH}_2\text{OH}
\end{array}
\rightleftharpoons
\begin{array}{c}
\text{CH}_3 \\
|\\
\text{C} = \text{O} \\
|\\
\text{C} - \text{OH} \\
||\\
\text{C} - \text{OH} \\
|\\
\text{H} - \text{C} - \text{OH} \\
|\\
\text{CH}_2\text{OH}
\end{array}
\xrightarrow[-\text{H}_2\text{O}]{+\text{Amine}}
\text{Melanoidins}
$$

Subsequent steps in the Maillard reaction

## FUNCTIONAL PROPERTIES OF SUGARS

The sweet taste of sugar is utilized in many food products in amounts ranging from minute to the major ingredient, as in candies. When sugar-containing products are heated to very high temperatures, as occurs in candy making, the degradation products that begin to form contribute additional flavor components.

Sugars contribute color to products that are heated either to a high temperature or for an extended period of time (Figure 8.2). Part of the color noted on the surface of cakes is due to some chemical breakdown of the sugar, and part is the result of the Maillard reaction (non-enzymatic browning) that also is responsible for much of the color that develops while caramels are boiling. Caramelization is the key process responsible for coloring brittles and toffee, which are heated to a much higher temperature than are caramels.

Depending on their concentration in a product, sugars can have considerable impact on the texture of various food products. Sugar syrups become increasingly viscous as the sugar content is increased by boiling water away. Cakes become increasingly tender as sugar content is increased until a critical maximum is reached. Volume also is increased with some increase in sugar if the sugar level does not become so high that the structure ruptures, and the cake falls (see Chapter 17).

The effects of sugar content on the mouthfeel of candy are discussed in the next section. Sugar stabilizes egg white meringues and also causes the foam to have smaller cells and a finer texture as a result of the need for increased beating. In starch-thickened puddings, sugar serves as a tenderizing ingredient. Custards and other protein-containing products with increased sugar levels need to be heated to somewhat higher temperatures to coagulate the protein than is necessary when less sugar is used.

## FOOD APPLICATIONS

### Crystalline Candies

**Crystalline candies**
Candies with organized crystalline areas and some liquid (mother liquor).

Candies that are easy to bite and have large areas of organized sugar crystals are categorized as **crystalline candies**. They are made by boiling sugar and water to concentrate the sugar syrup

**Figure 8.2** The thick strand of caramel being inspected by this worker in Sri Lanka is a rich brown color due to the high temperature reached during boiling of the sugar solution.

sufficiently to form a firm, crystalline structure when cooled. However, most recipes contain other ingredients, too. For example, corn syrup or cream of tartar, butter or margarine, and flavorings such as chocolate or vanilla are included to enhance the quality of the finished product. Even with these additional ingredients, crystalline candies appear deceptively simple to prepare. In fact, they illustrate some key chemical and physical principles of food preparation.

***Preparation.*** The quantity of sugar relative to the amount of liquid in crystalline candy recipes exceeds the amount that can be dissolved at room temperature (see Table 6.2). Consequently, the mixture feels gritty when stirred prior to heating. As the candy heats, the grittiness gradually disappears as more and more of the crystals dissolve. By the time the temperature of the boiling candy rises above 100°C (212°F), all of the sugar is in solution, and no crystals remain.

Water evaporates while the candy is boiling; therefore, the concentration of sugar in the solution increases gradually. This increase in sugar concentration causes a decrease in vapor pressure, and the boiling temperature rises slowly. The recommended test for determining when a candy is done is to boil the solution to the correct final temperature (Table 8.4). This temperature is correlated with the concentration of sugar, and thus a thermometer is used to indicate when the correct sugar concentration has been reached.

Crystalline candies vary slightly according to the type of candy, but most have a sugar concentration of about 80 percent or slightly greater. This concentration is achieved when the boiling temperature reaches 112°C (234°F). A slightly higher temperature means a higher concentration of sugar and a firmer candy; conversely, a lower temperature yields a softer one.

Acid, usually in the form of cream of tartar, commonly is an ingredient in crystalline candies. During the boiling period, acid promotes **inversion** (hydrolysis) of some of the sucrose molecules. The end products of inversion are equal amounts of glucose and fructose, referred to collectively simply as **invert sugar**.

The extent of inversion accomplished by the acid during the boiling period is directly related to the rate of heating. A candy that is boiling slowly undergoes more inversion

**Inversion**
Formation of invert sugar by either boiling a sugar solution (especially with acid added) or adding an enzyme (invertase) to the cool candy.

**Invert sugar**
Sugar formed by hydrolysis of sucrose; a mixture of equal amounts of fructose and glucose.

**Table 8.4**   Final Temperatures and Approximate Concentrations of Sugar in Selected Crystalline and Amorphous Candies

| Candy | Type | Final Temperature (°C) | Approximate Concentration of Sugar (percentage) |
|---|---|---|---|
| Fudge | Crystalline | 112 | 80 |
| Penuche | Crystalline | 112 | 80 |
| Fondant | Crystalline | 114 | 81 |
| Caramels | Amorphous[a] | 118 | 83 |
| Taffy | Amorphous | 127 | 89 |
| Peanut brittle | Amorphous | 143 | 93 |
| Toffee | Amorphous | 148 | 95 |

[a]The large amount of fat interferes with crystallization.

because it requires a longer time to reach the desired concentration of sugar than one boiling so vigorously that a large quantity of steam constantly is escaping from the candy.

This chemical change catalyzed by acid is significant in influencing the textural characteristics of the finished product. Extensive inversion causes crystalline candies to be somewhat softer than would be anticipated. Some inversion, however, is helpful in promoting a smooth texture. The mixture of different types of sugars (sucrose, glucose, and fructose) resulting from some acid hydrolysis makes it somewhat difficult for the crystallizing sugars to form the large aggregates that give some candies a sandy, gritty texture.

Inversion of sucrose by acid hydrolysis is but one of the means used to obtain a mixture of sugars and to promote a smooth texture in crystalline candies. Corn syrup actually is a mixture of carbohydrate compounds derived by hydrolyzing cornstarch. All corn syrups contain glucose and maltose plus some larger glucose polymers, but those made with high-fructose corn syrup (HFCS) also contain fructose because of the enzymatic conversion of some of the glucose during production. Any of the corn syrups will enhance the texture of crystalline candies because of the mixture of sugars they provide. The distinctive shapes of the crystals of different kinds of sugars interfere with ready alignment of crystals into large aggregates.

Fat also interferes with aggregation of sugar crystals in the finished crystalline candies. When cream is used in preparing candies, its fat is useful in promoting a smooth texture. This is one of the reasons that fudge is more likely to have a smooth texture than is a simple fondant made with water as the liquid and with no added fat. Not only is fudge commonly made using cream, but it always has a significant amount of fat from the chocolate in the recipe (chocolate is at least 50 percent fat). Traditionally, some butter is added to crystalline candies at the end of the boiling period. This addition has two advantages: it promotes a fine texture by interfering with crystallization, and it enhances flavor.

***Crystallization.***   A key factor in crystalline candy making is control of crystallization. When the boiling candy is removed from the heat, the solution is saturated. No more sugar could be held in solution at that temperature, but all that is present is in solution and not in crystals. This is a stable arrangement. As the solution cools, however, the amount that theoretically can be in solution is reduced. For example, 669 g of sugar can be dissolved in 100 milliliters of water at 115°C (239°F), but only 487.2 g can be dissolved at 100°C (212°F) and 260.4 g at 50°C (122°F) (refer back to Table 6.2).

Interestingly, it is possible to cool a solution that was saturated at the end of the boiling period without precipitating the crystals immediately. If considerable care is taken to prevent crystal formation, crystalline candies can be cooled to about 45°C (113°F) while maintaining all of the sugar in solution. At this point, the cooling syrup is quite viscous, and the dissolved sugar greatly exceeds the quantity that theoretically can be in solution. It is a supersaturated solution and is very unstable. Almost twice as much sugar is in solution as theoretically can be dissolved at the same temperature. The excess sugar will start to crystallize promptly if any nuclei for crystallization are provided.

The goal in making crystalline candies is to achieve a fine, smooth texture by controlling crystallization. Absolutely no nuclei should be available while the candy is cooling to the desired degree of supersaturation [a temperature of about 45°C (113°F)]. Care should be exercised to avoid the presence of sugar crystals on the sides of the pan in which the candy is cooling. Nothing should touch the surface at any time during the cooling period, and no movement should disturb the candy. Even adjustment of a thermometer in the cooling candy can be sufficient to start crystallization too soon.

When the desired degree of supersaturation is reached, beating is initiated to provide constant disruption of the crystals as they attempt to aggregate (Figure 8.3). The combination of the viscous solution at about 45°C (113°F) and the agitation results in crystallization of the excess sugar in such small aggregates that the finished product has a smooth, almost velvety feel on the tongue. This is the ideal situation.

If something happens to start crystallization before the candy achieves the desired degree of supersaturation, it is important to begin beating immediately and to continue until the candy solidifies. This agitation is essential to keep breaking aggregations into as small units as possible so that the texture will be fairly smooth when finished. Even with this effort, the texture will not be as smooth as when crystallization is avoided until the candy has become highly supersaturated (Figure 8.4).

When careful techniques to achieve supersaturation are combined with adequate beating, excellent crystalline candies can be produced. In addition, the presence of a variety of sugars and other interfering agents enhances quality by promoting a smooth texture. By selecting a recipe with interfering agents (cream of tartar and fat-containing ingredients) and by boiling candies containing an acid at a moderate rate to achieve an appropriate amount of inversion, candy makers can help ensure successful crystalline candy products.

***Ripening.*** Although they appear to be in a permanently solid form, crystalline candies actually are quite dynamic in nature. Continual dissolution and recrystallization of sugar crystals occur in these products during storage. When viewed under a microscope, the structure reveals some liquid (referred to as **mother liquor**) as well as many crystals in various-sized aggregations. As crystals dissolve into the mother liquor, other crystals form on existing aggregates. The small, individual crystals are fairly susceptible to going into solution and ultimately being recrystallized. Although this process is rather slow, it is relentless in causing a gradual change during storage. These changes are referred to as the **ripening** of the candy.

**Mother liquor**
Liquid in a seemingly solid crystalline candy.

**Ripening**
Changes that occur in crystalline candies when they are stored.

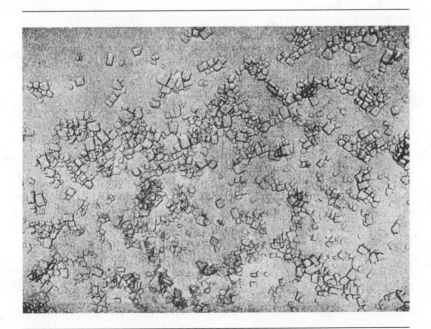

**Figure 8.3** Crystals from fondant cooked to 113°C (235°F) and cooled to 40°C (104°F) before being beaten until the mass was stiff and could be kneaded (X200).

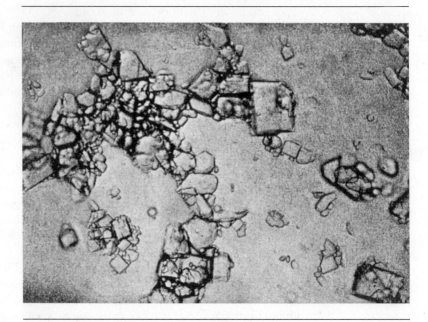

**Figure 8.4**   Crystals from fondant boiled to 113°C (235°F) and then beaten immediately and continuously until the mass could be kneaded (X200).

During the first few days of ripening, there is a bit of softening and smoothing of the texture as equilibrium is established. However, the texture of ripening candies gradually becomes coarser during storage. This developing grittiness is due to the increasing size of the crystal aggregates. Because there is no agitation while the new crystals are forming, they are attracted to existing crystals rather than forced apart mechanically. This transition is detectable over several days of storage. An adequate amount of interfering substances in the recipe helps retard undesirable changes caused by ripening (Figure 8.5).

Chocolates with cream centers are popular commercial candies that require several days of ripening before reaching their optimum texture. The difference between these soft-centered

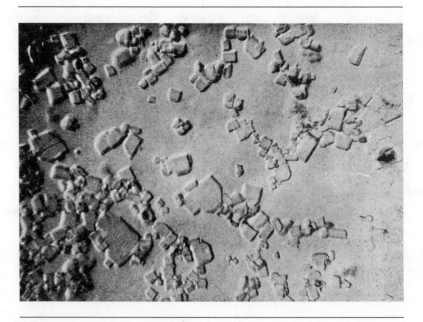

**Figure 8.5**   Crystals from fondant in Figure 8.3 after 40 days of storage show significant increase in crystal size (X200).

dipped chocolates and homemade candies during the ripening process is due to an ingredient added to the commercial candies. To be dipped easily during commercial production, the chocolates must have firm centers. **Invertase**, an enzyme that catalyzes inversion of sucrose during storage at room temperature, is added to the cooked and crystallized fondant center. The enzyme slowly inverts the sucrose during storage, resulting in a mixture of sucrose, fructose, and glucose in the fondant; this mixture crystallizes less readily than sucrose alone, causing the centers to soften. In this type of candy, the ripening process produces a creamier, smoother texture than is present before ripening.

*Evaluation.*   Crystalline candies should hold their shape when cut, yet be bitten easily. Their mouthfeel should be velvety smooth. The flavor should be characteristic of the type of candy and pleasing, with no suggestion of scorching.

Certain variables may cause the firmness of the candy to not be optimal. Commonly, failure to obtain the correct final temperature is the problem, for the thermometer must be read accurately. Too high a temperature means that the sugar concentration is too great, and the candy will be harder than it should be. Conversely, too low a final boiling temperature results in a candy that has too much water in relation to the sugar, causing the product to be soft and sticky. Either of these problems can be corrected by adding water to the candy and reboiling to the proper temperature, followed by controlled cooling and beating.

The hygroscopic (water-attracting) nature of sugar can cause a candy to be too soft when it is made on a rainy day. When hot candy is cooled on a damp day, the sugar syrup actually takes up moisture from the humid air. This added moisture is sufficient to cause the candy to be less firm than desired. In addition, low atmospheric pressure lowers the boiling point. To compensate for this problem, crystalline candies made on stormy days should be boiled to a temperature about 1F° higher than the recipe states.

An excessively slow rate of boiling can result in too much hydrolysis when acid is present. Extensive breakdown of sucrose into glucose and fructose can cause an increase in solubility sufficient to result in a very soft candy. This clearly is a case that proves that "more is not always better."

Too little beating of crystalline candies results in a coarse texture, but does not influence the firmness of the candy. Beating is effective in lightening the texture a bit by incorporating some air into the structure. It also causes these candies to become opaque and lighter in color as a result of the disruption of light transmission by the numerous fine-crystal aggregates developed during beating.

A gritty texture may result from addition of too few interfering substances, premature initiation of crystallization, and improper beating (starting too late or stopping too soon). Very rapid boiling allows so little time for inversion that little invert sugar is formed to impede the aggregation of sucrose crystals in the finished product. Omission of acid, corn syrup, or butter (or another source of fat) makes it comparatively easy for the sucrose crystals to aggregate and cause a gritty texture.

If crystallization begins while the candy is still very hot, the solution will be rather fluid. Crystals are able to move fairly easily through the hot candy to collect in large aggregates even when the candy is being beaten. When the candy is cooled to about 45°C (113°F) without any crystal formation, the supersaturated solution is so viscous that the crystals move slowly through the candy as they form. Beating is sufficient to interfere with the formation of large aggregations of sugar crystals.

## Amorphous Candies

Amorphous candies are boiled until they reach an appreciably higher temperature than is used for crystalline candies. In essence, this means that the concentration of sugar is far greater in amorphous than in crystalline candies, usually more than 90 percent (Table 8.4). This high concentration of sugar, often combined with significant quantities of interfering agents, produces an extremely viscous candy even when first taken from the heat. Such viscosity prevents any organization of sugar crystals, hence the term **amorphous candies**. Those that are hard are sometimes referred to as **amorphous glasses**.

**Invertase**
Enzyme that catalyzes the breakdown of sucrose to invert sugar (fructose and glucose).

**Amorphous candies**
Candies that lack an organized, crystalline structure because of their very high concentration of sugar or interfering substances.

**Amorphous glass**
Hard amorphous candy.

***Preparation.***    The basic ingredients in amorphous candies, just as in crystalline candies, are sugar and water; often, fat is included in the form of cream or butter. The key to preparing amorphous candies is to bring the boiling sugar syrup to the correct final temperature without scorching any portion of the product. Uniform heating is facilitated by the use of a heavy aluminum saucepan or other pan with uniform heating characteristics.

Accurate temperature readings are essential, but they may be difficult to obtain, particularly when a small quantity is being prepared. Because these candies are extremely viscous, air pockets often form between the candy and the bulb of the thermometer. Readings taken when some air is trapped next to the bulb are lower than the actual temperature, leading to overheating of the candy and increased risk of burning or scorching.

Extensive chemical changes occur in the sugar during the boiling of amorphous candies. The extremely high temperatures accelerate chemical breakdown and caramelization. The protein in the cream used in making some amorphous candies can combine with the sugars in the Maillard reaction to contribute to the noticeable browning that occurs in the later stages of boiling. The possible mechanism for the Maillard reaction was outlined previously in this chapter.

The intense heat involved in preparing amorphous candies is destructive to sugar molecules. When sugar molecules begin to undergo the extensive changes described earlier in this chapter, the flavor characteristics, as well as the color, change. The extent of change is influenced by the rate of heating and by the final temperature actually reached in the candy. More time is available for the chemical reactions to proceed when the rate of heating is comparatively slow than when it is more rapid. The color begins to become overly dark, changing from a pleasing golden brown to a fairly dark brown, or even black, particularly if the temperature rises too high.

Darkening color is a visible signal that flavor changes also are occurring. When the color begins to shift from golden brown to brown, the fragmentation of the original sugar molecules creates a jumble of smaller organic compounds and undesirable flavors. Of course, these breakdown products no longer carry a sweet taste, and the resultant candy tastes less sweet. This difference in sweetness tends to be masked when scorching or burning creates other flavors in the candy.

The extremely viscous nature of amorphous candies when they are removed from the heat prevents sugar crystals from forming in organized aggregates. In fact, caramels can be cut into separate pieces while they are still extremely hot, because they are too thick to flow. This is quite a distinct contrast to the physical properties of any of the crystalline candies. Taffy, with its high concentration of sugar, can still be pulled and twisted when it is cool enough to be held without burning the hands (Figure 8.6), but it is much too firm to permit easy movement and aggregation of sugar crystals to form a crystalline structure.

Toffee has such a high concentration of sugar and is so viscous when poured from the pan onto a sheet that it defies organization of crystals. In fact, it becomes so brittle that it can be cracked easily with a knife handle even while it is still very warm. Candies with such a high concentration are in the glassy state when cooled.

***Evaluation.***    Amorphous candies are evaluated on the basis of their color, flavor, and texture. Each type is quite unique in the textural characteristics desired, but the color and flavor evaluations are similar for all of them. These candies should have a pleasing, golden brown color that is uniform throughout, and the flavor should be pleasing, with no evidence of scorching. The exception in color evaluation is taffy, because pulling the hot taffy incorporates air into the candy, causing a difference in light refraction and a lighter color.

Caramels may range suitably from very sticky and chewy to somewhat rigid but still able to be bitten. If a sticky, chewy caramel is desired, the candy should be boiled to a temperature of about 120°C (248°F); firmer caramels should be boiled to a temperature of about 125°C (257°F). Taffy should be able to be pulled while hot, but not able to be bitten when cold. Brittles and toffees should be bitten with a bit of difficulty. The fact that these candies are in the glassy state can be demonstrated by hitting them with a knife handle, causing them to break into irregular pieces.

The problems noted in making amorphous candies result from errors in the cooking process, not in the handling of the candy during cooling. This is quite different from the crystalline candies, for which control of crystallization is the major problem. The primary factors to control

**Figure 8.6**    A Vietnamese lady pulls strips of taffy into slots in a tray to finish cooling before cutting into pieces.

when making amorphous candies are use of a heavy saucepan with excellent heat conducting properties, adequate stirring to maintain uniform heating throughout the candy, and accurate endpoint temperature (achieved by avoiding trapping air around the thermometer bulb).

## Syrups, Sauces, Jams, and Jellies

Love of a sweet taste has resulted in the production of many different sauces and syrups that are used to top everything from pancakes to elegant desserts. The primary function of sugars in syrups and sauces is to provide sweetness, but the viscosity of these sugar-containing solutions is influenced significantly by the concentration of sugars achieved during the "boiling down" phase in producing these items. Syrups produced in the home can be boiled until the desired flow properties are achieved by the evaporation of excess water. This also can be done commercially, but food additives, such as gums, often are used to achieve the right rheological characteristics quickly.

Sauces for desserts usually are sweetened with sugar, but the sugar concentration often is considerably lower than in syrup. This is possible because of the incorporation of another ingredient to aid in thickening the sauce. Frequently, starch serves as the thickening agent (see Chapter 9).

Jams and jellies are popular because of their sweet, fruity flavors. Sugar is essential for the sweet taste in most jams and jellies, but this ingredient also is vital for preserving the fruit and preventing spoilage. The high concentration of sugar essentially dehydrates microorganisms as a result of the unfavorable osmotic pressure created by the sugar (see Chapter 20).

## Baked Products

In yeast-leavened breads, sucrose is split and inverted by yeast invertase to provide glucose and fructose for fermentation. This fermentation is the source of the carbon dioxide that leavens the dough. High-fructose corn syrup (HFCS), corn syrup, and commercial dextrose (glucose) are other possible ingredients that can be used in bread dough for fermentation.

Interestingly, the crusts of breads made with sucrose or HFCS brown evenly to a fairly deep color, as contrasted with a lighter, more reddish brown (sometimes referred to as foxy-red) color that is characteristic of breads containing glucose (dextrose) or corn syrup. It is apparent from these differences that fructose contributes a somewhat different (fairly deep brown) color than that resulting from fermentation of glucose (foxy-red).

Sugar is present in large quantities in most cakes, causing several effects. Large amounts of sugar in cakes dilute the protein structure. The result is an increase in volume of the cake because of the somewhat higher coagulation temperature and the delayed setting of the structure (see Chapter 18).

Shelf life is extended by the presence of sugars. Sucrose interferes with microbiological growth by aiding in reducing water activity in a baked cake; invert sugar (equal amounts of glucose and fructose) is even more effective than sucrose in reducing water activity and inhibiting spoilage. The hygroscopic nature of sugars draws moisture from the air into a cake, which helps to lengthen shelf life.

Browning of the crust, the result of the Maillard reaction and some caramelizing, occurs most quickly when monosaccharides are contained in a cake (Figure 8.7). In fact, a cake

**Figure 8.7**    Even honey from a street festival in Serbia will cause cakes to brown faster because of its high fructose content.

made with HFCS may brown extremely fast because of the large amount of fructose and glucose. However, sucrose (a nonreducing sugar) needs to be inverted to fructose and glucose for the Maillard reaction to occur. Maltose and lactose are reducing disaccharides and, therefore, can undergo browning, although more slowly than the browning of monosaccharides.

Corn syrup can influence crumb color, as well as crust color of a cake. This is especially true in chocolate cakes; the acidity of corn syrup is sufficient to shift the chocolate color from a reddish overtone to a deep brown that lacks any trace of red. High-fructose corn syrup can result in a detectable browning of the interior of white cakes, to the detriment of the cake. By reducing the pH of the batter from its probable pH range between 7.1 and 7.5 to slightly less than 7.0, browning can be controlled because the Maillard reaction proceeds more slowly.

Cookies often have slightly less sugar than flour (by weight). Sucrose, the usual type of sugar in cookies, contributes to the spreading of the dough during baking. Dessert sugar causes considerably more spreading in a baking cookie than does granulated sugar. This effect is influenced by the amount of sugar syrup that forms in the cookie. Small crystals dissolve easily, promoting spreading. After baking, sucrose recrystallizes as the cookies cool, which contributes to a crisp surface. If cookies are made with more than one type of sugar (e.g., fructose and glucose), crispness is reduced because of reduced ease of crystal formation.

## CHOOSING SWEETENERS

### Types

The type of sweetener in a food affects the final product. The various roles that sugars are expected to perform in the food should be identified before deciding on the sweetener to use. Possible choices include sugars, syrups, liquid sugars, and various alternative sweeteners. After identifying appropriate sweeteners, the final choice can be based on factors such as cost and effectiveness.

***Sugars.***    Sucrose is by far the most common sugar available in crystalline form. Granulated sugar (from either sugarcane or sugar beets) is the type used for most applications, but superfine or dessert sugar is desirable for making meringues because of the ease of dissolving the fine crystals. Confectioners' sugar (powdered sugar) is pulverized sucrose crystals blended with cornstarch at a level of 3 percent to prevent caking of the sugar and also to help bind moisture in such products as uncooked icings.

Brown sugars are less refined than white sugars, with fewer of the impurities removed during processing (Figure 8.8). The impurities contribute flavors to these sugars; the darker the brown sugar, the more intense are the flavors of the impurities. Brown sugars are more acidic and higher in moisture than granulated sugar.

Terminologies used in identifying these various brown sugars may include **turbinado**, light, dark, **brownulated**, and raw. Even the so-called "raw" sugar has been refined to some extent to remove harmful substances. Turbinado sugar is the choice of those who seek a sugar they view as more healthful and natural, although such attributes are not documented. Brownulated sugar is convenient for those who wish to pour a bit of it onto cereal or other food. Selection of a brown sugar should be based on the flavor desired; the darker the color, the stronger will be the flavor. None of them has sufficient nutrients other than carbohydrate to make a viable contribution to the diet.

Fructose is available in some consumer markets, as well as in the food industry. The appeal of fructose is its potential for contributing greater sweetening power than sucrose. This is true in some applications, such as sweetening a beverage, but it is not so when fructose is used in baked products, apparently because of the presence of other ingredients. However, its comparatively slow rate of assimilation into the body makes this sugar of particular interest to diabetics.

The food industry uses certain other sugars to meet specific requirements. For instance, lactose (because of its relatively low hygroscopicity) sometimes is used as both a source of

**Turbinado sugar**
Light brown sugar from which only the surface molasses is removed, resulting in a mild flavor.

**Brownulated sugar**
Light brown sugar that can be poured.

**Figure 8.8**    Organic granulated and brown sugars are in markets today to meet the demand from shoppers who want to buy organic foods despite the price.

---

**Jaggery**
Form of brown sugar made in India from sugarcane or sap from the date palm.

**Muscovado**
Molasses-flavored, dark brown, somewhat sticky, sugar originating in Barbados.

**Demerara**
Light brown sugar made in Guyana from the first pressing of sugarcane.

---

## FOOD FOR THOUGHT: Brown Sugar Options

Consumers may be limited to a choice between dark and light brown sugar in their markets, but other versions have been developed in distant countries and may now be found in specialty markets. Recipe ingredients from India and countries in Southeast Asia may include **jaggery**, the traditional brown sugar in that part of the world (Figure C28). Actually, jaggery comes in two forms: one that is soft enough to spread and the other in a hard, crystalline form that may be shaped into a bar, ball, or cylinder. It is made from sugarcane or sap from date palm trees and has a somewhat distinctive flavor, which is particularly apparent in the versions that are a darker brown.

Gur, which usually is made from sago or coconut palms, is a similar brown sugar in that region of the world. Khandsari is an unrefined raw sugar ranging in color from golden yellow to brown; this brown sugar may be made in India by villagers or in small factories.

Other versions of brown sugar originated in the Western Hemisphere. **Muscovado** is a dark brown, molasses-flavored brown sugar with coarse crystals that is made in Barbados. This island, which is in the western part of the Atlantic Ocean just north of Venezuela, has a long history of sugarcane and muscovado production that began under the British.

A light golden brown sugar with somewhat sticky, large crystals is made in Guyana, a small country next to Venezuela on the northeastern shoulder of South America. This fine-grained brown sugar, called **demerara**, is light colored because it is made from the first crystallization of cane sugar processing, whereas the darker, stronger flavored muscovada is from the third crystallization. Mexico makes its own brown sugar called panela or piloncillo; this is similar to the jaggery of India.

some sweetness and an anti-caking agent to coat other sugars and to restrict their water absorption. Lactose helps keep candies and icings from becoming sticky on the surface. In contrast, invert sugar and glucose attract moisture and improve shelf life of baked goods.

***Syrups and Liquid Sugars.***   **Corn syrup** represents a mixture of carbohydrates, ranging from glucose to oligosaccharides and even dextrins of varying chain lengths. This familiar sweetener is produced by acid hydrolysis or a combination of acid and enzymatic hydrolysis, either of which results in a slightly acidic reaction.

**High-fructose corn syrup (HFCS)**, a particularly popular commercial product, is made by using an enzyme to convert some of the glucose in corn syrup into fructose. Isomerase, an enzyme produced by *Streptomyces*, is the enzyme used to convert as much as 90 percent of glucose to fructose, but usually the level ranges from 45 to 55 percent, depending on the conditions.

High-fructose corn syrup has a high **dextrose equivalent (D.E.)** of 65 D.E., which indicates that its sweetness, fermentability, browning reaction potential, and hygroscopicity (ability to attract moisture) are greater than those of regular corn syrup. However, HFCS is less viscous than traditional corn syrup, which usually has a sweetness of 38–49 D.E. High-fructose corn syrup and regular corn syrup differ in composition, with HFCS having approximately twice as high a concentration of mono- and disaccharides.

**Molasses**, commonly available as the sulfured or unsulfured liquid product, is a by-product of the production of sugar from sugarcane. Boiling of the sugarcane juice first results in the production of light molasses, an acidic liquid containing a maximum of 25 percent water. Subsequent boiling produces molasses darker than that from the first boiling, but still flavorful and palatable.

The final boiling yields blackstrap molasses, a product most commonly used in animal feeds. Sulfured and unsulfured molasses may contain as much as 70 percent sugar, largely sucrose plus a little glucose and fructose. Unsulfured molasses is reddish brown, in contrast to the sulfured version, which is light to dark brown.

Honey is a popular sweetener because of its distinctive flavor. The types of blooms from which the bees gather the nectar significantly influence the specific flavor (Figures C25 and C26). All types of honey are high in fructose. In fact, honey is the richest source of fructose in natural foods. The high fructose content promotes rapid browning when honey is used in baked products. Honey also contains a bit of sucrose and organic acids.

**Sorghum syrup** is made from grain sorghum by boiling down the juice to a syrup containing a maximum of 30 percent water. This syrup, like molasses, can vary from light to dark brown in color, and it has a distinctive flavor. It sometimes is used in various southern dishes and in some beers.

## Alternative Sweeteners

***Polhydric Alcohols.***   **Polyhydric alcohols (polyols)** are naturally occurring compounds in foods and are approved as generally recognized as safe (GRAS) additives. Even though they are not sugars, erythritol, lactitol, maltitol, sorbitol, and xylitol can contribute some sweetness to food products and have commercial applications when used in comparatively small amounts (Table 8.5).

**Table 8.5**   Some Polyhydric Alcohol Sweeteners

| Sugar | Polyol | Kcal/g | Comments/Applications |
| --- | --- | --- | --- |
| Erythrose | Erythritol | 0.2 | Anticariogenic; used in combination with other sweeteners |
| Lactose | Lactitol | 2.0 | Bulking agent with high-intensity sweeteners |
| Maltose | Maltitol | 2.1 | Chocolates |
| Mannose | Mannitol | 1.6 | Bulking agent; chewing gum (anticariogenic) |
| Sorbose | Sorbitol | 2.6 | Metabolized without insulin; baked goods, beverages |
| Xylose | Xylitol | 2.4 | Cooling mouthfeel; chewing gum (anticariogenic) |

**Corn syrup**
Sweet syrup of glucose and short polymers produced by hydrolysis of cornstarch.

**High-fructose corn syrup (HFCS)**
Especially sweet corn syrup made by using isomerase to convert some glucose to fructose.

**Dextrose equivalent (D.E.)**
Measure of the amount of free dextrose (glucose), which parallels glucose formation by hydrolysis of larger carbohydrate molecules; pure dextrose = 100. D.E.

**Molasses**
Brown syrup remaining after sugarcane juice has been boiled and some of its sugar removed during the refining of sugar.

**Sorghum syrup**
Syrup sweetener produced by boiling the juice of grain sorghum.

**Polyhydric alcohols (polyols)**
Alcohols with several hydroxyl groups, which enables them to be used as sweeteners—for example, xylitol and sorbitol.

Although they are somewhat less sweet than sucrose, polyhydric alcohols are useful sweeteners because they contribute fewer calories per gram and may be synergistic when used in combination with other sweeteners. Xylitol and sorbitol are the alcohol counterparts of the monosaccharide sugars xylose and sorbose, respectively. Maltitol is the polyhydric alcohol related to maltose, a disaccharide.

$$
\begin{array}{cc}
CH_2OH & CH_2OH \\
| & | \\
H-C-OH & H-C-OH \\
| & | \\
HO-C-H & HO-C-H \\
| & | \\
H-C-OH & H-C-OH \\
| & | \\
CH_2OH & H-C-OH \\
& | \\
& CH_2OH \\
\text{xylitol} & \text{sorbitol}
\end{array}
$$

These alcohols are absorbed much more slowly into the body than are sugars, a quality that has some merit for diabetics. However, the gas produced as a result of the prolonged microbiological action in the intestines can cause intestinal discomfort and diarrhea, thus limiting the usefulness of polyhydric alcohols as sweeteners for general applications. However, they are used in combination with other sweeteners in many food products.

Xylitol has been used effectively for sweetening chewing gum without introducing the cariogenic properties associated with sugars as sweeteners. This is a particularly important consideration because of the prolonged contact chewing gum has with teeth. A cooling sensation is noticed in the mouth when xylitol dissolves because solution of this alcoholic sweetener is an endothermic (heat-absorbing) reaction.

Sorbitol is useful as a humectant and bulking agent, as well as a sweetener. Its water-binding capabilities enable sorbitol to help limit mold growth, thus enhancing shelf life. These various attributes have resulted in sorbitol becoming a sugar alcohol that is used in a variety of food products, ranging from baked products to beverages.

Mannitol sometimes can be used as a bulking agent in products in which the hygroscopic nature of sorbitol is a detriment. In fact, mannitol is useful in powdered products because water is not attracted to this alcohol. Like xylitol, mannitol is well suited for use in sweetening chewing gums because it is not cariogenic.

**Saccharin**
Non-nutritive sweetener.

***Reduced or No-Calorie Sweeteners.***    Intense sweeteners (Table 8.6) are available for use in foods without adding many, if any, calories. Their use began with **saccharin** in the late nineteenth century. Originally, saccharin was important as a sweetener in foods for diabetics, but now much of the significance is its use as a noncaloric sweetener. Its disadvantage is the distinctive and strong bitter aftertaste that is so evident at comparatively high levels. In some food products, saccharin is combined with another sweetener, such as sorbitol, to yield a synergistic sweetening effect from small amounts of the sweeteners and with a greatly reduced problem of aftertaste from saccharin (Figure 8.9).

$$
\begin{array}{c}
C \\
C \quad C-C=O \\
| \quad | \quad \backslash \\
C \quad C \quad N-Na \\
C \quad C \\
| \\
SO_2
\end{array}
$$

saccharin

**Aspartame**
Very sweet, low-calorie dipeptide composed of phenylalanine and aspartic acid; used as a high-intensity sweetener.

**Aspartame** is a low-calorie sweetener resulting from the combination of two amino acids—aspartic acid and phenylalanine. People who have phenylketonuria must control

**Table 8.6**   Overview of Some Non- and Low-Calorie Sweeteners

| Sweetener | Trade Name | Sweetness[a] | Description | Applications |
|---|---|---|---|---|
| Acesulfame-K | Sweet-One® Sunett® | 200X | Soluble to 200°C (392°F), good with aspartame | Candies, baked Products, desserts, beverages |
| Alitame | Aclame™ | 2000X | Not hygroscopic, soluble | Candies, baked goods |
| Aspartame | NutraSweet® Equal® | 200X | Phenylalanine-aspartic acid dipeptide, heat sensitive | Beverages, dairy products, desserts |
| Cyclamate | | 30X | Sulfonated cyclohexylamine. Banned in the United States | Often combined with saccharin or aspartame in beverages and other applications |
| Dihydrochalcone | | 300–2000X | From citrus bioflavonoids | Baked goods, beverages, chewing gum |
| Isomalt | | | Derived from sucrose using an enzyme | Synergistic with sorbitol, browns slowly, 2 kcal/g |
| Neotame | | 7000–13,000X | Stable in baking; dipeptide phenylalanine not released | Beverages, dairy foods, cereals, cakes |
| Saccharin | Sweet'N Low® SugarTwin® | 300X | Metallic aftertaste | Beverages, tabletop sweetener |
| Stevia/Rebaudioside A (Reb A) | Truvia™ | 300X | From Stevia rebaudiana plant | Menthol; bitter aftertaste, limited use in world |
| Sucralose | Splenda® | 600X | Soluble, heat stable | Baked goods, tabletop sweetener, beverages |
| Tagatose | Gaio® Naturlose™ | Slightly > | Not hygroscopic, browns fast | Candies, cereals, beverages, ice cream |
| Thaumatin | Talin™ | 2000–3000X | Plant protein (Thaumatococcus danielli) from Sudan | Licorice aftertaste; no heat; chewing gum, beverages, jams, dairy products |
| Trehalose | Ascend™ | 0.5X | Preserves cell structure; heat stable | Dried fruits and vegetables, beverages, surimi |

[a]Compared with sucrose.

their intake of phenylalanine; therefore, package labels must carry a warning regarding phenylalanine. Aspartame, which is marketed as NutraSweet® or Equal® by NutraSweet Co., provides 4 calories per gram, but so little is needed for sweetening that aspartame affords a practical means of reducing calories in some food products (Figure 8.10).

The fact that aspartame is a dipeptide limits its applications in foods; heat alters the molecule sufficiently to cause a loss of sweetness. An encapsulated form of aspartame may overcome this difficulty. An acidic medium (pH 3–5) is especially favorable to aspartame use for sweetening. The soft drink industry has used aspartame widely in the production of diet drinks.

aspartame

**Figure 8.9**    Sweet'N Low® is a saccharin product that sometimes is used in place of sugar; equivalent amounts are listed on the box.

**Neotame**
Low-calorie dipeptide (aspartic acid and phenylalanine) sweetener.

**Neotame** is another aspartic acid-phenylalanine dipeptide low-calorie sweetener that is marketed by NutraSweet Co. The FDA approved neotame in 2002 for use in beverages and food. Like aspartame, neotame provides 4 calories/gram; the intense sweetness of both products means that little is needed. Actually, neotame provides even more sweetness than does aspartame. The net result is a reduction in calories in products in which these are used, either alone or in combination with other sweeteners.

**Acesulfame-K**
Low-calorie sweetener derived from acetoacetic acid.

**Acesulfame-K**, marketed as Sunett® or Sweet-One®, is the potassium salt of 6-methyl-1,2,3-oxythiazine-4 (3H)-one-2,2-dioxide and is derived from acetoacetic acid. Its sweetness is similar to that of aspartame (about 200 times that of sucrose), but acesulfame-K is stable to heat, which makes it suitable for use in baked products, puddings, and various other applications. Acesulfame-K is used widely in Europe, and its use in the United States was approved in 1992 by the Food and Drug Administration. Actually, acesulfame-K provides 4 calories per gram, but its great sweetness makes only a very small amount necessary. Its manufacturer, Hoechst Celanese of Somerville, New Jersey, recommends using 1 g of Sweet-One as a replacement for 2 teaspoons of sucrose to sweeten beverages, or 12 g per cup of sucrose in food preparation substitutions. Sorbitol can be used with acesulfame-K to round out flavors.

acesulfame-K

**Figure 8.10**  Various sugar substitutes vie with sugar to sweeten coffee or other beverages.

**Thaumatin** is a protein obtained from the fruit of katemfe (a West African plant) and is between 2,000 and 3,000 times sweeter than sucrose. Its limitations are the delayed perception of sweetness that it imparts and an aftertaste reminiscent of licorice. Despite these problems, thaumatin is used in the United States, Japan, and England and is marketed as Talin™.

**Cyclamate** is used as a sweetener in many parts of the world, but it has been banned from use in the United States by the FDA since 1970. There has been considerable controversy over the banning of cyclamate because its excellent properties—for example, 30 times as sweet as sucrose, no aftertaste, and no calories—make it desirable to use.

The tremendous consumer interest in diet foods has stimulated ongoing efforts to develop non-nutritive or low-calorie sweeteners. **Sucralose** is one of these products and is of considerable interest because it is 600–800 times sweeter than sucrose, is heat stable, and lacks any unpleasant aftertaste (Barndt and Jackson, 1990). Sucralose, approved by the FDA in 1998, has a sugar-like appearance and can be used in place of sugar in cooking and baking.

**Stevioside** is one of several steviol glycosides that can be derived commercially from *Stevia rebaudiana*, a South American plant. Rebaudioside A (also called Reb A) is another of the glycosides which is being marketed at the present time (Figure 8.11). The effective sweetening power varies somewhat with the different form of the glucoside, but is approximately 300 times sweeter than sucrose. Although used in Japan to sweeten beverages, the FDA has now granted similar approval in the United States, where it is marketed as Truvia™, Sun Crystals®, and Stevia in the Raw™.

**Isomalt** is made from sucrose, utilizing an enzyme (sucrose-glucosylfructose-mutase) to make isomaltulose, from which isomalt is formed. Although isomalt has only about 0.45–0.65 percent the sweetening power of sucrose, it acts synergistically as a sweetener when combined

**Thaumatin**
Natural sweetener from katemfe (a West African plant) with some aftertaste of licorice.

**Cyclamate**
Sweetener widely used in the world but banned in the United States.

**Sucralose**
Sweetener made from sucrose and containing three chlorine atoms.

**Stevioside**
Sweetener extracted from South American plant, suitable for tabletop sweetener.

**Isomalt**
Low-calorie sweetener produced from sucrose by enzyme action.

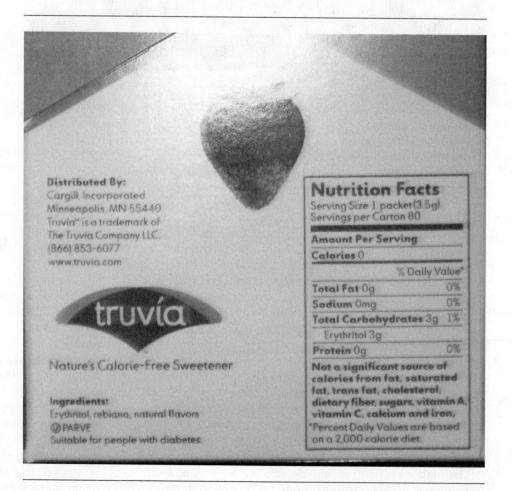

Sucralose                                          Stevioside

with sorbitol or some of the intense sweeteners. Isomalt is convenient to use because it can be substituted for sucrose on a 1:1 basis, and it has the advantage of providing only half as many calories (2 calories per gram) because it is only partially digested and absorbed in the small intestine. It can be substituted for sucrose in a wide array of products, although it does differ

**Figure 8.11**   Truvia™ is one of the brands marketing the sweetener Rebaudioside A (also called Reb A).

from sucrose in that it does not brown quite as readily and it is less hygroscopic. Isomalt is considered to be an excellent sweetener for future use by diabetics.

isomaltulose → α-D-glucopyranosyl-1,6-mannitol + α-D-glucopyranosyl-1,6-sorbitol

Isomalt

**Polydextrose** is a randomly bonded polymer of glucose, which is used extensively in food products as a bulking agent. The surprising thing is that this carbohydrate, which does not have a sweet taste despite its glucose heritage, is the source of only about 1 calorie per gram. This low-calorie contribution is the result of incomplete utilization of the compound in the body. Polydextrose is marketed by Pfizer Specialty Chemicals Group, New York, and it is identified by the trademark Litesse®.

**Polydextrose**
Polymer of glucose.

r = hydrogen/litesse/glucose/sorbitol/citric acid

Polydextrose (example of a molecule of randomly polymerized dextrose)

**Alitame** is an intense sweetener in development using L-aspartic acid, D-alanine, and an amine. Its intended use will be in beverages, frozen desserts, and other cold food, but not in hot foods because of possible off flavors. Pfizer, Inc., is seeking approval for alitame under the name Aclame™.

**Tagatose** is derived from dairy products, which allows this sweetener to qualify as GRAS. Its structure differs from fructose only at the fourth carbon. It provides 1.5 calories per gram

**Alitame**
Sweetener resulting from combining L-aspartic acid, D-alanine, and an amine.

**Tagatose**
Sweetener derived from dairy products and approved as GRAS.

because it is not absorbed completely and may result in some flatulence and laxation if consumed in excess. Arla Foods is marketing tagatose under the name Gaio® Tagatose.

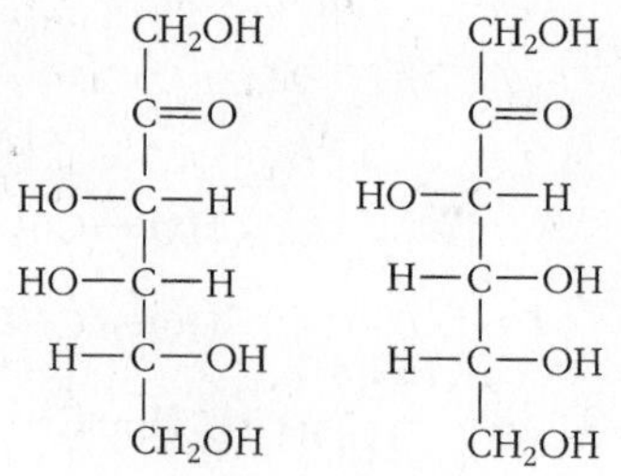

D-Tagatose        D-Fructose

**Trehalose**
Moderately sweet
ingredient used in some
foods to improve flavor
and/or texture.

**Trehalose** is found in a few foods naturally (e.g., honey and lobster) but commercially is produced from starch using enzymes. Despite the fact that it is only half as sweet as sucrose, it is used in a variety of foods, including surimi, beverages, and white chocolate. Trehalose is approved as GRAS.

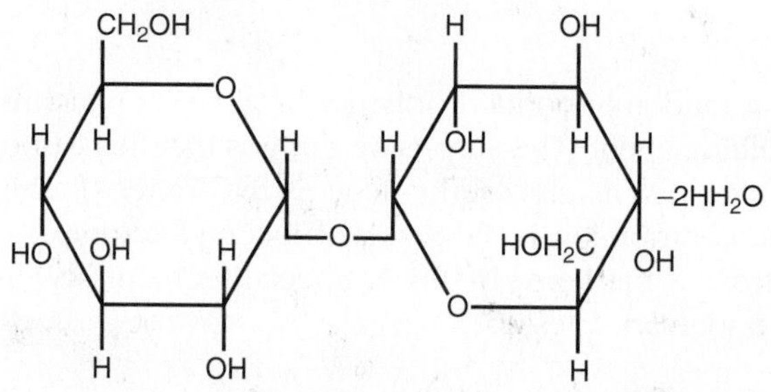

Some of the future sweeteners that may become important include miraculin, monellin, hernandulcin, and dihydrochalcones. Food and Drug Administration approval is required before a sweetener can be marketed in the United States. This is an area of food research that promises to be dynamic for a long time. At present, consumers can buy saccharin (Sweet'N Low®), aspartame (Equal®, NatraTaste Blue®, and NutraSweet®), sucralose (Splenda®), acesulfame-K (Sweet-One® and Sunett®), and stevioside or rebaudioside A (Truvia™, Sun Crystals®, and Stevia in the Raw™) in their local markets to use as tabletop sweeteners or in food preparation. Food manufacturers also may sweeten their products with several of the other sweeteners either alone or in various combinations to achieve the desired characteristics.

---

### FOOD FOR THOUGHT: Tagatose Enters the Market

In 2003, tagatose, a ketohexose, was included on the FDA's GRAS list. Immediately, tagatose began to be marketed by competing companies. Arla Foods, a European company, named its product Gaio® Tagatose, and Spherix, Inc., called its tagatose product Naturlose™. It debuted in the United States as an ingredient in 7-Eleven's Diet Pepsi Slurpee.

Commercially, Naturlose doubtless will find many applications because of its comparatively low-caloric content (1.5 kcal/g), its noncariogenic properties, and its limited impact on blood glucose levels. Much of the tagatose molecule is not digested until it reaches the large intestine, but some breakdown there produces gas and possible discomfort if large amounts of tagatose are eaten in a day. This sweetener also functions as a prebiotic by promoting production of butyrate and lactic acid bacteria in the large intestine.

Volume measurements for tagatose are approximately the same as for sucrose when it is used as a sweetener. Tagatose does not lose its sweetness when heated, but baked products brown more quickly than when sucrose is the sweetener. The impact of this new sweetener in the consumer market remains to be seen. Cost, availability, and its effect when used in making various baked products will be critical factors in its widespread use.

## SUBSTITUTING SWEETENERS

Functional and nutritional considerations may determine the choice that will be made in selecting one type of sweetener versus another. When the substitution involves replacing a liquid sweetener with a granular one, or vice versa, the adjustment in formulation of the product must include a change in the liquid as well as the sweetener. Table 8.7 lists feasible substitutions for various household sweeteners. Differences in color and flavor obviously will result in some of the substitutions suggested. Browning may be altered, too. For instance, the fructose in honey causes much more rapid browning than occurs with sucrose. Nevertheless, all of these substitutions will result in acceptable products because the functions of sugar can be provided, at least to a reasonable extent, by the substitutes suggested.

Food manufacturers have many possible sweeteners to consider when formulating their products. Specific nutritional considerations, perhaps for diet foods or for diabetic reasons, may make the choice of a non-nutritive sweetener suitable. However, that selection may trigger the need for other ingredients to perform the other functions sugar would handle. The tremendous difference in quantity that results when aspartame is substituted for sugar in a sweetened gelatin mix is evident if the aspartame-sweetened package contents are placed alongside the comparable sugar-sweetened mix. Table 8.8 reviews some sugar replacements.

**Table 8.7**  Substitutions for One Cup of Selected Household Sweeteners[a]

| Ingredient Stated in Recipe | Substitution Equivalent to 1 Cup of the Sweetener in the Recipe |
| --- | --- |
| Granulated sugar | 1 cup brown sugar, gently but firmly packed |
| Granulated sugar | 1 cup corn syrup, minus $\frac{1}{4}$ cup of liquid specified in recipe[b] |
| Granulated sugar | $1\frac{1}{3}$ cups unsulfured molasses, minus $\frac{1}{3}$ cup liquid specified in recipe, minus baking powder in the recipe, plus $\frac{3}{4}$ teaspoon baking soda |
| Granulated sugar | $\frac{2}{3}$ cup honey, minus $2\frac{2}{3}$ tablespoons liquid in recipe, plus $\frac{1}{16}$ teaspoon baking soda |
| Granulated sugar | $\frac{2}{3}$ cup honey, plus $2\frac{2}{3}$ tablespoons flour (if no liquid in recipe), plus $\frac{1}{16}$ teaspoon baking soda |
| Light brown sugar | $\frac{1}{2}$ cup dark brown sugar, plus $\frac{1}{2}$ cup granulated sugar |
| Turbinado sugar | 1 cup granulated sugar |
| Honey | $1\frac{1}{4}$ cups sugar, plus $\frac{1}{4}$ cup liquid |
| Corn syrup | 1 cup granulated sugar, plus $\frac{1}{4}$ cup liquid (same type of liquid as specified in recipe) |

[a]Products resulting from the suggested substitutions will exhibit color and flavor changes, according to the substitution being made.
[b]Substitute for no more than half of the sugar specified in the recipe.

**Table 8.8**  Guidelines for Using Sugar Replacements

| Sweetener | Substitution Equivalent | Brand Name | Recommended Uses |
| --- | --- | --- | --- |
| Saccharin | 1 packet = 2 tsp sugar | Sweet'N Low® | Hot or cold |
| Aspartame | 1 packet = 2 tsp sugar | Equal®, NutraSweet® | Cold beverages |
|  | 1 cup = 1 cup sugar | Equal® Spoonful | Limited baking |
| Neotame |  |  | Cold beverages |
| Acesulfame-K | 1 packet = 2 tsp sugar | Sweet-One®, Sunett® | Use hot or cold |
| Sucralose | 1 cup = 1 cup sugar | Splenda® | Hot or cold |
|  |  |  | Limited baking |
| Thaumatin |  | Talin® | Cold |
| Alitame |  | Aclame™ | Cold |
| Tagatose | 1 cup = 1 cup sugar | Gaio® Tagatose | Hot or cold |
| Trehalose | $\frac{1}{2}$ cup as sweet as sugar | Naturlose™ | Beverages, purees |
| Dihydrochalcones |  |  | Gum, toothpaste, candy |
| Glycyrrhizin |  |  | Candy (licorice) |
| Stevioside | $\frac{1}{2}$ cup = 1 cup sugar | Sun Crystals® | Soft drinks, gum |

## SUMMARY

Most of the sugar products on the market are sucrose derived from sugarcane or sugar beets and refined to the desired degree of purity. Fructose is gaining a market for special applications, particularly in commercial food products. Corn syrup (derived by hydrolysis of cornstarch and occasionally altered by isomerase to high-fructose corn syrup), molasses (from sugarcane processing), honey, and sorghum (from grain sorghum) are other sugar-containing products used as sweeteners.

Disaccharide sugars undergo some hydrolysis during heating in water, particularly when acid is present. With more severe heat treatment, the ring structure of sugar molecules splits, and a variety of acids and aldehydes form as the sugar molecules undergo caramelization. When in contact with proteins, reducing sugars can undergo the Maillard reaction. Caramelization and the Maillard reaction are evidenced by the gradual change in color to a golden brown and even to a deep brown or black with excessive overheating.

Crystalline candies gain their character from an organized crystalline structure, whereas amorphous candies are so viscous when hot that sugar crystals cannot aggregate in an organized fashion. The physical character of a candy is determined by the concentration of sugar and the presence of interfering substances, lower concentrations [indicated by a final temperature of about 112°C (233.6°F)] resulting in crystalline candies and higher concentrations [boiled to as high as 148°C (298°F)] in amorphous products.

Crystallization in crystalline candies needs to be controlled to achieve a fine texture, rather than the grainy quality associated with large crystal aggregates. The presence of a variety of sugars and the achievement of a high degree of supersaturation before crystallization is initiated during the cooling period are important to the control of crystal size. Adequate beating from the time the correct point of supersaturation is reached until the crystallizing candy becomes firm is an additional way to keep crystal aggregates very small.

During the first day after preparation of crystalline candies, the candy becomes slightly smoother and softer in texture, but ripening beyond this point results in increasingly grainy candy. However, invertase, which often is added to commercial chocolates with fondant centers, acts during storage to hydrolyze sucrose into invert sugar, creating a center smoother and softer than the original one.

Accurate temperature control and uniform heating are essential to the preparation of high-quality amorphous candies. Extremely high temperatures are reached as sugar concentration increases, and these high temperatures promote rapid degradation of the sugar. This can lead to scorching and burning. The high concentration of sugar at the end of the boiling period results in an extremely viscous syrup even when it is first removed from the heat. Formation of organized aggregates of sugar crystals is impossible in such a viscous mixture, hence the amorphous nature of these candies.

Saccharin (a non-nutritive sweetener used for many years) and aspartame (composed of two amino acids, L-aspartic acid and the methyl ester of L-phenylalanine) are sweeteners used when low-calorie products are the goal. Xylitol, sorbitol, and mannitol are sugar alcohols that are useful sweeteners in commercially produced foods.

The strong interest in reducing calories in foods has fostered the development and use of several low-calorie and no-calorie sweeteners. Saccharin, aspartame, stevia (rebaudioside A), neotame, tagatose, trehalose, acesulfame-K, thaumatin, sucralose, and isomalt are examples of sweeteners that are approved for use by the FDA. Cyclamates have been banned in the United States, but they are used in many parts of the world and may be approved for use at some future time in the United States as well.

Sugars can be substituted, but each type of sugar or sugar product has certain unique features. Saccharin is the sweetest of the non-nutritive sweeteners presently in use, being remarkably sweeter than sucrose, but plagued with a bitter aftertaste; aspartame also is much sweeter than sucrose, albeit distinctly less sweet than saccharin or the (banned) cyclamates. Fructose is the sweetest of the sugars and the most soluble, whereas lactose is the least sweet and least soluble of the common sugars. All of the sugars are hygroscopic, with fructose being considerably more hygroscopic than sucrose at room temperature, a characteristic that makes fructose of some interest in minimizing staling in baked products.

## STUDY QUESTIONS

1. Compare the recipes for candies from several cookbooks. What is the role of each of the ingredients? Are these same ingredients listed in comparable commercial candies? What additional or different ingredients are listed in the commercial candies?

2. Make a list of all of the sweetening products in a supermarket. What ingredients are included in each of these? How do the prices compare? Can these be used interchangeably?

3. Can fructose be used in place of sucrose in making candies? What differences, if any, might be noted?

4. Can aspartame be substituted for sucrose in making fudge? Explain the reasons for your answer.

5. Compare the applications and the results that would be expected if saccharin, aspartame, and sucrose were used in three different products containing sweeteners.

6. What differences would you expect to find when comparing use of regular corn syrup and high-fructose corn syrup in candies? Explain your rationale.

7. Describe the chemical changes that occur during the preparation of toffee.

8. Define saturated solution and supersaturated solution. How is a supersaturated solution prepared? Why is a supersaturated solution essential to the preparation of high-quality crystalline candy?

9. Can a lower-calorie fudge be made by reducing the amount of sugar in the recipe? Explain the rationale for your answer.

10. If a shortened cake is made with sugar and another is made with Splenda, what differences might be predicted in the two products?

11. True or false. Honey can be substituted for half the sugar in a recipe for fudge without causing a change in the end product.

12. True or false. Beating fondant as soon as it has been removed from the heat will improve the texture of the finished product.

13. True or false. It is necessary to remove a pan of candy from the heat as soon as it has reached a rolling boil.

14. True or false. Sugar influences the volume of a cake.

15. True or false. Splenda can be substituted for sugar in an equal amount.

## BIBLIOGRAPHY

Anonymous. 2009. Better reb-A flavor companions. *Food Product Design 19* (11): 72.

Anonymous. 2009. Streamline processing with dry sweeteners. *Food Product Design 19* (11): 74.

Backas, N. 2005. How sweet . . . and how hot it is! *Food Product Design 15* (2): 56.

Bakal, A. I. 1987. Saccharin functionality and safety. *Food Technol. 41* (1): 117.

Barndt, R. L. and Jackson, G. 1990. Stability of sucralose in baked goods. *Food Technol. 44* (1): 62.

Berry, D. 2009. Naturally sweet. *Food Product Design 19* (11): 84.

Blankers, I. 1995. Properties and applications of lactitol. *Food Technol. 49* (1): 66.

Campbell, L. A., et al. 1994. Formulating oatmeal cookies with calorie-sparing ingredients. *Food Technol. 48* (5): 98.

Chinachoti, P. 1993. Water mobility and its relation to functionality of sucrose-containing food systems. *Food Technol. 47* (1): 134.

Clark, J. P. 2004. Crystallization is key in confectionery processes. *Food Technol. 58* (12): 94.

Clemens, R. and Pressmann P. 2007. HFCS—a sticky matter. *Food Technol. 61* (12): 19.

Dea, P. 2004. Chewy confections. *Food Product Design 14* (6): 63.

Deis, R. C. 2005. Low glycemic foods—ready for prime time? *Food Product Design 15* (1): 79.

Deis, R. C. 2005. How sweet it is—using polyols and high-potency sweeteners. *Food Product Design 15* (7): 57.

Hollingsworth, P. 2002. Artificial sweeteners face sweet "n" sour consumer market. *Food Technol. 56* (7): 24–27.

Jacklin, S. 2004. Food manufacturers in the frontline against obesity. *Food Product Design. Functional Foods Annual.* (September): 15.

Knehr, E. 2005. Carbohydrate sweeteners provide sweet results. *Food Product Design 15* (May Supplement): 38.

Marshall, R. T. and Goff, D. 2003. Formulating and manufacturing ice cream and other frozen desserts. *Food Technol. 57* (5): 32.

Meyer, S. and Riha III, W. E. 2002. Optimizing sweetener blends for low-calorie beverages. *Food Technol. 56* (7): 42.

Nabors, L. O. 2002. Sweet choices: Sugar replacements for foods and beverages. *Food Technol. 56* (7): 28.

Nabors, L.O. 2007. Regulatory status of alternative sweeteners. *Food Technol. 61* (2): 24.

Ohr, L. M. 2006. Functional sweets. *Food Technol. 60* (11): 57.

Paeschke, T. 2003. Dropping calories, maintaining taste and functionality. *Food Product Design 12* (12): 32.

Prakash, I., et al. 2002. Neotame: Next generation sweetener. *Food Technol.* *56* (7): 36.

Pszczola, D. E. 2004. Confection: Sweet acronym. *Food Technol.* *58* (10): 50.

Pszczola, D. E. 2006. Synergizing sweetness. *Food Technol.* *60* (3): 69.

Shallenberger, R. S. 1998. Sweetness theory and its application in the food industry. *Food Technol.* *52* (7): 72.

Tragash, K. and Y. Tomiyama. 2005. Aspartame revisited. *Food Product Design 15* (7): 73.

Turner, J. 2003. Honey of an option. *Food Product Design 13* (4): 14.

White, J. S. 2010. Corn syrups: Clearing up the confusion. *Food Product Design 20* (1): 18.

## INTO THE WEB

*http://www.food-info.net/uk/colour/maillard.htm*—Discussion of browning due to the Maillard reaction.

*http://www.caloriecontrol.org/erythritol.html*—Information on erythritol.

*http://www.caloriecontrol.org/sweeteners-and-lite/polyols*—Summary of various polyols.

*http://chemistry.about.com/cs/5/f/bl031504a.htm*—Listing of relative sweetness of many sugars.

*http://www.caloriecontrol.org/sweeteners-and-lite/sugar-substitutes*—Background information on several sweeteners.

*http://search.oregonstate.edu/web/?query=sugar&site=food.oregonstate.edu*—Overview of sweeteners.

*www.aspartame.org*—Information on aspartame.

*http://pubs.acs.org/cen/whatstuff/stuff/8225sweeteners.html*—Article in *Chemical and Engineering News* on sugar substitutes.

*http://www.sweetenerbook.com/thaumatin.html*—Description of thaumatin.

*http://www.miraclefruitusa.com/*—Site about miracle fruit, the source of miraculin.

Cornstarch is a popular thickening agent; despite the prominent statement of 0 *trans* fats on this box, other brands also do not contain them.

# CHAPTER 9

# Starch

## Chapter Outline

## OBJECTIVES

After studying this chapter, you will be able to:

1. Describe the chemical components of starch and their physical organization.

2. Explain and differentiate between the processes of gelatinization, gelation, and retrogradation.

3. Identify the various starches and their potential choice as an ingredient in various food products.

4. Discuss the rationale for selecting various types of rice.

## INTRODUCTION

Starch is a complex carbohydrate that has been maligned by dieters in years past but now is considered to be a healthful source of energy. In food preparation at home, it is important as a thickening agent, but is only one of several polysaccharides used as thickeners in the food industry. This chapter examines several aspects of starch—its structure, functional properties, applications in food preparation, and types.

## STRUCTURE

Starch is a complex carbohydrate that consists of two fractions: amylose and amylopectin. The two fractions usually occur together in nature, so the overall behavior of a specific type is determined in large measure by the relative amounts of amylose and amylopectin. Each of the fractions has unique properties that contribute to the functionality of starch from various plant sources.

## Amylose

Amylose, the linear fraction, is characterized by 1,4-α-glucosidic linkages (see Chapter 7) and ranges in molecular weight from a few thousand to as large as 150,000. This large polymer of D-glucose is slightly soluble, a characteristic of importance in starch cookery. The actual length of amylose molecules varies considerably even within a single sample of starch, but generally cereal starches (e.g., corn and rice) have shorter, lighter amylose molecules than are found in potato and other starches from roots and tubers.

Although amylose is described as a linear molecule, individual molecules appear to form a loose, rather flexible coil when they are dispersed in a solution. This coiled arrangement allows iodine to be trapped within the helix when it is added to amylose, resulting in a blue color. Each six D-glucose units in the helix can bind one iodine atom. The iodine test is used to determine the presence of amylose in starch dispersions.

Starches from various sources differ in relative content of amylose (Table 9.1). Although it is possible to breed plants that produce starch containing essentially no amylose, the more common sources of starch range in amylose content from about 17 percent to around 30 percent. In general, the root and tuber starches contain somewhat less amylose than do those from cereals.

**Tapioca**
Root starch derived from cassava, a tropical plant.

Cornstarch typically ranges between 24 and 28 percent amylose; the range for wheat starch is similar (25–26 percent). However, the level in potato starch is somewhat lower (usually 20–23 percent). Of the starches commonly used, **tapioca** provides the least amylose (about 17 percent). At the other end of the spectrum, some special species of peas and corn have been bred with an amylose content as high as 75 percent.

## Amylopectin

Amylopectin, the other starch fraction, is also a polymer of D-glucose. However, the presence of 1,6-α-glucosidic linkages in addition to 1,4-α-glucosidic linkages (see Chapter 7) results in a different spatial arrangement. In contrast to the linearity of amylose, amylopectin molecules are **dendritic** (branching) because of the shift in direction of the D-glucose chain at each 1,6-α-glucosidic linkage. In essence, amylopectin molecules are linear for a span of about 10 to perhaps 25 or more glucose units, at which point a 1,6-α-glucosidic linkage occurs, causing the molecule to branch. Within a single amylopectin molecule, these branches are found frequently. Thus, an amylopectin molecule is described spatially as nonlinear, rather bushy, and dense.

**Dendritic**
Branching.

The molecular weight of amylopectin molecules is still a matter of conjecture, although this starch fraction clearly is one of nature's larger polymers. Researchers estimate the molecular weight can be as high as 500 million, with the lower end of the range being about 65 million. Despite the obviously far greater size of amylopectin compared with amylose molecules, amylopectin still has only one free aldehyde (reducing) group in each molecule.

**Table 9.1**    Amylose Content of Selected Starches

| Type | Percentage Amylose |
| --- | --- |
| *Root/tuber* | |
| Potato | 20 |
| Sweet potato | 18 |
| Arrowroot | 21 |
| Cassava (tapioca) | 17 |
| *Cereal* | |
| Corn | 28 |
| Waxy corn | <1 |
| High amylose corn | 70 |
| Wheat | 26 |
| Rice (long grain) | 22 |
| Waxy rice | 2 |

The difference in iodine-binding capability of the two starch fractions provides the basis for distinguishing the relative proportions of amylose and amylopectin in starch mixtures. If the iodine test is conducted on starch and a purplish-red color develops, amylopectin is present. This failure to show the blue color that iodine produces with amylose is due to the lack of a helical configuration within amylopectin molecules.

Typically, amylopectin is far more abundant in starches than is amylose. In root and tuber starches, amylopectin exceeds amylose content by approximately four times (ordinarily about 80 percent of the starch). Cereal starches are composed of around 75 percent amylopectin. However, genetic variations containing almost all amylopectin (e.g., waxy maize) have been developed and are of commercial significance. The relative proportions of amylopectin and amylose in starches are of considerable importance because of the different behaviors of these two starch fractions in cooked starch products.

## Starch Granule

In plants, starch is deposited in an orderly fashion in the form of granules (Figure 9.1). These granules, composed of amylose and amylopectin molecules, are made in the **leucoplasts** within the cytoplasm of the cells (see Chapter 10). Each granule consists of concentric layers of amylopectin molecules interrupted by some amylose molecules, which often are arranged in a somewhat organized manner within the layers (growth rings). From the very small inner layer, called the **hilum**, amylose and amylopectin molecules are deposited layer upon layer on the **starch granule** until the mature plant is harvested. Remarkably, the molecules within each layer are held together simply by hydrogen bonding, and similarly, the layers are bonded together by the same secondary bonding arrangement.

The physical nature of starch granules, as seen using X-ray diffraction and polarized light with polarized filters placed at right angles to each other, reveals a **birefringence** pattern that looks much like a **Maltese cross** because of the somewhat spherical shape and

**Leucoplast**
Plastid in the cytoplasm of plant cells; site of starch storage as granules.

**Hilum**
Innermost layer or the nucleus of a starch granule.

**Starch granule**
Concentric layers of amylose and amylopectin molecules formed in the leucoplasts that are held together by hydrogen bonding.

**Birefringence**
Refraction of light in two slightly different directions.

**Maltese cross**
A cross consisting of arms of equal length that terminate in a V-shape.

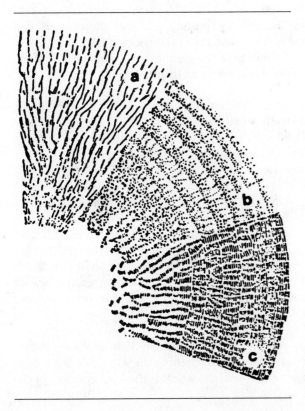

**Figure 9.1**    Diagram of ultra-thin sections of starch granule: (a) ordered radial arrangement of amylose, (b) amylose in amorphous region, and (c) amylopectin crystalline region. (Courtesy of D. R. Lineback and Baker's Digest. 1984. Vol. 58(2): 16. Diagram: P. Kassenbeck. Stärke. 1978. 30: 40.)

the crystalline areas. The unique birefringence patterns of starch granules are found only in the raw starch. When starch has been heated with water, the crystalline areas responsible for the original light refraction are altered, and the pattern of the Maltese cross can no longer be seen using polarized light.

## FUNCTIONAL PROPERTIES OF STARCH

For starch to be effective as a thickening agent in foods, it must be heated with water, which causes **gelatinization**, a physical change. If it is heated without water, it undergoes **dextrinization**, a chemical change. These processes are important in food preparation, for they determine the behavior of starch, its thickening ability in sols, and also its gel-forming properties.

## Gelatinization

**Gelatinization**
Swelling of starch granules and migration of some amylose into the cooking water when starch is heated in water to thicken various food products.

**Gelatinization** is a physical process that is unique to starches. The transformation that occurs when starch is heated in water is truly remarkable. The heat energy causes hydrogen bonds in the starch granules to break, which facilitates the entry of water into the granule and the shifting of some amylose molecules into the water surrounding the granules. Water continues to migrate into the granules, causing considerable swelling when the starch mixture is heated to the temperature range required for gelatinization. Because water is not compressible in the starch granule, the volume of the granule increases as more and more water enters and forms hydrogen bonds with the amylopectin and amylose molecules.

This bound water not only affects the viscosity of the starch mixture by increasing the physical size of the starch granules, but also reduces the amount of free water external to the granules. The tight organization of the starch granule is disrupted during the gelatinization process. The granules lose birefringence, and the paste becomes more translucent as amylose leaches out of the granule into the surrounding liquid.

**Pasting**
Changes in gelatinized starch, including considerable loss of amylose and implosion of the granule.

**Pasting** is another term to describe the changes that take place in starch as it heats in the presence of water. This is quite a descriptive word, for gelatinizing starch mixtures do begin to take on the qualities of a paste. Technically, pasting describes the changes that occur in the starch granule when heating is continued after gelatinization has taken place. These changes include loss of molecules of starch from the granule and complete loss of the granular organization, which may be described accurately as **implosion** of the granule.

**Implosion**
Violent compression.

***Temperatures.*** As heating continues beyond termination of the loss of birefringence, the granule continues to swell. In fact, most starches can be heated to 100°C (212°F) with little rupture of starch granules despite their vast increase in size as they bind increasing amounts of water. Amylose also leaches out of the granule during this phase of gelatinization; the shorter amylose molecules are the most likely to leave the granule because of their comparatively greater solubility.

At temperatures near 100°C (212°F), some of the granules begin to implode (compress inward) and fragment, a change that reduces viscosity. However, other granules may still be swelling and offsetting the effect of the few imploding granules. When a starch paste is held at 95°C (203°F) or higher for several minutes, most of the starch granules reach their maximum volume and begin to implode. This explains the decrease in viscosity of starch pastes with prolonged heating (Figures 9.2 and 9.3). It is important to remember that tapioca reaches maximum viscosity at a temperature about 20°C (68°F) lower than other starches and then begins to thin (Table 9.2).

***Type of Starch.*** The thickening ability of starches from different sources varies. Potato starch is far more effective than other starches as a thickening agent (Table 9.3). Root starches are somewhat more effective than cereal starches; wheat is the least effective of the

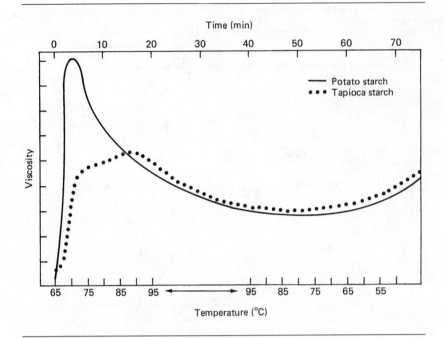

**Figure 9.2**    Comparison of the viscosity of potato and tapioca starch pastes during heating, holding, and cooling. (Based on data from T. J. Schoch and A. L. Elder. "Starches in the food industry." In *Users of Sugars and Other Carbohydrates in the Food Industry*. Adv. In Chem. Ser. 12, p. 24. 1955. American Chemical Society).

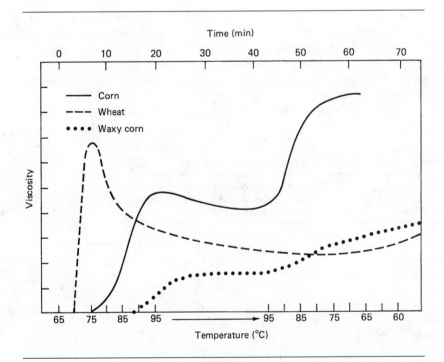

**Figure 9.3**    Comparison of the viscosity of corn, wheat, and waxy cornstarch pastes during heating, holding, and cooling. (Based on data from T. J. Schoch and A. L. Elder. "Starches in the food industry." In *Users of Sugars and Other Carbohydrates in the Food Industry*. Adv. In Chem. Ser. 12, p. 24. 1955. American Chemical Society).

**Table 9.2**    Temperature of Starch Pastes[a] at Maximum Viscosity and Temperature and Viscosity 20 Minutes after Reaching Maximum Viscosity

| Type of Starch | Temperature, Maximum Viscosity (°C) | Maximum Viscosity (g-cm) | 20 Minutes Later | |
|---|---|---|---|---|
| | | | Temperature (°C) | Viscosity (g-cm) |
| Potato | 90 | 105 | 95 | 49 |
| Waxy corn | 89 | 105 | 95 | 65 |
| Cross-linked waxy corn | 93 | 103 | 96 | 115 |
| Waxy rice | 87 | 110 | 96 | 106 |
| Waxy rice flour | 70 | 97 | 92 | 25 |
| Waxy sorghum | 91 | 108 | 96 | 65 |
| Tapioca | 71 | 108 | 94 | 59 |
| Arrowroot | 80 | 103 | 96 | 85 |
| Sorghum | 94 | 110 | 97 | 73 |
| Corn | 91 | 113 | 95 | 77 |
| Rice | 94 | 101 | 97 | 54 |
| Rice flour | 93 | 100 | 97 | 77 |
| Wheat | 92 | 105 | 94 | 62 |
| Wheat flour | 83 | 108 | 95 | 5 |

[a]The percentage of starch was varied so that pastes had similar viscosities when they reached maximum viscosity.

starches commonly available. Flour, which is used in the home, is even less effective than pure wheat starch because of its protein content. The waxy starches are more effective thickening agents in starch pastes than are their standard counterparts.

Increased translucence during gelatinization is particularly noticeable in the root starches. Potato starch and tapioca are much more translucent when gelatinized than are the cereal starches. Among the cereal starches, the gelatinized waxy (high amylopectin) starches are more translucent than their regular counterparts. However, regular cornstarch and, to a lesser extent, rice and wheat starches exhibit a distinct increase in translucence during gelatinization.

The texture of gelatinized starch pastes is of concern because of its influence on palatability. Ideally, a thickened starch paste will be smooth, but unmodified starches (especially root starches) tend to be mucilaginous. Those pastes that are the clearest (e.g., potato) are also the most stringy in texture. This is a major disadvantage in the use of potato starch as a thickening agent.

Similarly, tapioca-thickened pastes are much too mucilaginous in texture to be competitive with other starches unless the tapioca has been modified. This textural problem

**Table 9.3**    Comparative Thickening Ability of Various Starches

| Type of Starch | Comparative Amount Needed to Achieve Designated Viscosity of Hot [approximately 95°C (203°F)] Starch Paste |
|---|---|
| Potato | 1.96 |
| Waxy corn | 2.98 |
| Waxy rice | 3.13 |
| Waxy sorghum | 3.42 |
| Tapioca | 3.54 |
| Cross-linked waxy corn | 4.15 |
| Arrowroot | 4.37 |
| Sorghum | 4.66 |
| Corn | 4.90 |
| Waxy rice flour | 5.48 |
| Rice | 5.49 |
| Rice flour | 5.57 |
| Wheat | 6.44 |
| Wheat flour | 9.27 |

is the reason for processing tapioca into **pearl tapioca** or else modifying the starch chemically. Pearl tapioca consists of agglomerates of partially gelatinized starch dried into pellets so hard that they require overnight soaking for practical use in puddings and other products. When the partially gelatinized tapioca is dried in much finer pellets, the extended soaking problem is eliminated without sacrificing the improved texture. Finely ground minute tapioca is used often because it hydrates quickly.

Cereal starches are much less mucilaginous when made into gelatinized starch pastes than are the root starches (Table 9.4). This makes cornstarch an excellent choice as a thickening agent when a starch paste with limited translucence is acceptable. However, waxy cornstarch produces a distinctly stringy, albeit translucent paste. Fortunately, cross-linking during the manufacture of waxy starches alters the textural properties of the gelatinized starch pastes, and the desired smooth paste with excellent translucence can be made using cross-linked waxy cornstarch.

*Effect of Ingredients.*   When starch-thickened mixtures are incorporated into desserts, sugar usually is a prominent ingredient. The hygroscopic nature of all types of sugars causes the sugar used in the recipe to compete with the starch for the water needed for gelatinization. This competition partially explains the fact that gelatinization is delayed, and the final temperature required to achieve gelatinization is raised as the level of sugar in a starch mixture is increased. The other contributing factor appears to be the cross-linking between sugar and starch molecules.

Obvious effects of sugar include increased translucence and reduced paste viscosity and gel strength. Monosaccharides have less effect than do most disaccharides; curiously, maltose has less effect than either lactose or sucrose. The effect of sucrose is of particular interest because of its frequent use in puddings and pie fillings.

Occasionally, lemon juice or another acid ingredient may be a part of a starch-thickened product. The combination of acid and heat, particularly below pH 4, causes a hydrolytic reaction that begins to break down molecules of starch into slightly smaller molecules. The shorter molecules created by acid hydrolysis of starch move somewhat more freely in the thickening paste. The result is a gelatinized starch paste that is a little thinner than it would be if no acid were present.

If starch must be heated in the presence of acid (as is the case when thickening fruit juice for a pie filling), a rapid heating rate will keep the length of time for acid hydrolysis to a minimum and result in a thicker product than is produced when the filling is heated more slowly. Whenever possible, the acid should be added after gelatinization is complete so that the thinning caused by acid hydrolysis will be avoided. However, recipes do need to be formulated so that the liquid in lemon juice (or other similar acidic substances) is calculated as a part of the total liquid. This liquid added after gelatinization has occurred is not bound in the starch granules of the paste.

**Pearl tapioca**
Large pellets of partially gelatinized tapioca that are dried, resulting in a product that requires a long soaking period before use but yields a translucent, nonstringy paste.

**Table 9.4**   Characteristics and Qualities of Selected Starches

| Characteristic or Quality | Root and Tuber | | | Cereal | | | |
|---|---|---|---|---|---|---|---|
| | Tapioca | Potato | Arrow-root | Rice | Wheat | Corn | Waxy Corn |
| Percentage amylose (approx.) | 17 | 20–23 | | | 25–26 | 24–28 | 0 |
| Granule size (μ) | 12–25 | 100 | | 3–8 | 10–35 | 12–25 | |
| Temperature for maximum paste viscosity (°C) | 71 | 90 | 80 | 94 | 92 | 91 | 89 |
| Amount for comparable paste viscosity | 3.54 | 1.96 | 4.37 | 5.49 | 6.44 | 4.90 | 2.98 |
| Gel strength | 0 | 0 | 87 | 31 | 345 | 52 | 0 |
| Translucency | high | high | high | good | good | good | high |
| Texture | stringy | stringy | stringy | good | good | good | stringy |

*acid + sugar*

Both sugar and acid are ingredients in some starch-thickened products, such as a lemon pie. In such recipes, the amount of sugar used has a considerable effect on the behavior of the paste. If the sugar content is fairly low, increasing the amount of acid in sugar–starch mixtures causes an increase in the viscosity and a lower gelatinization temperature, but a rapid thinning of the starch paste when heating is continued. However, high sugar content apparently attracts so much of the water needed for gelatinization that the starch granules swell slowly. This delays the movement of molecules from starch granules and contact with the acid and so retards acid hydrolysis.

When it is necessary to gelatinize starch in the presence of both acid and sugar, sufficient heating is imperative to ensure complete gelatinization and maximum thickening. Careful attention is required so that the point of maximum thickening is recognized, and heating is halted before the rather rapid thinning that occurs once hydrolysis becomes the dominant factor. Use of a cross-linked starch is advised for commercial applications; cross-linked starches are more resistant to hydrolysis than are native starches.

Fats are a part of some formulations that are thickened with starch. Their presence results in a reduction in the temperature at which maximum gelatinization and viscosity occur. Milk proteins also lower the temperature required for maximum gelatinization of starch mixtures.

## Gelation

Gelatinized starch mixtures may exist as sols or gels (see Chapter 6). Hot starch pastes exhibit flow properties; starch is the dispersed solid phase, and water is the continuous phase. However, many starch pastes (sols) are converted to gels as they cool. Amylose molecules that have left the starch granules during the gelatinization process are free to move about in a paste. The paste loses energy gradually during cooling and the free amylose molecules move more slowly as energy is lost. Occasionally, one amylose molecule may happen to move close to another amylose molecule and a hydrogen bond forms between the two molecules. While the paste is still warm, this hydrogen bond may break readily and permit the two molecules to continue their separate journeys through the paste. Increasing opportunity for more stable hydrogen bonding arises as the amylose molecules move ever slower among the other amylose molecules and swollen starch granules.

**Gel**
Colloidal system in which a liquid is dispersed in a solid.

**Gelation**
The process of formation of a gel.

The gradual reduction in energy during cooling promotes the formation of hydrogen bonds between amylose molecules, with the result that frequently a starch paste is transformed into a gel. When there are enough amylose molecules to establish a continuous network joined by hydrogen bonds, the swollen granules become trapped in the amylose superstructure. This network of solids becomes the continuous phase of the newly formed starch **gel**, and the water is now the dispersed or discontinuous phase (Figure 9.4). This colloidal system, a gel, no longer has the flow properties associated with a sol. The process of forming a gel is **gelation**.

*starch types + gelatinization*

***Type and Concentration of Starch.***   After reading the preceding description of gelation, you should not be surprised to find that starches low in amylose (i.e., waxy starches) do not form gels in the concentrations normally used in making food products. If the level of starch is raised to 30 percent, the amylopectin molecules from the waxy starches can form a limited amount of hydrogen bonds and form a soft gel. This is of academic interest but is not practical in the food industry. When a gel is desired, other starches are used.

Unlike potato and tapioca starches, arrowroot starch is a root starch capable of forming at least a soft gel. In fact, it forms a much stronger gel than does rice starch (Table 9.5). Cornstarch forms a pleasingly firm gel, and wheat starch sets to a strong gel. Sometimes flours from cereals are used as thickening agents. Although wheat, rice, and even waxy rice flours form starch pastes effectively when gelatinized, they are comparatively poor in their gel-forming ability. A particular disadvantage of using wheat or other cereal flours as thickening agents is that they form rather opaque sols and gels because of their protein content.

**Figure 9.4**    Cornstarch pudding forms a strong enough gel when chilled so that it can be removed from a cup for service. The gel structure is weak enough to sag slightly, but still hold its basic outline.

***Extent of Heating.***    Gelation depends on the availability of free amylose molecules for hydrogen bonding to form the continuous network. These molecules become available when starch has been heated with water sufficiently for some of the hydrogen bonds within the granules to break and release some of the amylose into the surrounding water. For optimal gel strength, starch pastes need to be heated until enough amylose has been released, but not so much that the granules start to split apart into fragments.

With a moderate rate of heating, gelatinization is complete and amylose becomes available long before the granules are fragmented to any great extent. However, with

**Table 9.5**    Strength of Starch Gels Made from Various Starch Pastes of Approximately Equal Viscosity[a]

| | Gel Strength (g-cm) | |
| Starch | Cooled at Maximum Viscosity | Cooked 20 Minutes Beyond Maximum Viscosity Before Cooling |
| --- | --- | --- |
| Waxy corn | 0 | 0 |
| Waxy rice | 0 | 0 |
| Waxy rice flour | 0 | 0 |
| Potato | 0 | 0 |
| Tapioca | 0 | 0 |
| Cross-linked waxy cornstarch | 0 | 0 |
| Rice flour | 11 | 15 |
| Wheat flour | 26 | 88 |
| Rice | 31 | 30 |
| Arrowroot | 87 | 115 |
| Cornstarch | 52 | 142 |
| Thin-boiling cornstarch | 115 | 210 |
| Wheat | 345 | 440 |
| Thin-boiling wheat | 434 | 1440 |

[a]See Table 9.3 for the levels of dry starch used.
Adapted from Osman, E. M. and Mootse, G. Behavior of starch during food preparation. 1958. *Food Res.* 23: 554.

vigorous stirring and a prolonged heating period, considerable fragmentation does occur. This results in a pasty texture and a weakened gel structure.

***Agitation.*** For maximum gel strength, starch mixtures should be allowed to cool without disturbance. When hydrogen bonds form between amylose molecules that are in close proximity, they begin to provide the stable network needed for a strong gel. Agitation during the gelling period disrupts hydrogen bonds already formed and weakens the ensuing gel. For this reason, butter and flavorings added to puddings and pie fillings should be stirred in immediately after these products are removed from the heat, and the mixtures should then be allowed to cool undisturbed.

***Effects of Other Ingredients.*** Sugar and acid often are added to starch-containing products to alter their flavor, but these ingredients also influence the properties of starch gels. As noted in the discussion of gelatinization, both sugar and acid have a softening effect on the resulting gel. Sugar not only causes the gel to be more tender, but it also increases its translucence. The extent of softening caused by the acid is determined to a great extent by the presence of sugar, as well as by the rate of heating and the pH if the acid is added prior to gelatinization. If the acid is added before the starch is gelatinized and the pH is 4 or less, acid hydrolysis will result in a rather tender gel, the consequence of shorter amylose chains. When lemon juice or vinegar is added to a gelatinized starch paste prior to gelation, the softening observed is due primarily to the added liquid and not to hydrolysis.

Fats and proteins may influence gel strength too, in part as a result of the small reduction in the percentage of starch that occurs when other ingredients are added. However, the effect of egg yolk protein in starch mixtures is greater than can be explained by this rationale. When egg yolk is added to a gelatinized starch mixture, it must be heated sufficiently to coagulate the yolk proteins including **α-amylase** (a starch-digesting enzyme in the yolk). Then the thickened starch-protein mixture will gel on cooling. If the yolk proteins are not coagulated, the cooled thickened mixture does not gel Instead it becomes quite fluid because of α-amylase action, which breaks down starch and yields a product that is far less viscous than it was prior to addition of the yolks. From a practical perspective, careful coagulation of added yolks in a starch-thickened paste is essential to achieve the desired gel strength.

***Syneresis.*** Water is trapped within the starch gel. A layer of water is hydrogen bonded all along the individual amylose molecules, as well as to the surface molecules of the granules. Some water is bound within the starch granules. Additional free water (not actually bound to the starch) is trapped in the interstices within the gel structure.

As a gel ages, some of the amylose molecules draw together, and some water is squeezed out of the gel. In a similar fashion, water separates from a starch gel when the surface is cut and the trapped liquid is released from the areas that have been exposed by the cut. This loss of liquid from a gel is termed *syneresis* (see Chapter 6).

## Retrogradation

Although gels appear to be static after they are formed, some undergo considerable change. This change results from the breaking of some of the hydrogen bonds holding the gel together in a continuous network and the re-formation of other hydrogen bonds as the amylose molecules shift around within the gel. Over a period of time, there is a tendency for amylose molecules to orient themselves in crystalline regions. This more orderly alignment of amylose, termed **retrogradation**, is detected on the tongue as a somewhat gritty texture.

Amylopectin also participates in retrogradation, albeit at a slower rate than does amylose. The outer branches of the amylopectin molecules also are capable of forming some hydrogen bonds with other molecules in a starch gel. Obviously, retrogradation is undesirable and reduces the quality of the food in which it occurs. A starch-thickened pudding develops this textural quality if held in the refrigerator for a few days. A fairly common example is bread that develops a slightly harsh texture after several days of storage.

**α-amylase**
Amylose-digesting enzyme contained in abundance in egg yolk and to a lesser extent in egg white.

**Retrogradation**
Gradual increase of crystalline aggregates in starch gels during storage, the result of amylose molecules rearranging in an orderly fashion.

The crystalline aggregates of amylose that occur in retrogradation can be eliminated by gently heating the food containing the starch. The heat energy is sufficient to break the hydrogen bonds holding the amylose molecules together and allow them to move in the gel again. This change is detected readily when stale bread is tightly covered and reheated. The heated bread does not exhibit any harshness, but as the bread cools, the amylose molecules retrograde again into crystalline aggregates.

## Dextrinization

Occasionally, as when browning meat dredged in flour or making certain gravies and sauces, starch is heated without any water being added to the pan. Without water, the temperature rises rapidly beyond the maximum that can be reached when water is present. The high energy causes chemical degradation of starch, via a chemical reaction with the water that is naturally present in the flour. The amylose and amylopectin molecules split at one or more of the linkages between the glucose units. The result is formation of shorter molecules of varying lengths called dextrins (see Chapter 7). The chemical change that occurs, called **dextrinization**, is shown here.

**Dextrinization**
Hydrolytic breakdown of starch effected by intense dry heat, a chemical change that produces dextrins.

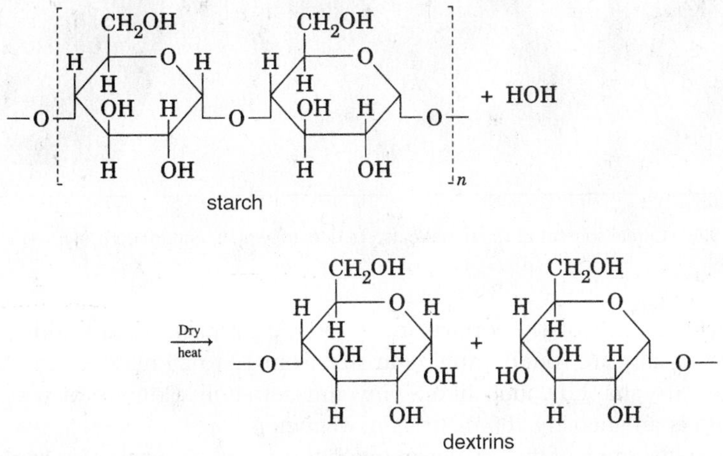

*dry heat*

starch

dextrins

## EXAMINING STARCHES

### Sources

Depending on the part of the plant used, food starches are classified as cereal, root, or tree starches. The specific cereals that are used as sources of starch include oats, rice, wheat, and corn (Figure 9.5). Although the shapes of the various grains or kernels differ from one cereal to another, all grains consist of three components—a bran covering, a small region of germ or embryo, and a major portion called the **endosperm**. The endosperm is the location of the starch granules within the individual grains of cereal.

**Endosperm**
Largest part of a cereal grain and the area where starch is deposited.

The granules of starch in each cereal have a distinctive shape, depending on the type of cereal (Figure 9.6). Cornstarch typically has polygonal granules ranging in diameter from about 12 to 25 microns. In contrast, rice starch granules are the smallest of the starches, usually only from 3 to 8 microns in diameter, but also polygonal in shape. Rice starch is particularly difficult to split from the protein with which it is agglomerated in the endosperm. A recently developed microfluidizer uses high pressure to break the two components apart so they can be separated into pure rice starch and protein. Surprisingly, wheat has two basic forms of granules: small spheres about 10 microns in diameter and larger disks about 35 microns across, significantly larger than the other cereal starch granules.

Root starches are produced commercially from the root of the cassava plant (also called manioc) and from the potato (actually a tuber). The starch from the cassava root is marketed

**Figure 9.5**    Cereal sources of starch: (a) oats, (b) rice, (c) wheat, and (d) corn.

as tapioca, which is particularly popular in its cooked form—tapioca pudding. Granules of tapioca starch typically are round, ranging in size from 12 to 25 microns. In contrast, potato starch granules are shaped much like a tiny mussel shell. Compared with other starch granules, potato is an amazing 100 microns in diameter.

The pith from the trunk of the sago palm tree is the source of yet another commercial starch. Sago starch granules are somewhat larger than those from potatoes and are elliptical in shape. Arrowroot and milo (also known as grain sorghum) are other sources of commercial starches.

Other plants rich in starch include mature legumes, such as dried beans. However, legumes are not processed to obtain starch for use in food preparation; instead, they are eaten as whole seeds because they contain protein, as well as other nutrients.

## Native Starches

Numerous choices are available when starch is required as an ingredient. Any of the grain, root, tuber, or tree starches mentioned in the preceding section are available from their plant source in a comparatively unmodified state. These **native starches** are frequently used as thickening agents in the home and also may be used in commercial food products. In addition, rice flour and wheat flour may be used for thickening, particularly in the home. These flours provide not only starch, but also some protein and other components of the grains from which they are milled.

Some other starches have unique properties because genetic research has led to new varieties of grains that differ from their cereal ancestors in the composition of the starch they contain. Some have been bred to be nearly 100 percent amylopectin. Such starches are **waxy starches**. The absence of amylose inhibits the formation of a gel structure when a waxy starch is used as a thickening agent; this characteristic makes waxy starches desirable in the preparation of fruit pies and other items in which a thickened, but ungelled consistency is desirable. Waxy cornstarch and waxy sorghum starches have unique applications in the food industry because of their chemical composition and consequent properties as thickening agents.

**Native starches** Starches that are used without being modified after they come from the plant.

**Waxy starch** Starch containing only amylopectin, the result of genetic research and breeding for this composition.

**Corn**        **Sorghum**        **Wheat**

**Oat**        **Barley**        **Rye**

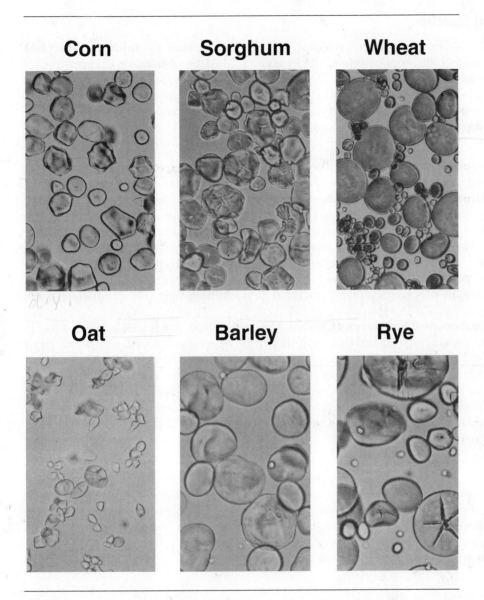

**Figure 9.6**    Photomicrographs of ungelatinized cereal starch granules (by M. M. McMasters. Courtesy of the Northern Regional Research Center, USDA-ARS. Peoria, IL and Jerold A. Bietz).

Genetic researchers also have developed starch sources that have an extremely high amylose content, sometimes as high as 75 percent. The remainder of the starch granule is amylopectin. Nevertheless, such high levels of amylose result in unique properties of food products utilizing **high-amylose starches**. The linear nature of the amylose molecules enables considerable hydrogen bonding to occur; as a result, the high-amylose starches can be made into thin films suitable for wrapping items such as candies. Because these **edible films** are made of this special starch, the wrappers can be eaten, resulting in a unique and definitely marketable item.

Research into developing edible films includes the use of varied starch sources, such as cassava, rice, and corn, as well as others. Impetus for such products is provided partially by the need to help developing countries produce agricultural products with value added beyond simply the basic crop. American food industries already are using a variety of edible films for diverse purposes such as wrappings for cough drops and candy to coatings for potatoes (to be deep-fat fried), fresh produce, and baked products. Such coatings provide edible protective barriers to enhance the quality of food when it reaches consumers.

**High-amylose starch**
Native starch resulting from breeding plants to produce starch that is exceptionally high in amylose and much lower than normal in amylopectin.

**Edible films**
Thin film of gelatinized starch that can be eaten; made by gelatinizing high-amylose starch.

## Modified Starches

**Modified starch**
Starch that has been altered from its native state by either physical or chemical means.

Today's sophisticated food industry has generated a demand for **modified starches** designed for specific applications. Changes accomplished by physical or chemical means can be useful in developing a variety of special starches with unique characteristics. Physical methods used to modify starches include extruding, drying by drum or spray, and heating. Chemical modifications include reactions with acids or other compounds to form esters or ethers to replace some hydroxyl groups.

*Pre-Gelatinized Starches.*   Heavy demands on time have created considerable need for time-saving products, and it was this priority that prompted the development of **pre-gelatinized starches**. These starches are examples of products in which a physical change modifies the characteristics of the native starch. Pre-gelatinized starches are cooked with water to gelatinize them and then are dehydrated after they become swollen. The result is a starch that swells to a desirable thickness when water is added; no heating is necessary. For ease in dispersing these starches, some manufacturers agglomerate these instantized particles as a final production step. Instant puddings are a popular example of the application of pre-gelatinized starch as a thickening agent in commercial food products.

**Pre-gelatinized starch**
Starch that has been gelatinized and then dehydrated; addition of water produces a thickened product.

Cold water–swelling starch (CWS) is a pre-gelatinized starch well suited for use in microwave cookery. Its characteristics are created by treating starch with an alcohol and a strong base at a high temperature and pressure to cause it to swell before it is dried. Another way to create a starch that swells in cold water is cooking the starch under controlled temperature and pressure and spray drying. Spray cooked starch is the name of the resulting product. These ingredients offer unique cold swelling properties and have all the properties of granular starch and do not require cooking.

**Cold water–swelling starch (CWS)**
Pre-gelatinized starch.

*Thin-Boiling Starches.*   Starch can be modified chemically to alter its properties for special applications in food products. Treatment of starch with hydrochloric or nitric acid in a controlled reaction results in acid hydrolysis. Depending on the extent of the hydrolytic reaction, the solubility of the starch increases and its thickening power diminishes. **Thin-boiling starches** are formed in this way. These starches contain many debranched molecules of amylopectin, the result of hydrolytic cleavage at the 1,6-$\alpha$-glucosidic linkage. This reaction facilitated by acid is presented here.

**Thin-boiling starch**
Debranched amylopectin molecules, produced by acid hydrolysis of starch, resulting in a starch that forms a thin sol when hot, but a strong gel when cold; useful for making gumdrops.

debranched amylopectin

The advantage of thin-boiling starches is exhibited in the commercial production of starch-thickened food items that need to be delivered into packages or molds through pipes. When hot, thin-boiling starches are fluid and pass easily through the delivery system into containers. Uniquely, these thin liquid starches form strong gels when cooled, apparently because of the greater ease of hydrogen bonding between molecules resulting from the reduced branching of some of the starch.

*Oxidized Starches.*    When starch reacts with sodium hypochlorite, the oxidation reaction produces **oxidized starches**. These starches are similar in behavior to acid-hydrolyzed starches. Oxidized starches have only limited use in processed foods despite the fact that they also are thin boiling and are capable of forming gels, albeit softer gels than those from acid treatment of starches.

**Oxidized starches**
Thin-boiling starches produced by alkaline (sodium hypochlorite) treatment, but forming only soft gels.

*Cross-Linked Starches.*    The tendency for starches to undergo retrogradation during storage of starch gels is an undesirable characteristic. Various **cross-linked starches** have been developed to minimize this problem. Formation of a cross-linked starch requires substitution of the hydroxyl groups on two different molecules within the same granule, a chemical change that increases molecular size. Such granules are less likely to rupture, which helps to retain viscosity, as well as minimize stringiness of starch pastes. However, paste clarity and stability during cold storage are reduced. Cross-linkages may be formed under alkaline conditions, often by use of either acetic anhydride (to form acetylated products) or succinic anhydride. Both acetate and succinate cross-linked starches are useful in formulating salad dressings and related products.

**Cross-linked starch**
Starch produced under alkaline conditions, usually in combination with acetic or succinic anhydride; notable as a thickener and stabilizing agent that undergoes minimal retrogradation.

Cross-linked starches are effective thickening agents in a variety of products because they form gels that undergo limited retrogradation. In addition, these products (even acid-containing ones) exhibit little change from temperature, shaking, or other agitation during storage.

Hydroxyalkyl starches are made by reacting starch with ethylene (shown here), which can lead to hydroxyethyl addition at carbons 2, 3, and 6 if the reaction is allowed to proceed for an extended period. The replacement occurs first at carbon 2. Hydroxyethyl starches are used to a limited extent in the food industry as a thickening agent.

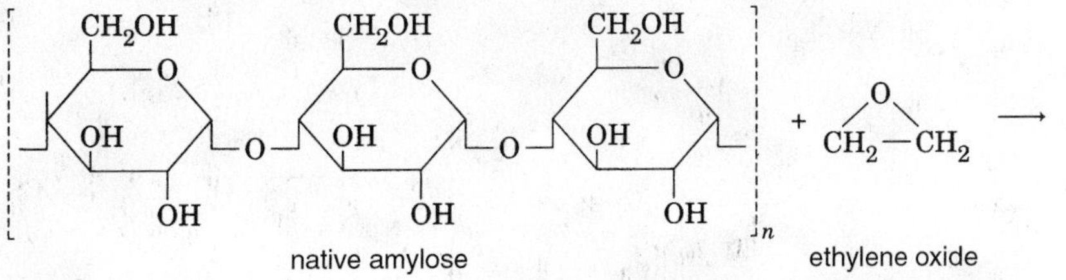

native amylose    ethylene oxide

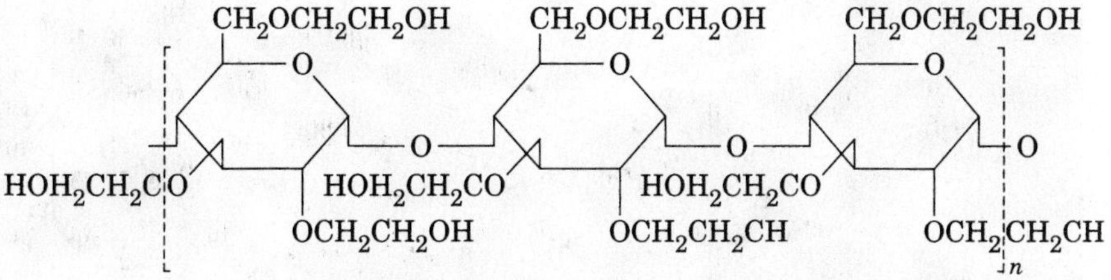

hydroxyethyl amylose (a cross-linked starch)

*Starch Phosphates.*    Yet another means of modifying starch chemically is esterification with phosphates, commonly sodium tripolyphosphate. **Starch phosphates** are valuable commercially because they increase the stability and improve the texture of starch pastes thickened with them. The repulsion between molecules containing the electrically charged phosphates accomplishes this action.

**Starch phosphates**
Starch derivative made by reaction with sodium tripolyphosphate or other phosphates to achieve a thickener with excellent stability and clarity.

The excellent clarity of starch products thickened with starch phosphates is one of the compelling reasons for their use in commercial food products. Another advantage of stabilized starches such as starch phosphates is their ability to reduce syneresis or weeping.

When a starch has been cross-linked and stabilized, its performance is outstanding: It forms smooth, nonstringy pastes and gels with little or no syneresis. Waxy starches benefit particularly from both cross-linking and stabilization processing during manufacturing.

***Spherical Aggregates of Starch Granules.***    Industrially, small starch granules and a little protein can be spray dried to form edible, spherical aggregates of a somewhat porous nature. This porosity is useful because flavorings can be entrapped in these granular spheres, which then serve as flavor carriers in selected food products. These aggregates can absorb a flavor or essence equal to as much as 60 percent of their weight. Without any protective coating, flavorings will be retained for several months. With a commercial tablet coating, flavors will be trapped for prolonged periods of storage.

## USES

Starch is used in a number of different food products because of its ability to thicken sauces and to bind fat. White sauces of varying viscosities (depending on their concentration of starch) form the basis for recipes ranging from cream soups to timbales. Thin white sauces are thickened with enough starch to create a soup that is about the same viscosity as cream. Medium white sauces and gravies have moderate flow properties but are distinctly thicker than a cream soup. Soufflés are made using a thick white sauce, one that flows just enough to be able to be folded into the egg white foam, yet not drain out of it before the structure sets during baking (Figure 9.7). Very thick white sauces are used as binding agents or edible pastes.

There are numerous other starch-thickened sauces. Some are used in the preparation of casseroles or as sauces over vegetables, while others may be served as dessert sauces and even in fruit pie fillings. When these sauces are hot, they will flow; but their viscosity will increase as they cool. In fact, some of them will form gels when they begin to approach

**Figure 9.7**    Soufflés are made by preparing a thick white sauce and folding it into an egg foam.

room temperature. When formulating starch mixtures, the temperature at which they will be served must be borne in mind so that the correct concentration of starch can be used to achieve the desired flow properties.

Starch-thickened puddings are familiar desserts. Some are made with cornstarch as the thickening agent, while tapioca may be the starch selected in others. Cream fillings in favorite pies such as banana cream, coconut cream, lemon, and chocolate are thickened with sufficient starch to form a gel after the filling has cooled.

Cereals of all types contain a considerable amount of starch, which means that part of the changes that occur during boiling of rice or other cereals will be gelatinization of starch. Gelatinization of starch is required in many recipes—for example, preparing oatmeal or other hot cereals for breakfast, boiling pasta, and making rice pilaf.

Commercially, starches are used in combination with some other component such as oil to serve as a fat replacer in appropriate products targeting the market for reduced fat and calories (see Chapter 12). Modified starches from potato, oats, rice, corn, wheat, and tapioca include ClearJel®, Amalean® I and II, Instant Sellar™, and N-Lite®, to name a few. Depending on the product, these modified starches may be used in conjunction with proteins, gums, or emulsifiers in making reduced-fat and fat-free sauces, processed meats, frozen desserts, salad dressings, and other food products.

---

## FOOD FOR THOUGHT: Resistant Starch

**Resistant starches** are not digested in the small intestine and, therefore, enter the large intestine intact. There they act as fiber to promote motility while also undergoing limited digestion to yield some energy. These attributes make them useful in formulating food products for the large market of consumers seeking reduced calories and more fiber in their diets.

Resistant starches are classified as RS1 (starch trapped in cells, as in coarsely ground grains or seeds and legumes), RS2 (native starch granules, as in raw potatoes, bananas, and waxy maize), RS3 (crystalline, nongranular starch, as in gelatinized and cooled potatoes), and RS4 (chemically modified or re-polymerized starch). During food processing, resistant starches may be changed sufficiently to become digestible, which reduces their effectiveness as fiber. Processing changes RS4 starches less than the other types. Fiber content analysis for labeling purposes may need to be done in a laboratory to ascertain such changes.

The K-State Research Foundation patented a process that researchers developed there to convert starch from grains (including corn, wheat, rice, and oats), tapioca, potatoes, and mung beans into resistant starch. The merits of this process were to reduce calories and increase fiber in products made with any of the resistant starches. Through licensing agreements, these special starches are being produced by MGP Ingredients, Inc. (various Fibersym™ starches from wheat and potatoes) and Cargill (ActiStar™ from tapioca). They are rather easily incorporated into baked products because they have low water-holding capacity, which helps prevent gumminess during mixing. Tapioca resistant starch has the additional advantage of contributing virtually no detectable flavor.

Novelose®, a resistant starch developed by National Starch Innovation, is identified on ingredient labels as maltodextrin and is an RS3 starch. It has a unique ability to contribute desired texture to crisp processed snacks and cereals while also contributing a significant amount of dietary fiber. Unlike some other fiber sources such as bran, Novelose does not contain fat and thus avoids potential flavor changes due to rancidity. Dough stickiness can cause manufacturing problems when various sources of fiber are included in the formulation, but Novelose takes up less water, and the dough containing it is easy to handle. Its small, white granules promote formation of comparatively light, even-textured baked products. This unique resistant starch, which contributes less than 3 calories/gram, is well suited for use in baked products to be marketed as high in fiber.

**Resistant starches**
Starches that are not digested until they enter the large intestine.

The Agricultural Research Service (ARS) of the USDA is sponsoring research on rice starch that has resulted in development of a slowly digested starch that may have potential in sport beverages, diabetic foods, and reduced-calorie or fat-free items. Another project sponsored by ARS was the development of Fantesk™, an emulsion of oil droplets encapsulated by starch by a special co-jet cooking process. Fantesk may have future applications in a wide range of reduced-fat products.

## CRITERIA FOR SELECTING A STARCH

Obviously, many different starches can be considered for inclusion in commercial food formulations. No single starch is ideally suited to the entire array of products that use starches. Instead, the starch or starches need to be well suited to the specific requirements of the item being formulated.

If a product is made for people seeking a low-calorie option, the starch needs excellent thickening and/or gelling capabilities for the amount of starch used. This enables the product to be thickened adequately with fewer calories than if a greater amount of another starch had to be used.

Mouthfeel is another possible consideration. A gummy or stringy feeling in the mouth may be quite evident in a sauce with soft ingredients but perhaps would not be noted if water chestnuts or other distinctive textural components were included in the sauce.

**Freeze-thaw stability** is the ability of a starch-thickened product to be frozen and thawed without undesirable loss of quality. Syneresis can be a problem when some products have been frozen and thawed. If the item is a stew or other dish that will be reheated and stirred, syneresis presents no problem, even when there has been a considerable amount of drainage. The liquid will all reunite in the sauce during the heating and stirring period. However, if a starch-thickened cream pie filling has been frozen and then exhibits considerable syneresis after thawing, the crust will be soggy, and the quality of the pie will be poor. In this situation, there is no opportunity to recombine the liquid.

Cornstarch and wheat starch have poor freeze-thaw stabilities. In contrast, waxy rice flour is a particularly useful starch for products that are to be frozen and thawed, because this type of starch shows little tendency to form crystalline areas. Apparently, the protein in waxy rice flour is partially responsible for its excellent freeze-thaw stability. Purified rice starch is better than many of the other starches for freeze-thaw stability, but it is not as good as waxy rice flour.

Unless starch-thickened products are to be frozen and thawed before use, syneresis is not important in selection of the type of starch. Also, freeze-thaw stability of the starch used in frozen starch-containing products that are heated and stirred before being served is not important because the crystalline nature of the product is eliminated by the heating.

Retrogradation is a concern in starch-thickened items, such as a pudding that is to be in refrigerator storage and eaten without reheating. Starches that retrograde easily will quickly develop a rather gritty texture that is noticeable unless the food is warmed sufficiently to break the bonds holding the crystalline structure.

Cross-linked starches resist forming crystalline areas in amylose-containing gels. For cross-linked starches to undergo freezing and thawing and still retain a stable gel structure, some of the hydroxyl groups on the amylose molecules must be either acetylated or propionylated. These slightly larger functional groups inhibit the closely ordered alignment of amylose molecules that causes the crystalline areas to form during and after freezing.

Viscosity of starch pastes and the firmness of gels are properties that need to be considered, particularly if special equipment demands need to be met during production of the item (e.g., gum drops). An example is the need in some operations for a starch sol with little viscosity so that it will flow readily through the pipes involved in the production setup.

Marketing stresses also may influence the strength of a starch gel that is needed in a pudding or pie filling. The flow characteristics of starch-thickened sauces and other sols are important to developing successful products. Products based on starch gels also require use of a starch that delivers the rigidity and other textural properties consumers expect.

**Freeze-thaw stability**
Ability of a starch-thickened product to be frozen and thawed without developing a gritty, crystalline texture.

**Table 9.6**    Advantages and Limitations of Selected Starches

| Starch | Advantage | Limitation | Use |
|---|---|---|---|
| Potato | Excellent thickening; translucent | Thins with long heating; stringy | Pudding |
| Corn | Translucent; satisfactory gel | Moderate thickening | Puddings, pie fillings, gravy, sauces |
| Waxy corn | Will not gel | Stringy | Fruit pie filling |
| High-amylose | Extrudes in thin films | Will not form traditional gel | Edible films |
| Pre-gelatinized | Swells without heat | Reduced thickening | Instant foods |
| Cold water–swelling | Improved texture, good stability | Reduced thickening | Microwavable sauces and entrèes |
| Thin-boiling | Very fluid sol; strong gel | Gummy texture | Gum drops |
| Oxidized | Thin-boiling | Soft gel | Stored starch products |
| Cross-linked | Little retrogradation | | |
| Phosphates | Little syneresis; nonstringy; clear | | With waxy starches |

These are some factors that need to be weighed when selecting a starch. Specific products may have refinements of these qualities that need to be examined carefully. In fact, in many instances, there may be a need to consider not only what kind of starch to use, but perhaps whether some other form of thickening would be preferable (see Chapter 10 for information on gums, which increasingly are replacing starches in some food applications). Table 9.6 provides information regarding some of the advantages and limitations of selected starches.

## RICE AND ITS STARCH

Rice has been the principal cereal in many Asian cultures for countless generations (Figures C29–C34), and now its popularity is gaining immensely as an alternative to potatoes as the starch source of choice in American menus, too. The known varieties of rice exceed 40,000 in number. Many of these are not of commercial importance, but about 20 are grown in this country.

Categorization of rice types is based on the length of the grain: long-grain rice measures between 6 and 7 mm, medium grain is 5–5.9 mm, and short grain is less than 5 mm. The starch in long-grain rice is higher in amylose than the other two lengths of grains. Since amylose absorbs less water than amylopectin during cooking, long-grain rice stays more fluffy and distinct than short- or medium-grain rice. These latter two absorb water more readily because of their higher amylopectin content; this water increases rupturing of granules and causes short- or medium-grain to be more moist and sticky than the long-grain rice.

Specialty rices are gaining a place in American cuisine and often are imported from various countries around the world (Figure 9.8). Waxy rices (virtually 100 percent amylopectin) are of two varieties in the United States. **Mochigome** is sticky and pasty when cooked. These characteristics are useful in making certain Oriental noodles and confections. **Calmochi** is the other waxy rice, and its characteristics are similar to those of mochigome. Either of these waxy rices, also called **sweet glutinous rice**, is suited for batter coatings used in frying and in making crisp pizza crusts. **Koshihikari** is a short-grain rice with a trace of sweetness that is prized in Japan for making sushi.

**Arborio** is a specialty rice that is a particularly absorptive, medium-grain rice well suited for preparing risotto and paella. **Basmati** has a remarkably long grain and a fragrance that carries into the flavor qualities of this rice from India. **Jasmine** also is a long-grain specialty rice that is appreciated for its aromatic reminder of the flower. Its uniqueness is that it stays soft in the refrigerator rather than undergoing the retrogradation typical of other long-grain rices when they are stored after cooking.

Some specialty rices are of various colors. For example, black japonica was created by breeding a black short grain and a reddish brown medium grain to create a dark rice with a

**Mochigome**
Waxy (high amylopectin), sticky rice.

**Calmochi**
Variety of rice that is sticky when boiled because of its high amylopectin content.

**Sweet glutinous rice**
General name for waxy, sticky rices; mochigome and calmochi are examples.

**Koshihikari**
Short-grain, sweet-flavored Japanese rice used to make sushi.

**Arborio rice**
Medium-grain rice used for making paella and risotto.

**Basmati rice**
Long-grain, aromatic rice.

**Jasmine rice**
Long-grain, aromatic rice; resists retrogradation in storage.

**Figure 9.8**    Among the various imported types are red (left) and black (right) rice.

trace of spiciness. Chinese black rice changes to a deep purple when cooked; its somewhat sweet taste makes it a good choice in desserts. Purple sticky (alias black Thai) sweet-tasting rice also is dark before cooking, but becomes an indigo blue after boiling. French red rice is grown in salty soil around Camargue, France. To be marketed as Camargue rice, red rice must be grown there because it has been designated by the European Commission as a Protected Geographical Indication.

## SUMMARY

Starch is composed of a linear fraction, amylose, and a highly branched fraction of large molecular weight, amylopectin. These fractions are deposited in granules in the leucoplasts of some cereal grains and roots and some other food sources; about 75–80 percent of the granule is amylopectin, and the remainder is amylose.

Gelatinization is the imbibition of water into the starch granules when the starch is heated in water; this is accompanied by some leaching of amylose molecules into the surrounding water. Different types of starch exhibit slightly different characteristics in their gelatinization, gelation, and retrogradation during storage. Sugar delays gelatinization and competes with the starch for water, resulting in decreased viscosity, more tender gels, and increased translucence. Acid causes some hydrolysis of starch when it is present during gelatinization, producing some loss in viscosity, especially below pH 4.

Gelation (formation of a gel) occurs when many starch pastes cool. The specific characteristics of starch gels are influenced by the type and concentration of the starch, the extent of heating, agitation, and the presence of acid and/or sugar. During storage, gels may exhibit syneresis, and some develop a gritty texture when crystalline areas form (retrogradation) as the amylose and amylopectin molecules organize themselves more tightly. Retrogradation may be a particular problem when starch gels are frozen and thawed.

Dextrinization is a chemical breakdown (hydrolysis) of starch molecules resulting from intensive dry heat. The shorter molecules have reduced thickening ability.

Native starches used in foods include cornstarch, wheat, rice, potato, tapioca, and waxy cornstarch. Modified starches perform special functions: pre-gelatinized starches reconstitute quickly to thicken foods without additional cooking; thin-boiling starches (treated with acid to reduce branching of amylopectin) are poured easily and allowed to set to form a strong gel; cross-linked starches form thickened starch gels and sauces that have a pleasing, non-gritty texture; some have a role as a fat replacer; and starch phosphates are effective thickening agents that form nonstringy pastes and gels with little, if any, syneresis.

The properties of the many varieties of rice are a reflection of their starch composition. Long-grain rice has comparatively high amylose content and, therefore, absorbs less water and is less sticky than the medium- and short-grain rices. Waxy (sweet glutinous) rice is extremely sticky because its starch is almost entirely amylopectin. Various specialty rices provide subtle flavors and colors.

## STUDY QUESTIONS

1. Select five recipes in which starch is used as a thickening agent. Identify the type of starch available to consumers that is best suited to each recipe and explain the reason for the choice. Similarly, select the type of starch available to the food industry that is best suited to a comparable food product and state the rationale for the selection.

2. Sketch a potato starch granule in its native state and then sketch it as it undergoes gelatinization. Write a description of the gelatinization process.

3. What difference can be predicted to exist between a starch-thickened pudding made with lemon juice and a comparable one prepared without lemon juice? Write the chemical reaction involved in the first product.

4. What is the difference between wheat starch and wheat flour? How does this difference influence the use of the two products in cookery?

5. Describe in detail the processes of gelatinization, gelation, and retrogradation.

6. Identify an appropriate type of rice for preparing each of the following: (a) side dish to be eaten with chopsticks, (b) sushi, (c) dessert, (d) rice casserole.

7. True or false. Gelation is another word for gelatinization.

8. True or false. The linear fraction of starch is amylose.

9. True or false. The linkages in amylose are $1,6\text{-}\alpha$-glucosidic.

10. True or false. *Dendritic* is a word used to describe the structure of amylopectin.

11. True or false. Gelatinization requires dry heat.

12. True or false. Potato starch will become less viscous if it is heated above 80°C.

13. True or false. Waxy cornstarch will not gel.

14. True or false. Gelatinization is a reversible action.

15. True or false. Retrogradation is a reversible action.

## BIBLIOGRAPHY

Berry, D. 2003. New times for cereals. *Food Product Design 13* (1): 35.

Berry, D. 2005. Through thick and thin. *Food Product Design 15* (2): 34.

Darling, K. 2009. Starch on the side. *Food Product Design 19* (11): 46.

Davis, R. C. 1998. The new starches. *Food Product Design 7* (11): 40

Decker, K. J. 2003. Souper mixes. *Food Product Design 13* (2): 37.

Decker, K. J. 2009. Getting sauced: Pasta sauce formulating secrets. *Food Product Design 19* (12): 18.

Duffy, R. 2001. *Specialty Rices of the World*. FAO and Science Publishers, Inc. Enfield, NH.

Fitt, L. E. and Snyder, E. M. 1984. Photomicrographs of starches. In *Starch, Chemistry and Technology. 2nd ed*. Whistler, R. L., et al., eds. p. 675. Academic Press. Orlando FL.

French, D. 1984. Organization of starch granules. In *Starch, Chemistry and Technology. 2nd ed*. Whistler, R. L., et al., eds. p. 183. Academic Press. Orlando, FL.

Kulp, K., et al. 1991. Functionality of carbohydrate ingredients in bakery products. *Food Technol. 45* (3): 136.

Kuntz, L. A. 2005. A starch that's hard to resist. *Food Product Design 15* (6): 19.

Meng, Y. and Rao, M. A. 2005. Rheological and structural properties of cold-water-swelling and heated cross-linked waxy maize starch dispersions prepared in apple juice and water. *Carbohydrate Polymers 60*: 291.

Pszczola, D. E. 1999. Starches and gums move beyond fat replacement. *Food Technol. 53* (8): 74.

Pszczola, D. E. 2001. Rice: Not just for throwing. *Food Technol. 55* (2): 53–59.

Pszczola, D. E. 2003. New ingredient developments are going with the grain. *Food Technol. 57* (2): 46.

Pszczola, D. E. 2006. Which starch is on first? *Food Technol. 60* (4): 51.

Skillicorn, A. 2003. The pie and pastry filling picture. *Food Product Design 12* (11): 91–107.

Warner, K. et al. 2001. Use of starch lipid composites in low-fat ground beef. *Food Technol. 55* (2): 36.

## INTO THE WEB

*http://www.starch.dk/isi/starch/glossary.asp*—Glossary of terms relating to starch.

*http://sci-toys.com/ingredients/starch.html*—Overview of starches.

*http://food.oregonstate.edu/learn/starch.html*—Discussion of starches and gelatinization.

*http://www.foodproductdesign.com/articles/1996/01/understanding-starch-functionality.aspx*—Overview of starch functionality.

*http://www.google.com/search?hl=en&client=firefox-a&rls=org.mozilla:en-US:official&q=starch+gelatinization&start=20&sa=N*—PowerPoint on starch behavior.

*http://www.google.com/search?hl=en&client=firefox-a&rls=org.mozilla:en-US:official&hs=sps&q=starch+gelatinization&start=10&sa=N*—PowerPoint on starch.

*http://www.palmtree-species.com/2008/05/extracting-starch-from-sago-palm.html*—Describes getting starch from the sago palm in New Guinea.

*http://eatingasia.typepad.com/eatingasia/2008/03/the-tree-of-lif.html*—Numerous photos of the making of sago palm starch.

*http://carillon.up.edu.ph/?p=416*—Possibilities for producing sago palm starch in Malaysia.

*http://ift.confex.com/ift/2004/techprogram/paper_24037.htm*—Formation of edible starch films.

*http://cerealchemistry.aaccnet.org/doi/abs/10.1094/CC-82-0131*—Report on permeability of edible starch films.

*http://www.google.com/search?q=edible+starch+films&ie=utf-8&oe=utf-8&aq=t&rls=org.mozilla:en-US:official&client=firefox-a*—Physicochemical properties of some edible starch films.

*www.aaccnet.org/cerealchemistry/abstracts/2001/0606-06R.asp*—Abstract of article, Effects of preparation temperature on gelation properties and molecular structure of high-amylose maize starch by E. Vesterinen, T. Suortti, and K. Autio to *Cereal Chemistry*, accepted 2001.

*http://apps3.fao.org/jecfa/additive_specs/docs/9/additive-0840.htm*—Information on modified starches.

*http://mysbfiles.stonybrook.edu/~ymeng/publications/Meng_CarboPolym_2005.pdf*—Article on cold-water-swelling and heated cross-linked waxy maize starch dispersions.

*http://www.aaccnet.org/meetings/2001/Abstracts/a01ma277.htm*—Paper on oxidized starches.

*http://www.cassava.org/Poland/Modification.pdf*—Overview of various starch products made in Thailand from cassava.

*http://www.preparedfoods.com/Articles/Feature_Article/BNP_GUID_9-5-2006_A_10000000000000234686*—Article on cross-linked starches.

*http://www.foodinnovation.com*—National Starch and Chemical Company product information.

Farmers' markets are a fine source of fresh vegetables and fruits.

# CHAPTER 10

# Vegetables and Fruits

## Chapter Outline

## OBJECTIVES

After studying this chapter, you will be able to:

1.  Describe the structures of fruits and vegetables.
2.  Identify the structural carbohydrates in plants and the changes they undergo.
3.  Discuss pigments and the changes that occur during various treatments.
4.  Review the various gums and their applications.

## INTRODUCTION

Plants provide considerable variety in the food supply, both in terms of gustatory aspects and nutrition. Fruits and vegetables are found in ever-greater abundance in the marketplace as world trade and nutritional awareness have prompted consumers to seek a variety of fruits and vegetables to optimize their health.

Fresh produce is available in most markets throughout the nation at any time of year. However, there still is a large market for both fruits and vegetables in cans, as frozen items, and even dried. Legumes, technically vegetables, are unique among vegetables. Cereals are grains and certainly differ from other types of plant foods, but serve as a mainstay of the diet. Various plants are also the sources of many of the fats and oils used today.

## A Different Meaning to Overhead

*Overhead* is the word grocers use to designate the cost of doing business, but workers who harvest coconuts in southern India and other regions where they are a commercial crop think of overhead as the place where they work (Figures C22 and C23). Unlike trapeze artists in a circus, these laborers climb sometimes as high as 100 feet up the trunk to cut the ripe coconuts loose to fall to the ground far below (Figure 10.1). Skilled workers climb the heights to determine which of the approximately 60 coconuts are ripe enough and cut them loose. Unlike most tree fruits, coconuts must be harvested all year long because flowering and fruit development occur continuously. Therefore, decisions regarding maturity need to be made on each bunch of coconuts.

The only thing protecting workers is a somewhat fragile-looking palm-frond tether attached to the tree, which hopefully will stop their fall if they slip. Unfortunately, falls

**Figure 10.1**    Harvesting coconuts is a daunting job that requires workers to climb up the trunk of the tree as high as 100 feet to the crown where the fruits forms in clusters. (Courtesy of Hema Latha.)

do happen in this very dangerous workplace, which historically suggests why most of the coconut harvesters in southern India used to be from the group called untouchables. However, efforts there to erase the caste system have led to some social mobility, which means that many men who previously had little chance to avoid picking coconuts now have other career options that are less dangerous. This important social change has resulted in a shortage of workers to harvest the fruit. The need for harvesting machinery to replace workers is spurring efforts to design equipment that can do the job successfully, but success is elusive because of the problem of determining which coconuts are ready for harvest.

The sea, as well as the land, is an important plant food source. Seaweed has played a prominent role in Japanese dishes for centuries. Now, gums obtained from seaweed are gaining in importance. This chapter examines these and other diverse plants and the foods obtained from them, including dietary supplements. The polysaccharides found in various plant foods and the structure and characteristics of the edible parts of plants are emphasized.

## STRUCTURE OF FRUITS AND VEGETABLES

Fruits and vegetables are composed of tissue systems and various types of cells. The overall framework or design differs from one specific type to another, yet the tissue systems have certain basic similarities. There obviously are differences between the type of cell found in a pear and that in a mango. However, noticeable differences also exist between the underripe, ripe, or overripe stages of maturity in the same type of fruit.

### Tissue Systems

The tissues in fruits and vegetables are designated in three different systems: dermal, vascular, and ground. The dermal tissue (skin or rind) is the protective covering of each portion of the plant. Within this dermal tissue is the vascular system, the system responsible for transport of fluids, nutrients, and waste products. The remaining inner portion is the ground system. Details of these systems vary with the specific plant.

The **dermal system** undergoes changes during development and maturation of the plant, yet the protective function remains constant. The layer of **epidermal cells** forms a thin, protective surface coating. Evaporation of moisture from the surface is minimized as a result of a layer of cutin and sometimes various waxes.

Potatoes and other tubers develop a somewhat cork-like protective coating, the **periderm**. Even more protection is developed in peas, which have strongly supported epidermal cells (supported by hemicellulose) and then a **hypodermal layer** with considerable intercellular spaces.

The **vascular system** is composed of two parts: the **xylem** and the **phloem** (Figure 10.2). The xylem usually is composed of elongated, tubular cells and is the portion of the vascular system that moves water. The phloem tissue is responsible for transporting organic matter in solution.

The remainder of the fruit or vegetable, which constitutes much of the edible portion, is the **ground system**. The bulk of the ground system consists of the parenchyma cells. Supporting tissues may contain collenchyma and sclerenchyma cells.

### Parenchyma Cells

The principal type of cell in fruits and vegetables is the **parenchyma cell**, which is the most abundant type of cell in the ground system. These cells are polyhedral, ranging from 11 to as many as 20 faces. The exact number of faces is specific to each species and so is the tightness of the fit between these cells. Intercellular space depends on the fit of the cells.

**Dermal system**
Outer protective covering on fruits and vegetables, as well as other parts of plants.

**Epidermal cells**
Layer of cells providing a continuous outer covering for fruits and vegetables.

**Periderm**
Layer of cork-like cells protecting vegetable tissues underground.

**Hypodermal layer**
Layer of cells beneath the epidermal cells.

**Vascular system**
System in plants that transports water and other essential compounds; composed of the xylem and phloem.

**Xylem**
The water transport system in plants; the tubular cells that move water.

**Phloem**
Portion of the vascular system that transports aqueous solutions of substances such as nutrients.

**Ground system**
Bulk of edible portion of plant foods.

**Parenchyma cells**
Predominant type of cell in the fleshy part of fruits and vegetables.

**Figure 10.2**    Cross section of beet. Note the vascular system (phloem and xylem) in enlarged granules appearing in an orderly arrangement in concentric rings.

**Middle lamella**
Region between adjacent cells that cements the cells together; composed mostly of pectic substances.

**Plasmalemma**
Thin membrane between the cell wall and the interior of the cell.

**Mitochondria**
Organelles in cells involved in respiration and other biochemical processes.

**Plastids**
Organelles in the cytoplasm that contain pigments or starch.

**Chloroplast**
Type of plastid containing chlorophyll.

**Chromoplast**
Type of plastid containing carotenoids.

**Leucoplast**
Type of plastid in which starch is formed and deposited in granules.

**Tonoplast**
Membrane separating the protoplasm from the vacuole in a parenchyma cell.

**Vacuole**
Portion of the cell containing most of the water, flavoring components, nutrients, and flavonoid pigments.

The parenchyma cells of potatoes fit together in a manner resembling the tight fit seen in ancient Incan walls, with the result that only about 1 percent of the volume of potatoes is due to intercellular air spaces. In contrast, apples do not have tightly fitting parenchyma cells. Consequently, apples have a much more open, slightly loose texture; about 25 percent of the volume of apples is intercellular air space, which is why they float.

A look at parenchyma cells begins with the **middle lamella**, the material between adjacent parenchyma cells that serves to hold the cells in a fixed position. The middle lamella is composed primarily of pectic substances, discussed in a later section.

The primary cell wall is composed of several complex carbohydrates, including cellulose, hemicelluloses, pectic substances, and noncellulosic polysaccharides. Between the cell wall and the interior of the cell wall is a thin membrane, the **plasmalemma** (Figure 10.3).

The protoplasm is found immediately within the plasmalemma of the cell. Within this are subcellular structures or organelles of several types: plastids, **mitochondria**, and the nucleus. The mitochondria are of great significance in fresh produce because of their involvement in respiration. Enzymes associated with them are responsible for catalyzing numerous biochemical reactions.

**Plastids** are of special interest, for it is in these structures that two types of pigments and starch are formed and stored. There are three types of plastids: chloroplasts, chromoplasts, and leucoplasts. **Chloroplasts** are the plastids in which chlorophyll is found. **Chromoplasts** contain the carotenoid pigments. It is in the **leucoplasts** that starch is made and stored in the form of granules. The relative proportion of the cell occupied by the protoplasm is reduced gradually as the plant matures, but the important functions that take place in the protoplasm continue throughout the lifetime of the cell.

Separating the protoplasm from the remaining interior of the cell is a thin membrane called the **tonoplast**. The vacuole inside the tonoplast increases in size during the maturation process and accounts for a large fraction of the cell interior at maturity.

The **vacuole** contains a variety of substances. This is the portion of the cell in which the numerous flavor components, including sugars, acids, and many other organic compounds contributing to the flavor are found. In addition, the flavonoid pigments (notably the anthoxanthins and anthocyanins) and some of the nutrients, including protein, are located in the vacuole. It also holds about 90 percent of the cellular water.

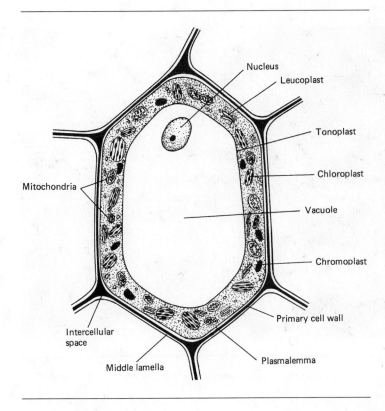

**Figure 10.3**   Diagram of a parenchyma cell.

## Collenchyma Tissue

Elongated cells such as those in the fibrous strands of celery are examples of **collenchyma tissue**. These cell walls are chewy and resistant to much softening when cooked. They are supporting tissue and contribute to the overall structure of the edible parts of fruits and vegetables containing collenchyma cells (Figure 10.4).

**Collenchyma tissue**
Aggregates of elongated collenchyma cells providing supportive structure to various plant foods, notably vegetables.

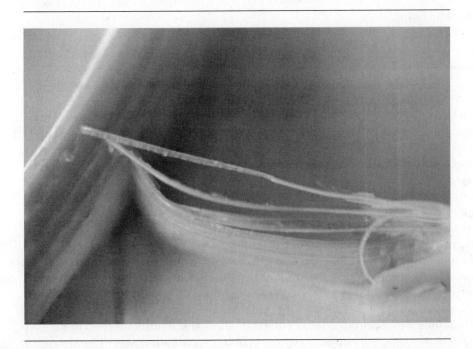

**Figure 10.4**   Collenchyma tissue in the fibers of celery is comprised of elongated collenchyma cells.

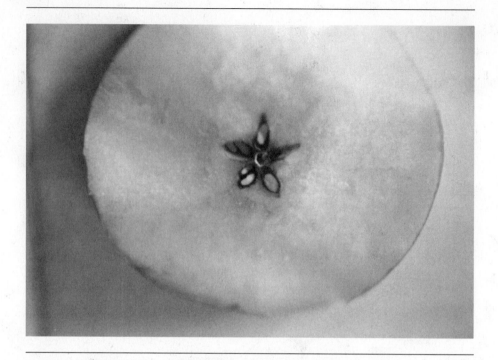

**Figure 10.5**    Sclereids (unique sclerenchyma cells) contribute to the somewhat granular texture noted in pears.

## Sclerenchyma Cells

**Sclerenchyma cells**
Unique supportive cells with a chewy, fibrous character.

**Sclereid**
Type of sclerenchyma cell that gives the somewhat gritty texture to pears and certain other fruits.

**Sclerenchyma cells** are woody cells that contribute a somewhat gritty texture to some plant foods. The unique texture of pears, for example, is the result of the presence of **sclereids**, a type of sclerenchyma cell. Both sclereids and the other type of sclerenchyma cells (often simply called fibers) have thick cell walls containing lignin, a wood-like compound. The fibers in asparagus and green beans are examples of the "fiber" type of sclerenchyma cell (Figure 10.5).

### Papayas and Sex

When one looks at a papaya tree quietly growing in a field, sex ordinarily is not a topic that comes to mind (Figure 10.6). In fact, who would even think that it had a sexual orientation? Surprisingly, some trees are male, others are female, and about one in five is a hermaphrodite. The unfortunate fact is that only hermaphrodites can be counted on to produce a good crop of fruit, because female trees must have pollen from a male if they are to bear fruit, and that pollen often does not arrive. Clearly, the best arrangement would be to plant only hermaphrodites, but these plants cannot be identified until they mature and flower. This means that farmers waste land and resources growing many unproductive plants just to produce some fruit.

This situation has caught the attention of scientists, and some researchers now are unwilling to let papayas decide their own fate. The first step was to identify the genome, which was drafted in 2008; then the X, Y, and $Y^h$ chromosomes became the focus to try to figure out a way to create the desired $X\,Y^h$ chromosome in seeds so that only hermaphrodite seeds can be planted. If scientists succeed in their quest, papaya farmers will be spared the problem of dealing with the sex of their crop.

**Figure 10.6**   Several papayas are growing near the top of this mature tree in Kenya.

# CARBOHYDRATE STRUCTURAL CONSTITUENTS

## Cellulose

**Cellulose** contributes to the structure of foods, although the pectic substances and hemicelluloses also are key compounds in determining the unique textural characteristics of specific fruits and vegetables. Cellulose is a glucose polymer, but unlike amylose, its glucose units are joined by 1,4-β-glucosidic linkages, as shown here. This difference in linkage is responsible for textural differences and also for the inability of people to digest cellulose as a practical source of energy. Cellulose molecules can aggregate into fibrils with a somewhat crystalline structure. Parallel clustering of these fibrils occurs in some fibrous vegetables.

cellulose

## Hemicelluloses

Although present in distinctly smaller quantities than cellulose, the **hemicelluloses** still are important structural components of the cell walls. Their chemical nature is quite heterogeneous. Unlike starch and cellulose, both of which are polymers containing glucose exclusively, the hemicelluloses contain a variety of sugars in their long chains. In fact, they even contain both pentoses and hexoses.

Xylose, a five-carbon sugar, combines with glucuronic acid in limited amounts to form a common hemicellulose polymer called **xylan**. Another of the pentoses, arabinose, is the primary constituent of the arabans, which are hemicelluloses that also contribute to the structure of plant foods. Xylan and araban are two of the hemicelluloses that are particularly common, yet they are present in much smaller quantities than cellulose.

xylose    arabinose    glucuronic acid

Hemicelluloses are matted together with the pectic substances to serve as a connection between the fibrillar cellulose in cell walls (Figure 10.7). An alkaline medium has a strong effect on hemicelluloses, and vegetables heated in cooking water to which soda is added quite quickly become flaccid and mushy. Even though hemicelluloses are present in much smaller amounts than cellulose, this change resulting from an alkaline medium quickly makes vegetables and fruits unacceptable. Furthermore, the destructive effect of soda on thiamine is sufficient to discourage the use of soda in cooking vegetables, especially legumes, which are good sources of thiamine.

## Pectic Substances

The **pectic substances** constitute a unique group of polysaccharides that are polymers of galacturonic acid, an organic acid in which the carbon external to the ring structure of galactose is combined with oxygen and hydrogen to form the organic acid radical rather than the alcohol radical of galactose.

galactose    galacturonic acid

**Figure 10.7** Many types of legumes (high in hemicelluloses, as well as in starch and protein) are sold in this Kenyan market.

The middle lamella between cells is made up of pectic substances, which change gradually to different forms during the maturation process. Significant textural changes accompany the transitions from one form to the next in the middle lamella. Pectic substances also are important in the primary cell wall, where they combine with the hemicelluloses to reinforce the structural contribution of cellulose.

***Definitions.*** Although there is no definite line of demarcation between types of pectic substances, specific terms designate the various forms. *Pectic substances* is the general term for any member of this family of polygalacturonic acid compounds. This term includes protopectin, pectin, pectinic acid, and pectic acid. Molecular weights range as high as 400,000, a figure indicative of the lengthy polymers involved.

**Protopectin** is the water-insoluble form of pectic substances occurring in immature fruits and, to a lesser extent, in vegetables. This pectic substance contributes significantly to the firm texture of unripe fruits. Essentially, protopectin is a very long polymer consisting of galacturonic acid units joined by 1,4-α linkages:

**Protopectin**
The form of pectic substances found in unripe fruits and some vegetables; a methylated, very long polymer of galacturonic acid.

**Pectins**
Galacturonic acid polymers in which most, if not all, of the acid radicals have been esterified with methanol; valued for gel-forming properties in making jams and jellies.

**Pectinic acids**
Galacturonic acid polymers in which between a fourth and a half of the acid radicals have been esterified with methanol; form of pectic substances formed as fruit begins to soften just a little.

protopectin

As fruit ripens, some demethylation and hydrolysis occur, seemingly randomly, along the protopectin molecules. When only a limited amount of degradation has occurred, the pectic substances are termed **pectins** (Figure 10.8). As additional demethylation occurs and hydrolysis continues, **pectinic acids** form.

pectinic acid

If a pectin molecule is esterified completely—that is, all organic acid residues are altered to methyl esters—the pectin will contain a little more than 16 percent methoxyl groups. Most pectins contain fewer than 16 percent methoxyl groups. In contrast to this are the **low-methoxyl pectinic acids (also referred to as low-methoxyl pectins)**, which are valued for their ability to form a gel structure with very little sugar, a characteristic of significance in making dietetic jams and jellies. Low-methoxyl pectinic acids contain methyl esters on no more than every fourth unit, and frequently only about one in eight of the galacturonic acid units has been esterified.

**Low-methoxyl pectinic acids (low-methoxyl pectins)** Galacturonic acid polymers in which only between an eighth and a fourth of the acid radicals have been esterified with methanol; pectic substances found in fruit that is just beginning to ripen.

low-methoxyl pectinic acid

**Figure 10.8** The albedo (white layer) in the skin of citrus is high in pectin.

Physical properties are modified as protopectin evolves into pectin and then pectinic acid. Of particular importance is the transition from a methylated, water-insoluble polymer (protopectin) to a shorter, methylated compound capable of being dispersed easily in water (pectin). Because of this change, it is feasible to heat fruit juices, pectin, and sugar and then let them cool to form pectin gels.

Not only are the cooking properties of pectins different from those of protopectin, but pectin has a significant effect on the texture of raw fruits. Demethylation and hydrolysis gradually occur during ripening, and protopectin is transformed into pectin. The fruits gradually soften from a very hard texture when green to a firm, but yielding texture in ripened fruits.

Pectinic acids form salts with calcium and some other ions. Of particular interest is the formation of gels utilizing low-methoxyl pectins and calcium ions. The salts formed when ions combine with the organic acid radicals in various pectins and pectinic acids are called **pectinates**. These pectinates are capable of playing an important structural role in forming gels.

As the degradation of pectinic acids continues, the molecules gradually become shorter and lose all of their methoxyl groups. These shorter polymers of galacturonic acid are designated as **pectic acid**. Pectic acid is found in overly ripe, very soft fruits and vegetables. This type of pectic substance has lost the gel-forming ability characteristic of the longer methyl esters of galacturonic acid polymers.

**Pectinates**
Compound resulting from the combination of pectinic acids or pectins with calcium or other ions to form salts that usually enhance gel-forming capability.

**Pectic acid**
The smallest of the pectic substances and one lacking methyl esters; occurs in overripe fruits and vegetables; incapable of gelling.

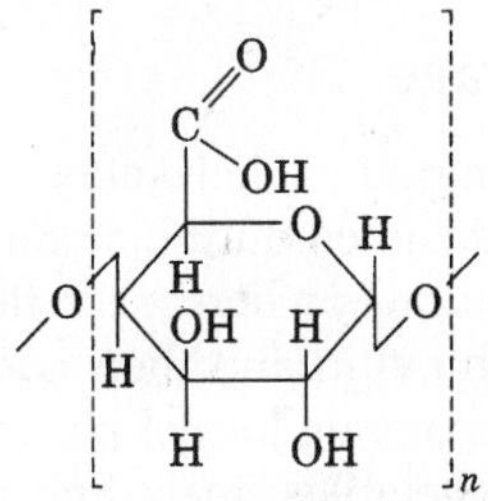

pectic acid

The transition from the long polymers of protopectin to somewhat shorter pectins and pectinic acids is a gradual one that is catalyzed by enzymes and organic acids. **Protopectinases** promote the shortening of the polymeric chains of protopectin to the shorter chains of pectins by the addition of a molecule of water at random locations between galacturonic units. This chain shortening by hydrolysis is essential to the development of the physical properties of pectins.

Yet another change, demethylation of protopectin, occurs gradually during the ripening of fruits and vegetables. The enzymes responsible for the splitting off of the methyl groups collectively are known as **pectinesterases**. Their action to remove methoxyl groups from the structures of the pectins and pectinic acids causes some loss of the gel-forming properties of the compounds and a distinct softening in the structure of the fresh fruit itself. The pectinesterases include pectin methoxylases, pectases, and pectin demethoxylases.

The reaction for demethylation of the galacturonic acid unit is:

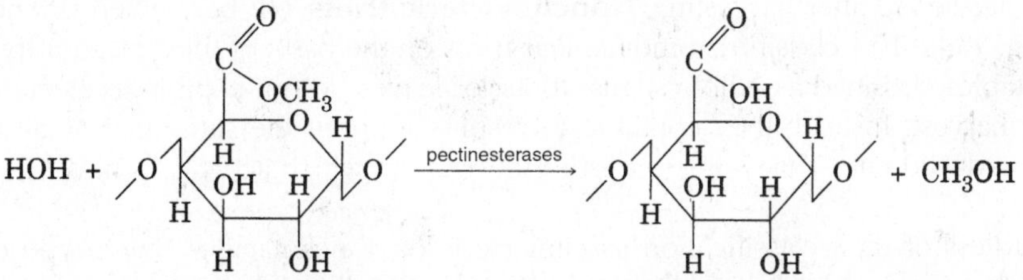

**Protopectinases**
Enzymes in fruits and vegetables capable of catalyzing the hydrolytic cleavage of protopectins to shorter chains of pectins.

**Pectinesterases**
Enzymes that de-esterify protopectin and pectin, a change that reduces gel-forming ability.

## Lignin

Unlike the other constituents of cell walls in plants, **lignin** is a noncarbohydrate polymer of many aromatic structures linked together to form an extremely large, complex molecule that gives a woody quality to plant foods. Although lignin itself does not contain amino groups,

**Lignin**
Structural component of some plant foods that is removed to avoid a woody quality in the prepared food.

one possible genesis of this diverse type of molecular structure is deamination of two amino acids, tyrosine and phenylalanine. Because of its tough and rigid texture, lignin is removed from any portions of fruits and vegetables during their preparation.

## Changes during Maturation

As fruits and vegetables mature, they gradually increase their content of cellulose, hemicelluloses, and even lignin. This increased structural support for cell walls causes fruits and vegetables, particularly the latter, to become less tender as they mature. Vegetables are much more likely than fruits to increase in lignin as they age, which explains the rather tough, woody texture of some vegetables. Lignin is more likely to be deposited in the xylem of the vascular system than in the phloem region of root vegetables, such as parsnips and carrots. This causes considerable toughening in these vegetables when they are harvested at maturity.

Pectic substances, as described in the previous section, undergo a series of chemical changes during the ripening of fruits. The overall changes are (1) hydrolytic cleavage catalyzed by protopectinase activity to produce shorter polymers and (2) de-esterification catalyzed by pectinesterases. The textural changes progress from the hard unripe fruits, through the softening of ripe fruit, to the mushy state characteristic of fruits that have ripened too much.

## Postharvest Changes and Storage

Virtually all synthesis of organic compounds halts after harvest, but numerous physiological changes continue in fruits and vegetables during storage. Bulbs, roots, tubers, and seeds become relatively dormant during storage, whereas the fleshy tissues of fruits and vegetables usually undergo ripening after maturation and then continue to **senescence**. Senescence occurs quite rapidly, with an accompanying loss of palatability. Certain types of biochemical activities occur in all fruits and vegetables, including respiration, protein synthesis, and changes in some constituents of cell walls.

**Senescence**
Accumulation of metabolic products, increase in respiration, and some loss of moisture in plant foods after maturation.

The rate of respiration is directly related to the perishable character of fruits and vegetables—that is, those that have a rapid rate of oxygen consumption and carbon dioxide production are the most perishable. Fortunately, the respiration rate of highly perishable fresh produce can be retarded appreciably by refrigeration, a measure that extends shelf life significantly. The rapid rate of respiration in corn results in increased starch deposition at the expense of sugar stores. The sugar in freshly harvested peas also is depleted quite rapidly as a result of the rapid respiration rate typical of the various legumes.

Respiration rate varies with the stage of maturity and ripening in many fruits, with the rate increasing to a maximum just prior to full ripening, the phase called the **climacteric**. Those fruits that exhibit this increase in respiratory rate just prior to senescence are termed **climacteric fruits**. They are distinguished by their ability to continue to ripen when they are harvested at the time that they are horticulturally mature, but not yet ripe. Peaches, pears, and bananas (Figure 10.9) are examples of climacteric fruits. Citrus fruits and grapes are familiar examples of fruits that are classified as nonclimacteric. Their respiration rate does not accelerate after harvesting. **Nonclimacteric fruits** are best when ripened before harvesting. Table 10.1 classifies some familiar fruits on the basis of their respiratory patterns.

**Climacteric**
Period of maximum respiratory rate just prior to the full ripening of many fleshy fruits.

**Climacteric fruit**
Fruit that continues to ripen after it has been picked—for example, bananas and peaches.

**Nonclimacteric fruit**
Fruit that needs to be harvested when ripe because it will not respire rapidly and ripen after picking—for example, grapes and oranges.

Vegetables classified as edible stems, roots, or leaves do not exhibit accelerated respiration after harvest. Instead, they continue to respire at about the same rate as at harvest or even at a reduced rate. This is in contrast to the tomato, which actually behaves as a climacteric fruit.

Regardless of their classification as climacteric or nonclimacteric, each type of fruit or vegetable has a temperature range over which storage is feasible for at least a short period of time. At the lower end of the range, storage can be done satisfactorily for a much longer time than at the upper end of the range. However, chilling injuries detrimental to the quality of produce occur if storage is at nonfreezing temperatures below the range for a particular fruit or vegetable. For example, potatoes held at a storage temperature of 4°C (39°F) accumulate sugars at the expense of starch content; apples stored at 3°C (37.9°F) gradually develop an internal browning and soft rot.

**Figure 10.9**    Bananas being transported to market by a young boy in Tanzania will continue to ripen because they are a climacteric fruit.

The atmosphere surrounding fresh produce also influences respiration rate. Lettuce needs at least 1 percent oxygen, and asparagus requires at least 5 percent. Sometimes carbon dioxide is added to retard deterioration of some fruits and vegetables during storage. The maximum level of carbon dioxide tolerated varies with the type of produce; strawberries are able to tolerate as much as 45 percent but apples are injured by carbon dioxide levels as low as 2 percent.

Enzyme levels frequently undergo change during storage of harvested fruits and vegetables because protein synthesis is a normal activity during storage and senescence. Enzymes observed to increase during ripening of fruits include lipase, pectic enzymes, invertase, chlorophyllase, and peroxidase. Researchers are trying to alter gene-directed senescence as a means of extending shelf life and reducing marketing losses in fresh produce.

**Ethylene gas** is credited with causing accelerated ripening and early senescence of fruits during storage and sometimes is dubbed the "ripening hormone." Formation of ethylene may be the result of oxidative decarboxylation of α-keto acid analogs of methionine. Storage of underripe fruit in an environment containing ethylene gas is a useful technique to speed ripening. In fact, clear plastic bubble-shaped containers with small holes at the bottom to help regulate gas and moisture levels create an ethylene-enriched environment in which fruits can ripen quickly.

Cell wall components undergo changes after harvest as a consequence of the action of various enzymes. The pectic substances in cell walls and the middle lamella undergo degradation as a result of the increasing levels of two types of enzymes: pectinesterases and

**Ethylene gas**
Gas produced in vivo that accelerates ripening of fruits. $H_2C = CH_2$.

**Table 10.1**    Selected Examples of Climacteric and Nonclimacteric Fruits

| Climacteric Fruits | Nonclimacteric Fruits |
| --- | --- |
| Apple | Cherry |
| Apricot | Citrus fruits |
| Avocado | Fig |
| Banana | Grapes |
| Peach | Melons |
| Pear | Pineapple |
| Plum | Strawberry |
| Tomato | |
| Tropical fruits, including papaya, mango, and passion fruit | |

**Polygalacturonases**
Pectic enzymes promoting degradation of pectic substances in avocados, pears, tomatoes, and pineapples.

**polygalacturonases**. The action of pectinesterases is valued in making apple and grape juices because the increased solubility of the degraded pectic substances, notably pectic acid, promotes a less cloudy beverage and increases the visual appeal of the juices.

Other enzymes include hemicellulases and cellulase. As a consequence of the reactions catalyzed by these enzymes, some sugars are released from the complex polysaccharides constituting the cell walls. The result is that ripening fruits increase in sweetness despite the fact that they may have little or no starch to serve as a potential source of sugar.

Another possible route for increasing sugar levels in some fruits is by the conversion of starch to sugars. This reaction is catalyzed by amylase. Invertase is the enzyme effective in converting sucrose in fruit into its component sugars, glucose and fructose.

**Phosphorylase**
Enzyme in potatoes that promotes sugar formation during cold storage.

Curiously, the amylase in potatoes held in storage does not appear to be responsible for most of the conversion of starch to sugars that takes place when the storage temperature is below 10°C (50°F). Instead, the active enzyme apparently is **phosphorylase** when the temperature is in the range between freezing and 10°C (50°F) or slightly warmer. Storage of potatoes above the active temperature range for phosphorylase is important if the detrimental effects of a high sugar content in potatoes (excessive browning in frying and too sweet a taste) are to be avoided.

In starch-containing vegetables such as legumes, potatoes (Figure 10.10), and carrots, starch synthesis may continue to occur if the storage temperature is approximately normal room temperature or slightly warmer. Although starch is desirable in potatoes, green beans and many other vegetables are considered more palatable when they retain a reasonable level of sugars rather than synthesizing starch. For vegetables in which some sweetness is desired, refrigerator storage is recommended.

## Textural Changes during Preparation

Corn and dried beans achieve part of their change in texture during cooking by taking up water as their starch is gelatinized. In contrast, spinach and other greens are softened visibly during steaming or other heat treatment; the cell walls become increasingly

**Figure 10.10**    Potatoes (whether white, red, or blue) should be stored at about room temperature to maintain their starch content, but green beans and fresh basil should be in the refrigerator.

permeable when they are heated, which causes loss of water and consequently loss of **turgor** in the cells. The leaves wilt, and enough water accumulates in the bottom of the pan to cook the greens.

Broccoli and carrots lose more water from their cells when cooked by microwave energy than when boiled conventionally. This difference appears to be at least partially responsible for the somewhat more rigid texture noted in some vegetables prepared in a microwave oven (Figure 10.11).

Loss of water from cells is but one of the changes taking place in vegetable cookery. The pectic substances undergo some chemical changes to become more soluble than they were in their original form as cementing substances between cells. These changes in some of the pectic substances help tenderize cooked vegetables. Hemicelluloses also become softer as vegetables are heated. Acid slows these changes during cooking, but an alkaline medium greatly accelerates softening.

Calcium can influence the texture of cooked vegetables. Considerable delay in softening occurs when calcium ions are able to react with the pectic substances in the middle lamella.

**Turgor**
Distension of the proto-
plasm and cell wall of a
plant by its fluid content.

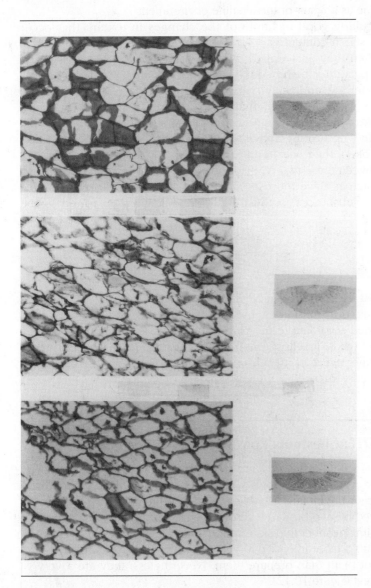

**Figure 10.11**　Cross section (X2) and parenchyma cells (X41) of the phloem of raw carrot (top); boiled conventionally (middle); cooked by microwave energy (bottom). (Reprinted from *Journal of Food Science*. 1975. 40: 1025. Copyright © Institute of Food Technologists. Courtesy of H. Charley.)

The resulting calcium pectinates or pectates precipitate and add rigidity to the structure. This phenomenon is utilized in the preparation of commercially canned tomatoes to achieve a firmer texture rather than shapeless stewed tomatoes.

Preparation of baked beans or other recipes for dried legumes that include molasses unfortunately also has the potential to illustrate the effect of calcium ions on vegetable texture. The recommended procedure is to soften the legumes to the desired extent before molasses is added. If this is not done, calcium ions from the molasses form insoluble calcium pectinates and pectates in the middle lamella and greatly delay softening.

Dried beans have a hard outer covering that requires some soaking in unsalted water prior to simmering until they are tender. Salt is not used in the soaking water because it makes the outer surface of the beans more resistant to water penetration during cooking. A 2-minute boiling period prior to soaking beans is recommended to inactivate enzymes and avoid possible souring during an overnight soaking in cold water. A quicker alternative is to boil the beans for 2 minutes, soak them in the hot water for an hour, and then simmer until done. The addition of 0.5 g ($\frac{1}{8}$ teaspoon) soda to 623 milliliters ($2\frac{2}{3}$ cups) of water reduces the softening time for cooking beans by about one-third. This saving in cooking time is offset by some loss of thiamine as a result of the alkaline environment.

Potatoes afford a particularly good example of the changes in texture that occur when a vegetable with high starch content is prepared. Some potatoes slough and seem to become slightly fluffy when they are cooked, whereas other types hold their shape well and remain rather compact after being cooked. The former are classified as mealy (non-waxy) potatoes; the latter are nonmealy (waxy). Mealy potatoes are characterized by large and numerous starch granules, particularly in the vascular parenchyma cells. These granules swell significantly during cooking; the pectic substances in the middle lamella become more soluble with heating and allow the cells to begin to separate a bit. When waxy potatoes are cooked, their cells tend to remain tightly associated. The exact reason for the difference between these two types of potatoes is not completely clear, but it may be that the waxy potatoes have more calcium ions available to form insoluble precipitates with the pectic substances in the middle lamella than are found in the mealy potatoes.

## CLASSIFICATION

Plant foods may be classified as fruits, vegetables, nuts, or herbs and spices. Fruits are subdivided into berries, citrus, drupes, grapes, melons, pomes, and tropical/subtropical fruits. Vegetables are subdivided according to the part of the plant that is eaten: bulb, root, tuber, leaves/stems, fruits, and seeds. It is apparent from these categories that there is room for a bit of confusion on whether tomatoes are a fruit or a vegetable; they are the fruit of the plant, but are not sweet like most fruits.

### TOPICAL NOTE: Chestnuts Roasting on an Open Fire

"The Christmas Song" starts with these words and conjures up warm, fuzzy images of holiday festivities, but the nut actually grows with other chestnuts inside a spike-covered ball, which protects it until it falls from the tree and eventually pops open. The nuts themselves are covered by two brown coverings, a somewhat brittle outer shell (pericardium) and a very clinging inner thin skin, which is hard to remove. These layers of protection make chestnuts a labor-intensive ingredient, which may explain why we are more apt to sing about them than prepare them. Nevertheless, they are a very interesting ingredient to meet (Figure 10.12).

Chestnuts of several varieties are grown in many parts of the world, including northern Europe, China, Korea, Japan, and the eastern United States. Although their size and flavor differ slightly depending on the variety, chestnuts are high in

**Figure 10.12**   Chestnuts need to be roasted and removed from the pericardium and inner skin before they are ready to eat.

carbohydrate, largely in the form of starch and some glucose and/or maltose. Because of their composition and bland flavor, they may be incorporated in a wide array of recipes including soups, stuffing, sauces, casseroles and other main dishes, and desserts.

The first step in their preparation is to cut an X through the pericardium to ease pressure and avoid an explosion when roasting or heating the nut. Then the chestnuts are heated about 3 minutes in boiling water or roasted about 20 minutes in an oven at 400°F or over a fire until the skin splits at the X. Finally, the pericardium and the inner skin are removed, leaving the nut with its wrinkled surface ready to eat or to use in chestnut dressing for the turkey or in other recipes.

## PIGMENTS

### Chlorophyll

The range of green hues available in foods enables the artistic chef to create some exciting monochromatic presentations. On a chemical basis, these variations are the result of rather minor alterations in quite complex molecules. The variations in green noted in green vegetables are possible because of two basic forms of chlorophyll, the varying ratios in which these are found, and the combination of chlorophyll with other pigments at times.

All chlorophyll-related compounds have the same basic **tetrapyrrole** structure, with connecting methyne bridges:

Chlorophyll molecules actually occur in two different forms, the difference being the functional group (designated as R in the structure) attached to one of the pyrroles. **Chlorophyll a**, the more abundant form in nature, has a methyl group at the R position.

**Tetrapyrrole**
Complex compound with four unsaturated, 5-membered rings (containing one nitrogen and four carbon atoms) linked by methyne bridges, resulting in a very large molecule with a high degree of resonance because of the extensive number of alternating double bonds.

**Chlorophyll a**
Blue-green, more abundant form of chlorophyll; the chlorophyll form in which the R group is a methyl group.

Note: In chlorophyll a, R = $CH_3$; in chlorophyll b, R = C$\begin{smallmatrix}O\\H\end{smallmatrix}$

*Pheophytin forms when magnesium (Mg) is replaced by hydrogen (H).

chlorophyll

phytyl group

The blue-green color of the "a" form of chlorophyll is due to the extensive resonance of the molecule and the presence of this methyl group.

The R group in the other form of chlorophyll, **chlorophyll b**, is an aldehyde group, which causes a yellowish-green color. The ratio of chlorophyll a to chlorophyll b varies depending on the specific plant. It even varies within the plant, as can be seen by the range of green colors within broccoli. In broccoli, chlorophyll a is definitely the dominant form in the blue-green florets, and chlorophyll b is the more dominant form in the yellow-green stalks.

Chlorophyll is a dominant pigment in plants, particularly in unripe fruits and some vegetables. However, other pigments also occur in these foods at the same time, and these pigments contribute to the color seen by the eye. During the ripening of fruits, the amount of chlorophyll present actually diminishes, and the other pigments begin to dominate to give the characteristic colors of the ripe fruit.

Some vegetables contain **chlorophyllase**, an enzyme that catalyzes the removal of the **phytyl** group, which is the complex alcohol esterified to the acid radical in the chlorophyll molecule. The phytyl group is responsible for the hydrophobic nature of chlorophyll. When chlorophyllase effects the removal of this group, the remainder of the chlorophyll molecule, now termed **chlorophyllide**, becomes water soluble. The action of chlorophyllase prior to the cooking of some fresh green vegetables is responsible for the slightly green appearance of the cooking water, as some chlorophyllide is dissolved in the cooking medium.

When vegetables containing chlorophyll are heated, a gradual color change occurs. First, there is an intensification of the bright green of chlorophyll as air is expelled from the tissues. This change unfortunately is followed by a slow transformation from bright green to an unattractive olive-drab color, the result of the elimination of the central magnesium ion and replacement with hydrogen to form pheophytin.

If the original chlorophyll molecule is the "a" form, the compound formed will be **pheophytin a**; similarly, chlorophyll b converts to **pheophytin b**. This conversion is facilitated by the presence of dilute acids, and these acids are released into the cooking water when vegetables are cooked (Figure C9). However, some volatile acids can be eliminated from the cooking medium if the vegetables are boiled without a cover.

Retention of chlorophyll also is favored by heating the water to boiling before adding the vegetable and by using a slight excess of water to dilute the acids released from the vegetable during cooking. These various precautions are of value when a vegetable is cooked longer than 5 minutes, because the breakdown of chlorophyll to pheophytin begins between 5 and 7 minutes after heating begins. By adding the vegetable to boiling water, the actual time required to tenderize the vegetable adequately is reduced, and chlorophyll retention is improved.

**Chlorophyll b**
Yellowish-green form of chlorophyll in which the R group is an aldehyde group.

**Chlorophyllase**
Plant enzyme that splits off the phytyl group to form chlorophyllide from chlorophyll.

**Phytyl**
Alcoholic component of chlorophyll that is responsible for its hydrophobic nature.

**Chlorophyllide**
Chlorophyll molecule minus the phytyl group; water-soluble derivative of chlorophyll responsible for the light-green tint of water in which green vegetables have been cooked.

**Pheophytin a and b**
Compounds formed from chlorophyll a and b in which the magnesium ion is replaced with hydrogen, altering the color to greenish gray for pheophytin a and olive green for pheophytin b.

Sometimes chlorophyll loses its magnesium and its phytyl group. This degradation results in a new compound called **pheophorbide**, which, like pheophytins, has an olive-drab color. Dill pickles and canned green beans provide examples of this change from chlorophyll to pheophorbide (shown here):

**Pheophorbide**
Chlorophyll derivative in which the magnesium and phytyl group have been removed; an olive-drab pigment.

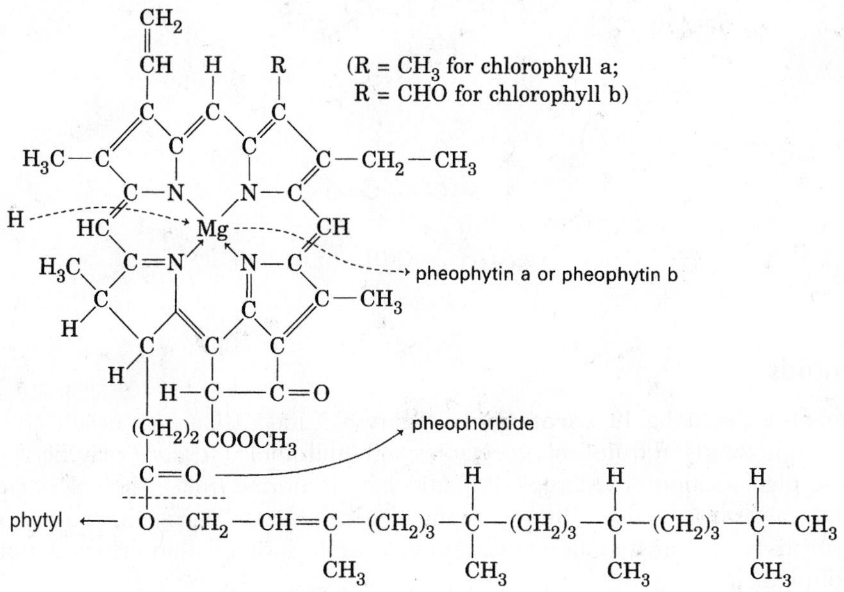

(R = CH$_3$ for chlorophyll a;
R = CHO for chlorophyll b)

→ pheophytin a or pheophytin b

→ pheophorbide

Processing vegetables affects chlorophyll. In the case of green vegetables, the blanching ordinarily done as a part of the preparation for freezing enhances the green because of the expulsion of air from the intercellular spaces. Fortuitously, this bright color is retained during freezing and the subsequent boiling prior to serving the vegetables. This excellent retention of chlorophyll results from the shortened cooking time and also the elimination of some of the organic acids during the blanching process, which results in a less acidic medium when the vegetables actually are cooked.

Canning is another means used for preserving green vegetables. The extremely long heating period and the high temperatures required for safe processing guarantee that chlorophyll in the fresh vegetable will no longer be present in the canned product. Instead, pheophytins and, ultimately, pheophorbide are formed, and the color is the typical olive-drab characteristic of canned green vegetables.

The comparatively easy destruction of chlorophyll during cooking and/or processing has generated some interest in finding ways of preserving the desirable green color. Addition of baking soda increases the retention of chlorophyll, but it has an unpleasant effect on texture if the pH rises above 7. There also is a pigment change when the pH exceeds 7, for the phytyl and methyl groups are eliminated, and a highly water-soluble compound, **chlorophyllin**, is the result.

**Chlorophyllin**
Abnormally green pigment formed when the methyl and phytyl groups are removed from chlorophyll in an alkaline medium.

Chlorophyllin has an unrealistic, bright green color. The undesirable textural change is extreme mushiness caused by breakdown of some of the hemicelluloses in the alkaline medium. By carefully controlling the pH so that there is only enough alkaline material present to neutralize the organic acids in a vegetable, the undesirable formation of chlorophyllin can be avoided. Addition of a small amount of calcium acetate or other calcium salt prevents the mushiness by blocking the breakdown of the hemicelluloses. These additions of alkali and a calcium salt are of interest in the food industry but are not feasible in home preparation of green vegetables.

Another area of research that is of interest, but not of practical value, is the addition of copper or zinc as a replacement for magnesium in the chlorophyll molecule. Such a replacement makes retention of the desirable green color possible. Unfortunately, these metals cannot be used because of problems with toxicity.

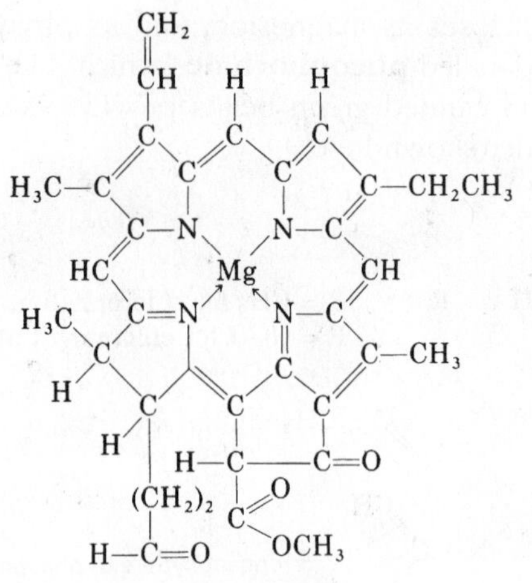

$$CH_2$$
$$CH$$

chlorophyllin

## Carotenoids

The pigments constituting the **carotenoids** (Figures C8 and C10) are especially colorful and important compounds in fruits and vegetables, although their presence may be masked by the intense pigmentation of chlorophyll, particularly in unripe fruits. The colors created by carotenoids range from yellow, through orange, to some reds. With such an array of colors, it is not surprising that many different compounds are included within the carotenoid classification of pigments.

Carotenoids are divided into two groups—the **carotenes** (some of which have potential vitamin A activity) and the **xanthophylls**. Both groups are composed of isoprene groups polymerized into larger molecules, usually containing at least 40 carbon atoms and the accompanying hydrogen. The carotenes contain only carbon and hydrogen; the xanthophylls are distinguished from the carotenes by the presence of at least one atom of oxygen.

The most prominent of the carotenes is β-carotene, a 40-carbon compound with a closed-ring structure at each end of the isoprenoid polymer chain. A closely related carotene (Table 10.2) is α-carotene, which is identical to β-carotene except that it has a closed ring at only one end of the isoprenoid chain. **Lycopene** is an example of a carotene that is acyclic (i.e., it has no ring structure in the molecule).

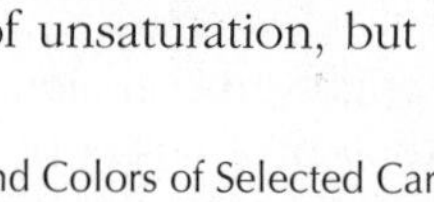

$$CH_3$$
$$CH_2{=}C{-}CH{=}CH_2$$
isoprene group

Lycopene has been studied extensively because of its possible benefit in protecting against some forms of cancer and coronary heart disease. Its antioxidant action and seeming interference in formation of low-density lipoproteins (LDLs) are thought to be significant in both diseases. Actually, lycopene (like other carotenoids) can exist in numerous isomeric forms because of its high degree of unsaturation, but the all-*trans* form accounts for about

**Table 10.2**  Structural Features and Colors of Selected Carotenoids

| Pigment | Structural Feature | Color |
|---|---|---|
| *Carotenes* | | |
| α-carotene | One closed ring | Yellow-orange |
| β-carotene | Two closed rings | Orange |
| Lycopene | No closed rings | Red |
| δ-carotene | No closed rings | Pale yellow |
| *Xanthophylls* | | |
| Lutein | One closed ring, 2—OH groups | Yellow |
| Zeaxanthin | Two closed rings, 2—OH groups | Orange |
| Cryptoxanthin | Two closed rings, 1—OH groups | Orange |

95 percent of the lycopene in foods. All-*trans* lycopene is quite stable during food preparation. The most common sources of lycopene in the diet are tomatoes and tomato-based products (e.g., spaghetti sauce, catsup, and pizza sauce), but such yellow fruits as apricots, papaya, and grapefruit augment lycopene consumption.

Most carotenoids are in the *trans* configuration at the double bonds, as can be seen in the structure of β-carotene:

β-carotene

This configuration gives an intensity of color because of the extensive resonance in the molecules resulting from the conjugated double bonds. Although the *trans* form is quite stable, the heat of cooking can cause some of the carotenoid molecules to transform into the *cis* form at one or two of the double bonds. This modification in the structure results in a molecule with a bend at the double bond in the *cis* configuration, in contrast to the same molecule with an all-*trans* configuration, which has a linear character.

The hue of carotenoids with a *cis* configuration is less intense than that of their counterparts with the *trans* configuration exclusively. The somewhat lighter color of carrots and other carotenoid-containing vegetables that is noted after cooking is an indication that some *cis* isomers have formed. This change is increased when a pressure saucepan is used because of the high temperature.

There is one example of a vegetable that undergoes the reverse transformation during cooking. Rutabagas naturally contain poly-*cis*-lycopene as a dominant pigment; however, some of the *cis* compound is shifted to form some all *trans*-lycopene, with the result that the orange pigment is intensified in the cooked vegetable.

Oxidation is responsible for some loss of color in fruits and vegetables containing carotenoids. The double bonds are susceptible to oxidation, particularly in dried foods. Blanching prior to dehydration is helpful in reducing the likelihood of oxidation. Apparently, blanching protects the carotenoids from oxidation by dissolving them in lipids in the cells, which is not possible until the lipids are freed from the protein with which they are complexed as lipoproteins.

Formation of both the xanthophylls and the carotenes occurs in the chromoplasts of the cells in plants. The exact forms of the carotenoids that are formed are determined genetically. Variations in the color of some plant foods, such as the yellow tomato, have been achieved through genetic manipulation. Regardless of the specific form of carotenoid, all carotenoids are naturally fat soluble. In some plants, it appears that the carotenoids do not remain in the chromoplasts but instead may migrate into the vacuole and dissolve in some of the lipid that is present. Still other molecules of the carotenoids may be found in crystalline form in the vacuole.

There are about 300 different carotenoid pigments, and these may occur in varying amounts and in varying locations within the cells; this helps explain the impressive color array possible in plant foods pigmented with carotenoids. Table 10.3 presents some examples of the types of carotenoids in fruits and vegetables.

## Flavonoids

The nomenclature for a group of closely related phenolic compounds composed of 2 phenyl rings and an intermediate 5- or 6-membered ring is widely varied. One way of categorizing this group of pigmented compounds is to call all of them **flavonoids** and then to subdivide

**Flavonoids**
Group of chemically related pigments usually containing 2 phenyl groups connected by an intermediate 5- or 6-membered ring.

**Table 10.3**    Carotenoids in Selected Fruits and Vegetables

| Plant Food | Pigment |
|---|---|
| *Fruits* | |
| Pink grapefruit | Lycopene, β-carotene |
| Watermelon | Lycopene |
| Peaches | Violaxanthin, cryptoxanthin, persicaxanthin, β-carotene, lycopene |
| Pineapple | Violaxanthin, β-carotene |
| Navel oranges | Violaxanthin, β-carotene |
| Italian prunes | Violaxanthin, β-carotene, lutein, cryptoxanthin |
| Muskmelons | β-carotene |
| *Vegetables* | |
| Yellow corn | Cryptoxanthin |
| Tomatoes | Lycopene, β-carotene |
| Red bell peppers | Capxanthin, capsorubin, β-carotene, violaxanthin, cryptoxanthin |
| Green bell peppers | Lutein, β-carotene, violaxanthin, neoxanthin |
| Carrots | α-carotene, β-carotene, γ-carotene, δ-carotene, lycopene |

**Anthoxanthins**
Phenolic compounds contributing white to yellow color and some flavor to plants; flavonoid pigment with no charge in the central ring.

**Anthocyanin**
Flavonoid pigment in which the oxygen in the central ring is positively charged.

the flavonoids into the **anthocyanins** and the **anthoxanthins**. Anthocyanins are highly pigmented, water-soluble pigments that range in color from red to purple to blue. The anthoxanthins are colorless or white and may change to yellow.

It is in the classification of the anthoxanthins that the confusion over terminology and categorization occurs. Sometimes the flavones are classified as one group of the flavonoids, while the other groups are identified as flavonols, flavones, chalcones, aurones, flavonones, isoflavonones, biflavonyls, leucoanthocyanins, and flavanols. For purposes of discussion here, anthoxanthins means any of the flavonoid compounds that are not classified as anthocyanins.

All of the flavonoid pigments are derived from the following basic structure:

The distinction between the anthocyanins and the anthoxanthins is found in the central ring between the two phenyl rings. The oxygen in the central ring carries a positive charge in the anthocyanins; in the anthoxanthins, the oxygen is uncharged.

*Anthocyanins.*    Anthocyanins, like the other flavonoids, are contained in the vacuole of plant cells where their solubility in water makes them disperse freely (Figures C11 and C12). These compounds are responsible for some of the most intense colors in plant foods, particularly in fruits. Cherries, red apples, various berries, blue and red grapes, pomegranates, and currants achieve their color appeal because of the predominance of anthocyanins. The red color in the skin of radishes and potatoes and the leaves of red cabbage is caused by anthocyanins, too.

Three of the prominent anthocyanins are pelargonidin, cyanidin, and delphinidin. As can be seen from their structures, these three compounds differ only in the number of hydroxyl groups on the right ring of the formula, yet the result is significant differences in color. Pelargonidin, with its single hydroxyl group, is red, whereas delphinidin is blue because of its three hydroxyl groups. Cyanidin has two hydroxyl groups and is intermediate in color.

**Anthocyanidin**
Anthocyanin-type pigment that lacks a sugar in its structure.

Technically, pelargonidin, cyanidin, and delphinidin are classified as **anthocyanidins** because they do not have a sugar complexed with them. When a sugar is complexed, the pigment becomes redder than it is as the anthocyanidin. The pigments containing a sugar are designated accurately as anthocyanins; however, this fine distinction commonly is not used, and all of the related compounds are grouped as anthocyanins. Other common anthocyanidins are malvidin, peonidin, and petunidin. In fruits and vegetables, one or more of these compounds may contribute to the overall color impression. Table 10.4 presents examples of foods colored by anthocyanidins.

• • • Figure C1
Root vegetables include turnip, celery root, and taro (left to right).

• • • Figure C2
Cross section of celery root (also known as celeriac).

• • • Figure C3
Artichokes on the plant ready for harvest.

• • • Figure C4
Cross-sectional view of artichokes.

• • • Figure C5
Green peppers arranged to the left and right of zucchini, eggplant, taro, and fresh coriander are ready for Kenyan shoppers at this outdoor market.

• • • Figure C6
Yellow bell peppers, fresh green beans, zucchini, and three colors of potatoes are topped off with fresh basil.

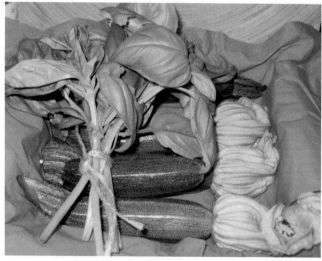

• • • Figure C7
A bunch of fresh basil rests on young zucchini still sporting the blooms from which they were formed.

• • • Figure C8
Kabocha squash, sometimes called Japanese pumpkin, is a winter squash with a flavor overtone of chestnuts.

• • • Figure C9
Chlorophyll turns yellowish green in acid (left) and bright green in alkali (right).

• • • Figure C10
Carotene, the dominant orange pigment in carrots, shows little effect from acid or alkali.

• • • Figure C11
Anthoxanthin, the pigment in cauliflower, bleaches whiter in acid (left) and turns somewhat yellow in alkali (right).

• • • Figure C12
Red cabbage, pigmented by anthocyanins, turns reddish in acid (right) and bluish in alkali (right).

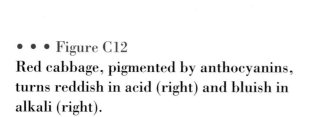

• • • Figure C13
White fish topped with a slice of beet documents the bright color of betalains, as the pigments drain quickly from the vacuole of beet cells.

• • • Figure C14
Rambutan, a prickly looking fruit from Southeast Asia, reveals a slippery white ball of pulp surrounding a seed after it is peeled.

• • • Figure C15
Passion fruit (left) is filled with numerous edible, soft seeds; mangosteen is another Southeast Asian fruit. It has a leathery outer shell encasing several sections of pulp, which sometimes surrounds seeds.

• • • Figure C16
Pomelo, the Chinese ancestor of grapefruit, has pink to reddish flesh on the interior and a thick greenish yellow, bumpy rind.

 Figure C17
Dragon fruit, with its bright red skin and speckled white interior, is a showy contrast to small Asian bananas.

• • • Figure C18
The chef explains that a banana blossom can be sliced for cooking and in salads, or a petal can be peeled off to serve as a liner under a salad.

• • • Figure C19
Papaya trees grow tall in the hot, wet climate of southern India.

 Figure C20
Green papaya salad features long strands of crisp, grated green papayas tossed with sliced cherry tomatoes, green onions, and a Thai vinegar dressing.

• • • Figure C21
A straw stuck into the slushy center of a green coconut is a favorite beverage to fight the thirst generated by the heat in Southeast Asia.

• • • Figure C22
Armed with his sharp knife, an agile man climbs the trunk of a tall coconut palm to harvest ripe coconuts in Pondicherry, India. (Courtesy of Hema Latha.)

• • • Figure C23
High atop the tree, the climber balances cautiously to harvest the mature coconuts. (Courtesy of Hema Latha.)

• • • Figure C24
Ripe coconuts are an important crop in Southeast Asia, but picking them is a dangerous occupation.

• • • Figure C25
Bees are essential for pollinating cherries and other tree fruit crops, a task that has the added benefit of producing honey. (Courtesy of Debra McRae.)

• • • Figure C26
Honey and its comb for sale at this street fair will add flavorful sweetness to toast and rolls.

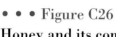

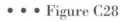

• • • Figure C27
Stalks of sugarcane are fed through these rollers to squeeze out the sweet juice preparatory to boiling it down and crystallizing it into jaggery (hard balls of unrefined cane sugar) in India.

• • • Figure C28
Balls of jaggery ready for market are so hard that cooks will need to grate them before adding to sweeten foods in India.

• • • Figure C29
Local villagers in Sri Lanka work together to transplant rice seedlings into flooded paddies where the crop will grow until mature and the rice can be harvested.

• • • Figure C30
A Japanese farmer and his wife work together to harvest rice from the small paddy they have created and tended in this mountainous terrain.

• • • Figure C31
Racks are used in this Japanese village to dry harvested rice in preparation for threshing to remove rice grains from the straw.

• • • Figure C32
Mochi, a Japanese favorite for New Year's, is made by pounding boiled sweet-glutinous rice into a very cohesive paste and shaping into cakes or balls.

• • • Figure C33
A Vietnamese lady prepares rice paper wrappers for spring rolls by spreading a thin rice-starch paste on a hot griddle and then hanging each wrapper on the edge of a rotating wheel before finally drying outdoors in the sun.

• • • Figure C34
Spring rolls served in Vietnam are the final destination for the wrappers made by the lady shown in C33.

• • • Figure C35
Cages are ready to be submerged just off the shore of Vietnam; pregnant prawns placed in them reproduce in very large numbers and ultimately yield an important cash crop of tiger prawns.

• • • Figure C36
Worker drops food down the tube into a cage housing tiger shrimp larvae released from the pregnant prawn.

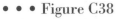

• • • Figure C37
The night's catch by Vietnamese fishermen is unloaded in the morning, sorted by type, and then iced for transport to market.

• • • Figure C38
Shrimp salad sporting a sailing mast made of a green onion rests on a petal from a banana blossom; the carrot rose provides a colorful accent.

• • • Figure C39
Fishermen at Cochin on the coast of the Indian Ocean drop their giant nets into the sea and then tip them back high above the water while their catch flops and jumps until emptied from the nets to be sold to waiting shoppers.

• • • Figure C40
Fishermen can preserve their catch by smoking fish for several hours on racks in a smokehouse.

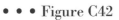 Figure C41
Giant ahi tuna is carved with a sword-like blade as the master carver begins his work at the Tsukiji Fish Market in Tokyo, Japan.

• • • Figure C42
The first large slab is filleted from a giant ahi tuna at the Tsukiji Fish Market in Tokyo, Japan.

• • • Figure C43
At salt pans along the southeastern coast of India, workers regulate moisture levels to optimize salt production; salt is a particularly important ingredient in very hot countries.

• • • Figure C44
Vines of pepper climb the trunks of trees scattered among tea bushes on an Indian plantation in southwestern India.

• • • Figure C45
Black peppercorn still attached to the vine, green cilantro leaves, and horseradish are ingredients used to great advantage in Indian cookery.

• • • Figure C46
Colorful spices and several kinds of legumes are on sale at this stall in Sousse, Tunisia.

• • • Figure C47
Red peppers of various varieties are essential ingredients in Hungarian recipes.

• • • Figure C48
Bark from the cinnamon tree is the source of cinnamon, a popular spice that is marketed as sticks or ground stick.

• • • Figure C50

The fruit ripening on this evergreen nutmeg tree at a plantation in southwestern India is valued as the source of two spices—mace and nutmeg.

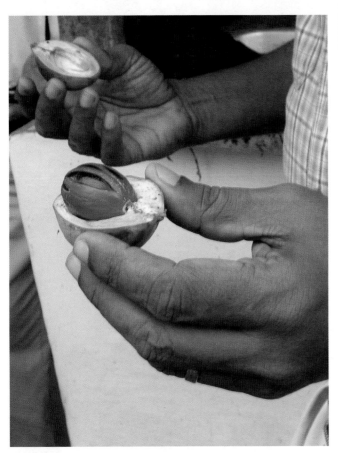

• • • Figure C52

Whole nutmeg contributes its optimal delicate flavor if grated or ground while preparing baked products; dried mace can be crumbled and added to impart an even more subtle flavor to similar products.

• • • Figure C49

Basil in a pot is a convenient herb that can be available fresh for a distinctive accent to salads and other dishes. Saffron strands (right), pistils from saffron crocus, are the most expensive spice and are valued for their orange-red color, pleasing aroma, and delicate flavor.

• • • Figure C51

The interior of the fruit contains the nutmeg seed; the red, wispy coating around the seed is mace, another spice.

 • • • Figure C53
Cardamom is a type of ginger that grows in the tropical climate of Southeast Asia.

• • • Figure C54
The pods that form on the panicle that grows along the ground contain the cardamom seeds that are used to flavor some baked products, including some Scandinavian cookies.

• • • Figure C55
Coriander seeds, turmeric, cumin, and peppercorns are among the ingredients drying in the sun before being ground into garam masala for flavoring Indian dishes. Dried hot chilies are often added to increase the heat level in the mouth.

• • • Figure C56
A Sri Lankan lady uses a mortar and pestle to grind her choice of spices to make garam masala.

• • • Figure C57
Chives can be snipped into short pieces to garnish and also add an overtone of onion flavor.

• • • Figure C58
Oregano is an herb used to enhance the flavor of a variety of dishes in provincial French cookery and other European and Mediterranean cuisines.

• • • Figure C59
Sage can be used fresh or dried to season poultry and dressing, as well as other meats.

• • • Figure C60
The flavor of thyme blends well to season soups and meat dishes.

• • • Figure C61

The aroma and flavor of mint add a distinctive, refreshing quality to salads and beverages. Its leaves also make an attractive garnish.

• • • Figure C62

Basil can be grown as a kitchen herb for convenience in highlighting flavors in salads and meat dishes.

• • • Figure C63

Butter lettuce grown hydroponically and marketed in a special container that keeps its roots damp retains its freshness somewhat longer than other types of lettuce, but is more expensive.

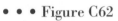

• • • Figure C64

Cacao pods contain the nibs that are roasted and made into cocoa and chocolate.

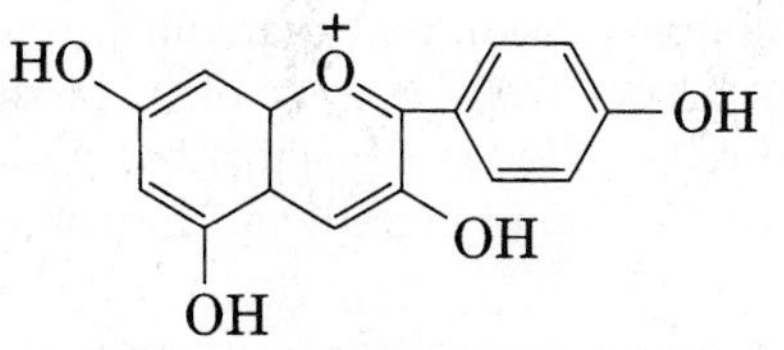

pelargonidin
(red)

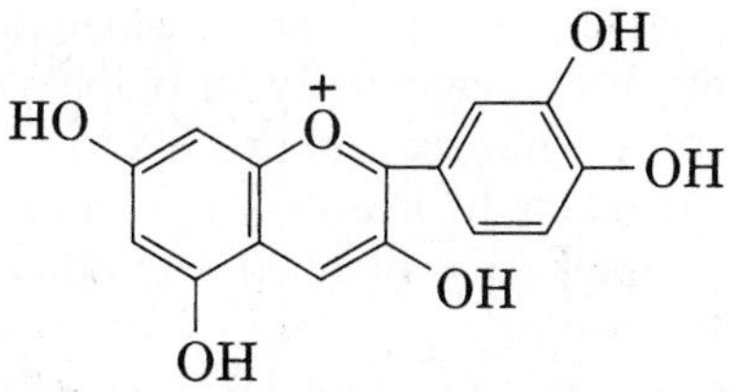

cyanidin
(reddish blue)

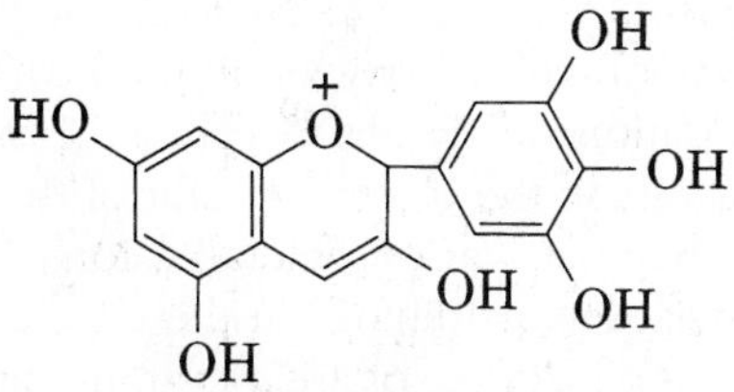

delphinidin
(blue)

Anthocyanidins often form a complex with glucose or some other sugar that may be present in the plant to form the anthocyanin. Whether present as an anthocyanidin or an anthocyanin, these compounds are highly sensitive to the pH of the medium in which they occur. The acidity of the cell in which these compounds form causes the molecules to have a positive charge on the oxygen atom (called an oxonium ion), as was shown earlier in the structures for pelargonidin, cyanidin, and delphinidin. This form, which is the common form at a pH of 3 or less, maintains or shifts the hue toward red. However, as the pH is raised toward a weak acid or even neutral solution, the positively charged oxygen form changes to the quinone form. The quinone form has a violet color.

In an alkaline medium, still another change takes place as a salt of the violet compound, called a color base, forms. The alkaline salt of the color base has a distinctly blue color. These dramatic changes in the color of foods highly pigmented by anthocyanins make it necessary to pay careful attention to pH in working with foods such as red cabbage and several others, including those identified in Table 10.4.

The actual susceptibility to gross changes in pigments as a result of shifts in pH varies with the type of pigment. Red cabbage exhibits unusually wild swings in color with a change in pH because of the presence of more than four hydroxyl groups on the anthocyanin molecule. To avoid the development of a highly unpalatable blue color, red cabbage frequently is cooked with the addition of slices of a tart apple to ensure that the pH is acidic. In contrast, strawberries show much less of a color change with a change in pH because their primary pigment has only three hydroxyl groups on the molecule.

**Table 10.4**    Examples of Anthocyanidins in Selected Foods

| Food | Anthocyanidin |
| --- | --- |
| Strawberry | Pelargonidin |
| Raspberry | Cyanidin |
| Cherry | Cyanidin and peonidin |
| Cranberries | Cyanidin and peonidin |
| Apple | Cyanidin |
| Orange | Cyanidin and delphinidin |
| Black currant | Cyanidin and delphinidin |
| Blueberry | Cyanidin, delphinidin, malvidin, peonidin, and petunidin |
| Grape | Cyanidin and petunidin |
| Peach | Cyanidin |
| Plum | Cyanidin and peonidin |
| Radish | Pelargonidin |
| Red cabbage | Cyanidin |

Color change needs to be considered when combining Concord grape juice with other ingredients. It is pigmented with delphinidin-3-monoglucoside and cyanidin-3-monogluco-side, which contain six and five hydroxyl groups, respectively, in their pigment structures. This abundance of hydroxyl groups causes the extreme color changes that may occur when Concord grape juice is blended with other juices or with other ingredients that influence the pH of the mixture.

Heat processing of fruits and vegetables containing anthocyanins also presents significant problems. Strawberry jam affords an example of the potential problems involved in retaining the desirable red color during prolonged shelf storage after the intense boiling required in production of the jam. A gradual change from the pleasing red to a dull reddish-brown occurs if factors such as a high pH, oxygen in the headspace, and/or a high storage temperature are present. Oxidation of the anthocyanins in the jam is promoted by these conditions and results in a change in the observed color of the pigments.

In addition to heat and oxygen, various metallic ions can cause undesirable color changes in anthocyanins. Special enamel linings in the cans used for heat-processed foods prevent metallic interactions during storage of anthocyanin-containing fruits and vegetables. Foods pigmented with anthocyanins must not contact iron, aluminum, tin, and copper ions in order to maintain desirable colors. Unusual colors, ranging from green to slate blue, develop when anthocyanins contact these metals. The presence of ascorbic acid and copper or iron accelerates the oxidation and undesirable color changes of anthocyanin compounds. The disastrous color changes that occur when metallic ions interact with anthocyanins underline the importance of avoiding contact with copper and iron, which can occur easily if worn utensils or knives other than stainless steel are used in preparing these foods.

Enzymes also can cause detrimental changes in anthocyanin pigments. Anthocyanase is an enzyme that can catalyze reactions that result in the loss of color of anthocyanins. In a product such as strawberry jam, such a change is undesirable, but in the production of white wines, it can be helpful. Other enzymes of interest in reactions with anthocyanins include peroxidases, phenolases, and glycosidases. Peroxidases and phenolases that are naturally present in some fruits and vegetables can catalyze oxidative reactions that result in less desirable colors.

Glycosidases split the sugar from an anthocyanin to form a very unstable anthocyanidin. Mild heat treatment is sufficient to inactivate these enzymes and eliminate this potential problem of color retention in produce pigmented by anthocyanins.

**Betalains**
Two groups of pigments (betacyanins and betaxanthins) that contribute the anthocyanin-like color to beets but differ chemically from the anthocyanins.

**Betacyanins**
Group of betalains responsible for the reddish-purple color of beets; not an anthocyanin, but behaves colorwise in the same fashion.

**Betaxanthins**
Type of betalain that contributes a delicate yellow color.

**Tannins**
Term sometimes used to designate plant phenolic compounds.

***Betalains.***    Beets contain **betalains**, pigments that are closely related to anthocyanins but are not actually categorized as such. The two groups of betalains (both of which contain nitrogen) are **betacyanins** and **betaxanthins**. The distinctive color of beets is derived largely from the presence of the betacyanins; **betaxanthins** contribute a somewhat yellow pigment.

Among the water-soluble pigments within the betacyanin group are betanidin and betanin. Although these pigments are held tightly within cells in the raw vegetable, they diffuse rather rapidly into the cooking water, resulting in the highly pigmented water associated with boiling beets. This problem is aggravated by cutting into beets prior to cooking, a practice that leads to dull coloration in the boiled product.

Betacyanins are sensitive to the pH of the medium in which they are located and, in fact, undergo color changes parallel to those noted previously for the anthocyanins. In other words, an acidic medium promotes a reddish color, whereas a neutral or somewhat alkaline pH brings out the blue of the pigment. The betacyanins are the subject of considerable interest as the food industry turns increasingly to natural food colorants.

***Anthoxanthins (Tannins).***    The compounds included in this discussion of anthoxanthins (sometimes called **tannins**) are colorless or white to yellow, depending on the pH of the medium in which they occur. Compared with other pigments in fruits and vegetables, anthoxanthins make only a slight contribution to the beauty of these foods. However, they are responsible not only for the color of some vegetables, such as cauliflower, but also for some flavor overtones. In white or light-colored vegetables, an acidic pH has a bleaching effect, causing cauliflower to become quite white. At a pH above 7, the anthoxanthins

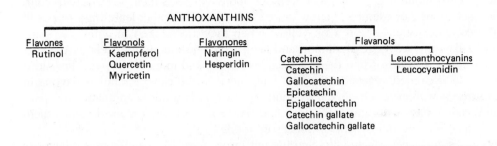

**Figure 10.13**   Subdivisions of anthoxanthin (type of flavonoid) pigments and specific pigments within the various subdivisions.

change to a distinctly noticeable yellow, although this yellow is much less brilliant than the yellow of the xanthophylls in the carotenoid pigment group.

Flavones and flavonols are anthoxanthins that are especially abundant in fruits and vegetables (Figure 10.13). As the structures for kaempferol, quercetin, and myricetin (three of the most abundant flavonols) show, these anthoxanthins are higher in oxygen than are their closely related compounds, the anthocyanidins pelargonidin, cyanidin, and delphinidin. This difference in oxidation appears to account for the significant difference in coloration provided by these similar compounds.

kaempferol

quercetin

myricetin

Isoflavone is an anthoxanthin in soybeans that is of considerable interest at the present time because of its apparent role in reducing LDL and total serum lipids when soybeans and various soy products are prominent in the diet. The structure of isoflavone (of importance also for the creamy color of soybeans) is:

Isoflavone

The solubility of flavones and flavonols in water can be seen; if the water in which the cauliflower or other appropriate plant food is cooked is slightly alkaline, it will have a distinctly yellow color to it. On the acid side of neutral, this coloration is not evident because of the colorless nature of flavones and flavonols in an acidic pH.

**Catechins**
Subgroup of flavanols,
including catechin,
gallocatechin, epicatechin,
epigallocatechin,
catechin gallate, and
gallocatechin gallate;
important to color and
flavor of tea.

**Leucoanthocyanins**
Flavonoid pigments that
are a subgroup of the
flavanols and that are often
termed *proanthocyanidins*.

One of the confounding problems of flavones and flavonols is their propensity to complex with metal ions. For example, yellow onions and spinach or other leafy vegetables will cause the cooking water to turn a bit yellow when they are cooked in aluminum pans because the flavones scavenge aluminum and form a flavone–aluminum chelate. A similar problem is noted with iron when these vegetables are cooked in cast-iron pans. Reactions also can occur if canned vegetables interact with metal in the cans. Rutinol, a flavone in asparagus, reacts with iron to form a ferric-rutin complex that produces an unattractive dark color. In fact, the ease with which flavones and flavonones can be oxidized makes them potentially important as antioxidants in processed foods.

Flavanones and flavanols are slightly different in chemical structure from the flavones and flavonols because they do not have the double bond in the middle ring that is found in quercetin and related compounds. This difference in structure is evidenced in the two subgroups of flavanols: **catechins** and **leucoanthocyanins**. Catechin, gallocatechin, epicatechin, epigallocatechin, catechin gallate, and gallocatechin gallate are included in the flavanols designated as catechins. Of particular interest is gallocatechin gallate because it may help block cholesterol absorption. The structures of the leucoanthocyanins differ from those of the catechins by only one hydroxyl group, as can be seen by comparing the structures of catechin and leucocyanidin.

catechin          leucocyanidin

Pears and white potatoes sometimes develop a pinkish color in their cut surfaces when they stand for awhile after being peeled or sliced. Apparently this color change is caused by the conversion of the proanthocyanin to a pigmented and closely related compound, cyanidin. In canned pears, this color transformation is the result of too much heat processing—that is, too high a temperature or too long a time before the heated product is cooled.

The anthoxanthins, which are all polyphenolic compounds, may undergo browning or blackening reactions when they are bruised, cut, or exposed to air for a period of time. This darkening is attributed to the action of a group of enzymes, the **polyphenoloxidases**. These enzymes are capable of catalyzing oxidation reactions in foods such as mushrooms, potatoes, bananas, pears, peaches, cherries, apricots, avocados, sweet potatoes, eggplant, and apples.

**Polyphenoloxidases**
Group of enzymes capable of oxidizing flavonoid (polyphenolic) compounds to cause browning or other discoloration of bruised or cut surfaces in fruits and vegetables containing these compounds after harvest.

Tyrosine, chlorogenic acid, the various catechins, and several mono- and dihydroxyphenols are among the many compounds that can serve as substrates for oxidation by polyphenoloxidases to cause browning or other discoloration in these foods. The reaction in bananas and white potatoes often leads first to the formation of dihydroindole-quinone, a reddish compound sometimes called dopachrome. Subsequently, the chemical transition continues until finally the grayish-black color of melanin is seen. Eliminating oxygen or using an antioxidant blocks undesirable changes in color.

Polyphenoloxidase activity can be retarded greatly in fruits and vegetables by cold storage, for low temperatures reduce the effectiveness of these enzymes. When the fruit or vegetable has cut surfaces, it is possible to add an acid to inhibit enzyme action. Citric acid often is used because it is so readily available from lemon juice and other citrus fruits, but ascorbic acid is another useful acid. Sulfur in the form of sulfur dioxide gas or in a sulfite, bisulfite, or metabisulfite solution also is used commonly, particularly in the drying of apricots and apples. A sodium chloride solution is effective but is not used often because of the excessively salty flavor introduced by this procedure.

Ordinarily, the action of polyphenoloxidases is undesirable because of the detrimental darkening of the fruit or vegetable. There is one notable exception to this, and that is in the production of oolong and black teas, in which the green leaves are deliberately wilted and rolled to bruise the tissues and bring the epigallocatechin gallate and epigallocatechin in contact with the polyphenoloxidases. The resultant reaction forms two orange-pigmented compounds, theaflavin gallate and theaflavin, the two compounds largely responsible for the pleasing, dark orange-yellow color of brewed oolong and black teas. Actually, the final pigmentation is from the **thearubigens** that are oxidized from theaflavin gallate and theaflavin.

**Thearubigens**
Dark orange-yellow compounds formed when polyphenoloxidases oxidize epigallocatechin gallate and epigallocatechin to theaflavin gallate and theaflavin for ultimate oxidation to thearubigens in oolong and black teas.

## COLORING AGENTS IN FOOD PRODUCTS

The strong interest in natural foods has added emphasis to the use of colors derived from plant sources to enhance the color in a wide range of food products (Table 10.5). Federal regulations do not permit identification of coloring agents from plants as "natural." Instead, these agents are designated as "exempt," accompanied by a statement "color added with ___" or ___ "color." This terminology for colors from plant sources is in contrast to the chemical coloring agents that are identified as "certified."

Certified colors (approved by FDA) include FD&C Blue No. 1 and 2 (dye and lake), FD&C Green No. 3 (dye and lake), FD&C Red No. 3 (dye), FD&C Red No. 40 (dye and lake), FD&C Yellow No. 5 and 6 (dye and lake), Orange B (restricted uses), and Citrus Red No. 2 (restricted uses). All of these are chemical compounds. **Dyes** are water-soluble chemical coloring agents that are available as powders, granules, liquid, or other forms. **Lakes** are used with fats and oils because they are insoluble in water.

**Dye**
Water-soluble chemical coloring agent certified for use in coloring foods.

**Lake**
Water-insoluble chemical coloring agent certified for use in coloring foods.

## FLAVOR

The flavors of fruits and vegetables are extremely important to their acceptance in the diet and have been studied extensively. Considerable research in flavors has revealed the highly complex nature of the total combination of chemical compounds that blend together to give the specific flavor impressions that distinguish each type of fruit and vegetable from all others.

The overall flavor impression is the result of the tastes perceived by the taste buds in the mouth and the aromatic compounds detected by the epithelium in the olfactory organ in the nose. Through the combination of these senses, most people can detect distinctive differences from one fruit or vegetable to another. The compounds detected on the tongue are those that have a sweet, sour, salty, or bitter character. In fruits and vegetables, this means that sugars, acids, salts, and bitter quinone-like compounds are tasted while the food is chewed in the mouth.

Several sugars are found in various fruits and vegetables, fruits commonly having a far higher sugar content than do vegetables. Sweetness may result from the presence of glucose, galactose, fructose, ribose, arabinose, and xylose, as well as other sugars. Glucose, the most abundant of the sugars, may be found in the free sugar form or may be in phosphate esters or other forms.

Organic acids also contribute significantly to sour perceptions on the tongue and, consequently, to the overall flavor of both fruits and vegetables. The common organic acids in fruits are malic, citric, tartaric, and oxalic; those in vegetables include the ones identified as common in fruits plus isocitric and succinic acids. This is not an exhaustive list, but it does include the

**Table 10.5**   Some Exempt Coloring Agents Used in Food Products

| Food Source | Color | Stability |
|---|---|---|
| Turmeric root | Bright yellow | Heat stable |
| Annatto; seed (Caribbean shrub) | Orange-yellow | Stable at pH > 4 |
| Carmine/cochineal; insect in cactus | Orange-yellow, purple | Light stable; stable over wide pH range |
| Red cabbage, grapes, purple carrots, elderberries | Bright red, magenta | Stable at pH 2.5–4; > 4 shift toward purple; fade with extreme heat |
| Red beets | Red beet | Violet < pH 3; red pH 3–7; browns if heated |
| Red pepper pod | Paprika red | Stable to heat and light |

most abundant organic acids in both fruits and vegetables. Although vegetables contain a somewhat greater array of acids than do fruits, the content is not as great. This comparatively low acid content of vegetables explains the distinction between the use of water bath canning for fruits and the requirement for pressure canning of vegetables (see Chapter 20).

All fruits and vegetables naturally contain at least a small amount of salt, which is detected in the overall taste impressions contributing to flavor.

The flavonones, previously discussed because of their role in pigmentation, also contribute to taste sensations. Hesperidin and naringin, compounds found in the peels of oranges and grapefruit, respectively, have been studied extensively as possible sweetening agents because they become intensely sweet when their molecules are modified slightly by opening the central ring. However, they are not sweet in their unaltered form. Several compounds quite closely related to hesperidin and naringin are found in citrus peels; some of these are quite bitter.

Astringency is the feeling of puckering that occurs in the mouth when certain compounds, notably flavanols, are present. Although astringency is not actually a part of the flavor, this puckering sensation in the mouth blends with flavor perceptions in influencing acceptance of the particular fruit or vegetable containing these flavanols. This characteristic astringency is considered to be desirable in limited amounts in many foods, including wines, some fruit juice ciders, and some fruits.

When bananas and persimmons ripen, the excessive quantities of flavanols found in the unripe fruit are reduced. Temperature influences the amounts of these flavanols. A warm and clear environment during ripening of peaches results in a lower content of polyphenols than is developed when cooler, cloudy weather occurs. The lower polyphenolic content imparts a reduced astringency.

Volatile compounds are at least as important as those detected by the taste buds on the tongue. Several organic compounds contribute significantly to the aroma and flavor of fruits and vegetables because of their volatility. Included among these volatile compounds are esters, aldehydes, acids, alcohols, ketones, and ethers. No single type of compound stands out as the key to the aroma and flavor of specific fruits and vegetables. Analysis by gas–liquid chromatography reveals an impressive assortment of organic compounds in the aroma of different fruits and vegetables. Table 10.6 presents a few of the important volatile components of fruits.

Two types of vegetables have strong flavors resulting from the presence of various sulfur-containing compounds. **Allium** is the genus that includes onions, chives, garlic, and leeks. Brussels sprouts, broccoli, cabbage, rutabagas, turnips, cauliflower, kale, and mustard are members of the family **Cruciferae**, which also contains prominent sulfur compounds.

**Alliin** is a key compound from which a strong and pungent odorous compound forms when a garlic clove is crushed. Although the parent compound of a highly aromatic substance, alliin actually is odorless. It is not until **alliinase**, an enzyme found in garlic, comes in contact with alliin that diallyl thiosulfinate forms. Diallyl thiosulfinate is quite unstable and is quickly converted to **diallyl disulfide**, the compound considered to be the most important one in garlic's (and other *Allium* vegetables') aroma. Various other di- and trisulfides, hydrogen sulfide, and propanethiol are among the aromatic compounds identified in onions and other *Allium* vegetables.

***Allium***
Genus including onions, chives, garlic, shallots, and leeks; unique for its sulfur-containing flavor compounds.

***Cruciferae***
Family of vegetables (including brussels sprouts, cabbage, rutabagas, turnips, cauliflower, kale, and mustard) flavored with sulfur-containing compounds that differ from those found in *Allium* vegetables.

**Alliin**
Odorless precursor in garlic that ultimately is converted to diallyl disulfide:

$$CH_2=CH-CH_2-S-CH_2-CH-COOH$$
$$\overset{|}{O} \qquad \overset{|}{NH_2}$$

**Alliinase**
Enzyme in garlic responsible for catalyzing the conversion of alliin to diallyl thiosulfinate, the precursor of diallyl disulfide.

**Diallyl disulfide**
Key flavor aromatic compound from garlic:

$$CH_2=CH-CH_2-S$$
$$CH_2=CH-CH_2-S$$

**Table 10.6**    Some Volatile Flavor Components in Selected Fruits

| Fruit | Volatile Flavor Component |
| --- | --- |
| Apple | Esters; alcohols; aldehydes; ketone; acids, including hexanal; ethyl 2-methyl butyrate |
| Banana | Alcohols; esters, including amyl acetate, isoamyl acetate, butyl butyrate, and amyl butyrate |
| Cranberries | Benzaldehyde; benzyl and benzoate esters |
| Orange | Acetaldehyde, ethanol, limonene, ethyl esters, linalool; β-terpincol |
| Peach | Benzaldehyde, benzyl alcohol, γ-caprolactone, γ-decalactone |
| Pear | Esters of 2,4-decadienoic acid, especially the esters of ethyl, n-propyl, and n-butyl |
| Pineapple | p-Allyl phenol, γ-butyrolactone, γ-octalactone, acetoxyacetone, methyl esters of β-hyroxybutyric and β-hydroxyhexanoic acids |
| Strawberries | Methyl and ethyl acetates, propionates, and butyrates |

**Table 10.7**   Some Flavor Components of Selected Vegetables

| Vegetable | Flavor Component |
|---|---|
| Celery | Phthalides, *cis*-3-hexen-1-yl pyruvate, diacetyl |
| Cucumber | Nona-2,6-dienal, non-2-enal, hex-2-enal, propanal, ethanal, hexanal |
| Beans | Methanol, acetone, hydrogen sulfide, methanethiol |
| Corn | Methanol, acetone, hydrogen sulfide, methanethiol |
| Parsnips | Dimethyl sulfide, methanol, acetone, hydrogen sulfide |
| Peas | Methanol, acetone, dimethyl sulfide |

The eye-irritating substance in onions is an acid derived specifically from (+)-S-(prop-1-enyl)-L-cysteine sulfoxide as a result of the action of an enzyme. The irritant is **propenylsulfenic acid**, which decomposes rather rapidly. This compound and many of the sulfur compounds contributing flavors to various members of the *Allium* genus are quite volatile and escape during cooking unless the cooking time is short.

Isothiocyanates are among the sulfur-containing compounds found in members of the *Cruciferae* family. **Sinigrin**, an isothiocyanate found in cabbage and some other *Cruciferae*, can be broken down by myrosinase A, an enzyme, to yield a highly pungent compound, allyl isothiocyanate. When vegetables from this family are cooked, hydrogen sulfide is produced, with the production accelerating when cooking extends beyond 5 minutes. Cooked cauliflower yields about twice as much hydrogen sulfide as does cooked cabbage. Broccoli also contributes more hydrogen sulfide than does cabbage. Dimethyl sulfide is a key volatile compound produced when cabbage boils. Rutabagas also release hydrogen sulfide and dimethyl sulfide when boiled.

Sulfur-containing compounds are not the only source of flavors in vegetables. Other flavor components include short-chain aldehydes, esters, and other volatile organic compounds, a few of which are identified in Table 10.7.

**Propenylsulfenic acid**
Compound in onions causing eye irritation and tears:

$$CH_3—CH=CH—\overset{O}{\underset{\uparrow}{S}}—H$$

**Sinigrin**
Potassium myronate, an isothiocyanate glucoside in cabbage:

$$CH_2=CH—CH_2—C\overset{S—C_6H_{11}O_5}{\underset{N—O—SO_3K}{}}$$

# FIBER

## Categories of Fiber

Fiber is the combination of materials in foods that cannot be digested readily. Plant foods are valued in the diet as potentially outstanding sources of fiber. Although the word sounds simple, several different compounds comprise dietary fiber. Nutritionists often categorize fiber as soluble and insoluble fibers. Table 10.8 shows the division of fiber components according to solubility.

Obviously, both soluble and insoluble fibers are found in the same food. However, the relative proportions vary from one plant food to another. Soluble fibers appear to be digested to a limited extent to provide calories to the body, while insoluble fibers are excreted undigested, thus providing stool bulk, but not energy. Soluble fibers may be of some benefit in reducing serum cholesterol levels. Insoluble fibers are possibly beneficial because they speed the transit time, promoting excretion of waste from the body.

Citrus fruits, oats, and legumes are some of the foods that are especially good sources of soluble fibers, while wheat, rice, and many vegetables are valued for their content of insoluble fibers.

A soluble fiber in oats, β-glucan, has been shown to be effective in reducing cholesterol levels. This finding resulted in FDA approval to state on labels of food products containing

**Table 10.8**   Dietary Fiber—Soluble and Insoluble

| Soluble Fiber | Insoluble Fiber |
|---|---|
| Gums (guar, locust bean, gum arabic, gum tragacanth, gum ghatti, gum karaya, alginates, agar, carrageenan, xanthan, gellan, cellulose gums) | Cellulose Hemicellulose Lignin |
| Pectic substances | |

whole oats that soluble fiber from whole oats, when included as part of a diet low in saturated fat and cholesterol, may reduce the risk of CHD (coronary heart disease) by lowering blood cholesterol levels. The water-soluble gum from psyllium seed husks has been found to have a similar effect in lowering cholesterol; this allows products containing psyllium to make a similar health claim. Spanish psyllium, an annual that is native to the eastern United States, is a source of this gum.

## Bran in Food Products

As emphasis on fiber in the diet has increased, the food industry has rallied its efforts to develop numerous products with increased fiber content. An example is a fat replacer made with oat bran, flavorings, and seasonings suitable for incorporation with ground meat. This bran product can be blended with low-fat ground meats to produce a cooked patty containing only 10 percent fat. Because of the water retention by the fiber in the bran, the finished patty is juicy and flavorful and has an acceptable mouthfeel.

Breads and other baked products can be made with bran added up to a level of about 15 percent of the weight of flour. One effect of added bran is a decrease in volume. The rather sharp character of the bran particles cuts some of the gluten strands during mixing, which interferes with maximum stretching of the strands during baking. Bran aids in retaining moisture in bread during storage.

### FOOD FOR THOUGHT: Reducing Waste

Sugarcane is valued because of the sugar produced from it (Chapter 8), but the residual pulp remaining after the sugar has been removed creates an expensive disposal problem (Figure 10.14). The answer previously has centered on creating biodegradable

**Figure 10.14**    Piles of waste remain after the juice has been pressed from sugarcane whether at this rural factory near Mysore in southern India or in large commercial operations.

products, but a new product is emerging that may resonate in a marketplace that is becoming increasingly concerned about the environment.

Cane and paper pulp in a 9:1 ratio, water, and some chemicals are mixed together before being molded under steam and pressure into disposable plates, cups, and containers that can hold hot coffee, are resistant to fats and oils, and can be used in a microwave oven. The result is biodegradable products for an existing market, but with a slightly higher price tag than its Styrofoam competition that does not degrade. The other bonus is reducing waste from sugar processing and converting that problem into an income source. Clearly, this is a win–win solution.

## GUMS

**Gums**, as the word is used in food technology, are complex hydrophilic carbohydrates found in plants. Their hydrophilic nature explains why they are called hydrocolloids. Despite the fact that they are composed of a large number of monosaccharide units linked together by glycosidic linkages, they undergo little digestion in the small intestine. For this reason, they are classified as soluble fiber. Their monosaccharide components are more varied than is true for the energy-contributing disaccharides and the glucose polymers (dextrins and starch). Two characteristics of gums are particularly noteworthy: (1) a remarkable ability to attract and bind water, and (2) very limited caloric contribution due to limited digestion and absorption in the body, mainly in the large intestine.

Attention is focused on gums as possible additives in foods today because of the potential nutritional and health benefits that may accrue when they are incorporated in food products. The water-holding capacity of gums increases stool bulk until some digestion occurs in the large intestine. Even then, the insoluble fibers (cellulose, hemicelluloses, and lignin) still speed transit time of food waste through the remainder of the gastrointestinal tract to excretion.

The fact that gums are digested and absorbed quite inefficiently compared with sugars and starch makes them useful in food products aimed at weight-conscious consumers. Agar, guar gum, gum arabic, and pectin appear to provide somewhere between 1 and 3 calories per gram; carrageenan and xanthan gum probably provide, at the most, 1 calorie per gram. Evidence regarding the anticarcinogenic nature of gums is not yet definitive, but there is the possibility that gums may help protect against cancer.

### Functions

Gums are valued as thickening agents; their comparatively lower potential caloric value may be an important reason for a food product to be formulated with a gum rather than with starch to achieve the desired viscosity. Flour and sugar also can be used as thickening agents, but these common ingredients add 4 calories per gram, whereas gums can accomplish essentially comparable thickening with a significant reduction in calories.

The most dramatic calorie savings are afforded when gums are used to replace fats and oils, because fats contribute 9 calories per gram. For instance, gums can be used in salad dressings to provide appropriate thickening in place of some of the oil ordinarily included in the formula. Ice creams and baked products, especially cakes, may utilize them to enhance mouthfeel when fat levels are reduced. In cakes, shelf life is extended, staling is delayed, cell structure is enhanced, and volume is increased by the use of appropriate gums. In ice cream, large aggregates of ice crystals are blocked when they are added, resulting in a fine texture. Gums sometimes are used to increase the fiber content of a beverage or food and enhance its nutritional merits.

### Sources

Carbohydrate gums are derived from seeds, plant exudates, seaweed extracts, and microorganisms. These, as a group, are hydrophilic polysaccharides. The seed gums of primary

## FOOD FOR THOUGHT: Inulin, Ingredient in Functional Foods

Inulin is a complex fructooligosaccharide available from the roots of chicory. It is an ingredient of interest in formulating functional foods. This soluble fiber is not digested until it reaches the large intestine, where some bacterial digestion occurs. This prebiotic fiber has been shown to promote absorption of calcium and magnesium and decrease cholesterol and serum lipids. Inulin stimulates *Lactobacilli* and *Bifidobacteria* in the colon, which helps to combat possible harmful microorganisms and to promote health.

Inulin is a versatile food ingredient. It lacks flavor, so it can be incorporated into products without introducing detrimental overtones. It contributes to a pleasing mouthfeel when used in table spreads, yogurts, frozen desserts, and dressings. It can be added to breads and other baked products at levels high enough to be of merit in increasing fiber intake. Various food ingredient companies are marketing inulin, and it is finding a ready market as food products are developed with decreased fat and sugar to meet the demand for fewer calories.

commercial significance come from guar and locust bean (carob), both of which are classified as legumes. Gum arabic, tragacanth, karaya, and ghatti are the primary gums from plant fluids (actually the oozing liquid from cuts in tropical trees). Of these, gum arabic is used most commonly. The red and brown algae, familiarly known as Irish moss and giant kelp, provide agar, carrageenan, and algin gums.

Recently, some microorganisms have been used to produce gums. The most successful of these fermentation gums is xanthan gum, which results when *Xanthomonas campestris* ferments glucose in the presence of some trace elements. The potential for developing other gums of commercial value through the use of microorganisms may be great, and work is proceeding in this direction.

### Seaing a New Vegetable?

Kelp has long been harvested for use as a gum in commercial food products, but who in the United States would ever think that it had any potential as a vegetable? Although Asians have long had the tradition of using dried seaweeds of various types as ingredients in many different dishes, seaweed is not found in the typical American kitchen. That may change if a Maine company (Ocean Approved) is able to gain consumer acceptance of its kelp, which is grown and harvested from the sea just off the coast at Portland, Maine.

Although kelp can be harvested in several spots along the state's rugged coast, commercial quantities require farming of a controlled space of the ocean. Kelp farming begins with seedlings that are transplanted to long lines that extend upward from the ocean bottom. By the time the crop is ready for harvest, fronds will have grown to a length of 6–8 feet.

Fresh kelp is made ready for the market by parboiling to kill bacteria that might be present, cutting it into narrow strips the size of spaghetti, or even fettuccine, and packaging for freezing. *Voilá*, a new, slightly crunchy vegetable is on its way to American kitchens, where it may add a bit of excitement to a raw salad. Other ideas for its use include as an ingredient in soups or main dishes, or even as a replacement for pasta. Kelp is just one type of seaweed that is potentially available for inclusion in salads, soups, and even in breads. Alaria, dulse, and sea lettuce may also find their way to your table.

## Chemical Composition

The various gums have unique structures, but they have in common that they are polysaccharides in which glucose usually is absent. Galactose is common to seed, plant exudate, and seaweed gums (with the exception of algin).

mannose                galactose

*Seed Gums.*   Both guar and locust bean gums contain only two types of monosaccharides: mannose and galactose. Guar and locust bean gums are composed of mannose units linked together by a 1,4-$\beta$ linkage, with the galactose being linked by a 1,6-$\alpha$ linkage at regular intervals. In the structures that follow, note that guar gum has twice as much galactose as does locust bean gum. Locust bean gum requires heat for maximum hydration, whereas guar gum hydrates in cold water.

guar gum

locust bean gum

*Plant Exudates.*   The gums from plant exudates (Figure 10.15) are structurally less orderly than the seed gums, and they are likely to have other compounds associated with them, rather than being the pure polysaccharide polymer. Because of this variability, the plant exudates may present problems in food production. These gums naturally contain hexuronic acids, most of which normally are neutralized by the presence of calcium or other ions.

**Figure 10.15** Sap exuding from the acacia tree is the source of gum arabic. (Courtesy of Colloides Naturels, Inc.)

Actual structures of the plant exudate gums cannot be presented because of the great variability of contaminants. However, general observations regarding their primary constituents can be made. The sugars and hexuronic acids found in plant exudate gums are as follows:

- Gum arabic—galactose, arabinose, rhamnose, galacturonic acid, and the methyl ester of galacturonic acid
- Tragacanth—xylose, arabinose, galactose, fucose, rhamnose, and galacturonic acid
- Karaya—galactose, rhamnose, glucuronic acid, and galacturonic acid

***Seaweed Extracts.***    Seaweed extracts (Figure 10.16) are rich in galactose, often with sulfate esters. Agar has two fractions: agarose and agaropectin. Agarose is the linear fraction and is composed of β-D- and α-L-galactose; agaropectin contains sulfate esters and glucuronic acid, as well as the basic components of agarose. Agar is noted for its strong and transparent gels, which are reversible when heated and then re-form when chilled.

Carrageenan occurs in various fractions, some of which form gels and some of which do not gel. The important quality of carrageenan is its ability to interact with protein to aid in the stabilization of various milk products, notably ice creams, process cheeses, and chocolate milks. This gum also works effectively in concert with other gums because of its propensity for cross-linking with them. Carrageenan structurally can be characterized as predominantly sulfate esters of galactose.

**Figure 10.16**   Giant kelp (a commercial source of algin) being harvested from the sea. (Courtesy of Kelco Division of Merck and Company, Inc.)

Algin is used to form both gels and films. It is a gum with mannose and guluronic acid as its principal components and with numerous salts resulting from the presence of sodium, potassium, and ammonium ions.

***Microbial Exudates.***   Fermentation-produced polysaccharides are yet another type of gum. Xanthan gum is a microbial exudates gum produced by *Xanthomonas campestris*. Glucose and other carbohydrates serve as the substrates for the fermentation. The chemical structure of xanthan gum is based on a chain of glucose units (1,4-β-glucosidic linkage) and a side chain consisting of two mannose and a glucuronic acid unit attached on alternating glucose unit.

Curdlans is another gum produced by microorganism fermentation. *Alicaligenes faecalis* var. *myxogenes* produces curdlan, which is a polysaccharide in which glucose units are joined by 1,3-β-glucosidic linkages. Curdlan was approved by the FDA in 1996 and is marketed as Pureglucan®. It does not dissolve in water, but it can be suspended in water and heated to 80°C (176°C) to form a gel that is stable under freezing and retorting. When a suspension of curdlan is heated to 55–60°C (131–140°F) and then cooled below 40°C (104°F), a gel forms that behaves much like carrageenan.

***Synthetic Gums.***   Cellulose serves as the starting material for the production of some synthetic gums. Carboxymethyl cellulose (CMC) is the most prominent of these. Its chemical name is sodium carboxymethyl cellulose; it sometimes is designated simply as cellulose gum. The structure is cellulose that has undergone a chemical reaction to add a carboxymethyl group.

## Applications

Because of the potential nutritional benefits of adding fiber to various food products, many foods currently are formulated with gums added to replace part or all of traditional functional ingredients such as starch or fat. Gum arabic is a soluble fiber that is well suited to liquid foods, such as soups, because a significant quantity of fiber can be added without causing too much of an increase in viscosity. Carboxymethyl cellulose is another soluble fiber that may be used alone or in tandem with gum arabic to increase fiber content of a food. The high degree of solubility of both gum arabic and CMC makes it possible to use comparatively high levels without developing a gritty texture in liquid foods.

Guar gum and locust bean gum are two seed gums that can replace up to at least half of the starch used for thickening, thus reducing calories in the food. The added advantage is that the fiber has some nutritional benefits. If enhancing fiber content is an objective, a soup that is formulated with 4 percent starch can be made by using half as much starch (2 percent) and adding guar gum and CMC at a level of 1 percent each.

Guar gum is useful as a replacement for as much as 10 percent of the flour in some baked products. Biscuits in which guar gum replaced 10.35 percent of the flour resulted in a satisfactory product, and fiber increased by 9.5 g per 100 g. Guar gum can replace flour in cakes satisfactorily at up to 5.5 percent; replacement in bread can be done only up to a level of 3 percent.

Xanthan gum is used successfully as a thickener in salad dressings. Some type of starch also is included to obtain the desired texture in low-calorie, low-oil, or no-oil dressings. Carrageenan often is the gum of choice in making smooth ice creams with reduced fat levels. Guar gum, CMC, and locust bean gums may be used to give a firmer body to ice creams. Yet another gum—gum ghatti—is an effective emulsifying agent when used in conjunction with lecithin in the production of butter-containing syrups. Table 10.9 summarizes the characteristics and applications of these and other gums.

### FOOD FOR THOUGHT: Textural Polysaccharides

Different starches and various polysaccharide gums are widely used in the food industry to help define the texture of products ranging from liquids such as soups and sauces to bakery products and snack foods. Product designers face a complex challenge in formulating a product that meets a range of requirements. They often employ one or more starches or gums to achieve a texture that (1) is acceptable to consumers, (2) meets processing requirements, (3) releases or binds flavors appropriately, (4) has good shelf stability, and (5) is visually appealing.

Gums and starches are hydrocolloids and thus attract or bind water to influence the viscosity and flow properties of liquid products and batters. Depending on the other ingredients and the treatment during preparation and processing, these hydrocolloids may form a sol, gel, foam, and/or film to significantly affect the textural properties of a food. For instance, gum Arabic can be used to stabilize an emulsion and also contribute to mouthfeel that simulates the coating effect of fat. In contrast, xanthan gum in a salad dressing is a shear-reversible hydrocolloid that is thin enough so that the dressing can be shaken to blend ingredients, yet the dressing subsequently will cling to the greens.

Product designers need to consider the cost of ingredients. Starches are of interest as textural agents because they cost considerably less than the various gums. Sometimes more than one gum may be an economy because some combinations of gums are synergistic, requiring a smaller total gum content than if only one gum were used. Some other combinations may be additive in their effect, so more of each gum is required for the desired effect.

**Table 10.9**    Sources, Characteristics, and Applications of Selected Gums

| Gum | Source | Characteristics | Applications |
|---|---|---|---|
| | | **Seed Gums** | |
| Locust bean (carob seed) gum | Seed of evergreen tree (*Ceratonia siliqua*) (Mediterranean) | Insoluble in cold water; viscous at 95°C (203°F); nongelling alone, gels with xanthan gum | Stabilize ice cream, bologna, cheese, sauces, processed meats |
| Guar gum | Endosperm of guar plant (*Cyamopsis tetragonolobus*) (India, Pakistan) | Nongelling; stabilizing; increasing viscosity; water binding; gels with agar and κ-carrageenan | Desserts, baked products, ice cream stabilizer, sauces, soups, salad dressings |
| | | **Plant Exudates** | |
| Gum arabic | Sap of *Acacia* tree (Sudan) | Dissolves in hot or cold water; limited viscosity; readily soluble; emulsifying agent | Candies to retard crystallization, flavor fixative, soft drinks; beer (foam stabilizer) |
| Gum ghatti | Exudate from *Anogeissus latifolia* tree (India) | Nongelling; disperses in hot or cold water; substitute for gum arabic | Butter-containing syrups |
| Gum karaya | Dried exudate from *Sterculia* tree (India) | Low water solubility; swells greatly in water; boiling reduces viscosity; ropy in alkali; stabilizes foams | French dressing, sherbets and ices, meringues, bologna |
| Gum tragacanth | Exudate of *Astragalus* (leguminous) bush (Iran, Syria, Turkey) | High viscosity; cold-water soluble; heat stable; water binding; soft gels; emulsifier | Salad dressings, sauces, catsup, relishes, fruit fillings, ice cream, confections |
| | | **Seaweed** | |
| Alginates | Brown seaweeds (*Macrocystis pyrifera*, etc.) (U.S., U.K., Japan, Canada) | Irreversible gels with calcium ions in cold water; emulsifier; thickener; binding agent | Salad dressings, lemon pie filling, fruits for baking, meringues, fish coating |
| Agar | Red seaweeds (*Gelidium cartilagineum*, etc.) (Japan, Mexico, Portugal, Denmark) | Strong gels; insoluble in cold, but very soluble in hot water; gels at 35–40°C (90–104°F); melts at 85°C (185°F) | Culture medium, stabilizer in puddings, meringues, pie fillings, cheese, icings, sherbets |
| Carrageenan (Irish moss) | Red seaweeds (*Chondrus crispus*, etc.) (Iberia, North Atlantic, Japan) | Lambda fraction is nongelling; iota and kappa fractions gel (iota with Ca ions, kappa with K ions); heat-reversible gels; protein binding | Pet food (meats with gravy), low-sugar jams and jellies, low-fat or nonfat salad dressings, chocolate milk, puddings, cheese analogs, bakery fillings and icings |
| | | **Microorganism Gums** | |
| Xanthan gum | Culturing of *Xanthomonas campestris* (bacterium associated with rutabagas) | Very soluble in hot or cold water; high viscosity; stable to heat, pH, and enzymes; viscosity unaffected by temperature; thickening; stabilizing; suspending; good freeze-thaw stability | Frozen doughs, meringues, ice cream, tomato sauce, fruit gels |
| Curdlan | Culturing of *Alicaligenes faecalis* var. *myxogenes* | Water insoluble; stable to freezing, retorting if heated to 80°C (176°F); behaves like carrageenan if heated to 60°C (140°F) and cooled to 40°C (104°F) | Sausage, ham, and other meat products, Oriental noodles, sponge cake, cheese products, yogurt, low-fat sour cream |
| Gellan gum | Culturing of *Pseudomonas elodea* | Strong gels; heat to dissolve; gels with cations upon cooling (sets at 20–50°C (68–122°F) melts at 65–120°C (149–248°F); stable to acid and heat | Icings, nonstandardized jams and jellies |
| | | **Synthetic Gums** | |
| Cellulose gum (sodium carboxymethyl cellulose or CMC) | Alkaline conversion of cellulose into ethers | Water soluble; binds water; thins with heat, thickens when cooled; gels | Bulking agent in low-calorie foods, pie fillings, soya protein products |
| Microcrystalline cellulose (MCC) | Acid hydrolysis of cellulose | Thixotropic gels; stable to acids; stable over temperature range; increases film strength; stabilizer | Icings, partial starch replacement, chocolate drinks, oil replacement in emulsions |
| Methylcellulose (MC) | Chemically derived from cellulose | Soluble at cool temperatures; gels at high temperatures | Batters for coating fried foods, cream soups, sauces |
| Hydroxypropy-lmethyl-cellulose (HPMC) | Chemically derived from cellulose | Soluble at cool temperatures; gels at high temperatures | Batters for coating fried foods, cream soups, sauces |

## SUMMARY

Fruit and vegetable tissues consist of the dermal, vascular, and ground systems. Parenchyma cells, the predominant type of cell in the fleshy portions of fruits and vegetables, contain several structural features, including the vacuole, chromoplasts, chloroplasts, and leucoplasts in which the various pigments and starch are produced. Structural components include cellulose, hemicelluloses, pectic substances (protopectin, pectin, pectinic acid, and pectic acid), and lignin.

During maturation and following harvest, many changes occur in fresh produce. These changes are the result of enzyme action and respiration.

The pigments in fruits and vegetables include chlorophyll, the carotenoids, and the flavonoids (anthocyanins, betalains, and anthoxanthins). Color changes can occur as the result of heat, changes in pH, oxidation, and enzyme action. Chlorophyll and the flavonoids are more susceptible to change than are the carotenoids. Flavors are due to a variety of organic compounds, including aldehydes, ketones, acids, sulfur-containing compounds, esters, alcohols, ethers, and sugars.

Some gums are complex carbohydrates that are used as additives to modify the properties of many manufactured foods today. They may be used as thickening agents, fat replacements, and texturizing agents. The seed gums include guar and locust bean; plant exudates are gum arabic, gum tragacanth, karaya gum, and gum ghatti. Red and brown seaweeds are the source of agar, carrageenan, and alginates. Xanthan and gellan gums are the product of microorganisms. Cellulose is modified to make cellulose gum (CMC), microcrystalline cellulose (MCC), methylcellulose (MC), and hydroxypropylmethyl cellulose (HPMC). Bran from cereals is a useful source of fiber, which can be incorporated into reduced-fat ground meat products, as well as into breads and other baked products.

## STUDY QUESTIONS

1. Describe the tissue systems and the types of cells contained in the edible parts of plants. Name the various structural features.

2. What are the structural constituents of plant foods? Identify their chemical structures.

3. What changes do the pectic substances undergo during maturation, and how do these changes influence their behavior in food products?

4. What changes occur after harvest, and how can the storage environment be controlled to optimize shelf life?

5. What is chlorophyllide, and how is it formed?

6. What is the key difference between the anthoxanthins and the anthocyanins? How does this influence color?

7. What is polyphenoloxidase and what happens when it is present in fresh produce containing anthoxanthins? How can this problem be minimized?

8. Compare the flavoring compounds in fruits with those in onions and with those in cabbage.

9. Find some food product labels that list gums as ingredients. What functions are performed by each of the gums?

10. Why are gums important ingredients in commercial food products, but not in food produced in the home?

11. True or false. When boiling vegetables, adding acid to water causes them to soften more quickly.

12. True or false. The pH of the cooking water when boiling red cabbage influences the color.

13. True or false. The type of pectic substance in a fruit changes during ripening.

14. True or false. Sclereids are unique cells in pears that contribute a gritty quality to the flesh of the fruit.

15. True or false. Ethylene gas sometimes is used to prevent spoilage of fruits during storage.

## BIBLIOGRAPHY

Anscombe, A. 2003. Regulating botanicals in food. *Food Technol.* 57 (1): 18.

Arscott, S.A. and Tanumihardjo, S.A. 2010. Carrots of many colors provide basic nutrition and bioavailable phytochemicals acting as a functional food. *Comp. Rev. in Food Science and Food Safety* 9 (2): 1541.

Barrett, D. M. and Theerakulkait, C. 1995. Quality indicators in blanched, frozen, stored vegetables. *Food Technol.* 49 (1): 62.

da Silva, J. A. L. and Rao, M. A. 1995. Rheology of structure development in high-methoxyl pectin/sugar systems. *Food Technol.* 49 (10): 70.

Deis, R. C. 2001. Dietary fiber: A new beginning. *Food Product Design 11* (9): 65–79.

Draughon, F.A. 2004. Use of botanicals as biopreservatives in foods. *Food Technol. 58* (2): 20.

DuBois, C. M., et al, eds. 2008. *World of Soy.* University of Illinois Press. Urbana, IL.

Duxbury, D. 2004. Dietary fiber: Still no accepted definition. *Food Technol. 58* (5): 70.

Foster, R. J. 2004. Fruit's plentiful phytochemicals. *Food Product Design: Functional Foods Annual* (September): 75.

Foster, R. J. 2005. Curious cloves. *Food Product Design 15* (3): 18.

Grenus, K. 2005. Focus on fiber. *Food Product Design 15* (3): 92.

Hazen, C. 2006. New fiber options for baked goods. *Food Product Design 15* (10): 80.

Homsey, C. 2002. Formulating natural products. *Food Product Design 11* (12): 33–57.

Hoover, D. G. 1997. Minimally processed fruits and vegetables: Reducing microbial load by nonthermal physical treatments. *Food Technol. 51* (6): 66.

Katz, P. 1998. That's using the old bean. *Food Technol. 52* (6): 42.

Katz, P. 2002. The fine line between functional foods and supplements. *Food Product Design 12* (3): 101.

Klahorst, S. J. 2004. Nutrigenomics: Window to the future of functional foods. *Food Product Design: Functional Foods Annual* (September): 5.

Kuntz, L. A. 2002. Formulating by gum, pectin, and gelatin. *Food Product Design 12* (3): 63.

Lauro, G. J. 2002. Color, naturally. *Food Product Design 12* (5): 141–156.

Luff, S. 2002. Phytochemical revolution. *Food Product Design: Functional Foods Annual* (September): 77.

Luff, S. 2005. Better ingredients through biotechnology. *Food Product Design 14* (10): 91.

Miraglio, A. M. 2003. Fiber in the morning. *Food Product Design 13* (4): 131.

Miraglio, A. M. 2005. Fruits for better health. *Food Product Design 15* (3): 115.

Miraglio, A. M. 2006. Beyond lycopene. *Food Product Design 15* (10): 77.

Nguyen, M. L. and Schwartz, S. J. 1999. Lycopene: Chemical and biological properties. *Food Technol. 53* (2): 38.

Ohm, L. M. 2004. Fortifying with fiber. *Food Technol. 58* (2): 71.

Ohr, L. H. 2003. Botanically speaking: Nutraceuticals and functional foods. *Food Technol. 57* (1): 65–67.

Ohr, L. M. 2004. Nutraceuticals and functional foods. *Food Technol. 58* (2): 71.

Ohr, L. M. 2006. Functional foods sneak peek. *Food Technol. 60* (5): 95.

Okezie, B. O. 1998. World food security: Role of postharvest technology. *Food Technol. 52* (1): 64.

Palmer, S. 2005. Cancer-fighting foods. *Food Product Design 15* (4): 93.

Pszczola, D. E. 2006. Fiber gets a new image. *Food Technol. 60* (2): 43.

Pszczola, D. E. 1997. Curdlan differs from other gelling agents. *Food Technol. 51* (4): 30.

Sapers, G. M. 1993. Browning of foods: Control by sulfites, antioxidants, and other means. *Food Technol. 47* (10): 75.

Schauwecker, A. 2005. To preserve and protect. *Food Product Design 15* (3): 67.

Schildhouse, J. 2002. Flax facts. *Food Product Design: Functional Foods Annual* (September): 102.

Schultz, M. 2005. Xanthan and foods: Bonded for life. *Food Product Design 15* (7): 17.

Sideras, G. M. 2006. Sweet fruit meets savory. *Food Product Design 16* (2): 52.

Sloan, A. E. 2000. Top ten functional food trends. *Food Technol. 54* (4): 33–62.

## INTO THE WEB

*http://www.timesonline.co.uk/tol/news/world/asia/article 6804512.ece*—Article about the contest in Kerala for a machine capable of doing the jobs done by skilled workers who harvest coconuts.

*http://www.latimes.com/features/food/la-fo-quince28-2009oct28,0,5254414.story*—Story on nutrient levels in carrots of various colors.

*http://www.ctahr.hawaii.edu/oc/freepubs/pdf/F_N-5.pdf*—Information on the sexes of papaya.

*http://www.chestnuts.us/?gclid=CKKn8K7qxJ4CFRgbaw odtm9_ow*—Chestnut farming information.

*http://jn.nutrition.org/cgi/content/full/129/7/1402S*—Information on inulin.

*http://www.oceanapproved.com/nutrition/*—Information on kelp as a vegetable.

*http://seaveg.com/*—Information on several sea vegetables.

*http://www.cpkelco.com/products-xanthan-gum.html*—Information on xanthan gum.

*http://www.wiley-vch.de/books/biopoly/pdf_v05/bpol5006_135_144.pdf*—Information on curdlan.

*http://www.codexalimentarius.net/gsfaonline/additives/details.html?id=51*—Codex information on sodium carboxymethyl cellulose.

# 4

# Lipids

Oil palm plantations in Malaysia supply a considerable amount of oil to the world market.

# CHAPTER 11

# Overview of Fats and Oils

## Chapter Outline



## OBJECTIVES

After studying this chapter, you will be able to:

1. Describe the chemical structures of edible lipids.
2. Discuss the importance of the crystalline nature of solid fats.
3. Explain the development of rancidity.
4. Trace the changes that occur in fats and oils as a consequence of heating.

## CHEMISTRY

**Lipids**, like carbohydrates, are organic compounds composed of carbon, oxygen, and hydrogen. However, relatively little oxygen is present; hydrogen comprises a much larger proportion in lipids than in carbohydrates. This difference in composition accounts for the large difference in the energy value: lipids provide 9 kilocalories per gram compared with 4 kilocalories per gram for carbohydrates.

The two key components of **simple fats** (the lipid class of greatest significance in food preparation) are **glycerol** and **fatty acids**. In these compounds, glycerol and a fatty acid(s) are linked together to form an ester, as shown here.

glycerol    fatty acid    simple fat (monoglyceride)

**Esterification of glycerol and a fatty acid**

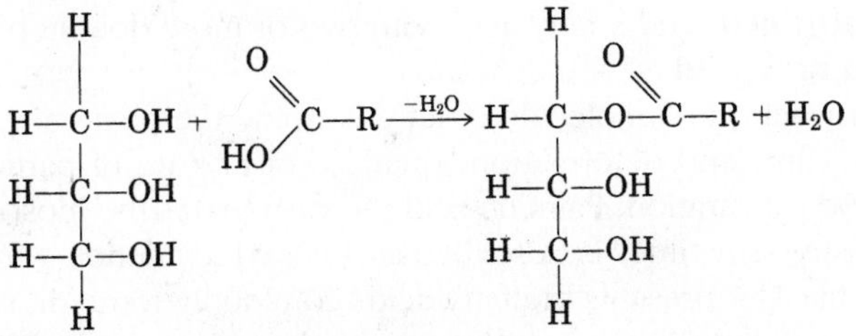

**Lipids**
Nonpolar, water-insoluble compounds composed of carbon, hydrogen, and a small amount of oxygen.

**Simple fats**
Lipids comprised only of glycerol and fatty acids.

**Glycerol**
Polyhydric alcohol containing three carbon atoms, each of which is joined to a hydroxyl group.

**Fatty acid**
Organic acid containing usually between 4 and 24 carbon atoms.

## Glycerol

As the preceding reaction shows, glycerol actually has three hydroxyl (alcohol or —OH) groups, each of which can be esterified with a fatty acid. The polyhydric nature of glycerol permits the formation of a wide range of simple fat molecules because each of the hydroxyl groups can esterify with a different fatty acid. This makes the range of possible molecules of simple fats extremely large. Researchers have found that food fats contain a relatively large and somewhat varying content of the different fatty acids. However, the glycerol molecule remains a constant structural feature.

## Fatty Acids

The fatty acids in simple fats are organic compounds with the characteristic carboxyl (-COOH) group identifying them as acids. They ordinarily contain an even number of carbon atoms. The smallest of these compounds is acetic acid (two carbon atoms), and the largest occurring with any regularity in foods is arachidonic acid (20 carbon atoms).

The structure of a fatty acid appears here, using oleic acid as an example:

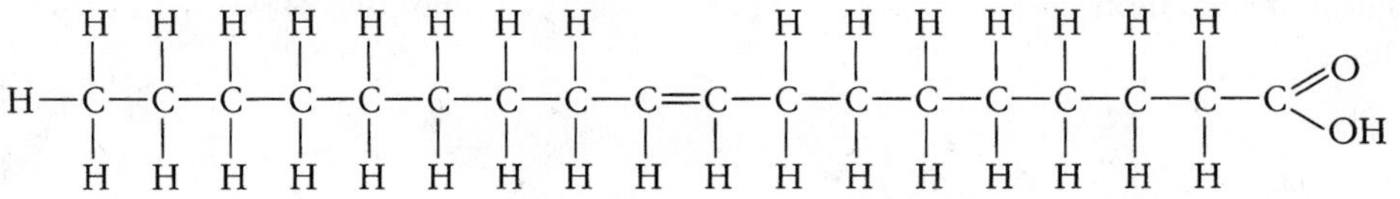

oleic acid

To facilitate identification of different carbon atoms in the structure, the carbon in the carboxyl group is numbered carbon 1, and the remaining carbons are numbered sequentially. This numbering system would be applied to the carbon atoms in oleic acid as follows:

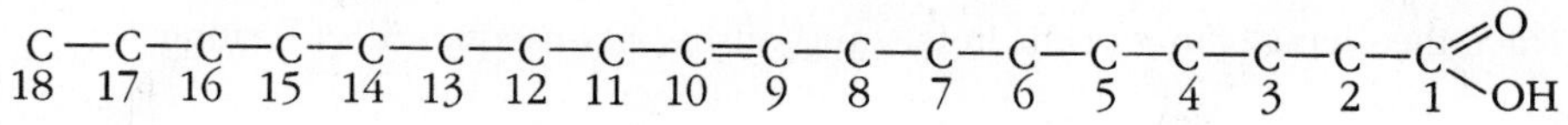

This representation, while accurate, is cumbersome; the total size of the molecule is only known when each of the carbon atoms is counted. A somewhat simpler system retains the accuracy while speeding identification simply by enclosing the $CH_2$ representation in parentheses and indicating the number of carbon atoms by use of a subscript. By this system, oleic acid can be written as follows:

$$CH_3(CH_2)_7CH = CH(CH_2)_7COOH$$

**Saturated fatty acids**
Fatty acids containing all of the hydrogen atoms they can possibly hold.

**Mono- and polyunsaturated fatty acids**
Fatty acids with one (mono) or two or more double bonds (polyunsaturated).

Some fatty acids are called **saturated fatty acids** because they contain all the hydrogen atoms it is chemically possible for them to have. Others contain one or more double bonds where more hydrogen can be added to the molecule. A fatty acid containing one double bond is **monounsaturated**, and a fatty acid with two or more double bonds is classified as a **polyunsaturated fatty acid**.

The specific fatty acids in a molecule of fat determine the physical characteristics of the fat. The rheological (flow and deformation) qualities of fats are of particular interest when they are used in food preparation. Palmitic acid (16 carbons) is the most abundant saturated fatty acid. Its chemistry sometimes is described as 16:0, which denotes 16 carbon atoms and 0 double bonds. Table 11.1 presents the fatty acids commonly found in simple fats in foods.

The most abundant monounsaturated fatty acid is oleic acid, which is described chemically as 18:1 (18 carbon atoms and 1 double bond). A more complete chemical description, 18:1(n-9), indicates that the double bond is at the 9th carbon, starting from the methyl carbon. This terminal carbon, which is at the end furthest from the carboxyl end, is designated as the omega (ω) carbon.

**Table 11.1**   Some Fatty Acids Occurring in Foods

| Common Name | Carbon Atoms | Double Bonds | Approx. Melting Point (°C) | Structure |
|---|---|---|---|---|
| Butyric | 4 | 0 | −8 | $CH_3(CH_2)_2COOH$ |
| Caproic | 6 | 0 | −4 | $CH_3(CH_2)_4COOH$ |
| Caprylic | 8 | 0 | 16 | $CH_3(CH_2)_6COOH$ |
| Capric | 10 | 0 | 31 | $CH_3(CH_2)_8COOH$ |
| Lauric | 12 | 0 | 44 | $CH_3(CH_2)_{10}COOH$ |
| Myristic | 14 | 0 | 54 | $CH_3(CH_2)_{12}COOH$ |
| Palmitic | 16 | 0 | 64 | $CH_3(CH_2)_{14}COOH$ |
| Palmitoleic[a] | 16 | 1 | 1 | $CH_3(CH_2)_5CH = CH(CH_2)_7COOH$ |
| Stearic | 18 | 0 | 70 | $CH_3(CH_2)_{16}COOH$ |
| Oleic[a] | 18 | 1[b] | 14 | $CH_3(CH_2)_7CH = CH(CH_2)_7COOH$ |
| Elaidic[a] | 18 | 1[c] | 44 | $CH_3(CH_2)_7CH = CH(CH_2)_7COOH$ |
| Linoleic[a] | 18 | 2 | −5 | $CH_3(CH_2)_4CH = CHCH_2CH = CH(CH_2)_7COOH$ |
| α-linolenic[a] | 18 | 3 | −11 | $CH_3(CH_2CH = CH)_3(CH_2)_7COOH$ |
| Arachidic | 20 | 0 | 76 | $CH_3(CH_2)_{18}COOH$ |
| Arachidonic[a,d] | 20 | 4 | −50 | $CH_3(CH_2)^4(CH = CHCH_2)_4(CH_2)_2COOH$ |

[a]Unsaturated fatty acid.
[b]Double bond is *cis* configuration.
[c]Double bond is *trans* configuration.
[d]The systematic name is 5, 8,11,14-eicosatetraenoic acid.

Health benefits associated with some polyunsaturated fatty acids focus attention on omega-3 and omega-6 fatty acids. Linoleic acid can be described chemically as 18:2(n-6) when identifying its structure by omega category or *cis*-9, *cis*-12-octadecanoic acid, a description that indicates 18 carbon atoms and double bonds in the *cis* configuration at carbons 9 and 12 (counting from the acid carbon). It is an omega-6 fatty acid.

Alpha-linolenic acid is an omega-3 fatty acid consisting of 18 carbon atoms and double bonds at carbons 9, 12, and 15; its chemical description is 18:3(n-3) or *cis*-9, *cis*-12, *cis*-15-octadecatrienoic acid. It is of particular importance because it is a precursor of 20-carbon **eicosapentanoic acid (EPA)** and 22-carbon **docosahexanoic acid (DHA)**, both of which are highly unsaturated (5 and 6 double bonds, respectively) omega-3 fatty acids that are biologically active.

The double bonds in naturally occurring polyunsaturated fatty acids usually are in the *cis* form and have a -$CH_2$ group between the carbons at the double bonds; hence, they are designated as nonconjugated fatty acids. However, **conjugated linoleic acid (CLA)**, a fatty acid that is abundant naturally in milk fat (the result of microbial action in the rumen), consists of double bonds on alternate carbon atoms, with one in the *cis* form and one in the *trans* form. Despite the fact that CLA has a *trans* double bond, it is not considered to be a *trans* fat for labeling purposes because FDA specifies only nonconjugated *trans* double bonds need to be reported on the nutrition label. Actually, CLA may have some health benefits.

Of special interest in Table 11.1 is the **melting point** of the various fatty acids. The melting point, which is the temperature at which the fatty acid is transformed from a solid to a liquid, is significant in determining specific applications in food preparation.

The flow properties and ability to solidify into crystalline form that fatty acids exhibit are significantly influenced by two factors: (1) their chain length and (2) the degree of unsaturation. As Table 11.1 shows, short-chain fatty acids have lower melting points than do long-chain fatty acids. This means that long-chain fatty acids are more likely to be solid at mixing temperatures than are short-chain fatty acids. When fats are hard (have high melting points) at room temperature, they may be poorly suited for specific preparations, such as creaming a shortened cake mixture.

**Eicosapentanoic acid (EPA)**
Biologically active omega-3 fatty acid with 20 carbon atoms and 5 double bonds.

**Docosahexanoic acid (DHA)**
Biologically active omega-3 fatty acid containing 22 carbon atoms and 6 double bonds.

**Conjugated linoleic acid (CLA)**
Fatty acid [18:2(n-6)] in milk fat that has one *cis* and one *trans* double bond on alternate carbons; not considered to be a *trans* fat.

**Melting point**
The temperature at which the crystals of a solid fat melt.

**Unsaturation**
Lack of hydrogen relative to the amount that can be held, a situation characterized by a double bond between two carbon atoms in the fatty acid chain.

**Stearic acid**
Saturated 18-carbon fatty acid.

**Oleic acid**
Monounsaturated 18-carbon fatty acid.

**Linoleic acid**
Essential fatty acid (18 carbons) containing two double bonds.

**Linolenic acid**
Fatty acid (18 carbons) containing three double bonds.

***Cis* configuration**
The hydrogens attached to the carbon atoms on either end of the double bond are from the same orientation and cause a lower melting point.

***Trans* configuration**
The hydrogens attached to the carbon atoms on either end of the double bond are from opposite directions and cause a higher melting point.

Another factor that affects melting points of fatty acids is the degree of **unsaturation**. Saturated fatty acids (acids holding all of the hydrogen possible) have higher melting points than do their counterparts containing a double bond (Figure 11.1). The higher the degree of unsaturation (i.e., the more double bonds in a fatty acid), the lower the melting point. This can be seen by comparing the 69.6°C (157.3°F) melting point of **stearic acid** (saturated 18-carbon fatty acid), the 14°C (57.2°F) melting point of **oleic acid** (18-carbon fatty acid with one double bond), the –5°C (23°F) melting point of **linoleic acid** (18-carbon fatty acid with two double bonds), and the –11°C (12.2°F) melting point of **linolenic acid** (also an 18-carbon fatty acid, but with three double bonds). This is why linoleic and linolenic acids remain fluid in the refrigerator.

At double bonds, the configuration is either *cis* or *trans*. If the double bond is in the *cis* form, the melting point is appreciably lower than when the comparable molecule has a double bond in the *trans* form. Table 11.1 illustrates this; the melting point for oleic acid (*cis* form) is 14°C (57.2°F) and for elaidic acid (*trans* form) is 43.7°C (110.7°F). Because both of these fatty acids contain 18 carbon atoms and one double bond, the difference in melting point is attributable to the difference in configuration at the double bond.

$$
\begin{array}{cc}
\text{cis} & \text{trans} \\
\text{(oleic m.p. = 14°C)} & \text{(elaidic m.p. = 43.7°C)}
\end{array}
$$

The *trans* configuration is able to form van der Waals forces with other molecules fairly easily because of its linearity, which permits molecules to approach each other closely. The *cis* form, on the other hand, is somewhat angular because of the bend in the molecule imposed by the orientation at the double bond. This makes it slightly more difficult for other molecules to align themselves closely enough for van der Waals forces to immobilize the molecules into a crystalline framework. Consequently, crystallization of the *cis* form requires

**Figure 11.1**  Rapeseed blooms as a prelude to making its oil-rich seeds that are the source of canola oil, an oil that is valued for its mono- and polyunsaturated fatty acids.

that considerable energy be removed from the system by cooling the mixture to a low temperature so that the molecules move slowly enough for van der Waals forces to begin to lock some molecules together, at least briefly.

## STRUCTURES OF FATS IN FOODS

Fats in foods can be classified on the basis of the number of fatty acids esterified to the glycerol molecule. If one fatty acid is esterified, the compound is designated as a **monoglyceride**. Chemists refer to monoglycerides as **monoacylglycerides**, which accurately identifies the presence of the $-COO^-$ of one fatty acid to form an ester with one $-OH$ of the glycerol. The fatty acid may be on one of the terminal carbon atoms or on the center one. The configurations possible are shown here. The R group represents the carbon chain of the fatty acid exclusive of the carboxyl group.

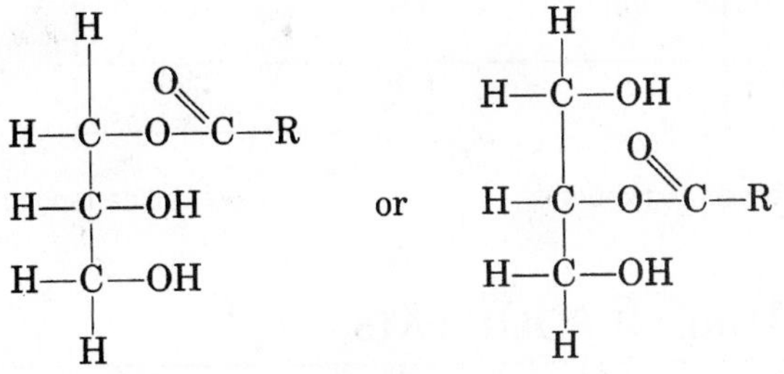

**Monoglyceride**
Lipid consisting of one fatty acid esterified to one of the hydroxyl groups of glycerol; synonym for *monoacylglyceride*.

**Monoacylglyceride**
Chemical name used to clarify the ester formed with one fatty acid and glycerol.

forms of monoglycerides

**Diglycerides**, also called **diacylglycerides**, contain two fatty acids esterified on one glycerol molecule. Esterification of the two fatty acids may be on the two terminal carbon atoms of glycerol or on two adjacent carbon atoms.

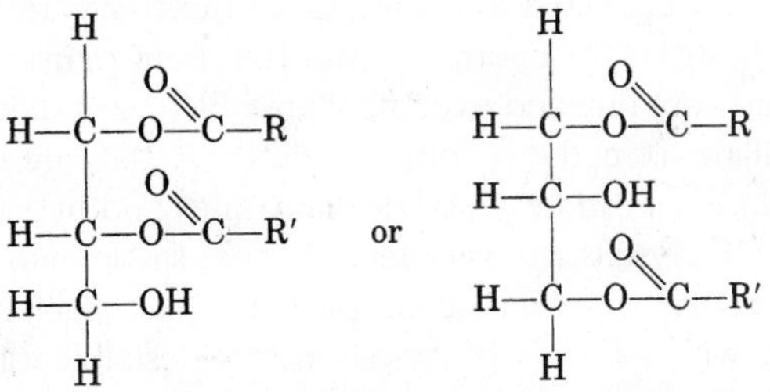

**Diglyceride (diacylglyceride)**
Simple fat containing two fatty acids esterified to glycerol.

forms of diglycerides

**Triglycerides** or **triacylglycerides**, the most common form of food fats, are composed of three fatty acids esterified to the glycerol molecule. This is the maximum number of fatty acids that can exist in the compound because there are no additional hydroxyl groups to which a fatty acid can be esterified. The general formula for this structure is:

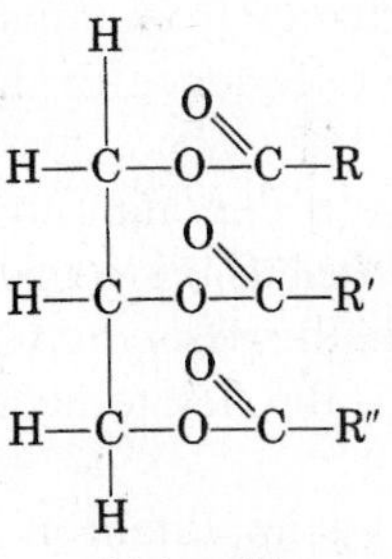

**Triglyceride (triacylglyceride)**
Simple fat containing three fatty acids esterified to glycerol; the most common form of simple fat.

triglyceride

In the preceding structures, the importance of the fatty acid moiety may be minimized by the use of the symbol R to represent any chain length. This is somewhat misleading because the R group for the most common fatty acids found in fats contains 16 or 18 carbons [palmitic (16 carbons, 0 double bonds), palmitoleic (16 carbons, 1 double bond), stearic (18 carbons, 0 double bonds), oleic (18 carbons, 1 double bond), and linoleic (18 carbons, 2 double bonds)]. These 16- and 18-carbon chains dominate the overall structure of the fat molecules, for they are significantly larger than the 3-carbon chain of glycerol.

Although the structural formula for fat molecules usually is written in the form of a capital letter E, with all three fatty acids seeming to extend in the same direction from the glycerol, steric hindrance precludes this from actually occurring. Instead, the three fatty acids extend spatially in different directions. The tuning fork arrangement is postulated to be one of the feasible configurations. In the diagram that follows, the glycerol moiety is represented by the vertical line and the three fatty acids by horizontal lines. A chair arrangement also has been suggested as being representative of the actual spatial arrangement.

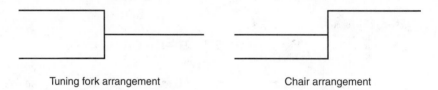

Tuning fork arrangement                    Chair arrangement

## CRYSTALLINE NATURE OF SOLID FATS

Fats high in polyunsaturated fatty acids are fluid (oils) at room temperature, but many food fats are in solid form at refrigerator and room temperatures. Although these fats appear to be a solid mass, they actually are a mixture of crystals of fat in oil. The nature of the crystals influences the usefulness of these solid fats in food preparation.

When melted fats cool, the molecules gradually move more and more slowly. Occasionally, a molecule links to another by means of van der Waals forces, and crystals begin to form. The tuning fork configuration is well suited to alignment of molecules to form a crystalline matrix.

**Alpha (α) crystals**
Extremely fine and unstable form of fat crystals.

**Beta prime (β′) crystals**
Very fine and reasonably stable fat crystals.

**Intermediate crystals**
Slightly coarse fat crystals that form when β′ crystals melt and recrystallize.

**Beta (β) crystals**
Extremely coarse and undesirable fat crystals.

Fat crystals may be in any of four forms: **alpha (α)**, **beta prime (β′)**, **intermediate**, and **beta (β)**. The α crystals are very fine and extremely unstable. They quickly melt and recrystallize into the next larger crystalline form, the β′ form. When fat crystals are in the β′ form, the fat has an extremely smooth surface, as can be noted in margarines and hydrogenated vegetable shortening of high quality. The β′ crystals are considerably more stable than are the α crystals; in fact, they are stable enough to survive the marketing process unless subjected to high temperatures during storage. For baking with solid fats, β′ crystals are the desirable form. Their presence aids in promoting a fine texture in the finished product (Figure 11.2).

Intermediate crystals give a somewhat grainy appearance to a fat and are not recommended for use. They may form if a fat is stored at a warm temperature. Under this condition, β′ crystals melt gradually and then recrystallize into the larger, coarser intermediate form (Figure 11.3).

The coarsest crystalline form is the extremely stable β form. When a fat melts completely and then recrystallizes without being disturbed, beta crystals form. This process can be seen when a small amount of fat is melted and cooled slowly without being stirred. One thing that is apparent is the gradual change from transparent melted fat to the opaque solid fat. This increase in opacity is the consequence of the crystals forming and refracting the light differently (Figure 11.4).

Controlling crystal size in solid fats is important from the perspective of visual appeal of the very smooth surface of the fat, as well as performance in baked products. The β′ crystals (not the intermediate or the β form) are the desired crystals. During commercial production of shortenings, agitation during cooling is the key to achieving the necessary control of crystal formation.

Storage at cool temperatures also is important in the marketing chain if the β′ crystals are to reach consumers. If warehouses or shipping containers are allowed to get extremely

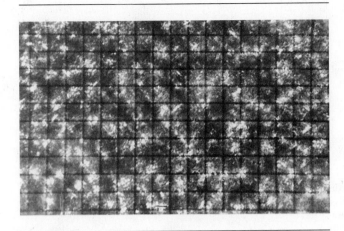

**Figure 11.2**    Photomicrograph of beta prime (β′) fat crystals in polarized light (×200). Grid lines represent 18 microns.

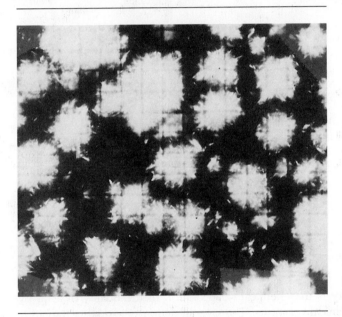

**Figure 11.3**    Photomicrograph of intermediate fat crystals in polarized light (×200). Grid lines represent 18 microns.

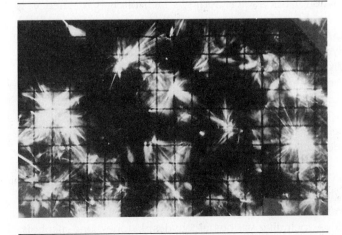

**Figure 11.4**    Photomicrograph of beta (β) fat crystals in polarized light (×200). Grid lines represent 18 microns.

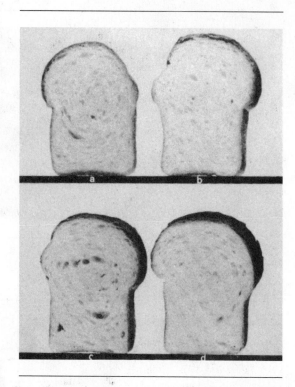

**Figure 11.5** Crystal size influences the texture of breads made with lard containing crystals of varying size: (a) control with no lard; (b) beta prime (β′) to intermediate crystals; (c) beta (β) crystals, and (d) beta prime (β′) crystals. Note the increased volume and closer grain when β′ crystals are present, as compared with the effect of β crystals. (Courtesy of J. G. Mahdi, Kansas State University and Baker's Digest. In Mahdi, J. G., et al. 1981. Effect of mixing atmosphere and fat crystal size on dough structure and bread quality. *Bakers Digest* 55 (2): 28.

warm, the quality of the fats they house begins to be modified as the β′ crystals melt, and the fat recrystallizes in the form of intermediate or even β crystals (Figure 11.5). Fat crystals can slowly gather around a nucleus to form spherulites, which are symmetrical superstructures of single fat crystals in concentric rings. Spherulites contribute a very coarse texture to the solid fat.

**Spherulite**
Spherical crystalline body of radiating crystal fibers.

## CHEMICAL DEGRADATION

### Rancidity

**Rancidity**
Chemical deterioration of a fat caused by the uptake of oxygen (oxidation) or water (hydrolysis).

**Oxidative rancidity**
Development of off flavors and odors in fats as a result of the uptake of oxygen and the formation of peroxides, hydroperoxides, and numerous other compounds.

**Free radical**
Unstable compound containing an unpaired electron.

**Peroxide**
Compound with oxygen attached to oxygen.

**Hydroperoxide**
Compound containing a —O—O—H group.

*Oxidative Rancidity.* **Rancidity** is the chemical deterioration of the quality of a fat by either oxidative or hydrolytic chemical reactions. The overall action of oxidative rancidity involves the uptake of oxygen at a double bond in an unsaturated fatty acid in a fat. When fats are exposed to oxygen, the double bond can be broken so that oxygen can then become a part of the molecule.

**Oxidative rancidity** begins when a free radical forms, which often is initiated in a polyunsaturated fatty acid, such as linoleic acid. Its structure is able to form a **free radical** at any of three carbons (9, 11, or 13). The free radical then combines with two oxygen atoms to form a **peroxide**. Subsequently, one hydrogen atom is removed from another unsaturated fatty acid, and that hydrogen is added to complete the formation of the **hydroperoxide** on the first fatty acid.

Unfortunately, a new free radical is formed when that hydrogen atom is removed from the second fatty acid. In turn, two oxygen atoms are added to this second free radical; now yet another fatty acid is stripped of one hydrogen atom to finish formation of the second hydroperoxide, leaving yet another free radical. This process is autocatalytic—that is, it is self-perpetuating.

Oxidative rancidity requires oxygen, but it is facilitated by the presence of certain metals (e.g., iron and copper) and by light and/or warm temperatures. The objectionable flavor changes associated with oxidative rancidity come from the formation of a wide range of compounds from the hydroperoxides, including various aldehydes, ketones, and alcohols.

The early phase in which this oxygen uptake begins to occur is referred to as the *induction period*. The fact that the oxidation reaction keeps generating additional free radicals to maintain the momentum of the formation of peroxides and hydroperoxides explains why this oxidation is called **autoxidation**.

The rate of oxidation accelerates markedly during the propagation period that follows the induction phase. One free radical formed during the induction period may lead to the formation of a hundred peroxide molecules before two peroxy radicals collide and termination occurs.

Storage in tightly closed containers in a cool, dark place helps slow the onset and continued development of oxidative rancidity. However, some oxygen still remains in the headspace of the closed container, and eventually oxidation will begin. The problems of oxidative rancidity can be delayed considerably by the addition of antioxidants to products high in unsaturated fatty acids, for these fats are particularly susceptible to oxidation.

**Antioxidants** protect against the development of oxidative rancidity by providing a hydrogen atom from their own molecule to react either with a free radical of a fatty acid or with a peroxide that has already formed. The addition of the hydrogen (from the antioxidant) to the free radical results in the return to the original fatty acid. If the hydrogen unites with the peroxide, a stable hydrogen peroxide is formed.

In both of these cases, fairly stable compounds result, and the autocatalytic continuation of oxidative rancidity is blocked. The free radical that is formed by the donation of the hydrogen atom from the antioxidant is not able to participate in the formation of a free radical in the fatty acid.

Several synthetic antioxidants are used extensively as additives to help retard oxidative rancidity in fats and oils (Table 11.2). **Tertiary-butylhydroquinone (TBHQ)** is particularly effective in both vegetable oils and most other fats. It is well suited for use not only in the fats and oils themselves, but it also retains its antioxidant capability in baking or frying. **Butylated hydroxyanisole (BHA)** is an effective antioxidant in animal fats used for baking. **Butylated hydroxytoluene (BHT)** also is rather effective in animal fats, but not in vegetable oils. **Propyl gallate (PG)** is somewhat useful as an antioxidant in vegetable oils, but its action is enhanced synergistically by combining it with BHA and BHT.

Some other substances act as scavengers to combine directly with copper or iron that might be present, thus blocking the ability of the metal to catalyze the oxidation of the fat or oil. Additives that are effective as sequestrants or scavengers include ethylenediaminetetraacetate (EDTA), ascorbyl palmitate, ascorbic acid, and citric acid. Some compounds (e.g., ascorbic acid) can interact to be oxidized themselves before the fatty acids undergo oxidation. Tocopherols, tocotrienols, polyphenols, and carotenoids are other examples of natural antioxidants which are effective in retarding oxidative rancidity because of their ease of oxidation.

**Autoxidation**
Oxidation reaction capable of continuing easily with little added energy.

**Antioxidant**
Compound that can retard oxidative rancidity by providing hydrogen to block formation of free radicals in fatty acids or by scavenging metal or oxygen.

**Tertiary-butylhydroquinone (TBHQ)**
Antioxidant often added to animal fats used in baking and frying.

**Butylated hydroxyanisole (BHA)**
Antioxidant effective in animal fats used in baking.

**Butylated hydroxytoluene (BHT)**
Antioxidant used to retard oxidation in animal fats.

**Propyl gallate (PG)**
Antioxidant somewhat effective in vegetable oils, often used in combination with BHA and BHT.

**Table 11.2    Some Chemical and Natural Antioxidants and Their Uses**

| Antioxidant | Action/Characteristics | Applications |
|---|---|---|
| EDTA[a] | Slow oxidation by metals | Vegetable oil-containing foods |
| Citric acid | Chelate metals in meat | Meats |
| Phosphates | Complexes with metal ions | Meats |
| BHA[b] | Survives baking and frying | Foods containing animal fats |
| BHT[c] | Survives baking and frying | Foods containing animal fats |
| TBHQ[d] | Survives frying temperatures | Vegetable oil-containing foods |
| Propyl gallate | Heat sensitive | Vegetable oil-containing foods |
| Tocopherols | Can add with vitamin C, etc. | Foods containing animal fats |
| Rosemary | Delay free radical formation | Meats, irradiated ground beef |
| Thyme, oregano | Avoid warmed-over flavor | Comminuted poultry, meat, fish |
| Dried plums | Retard lipid oxidation | Sausage and other ground meat |
| Honey | Darker is more effective | Ground turkey |

[a]Ethylenediaminetetraacetic acid
[b]Butylated hydroxyanisole
[c]Butylated hydroxytoluene
[d]Tertiary-butylhydroquinone

## FOOD FOR THOUGHT: Sunlight and Olive Oils

Virgin olive oils are prized for their delicate flavors, which result from their complex content of assorted organic compounds. Consumers have many choices of brands and qualities of olive oil. Each of these products will have slightly different quantities of volatile and flavor compounds, which accounts for the subtle differences in these oils.

Olive oil undergoes oxidative changes during storage as a consequence of lipoxygenase activity. Exposure to sunlight for a few days causes significant changes in the amounts of various compounds, resulting in undesirable flavor changes. For example, autoxidation of linoleic acid (18:2) leads to greatly increased amounts of trans-2-heptanal, 1-octene-3-ol, trans-2-octenal, and trans-2-nonenal, but decreased amounts of 2,4-decadienal. These changes contribute to the flavor changes as olive oil becomes rancid. Deteriorative chemical changes occur less quickly if the oil is stored in a dark, cool place.

**Lipolysis**
Reaction of a molecule of water with a fat molecule to release a free fatty acid in the presence of lipase or heat.

**Lipase**
Enzyme that catalyzes the hydrolysis of fat to yield free fatty acids and glycerol.

**Hydrolytic rancidity**
Lipolysis (hydrolysis) of lipids to free fatty acids and glycerol, often catalyzed by lipases.

*Hydrolytic Rancidity.*   The hydrolytic reaction in which free fatty acids are split from the glycerol in fat molecules, **lipolysis**, is shown here. The uptake of a molecule of water usually is promoted either by the action of **lipase** or by heat. Although some time is required for the complete liberation of a molecule of glycerol from a triglyceride, eventually one triglyceride molecule undergoes lipolysis to yield three free fatty acid molecules and one molecule of glycerol. These free fatty acids may seriously alter the aroma and flavor of a fat or oil in which lipolysis has occurred. This breakdown is termed **hydrolytic rancidity**.

Lipolysis

## Reversion

**Reversion**
Development of an off flavor (beany or fishy) in soybean, rapeseed, or various fish oils as a result of a reaction involving only very minor amounts of oxygen.

Another deteriorative change in fats occurs with only a small amount of oxygen present and apparently is the result of oxidation of some of the linoleic and linolenic acids (18:2 and 18:3, respectively) in oils. This deteriorative change, which results in development of off odors and off flavors, is called **reversion**. Flavors that develop as a result of reversion are often described as "fishy" or "beany" and are definitely detrimental to quality.

The oil most commonly plagued by this problem is soybean oil, but rapeseed oil and various fish oils also are very susceptible to reversion. Some additional flavor problems may be generated by heat reversion, a reaction apparently stemming from changes in glycerol. At the present time, the fact that reversion occurs is recognized as a significant problem when working with products containing susceptible oils, but research has not yet led to a solution.

## Effects of Heat

**Smoke point**
Temperature at which a fat or oil begins to emit some traces of smoke.

*Smoke Point.*   Fats and oils can be heated to extremely high temperatures, but eventually there is so much energy that the molecules begin to degrade. When fats and oils begin to degrade, a slight amount of smoke will appear. This phenomenon is called the **smoke point**. It occurs at high temperatures (Table 11.3) but varies with the substance being

**Table 11.3**    Smoke Points of Fats and Oils[a]

| Fat/Oil | Smoke Point | |
|---|---|---|
| | (°C) | (°F) |
| Safflower oil | 267 | 513 |
| Soybean oil | 248 | 478 |
| Canola oil | 243 | 470 |
| Corn oil | 242 | 468 |
| Palm | 235 | 455 |
| Peanut oil | 234 | 453 |
| Sunflower oil | 232 | 450 |
| Sesame seed oil | 226 | 438 |
| Grapeseed oil | 216 | 420 |
| Olive oil, extra virgin | 191 | 375 |
| Lard | 183–205 | 361–401 |
| Shortenings | 180–188 | 356–370 |
| Butter | ~177 | ~350 |

[a]Smoke points drop during prolonged heating.

heated and with its previous history (exposure to heat, water, and food particles). If fats and oils are heated well beyond the smoke point, they reach a flash point and can ignite.

The weakest linkage in the structure of a fat is the ester linkage joining the fatty acid(s) to the alcohol (glycerol). With the uptake of a molecule of water, lipolysis occurs, releasing a free fatty acid (shown in the equation for lipolysis). If the molecule is a monoglyceride, glycerol will be the other product of lipolysis. However, diglycerides and triglycerides also undergo lipolysis, with the reaction continuing until all the fatty acids have been removed and glycerol is formed. The temperature for such reactions to occur usually is 190°C (374°F) or higher.

Deteriorative changes do not stop there. Continued heating will cause the removal of two molecules of water from the glycerol, which results in the formation of an unusual aldehyde called **acrolein**.

**Acrolein**
A highly irritating and volatile aldehyde formed when glycerol is heated to the point at which two molecules of water split from it.

$$\begin{array}{c} H \\ | \\ H-C-OH \\ | \\ H-C-OH \\ | \\ H-C-OH \\ | \\ H \end{array} \xrightarrow[\text{Heat}]{-2H_2O} \begin{array}{c} H-C=O \\ | \\ H-C \\ \| \\ H-C-H \end{array} + 2H_2O$$

glycerol                                   acrolein

Acrolein is almost immediately vaporized, causing the fat to smoke. Acrolein-containing smoke is extremely irritating to the eyes and respiratory passages. Fats that have undergone sufficient lipolysis so that the glycerol content is fairly high exhibit a rather serious drop in smoke point, which rapidly approaches 190°C (374°F), the usual temperature for frying. This causes considerable discomfort for chefs when these are used for deep-fat frying.

The smoke point of a fat is not a specific temperature. As a fat is used at high temperatures over a period of time, its smoke point will gradually drop until the flavor and appearance of the foods fried in it are unacceptable. Different oils used for frying naturally have somewhat different original smoke points, but the point is sufficiently above 190°C (374°F) that most oils (with the exception of olive oil) can be used satisfactorily for a reasonable length of time before the smoke point drops to frying temperature or below.

Shortenings or other fats that have monoglycerides present quickly are altered so that the smoke point is too low for acceptable fried products. This result is not surprising because only one fatty acid has to be removed before the glycerol is available to begin forming acrolein.

**Polymerization.**    The free fatty acids formed as a result of lipolysis follow quite a different course from that of the changes in glycerol. Instead, these free fatty acids undergo additional chemical modification as they continue to be held at frying temperature. No single route of reaction has been identified for the free fatty acids that are liberated in very hot fat. Instead, it appears that several different reactions probably occur as they are coupled together into polymers of varying lengths or sometimes transformed into smaller molecules.

**Polymerization**
Formation of a variety of polymers, including simple dimers and trimers, when free fatty acids are subjected to intense heat for a long period during frying.

Fatty acids containing at least one double bond are particularly susceptible to **polymerization**. Through formation of new carbon-to-carbon bonds, dimers (two fatty acids) and trimers (three fatty acids) of a cyclic nature begin to evolve from the free fatty acids. Oxygen-to-carbon bonds may lead to oxidative polymerization when some hydroperoxides of polyunsaturated fatty acids are found in the fat used for frying.

Although the specific compounds formed during prolonged heating of fats doubtless are extremely varied, clearly the resulting compounds are appreciably larger than the original free fatty acids. These larger polymers of the free fatty acids increase the viscosity of the hot fat appreciably. In fact, the increased viscosity is evident to even the casual observer and is a sign that the fat is altered significantly in its chemical composition. Darkening of the color of the fat during extended use in frying is yet another indication of loss of quality.

**Acrylamide**
Carcinogen formed from natural sugars and asparagines in starchy fried foods and also in baked products.

**Acrylamide.**    Acrylamide (known to chemists as ethylene carboxyamide or 2-propenamide) is a monomer formed in foods containing starch, some sugar, and asparagine (an amino acid) when they are cooked at high temperatures. The structure of acrylamide is as follows:

$$H_2C=C-C=O$$
$$\overset{|}{\underset{NH_2}{}}$$

Acrylamide

This substance was not a subject of much interest in the food world until 2002, when Swedish researchers (Tareke, et al., 2002) reported finding high levels of acrylamide in staple foods, including rice and potatoes. Although acrylamide is recognized as a carcinogen in laboratory animals, the significance for humans of the levels reported by the researchers is the subject of considerable research today.

Recognition of this possible health concern has prompted governmental efforts to attempt to establish guidelines. At the present time, the levels of acrylamide that are consumed by most people are not thought to be dangerous.

Acrylamide is not naturally present in foods, but it can form at high temperatures in foods containing the essential components. Potatoes and baked products not only contain starch, natural sugars, and some proteins (specifically, the amino acid asparagine), but they usually are fried or baked at high temperatures [at least 163°C (325°F) and often much hotter], which creates some acrylamide. Baked products develop increasing amounts as they gradually brown, and so do potato chips and French fries subjected to very high temperatures [190°C (374°F)] during frying. Roasting coffee beans also form some acrylamide as they darken. In these and any other foods naturally containing some sugar and asparagine, increased browning creates higher levels of acrylamide.

One means of helping to minimize acrylamide formation is to soak French fries in water for half an hour before blotting them dry and then frying. Also, frying to a light golden brown rather than a deeper color minimizes the level. Similarly, baked products will form less acrylamide if they are removed from the oven when they are starting to brown, but are not yet dark. If people are particularly concerned about acrylamide levels, they can opt to eat boiled or microwaved potatoes. That option has the added advantage of lowering fat intake. French-roast coffee has a higher level than those that are lightly roasted.

## SUMMARY

The physical behavior of fats and oils is determined by the fatty acids that are esterified to glycerol to form monoglycerides, diglycerides, and triglycerides, which are the common forms of fats used in food preparation. Carbon chain length and degree of saturation determine the melting point of fatty acids. These fat molecules are found in foods as oils or as crystalline fats, with the crystal size ranging from alpha to beta prime, then on to intermediate, and finally to beta crystals. Beta prime is the preferred size for food applications.

Rancidity in unsaturated fatty acids can develop as a consequence of oxidation, a process that can be hastened by metals and/or warm temperatures. Antioxidants act as deterrents to oxidation of fats. In addition to oxidative rancidity, fats may undergo hydrolytic rancidity, which occurs when fatty acids are split from glycerol by the addition of a molecule of water at the reaction site.

Reversion is the development of beany or fishy aroma and flavor as the result of oxidation of some linoleic (18:2) and linolenic (18:3) acids in an oil.

Fats heated to high temperatures for an extended period will begin to smoke as glycerol and ultimately acrolein are formed. This is possible because of the release of free fatty acids from mono-, di-, and/or triglycerides in intense heat. The free fatty acids polymerize as heating continues. Fats need to be heated to temperatures below their smoke points to avoid this degradation. Acrylamide forms from sugar and asparagines at high temperatures when starchy foods are fried or baked.

## STUDY QUESTIONS

1. Write the chemical reaction for the esterification of glycerol with a molecule of stearic acid, one of oleic acid, and one of butyric acid.

2. Identify three factors that influence the melting point of a fatty acid and describe the effect of each factor.

3. Why is the crystalline form of a fat important in making shortened cakes?

4. Is the crystalline form of a fat of concern when selecting a fat for frying? Explain your answer.

5. Carefully describe the development of oxidative rancidity and of hydrolytic rancidity. Be sure to include chemical formulas and reactions to clarify each of the processes.

6. Describe chemically the degradative changes that an oil undergoes when it is being used for frying over an extended period.

7. True or false. Olive oil has more polyunsaturated fatty acids than lard does.

8. True or false. A short-chain fatty acid has a higher melting point than one with a long chain.

9. True or false. A double bond raises the melting point of a fatty acid.

10. True or false. A solid fat with beta crystals gives better results in cakes than one with beta prime crystals.

11. True or false. When a molecule of acrolein is formed, one molecule of water is formed.

12. True or false. Oxidative rancidity is accelerated by the presence of some water.

13. True or false. Hydrolytic rancidity requires oxygen.

14. True or false. Frying a potato can cause the formation of acrylamide.

15. True or false. Oxidative rancidity occurs at the ester linkage between glycerol and a fatty acid.

## BIBLIOGRAPHY

Allison, D. B., et al. 1999. Estimated intakes of *trans* fatty and other fatty acids in the US population. *J. Am. Diet. Assoc.* 99 (2): 166.

Berry, D. 2003a. Fat that's fit for oven and fryer. *Food Product Design 12* (11): 53–67.

Berry, D. 2003b. Fats' chance. *Food Product Design 12* (3): 74–88.

Berry, D. 2005a. Designer lipids. *Food Product Design 14* (12): 118.

Berry, D. 2005b. Lowering LDLs. *Food Product Design 15* (6): 71.

Bren, L. 2003. Turning up the heat on acrylamide. *FDA Consumer 37* (1): 10–11.

Clark, J. P. 2005. Fats and oils processors adapt to changing needs. *Food Technol. 59* (5): 74.

Coughlin, J. R. 2003. Acrylamide: What we have learned so far. *Food Technol.* 57 (2): 100.

Duxbury, D. 2005a. Analyzing fats and oils. *Food Technol.* 59 (4): 66.

Duxbury, D. 2005b. Omega-3s offer solutions to *trans* fat substitution problems. *Food Technol.* 59 (4): 34.

Fortin, N. D. 2005. Fats in the fast lane. *Food Product Design 14* (12): 148.

Gerdes, S. 2003. You gotta have heart-healthy ingredients. *Food Product Design 13* (2): 113–129.

Giese, J. 2002. Acrylamide in foods. *Food Technol. 56* (10): 71–72.

Hazen, C. 2005a. Understanding fats and oils today. *Food Product Design 14* (11): 38.

Hazen, C. 2005b. Trans-formulation alternatives. *Food Product Design 15* (5): 71.

Hicks, K.B. and R.A. Moreau. 2001. Phytosterols and phytostanols: Functional food cholesterol busters. *Food Technol. 55* (1): 63.

Juttelstad, A. 2004. Marketing of *trans*-fat-free foods. *Food Technol. 58* (1): 20.

Klapthor, J. N. 2002. Providing caution in the face of acrylamide. *Food Technol. 56* (6): 129.

Kuntz, L. A. 2001. Fatty acid basics. *Food Product Design 11* (8): 93.

Kuntz, L. A. 2002. Designer fats for bakery. *Food Product Design 12* (8): 55.

Kuntz, L. A. 2005. *Trans*-lating formulas. *Food Product Design 15* (7): 14.

List, G. R. 2004. Decreasing *trans* and saturated fatty acid content in food oils. *Food Technol. 58* (1): 23.

Luff, S. 2004. Ascendancy of omega-3s. *Food Product Design: Functional Foods Annual* (September): 67.

Luff, S. 2005. Better ingredients through biotechnology. *Food Product Design 14* (10): 91.

Marangoni, A. G. and Hartel, R. W. 1998. Visualization and structural analysis of fat crystal networks. *Food Technol. 52* (9): 46.

Ohr, L. M. 2005. Functional fatty acids. *Food Technol. 59* (4): 63.

Palmer, S. 2005. New current for tropical oils. *Food Product Design 15* (2): 87.

Pszczola, D. E. 2001. Antioxidants: From preserving food quality to quality of life. *Food Technol. 55* (6): 59.

Pszczola, D. E. 2004. Fats: In *trans*-ition. *Food Technol. 58* (4): 52.

Remig, V., et al. 2010. *Trans* fats in America: Review of their use, consumption health implications, and regulation. *J. Amer, Dietet. Assoc. 110* (4): 585.

Sloan, A. E. 2005. Time to change the oil? *Food Technol. 59* (5): 17.

Spano, M. 2010. Heart health and fats. *Food Product Design 20* (3): 22.

Tareke, E., et al. 2002. Analysis of acrylamide, a carcinogen formed in heated foodstuffs. *J. Agric. Food Chem.* 509170: 4998.

Tarrago-Trani, M. T., et al. 2006. New and existing oils and fats used in products with reduced *trans*-fatty acid content. *J. Am. Diet. Assoc. 106* (6): 867.

Zapsalis, C. and Beck, R. A. 1985. *Food Chemistry and Nutritional Biochemistry.* Wiley. New York.

# INTO THE WEB

*www.macnutoil.com*—Information about macadamia nut oil.

*www.canolainfo.org*—Information on canola oil compared with other oils.

*http://pubs.acs.org/doi/abs/10.1021/cg0706159*—Paper on transition from $\beta'$ to $\beta$ crystals in spherulites of trilaurin.

*http://www.springerlink.com/content/ar10243510519554/*—Paper on reversion in soybean oil.

*http://www.iseo.org/*—Site of Institute of Shortenings and Edible Oils.

*http://www.fda.gov/Food/FoodSafety/FoodContaminantsAdulteration/ChemicalContaminants/Acrylamide/ucm053569.htm*—FDA Web site about acrylamide.

*http://www.foodrisk.org/acrylamide/web_sites.cfm*—International information on acrylamide.

*http://www.sciencedaily.com/releases/2008/03/080306075222.htm*—Report on reduction of acrylamide by soaking before frying French fries.

*http://oehha.ca.gov/Prop65/acrylamideqa.html*—Information on California's legislative efforts regarding acrylamide.

Extra virgin olive oils are a speciality of this olive oil cooperative in Krista, Greece, on the island of Crete.

# CHAPTER 12

# Fats and Oils in Food Products

## Chapter Outline

## OBJECTIVES

After studying this chapter, you will be able to:

1. Describe the steps involved in manufacturing and modifying food fats.

2. Explain tests to determine the quality of fats and oils.

3. Discuss selection of specific fats and oils for different applications in food preparation.

4. Identify fat replacements and their applications.

## INTRODUCTION

Oils used in food preparation may be produced from various plant sources: seeds from such crops as cotton, corn, safflower, soy, sesame, and grape; in addition, walnuts, hazelnuts, peanuts, olives, and even avocados are commercial sources of oil. Fats are obtained from dairy cattle, cattle, and swine. Depending on the source and the product being made, various manufacturing steps will be completed before they are marketed. Technologists have developed a wide array of fats and oils to meet a range of consumer needs and expectations. There are numerous choices that can be made to obtain the desired results when cooking and baking.

## STEPS IN MANUFACTURING FOOD FATS

### Extraction

**Rendering**
Removing fat from animal tissues by either dry or moist heat.

The first step in manufacturing food fats is to remove lipids from their natural food sources (Figure 12.1). The process used for lard from swine and tallow from cattle is **rendering**, either wet or dry. Wet rendering, the more common technique for preparing edible fats, uses steam under pressure to heat the tissues and fat to at least 90°C (195°F). Then hot, fluids fats are separated from the water. Antioxidants frequently are added to retard the development of rancidity. Dry rendering is done by simply heating the tissues and collecting the melted fat as it drains and is finally squeezed from the residue.

Oils are removed from plants by applying pressure with or without adding heat. A mechanical or screw-type press can be used to express oils from appropriate seeds (e.g., sesame), nuts, or plant tissue (olive, for instance), a procedure termed **cold pressing** (Figure 12.2). Cold pressing is desirable because of the high quality of the oil that can be extracted in this manner. However, cold pressing does not extract as high a portion of the oil in the seeds as does hot pressing.

**Cold pressing**
Mechanical pressing of appropriate seeds, plant tissue, or nuts to express oil without heat, resulting in an oil of excellent purity.

**Hot pressing**
Using steam or hot water to heat plant seeds to about 70°C (158°F) to facilitate extraction of lipids from the seeds, a process that also extracts some gums, off flavors, and free fatty acids.

In **hot pressing**, steam is used to warm the tissues to about 70°C (158°F), and then they are pressed to remove the oil. The higher temperatures reached in hot pressing result in a somewhat lower quality product because of the presence of some gums, possible off-flavor overtones, and free fatty acids. Fortunately, a subsequent degumming operation can remove the extraneous gums acquired during hot pressing. Solvent extraction conducted on oil seeds after pressing can isolate additional lipids. Another extraction technique involves the application of carbon dioxide at pressures as high as 8,000 psi at warm temperatures [at least 31°C (88°F)].

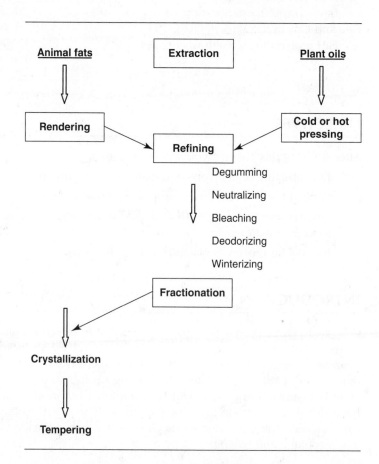

**Figure 12.1**    Steps in manufacturing fats and oils.

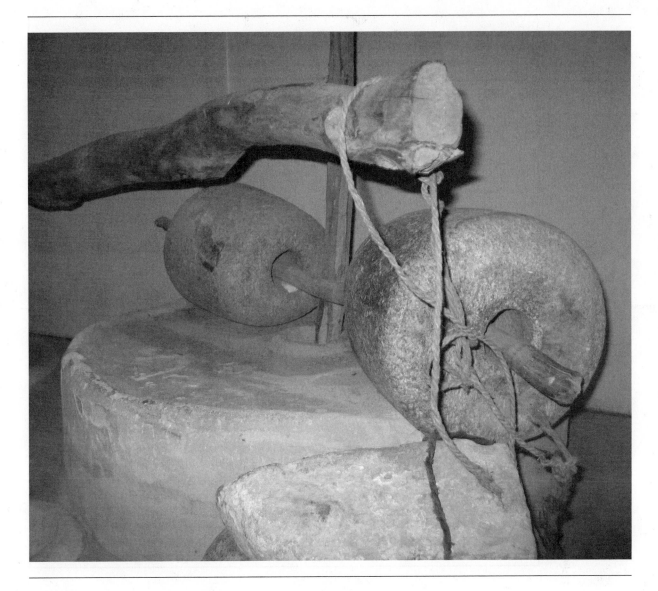

**Figure 12.2**   Oil can be pressed from olives using a press such as this one in Libya.

## Refining

Undesirable constituents may occur in fats after extraction, particularly if rendering was done. Gums, lipoproteins, lecithin, ketones, and aldehydes are just some of the compounds that may appear in fats at this stage of manufacturing. Several steps can help to refine these fats to the high level of purity needed to move them through marketing channels and into the kitchen.

**Degumming** and **neutralizing** procedures can remove gums and free fatty acids. A hot steam distillation procedure separates the relatively volatile free fatty acids from the fat molecules, one method of neutralizing the fat. Steam also can remove gums that may be present, a process called *degumming*. The use of dilute phosphoric acid is an alternative means of degumming.

**Bleaching**, the refinement process that removes undesirable coloring and flavoring contaminants from fats and oils, can be accomplished by several means, such as filtering through activated charcoal. A related step is **deodorizing**, a process accomplished by steam distillation. Deodorizing is particularly important to the production of high-quality coconut

**Degumming**
Separating natural gums from extracted fats; an important step in refining fats and oils.

**Neutralizing**
Removing free fatty acids from fats and oils; a step in refining.

**Bleaching**
Refining step in which coloring and flavoring contaminants are removed from fats, often by filtration through active charcoal or other suitable substrate.

---

### FOOD FOR THOUGHT: Olive Oil Quality

Olive oils are classified commercially on the basis of oil source and processing techniques; extra virgin olive oil and virgin olive oil contain only the oil obtained from the first mechanical pressing of olives. No oil can be added from other sources, and no heat or solvents can be used in preparing either of these top-grade olive oils. Differences in flavor, odor, and color can be observed and used as criteria in conducting sensory evaluations to assess palatability and quality of olive oils from various regions.

The price of olive oils varies greatly from one brand to another, with top brands commanding high prices in gourmet markets. Because of market competition, olive growers and olive oil producers constantly are striving to improve their products.

The European Union has developed a profile chart for expert tasters to use in evaluating olive oils from various olive groves from Spain to Greece. Acceptable perceptions include fruity olive taste, overtones of apple or other fruits, herbs, bitter, hot, spicy, and sweet; these are scored individually to create a flavor profile. Negative scores are given for overtones of sour, metallic, mold, and/or rancid.

The characteristics of oil from a particular harvest are influenced by many factors, including rainfall, soil, temperatures during the growing season, and the precise maturity of the olives at the time of harvest. The combination of these factors creates the olive oil available from that grove that particular season. Obviously, each year there will be variations because of weather conditions and harvesting decisions.

This variability adds to the importance of expert tasters to aid in maintaining quality standards in extra virgin olive oil. Members of a sensory panel need to be able to score olive oils to develop a flavor profile, and they also should detect if a sample is musty, rancid, or otherwise unpleasant.

Aficionados of various brands of extra virgin olive oil praise the sensory qualities of their favorites. In reality, the sensory qualities of olive oil that connoisseurs seek come from the combination of many different chemical compounds; growing and harvesting conditions determine relative amounts of these compounds. Fatty acids, sterols, and some alcohols influence the mouthfeel of olive oils, while polyphenols, aldehydes, ketones, esters, and organic acids are among the compounds contributing to flavor.

---

**Deodorizing**
Using steam distillation or other suitable procedure to remove low molecular weight aldehydes, ketones, peroxides, hydrocarbons, and free fatty acids that would be detrimental to the aroma and flavor of fats.

**Winterizing**
Refining step in which oils are chilled carefully to precipitate and remove fractions with high melting points that would interfere with the flow properties of salad dressings or other products containing the oils.

and palm kernel oils. Conversely, olive oil is not deodorized because its natural aroma is considered highly desirable.

**Winterizing** is a refining technique used to remove lipid fractions with melting points high enough to cause them to become a solid at refrigerator temperatures. If a fat is chilled to crystallize the fractions that would precipitate during storage in the refrigerator and the precipitated fats are removed by filtration, the resulting oil can be stored in the refrigerator and yet be poured as soon as it is removed from the cold. Winterized oils are used in salad dressings or other pourable sauces. The exception is olive oil. This unique oil is not winterized because important flavorful oils would be removed.

## Fractionation

Tropical oils are naturally high in saturated fatty acids, which means that they have comparatively high melting points and a firmer texture at room temperature than the other commercial sources of oils. These oils used to be considered less healthful than those with lower levels of saturated fatty acids. However, the emphasis on reducing intake of *trans* fats because of their negative impact on heart health has led to increased utilization of palm and palm kernel oils in the manufacture of commercial fats.

---

**FOOD FOR THOUGHT: Refining and Smoke Point**

Unrefined oils are detectably different from comparable refined oils. The flavor and aroma of unrefined oils are distinctive, whereas those of refined oils typically are bland. A particularly important difference is the significantly lower smoke point of unrefined oils. In unrefined safflower oil, the smoke point is only 107°C (225°F). Unrefined peanut and soybean oils have a smoke point of 160°C (320°F), and unrefined sunflower oil starts to smoke at only 107°C (225°F). Unrefined oils clearly are not acceptable for use in frying because of their low smoke points.

Oils used for deep-fat frying need a smoke point that is well above 190°C (375°F) to minimize the formation of harmful compounds during frying. Fortunately, the smoke points of the refined plant oils available in markets usually are well above this temperature. For example, refined safflower oil has a smoke point above 260°C (500°F), and sunflower, soybean, and corn oils have a smoke point of 232°C (450°F); they are all well suited for frying.

---

Palm oil can be separated into fractions with different physical properties and varying ratios of fatty acids by **fractionation**. Fats can be separated by careful temperature control and removal of fat crystals that form. The oils left behind have lower melting points. The crystalline fat can be blended with the amount of oil needed to create fats with the desired physical properties. This oil also has the advantage of being free of or low in *trans* fats because the oil was not hydrogenated.

**Fractionation**
Process of separating oils into fractions using controlled temperature to crystallize fatty acids with high melting points and separate them from oils with lower melting points.

## Crystallization of Fats

The final phase in the manufacture of fats is crystallization of the warm, fluid fat. The comparative stability and fine crystal size of β′ crystals make this type of crystal the goal in the final product. Careful control of temperature during cooling, combined with an appropriate amount of agitation, helps achieve a smooth fat, with β′ crystals being the predominant crystal size. Without such controls, these small needle-shaped crystals (about 1 μm long) are not able to form readily. Instead, the large β crystals (usually ranging between 20 and 45 μm) precipitate to give a coarse texture. Hydrogenated cottonseed oil often is included in the manufacture of fats because of its propensity to form the desired β′ crystals. In the manufacture of shortenings, flakes of cottonseed oil and tallow frequently are added to promote formation of β′ crystals.

In the manufacture of special fats for use in the confectionery industry, fats undergo *tempering* to yield a product with a mixture of crystal shapes (polymorphs). **Tempering** is a process in which temperature is carefully controlled by removing the heat as it is released when liquid fats crystallize (the heat of crystallization). By this means, the fat is held at a specified temperature for the time required for the crystals to form and to stabilize in the favored crystal form. Tempered fats can be stored with some variations in temperature, sometimes more than 25°C (78°F), and still retain their textural qualities.

**Tempering**
Removing heat resulting from crystallization of fats and maintaining a selected temperature to promote the formation of stable, desirable crystals.

Chocolate needs to be tempered to promote the formation of stable fat crystals. Otherwise, the smaller crystals in the chocolate melt and then recrystallize in coarse β crystals, which appear as somewhat discolored, granular areas on the surface of chocolate when it has become a little warm or has been stored for a long time. This unique and rather unattractive surface appearance is called **bloom** and definitely is to be avoided.

**Bloom**
Granular-appearing, discolored areas on the surface of chocolate; the result of melting of less stable crystals and recrystallization as β crystals on the surface.

## Quality Determinations

***Chromatographic Analyses.*** The heterogeneity of lipids leads to the use of chromatographic analyses to determine the relative amounts of key components of a fat. Both gas–liquid chromatography and high-pressure liquid chromatography are appropriate and practical analytical tools. To volatilize the sample for gas–liquid chromatography, fatty acids commonly are converted to their methyl esters. With high-pressure liquid chromatography, even trace amounts of free fatty acids can be detected.

***Iodine Number.***   Fatty acids containing double bonds can take up iodine at the points of unsaturation. The weight of the iodine held in this way indicates the amount of unsaturation in a fat and is referred to as the iodine number or iodine value.

***Peroxide Value.***   A related test measures the oxidation of potassium iodide in the presence of a fat to determine the peroxide value of the fat. This value indicates the deterioration of a fat due to oxidative rancidity, as measured by the formation of peroxides.

***Free Fatty Acid Content.***   By titrating a fluid fat sample with a standard sodium hydroxide solution, the free fatty acid content of the fat can be determined. This value rises as hydrolytic rancidity develops in the fat during storage or use.

***Standardized Testing.***   Methods for testing fats and other lipids have been developed and reviewed extensively by several groups, including the American Oil Chemists' Society (AOCS), the Association of Official Analytical Chemists (AOAC), the Codex Alimentarius Commission (CAC), and the International Union for Pure and Applied Chemistry (IUPAC).

## CHEMICAL MODIFICATIONS

Chemical methods can be applied to alter the physical properties of fats and oils and tailor products to have altered melting points and enhanced plasticity or spreadability for use in specific products. Hydrogenation, interesterification, and intraesterification (sometimes in conjunction with enzymes) are techniques being used to develop a wide variety of fats and oils for commercial use. The push to eliminate *trans* fat has driven a considerable amount of work in this area.

### Hydrogenation

**Hydrogenation**
Addition of hydrogen to an unsaturated fatty acid in the presence of a catalyst to reduce the unsaturation of the molecule and raise the melting point.

**Hydrogenation** is part of the manufacturing process of a wide range of fat products because this reaction alters the melting points of fatty acids by increasing their saturation with hydrogen. In hydrogenation reactions, a catalyst (commonly nickel) is present in conjunction with hydrogen gas and oil in a heat-controlled environment. Under these conditions, hydrogen atoms react with unsaturated fatty acids at the points of unsaturation, as shown here.

L → S = margarines + shortening

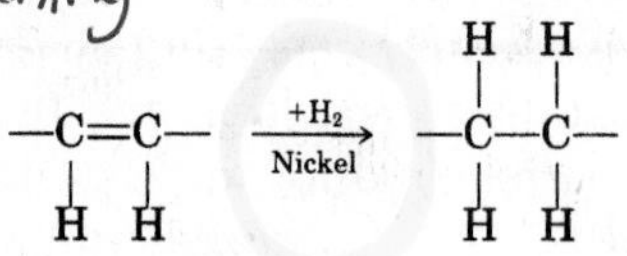

$$-C=C- \xrightarrow[\text{Nickel}]{+H_2} -\underset{\underset{H}{|}}{\overset{\overset{H}{|}}{C}}-\underset{\underset{H}{|}}{\overset{\overset{H}{|}}{C}}-$$

Hydrogenation occurs more readily on fatty acids with two or more double bonds than on those with only one. However, hydrogenation does occur on molecules with one double bond when most of the diene and polyene molecules have been modified to the corresponding monoene fatty acid.

Through the use of hydrogenation reactions, vegetable oils can be modified from liquids to solids, a change that makes these former oils suitable for use as margarines and shortenings. This process also is used to modify peanut butter from its original state (in which it separates to a concentrated solid mass and a layer of oil) to a spread that remains homogeneous, even during extended shelf storage.

One undesirable result of the hydrogenation process is that some unsaturated fatty acids undergo isomerization, resulting in the formation of some double bonds in the *trans*

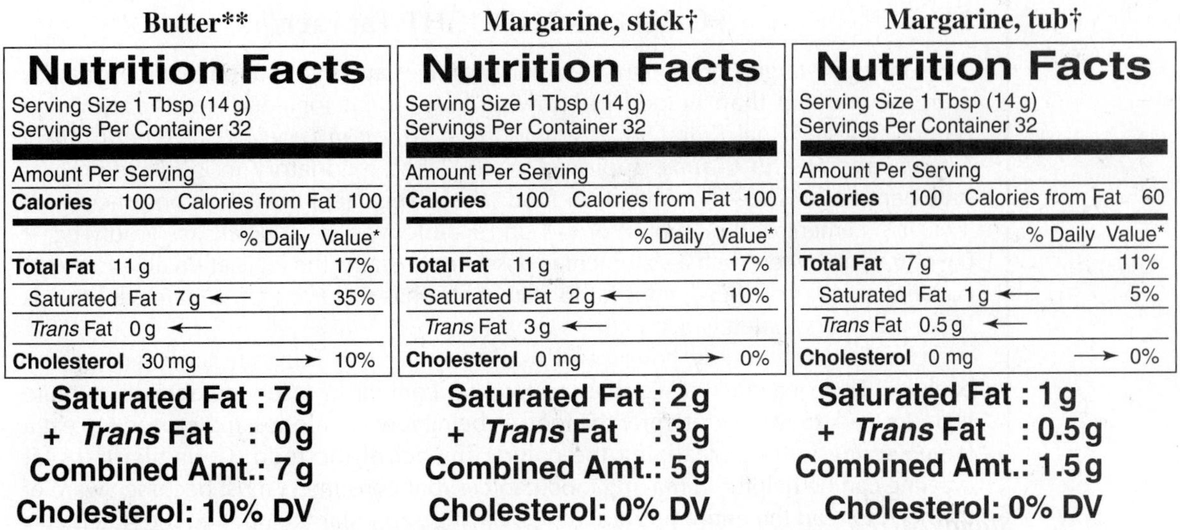

**Butter\*\***

## Nutrition Facts

Serving Size 1 Tbsp (14 g)
Servings Per Container 32

Amount Per Serving

**Calories**  100   Calories from Fat  100

% Daily Value\*

**Total Fat**  11 g                                         17%

  Saturated Fat  7 g  ←                          35%

  *Trans* Fat  0 g  ←

**Cholesterol**  30 mg              →           10%

### Saturated Fat : 7 g
### + *Trans* Fat   : 0 g
### Combined Amt.: 7 g
### Cholesterol: 10% DV

**Margarine, stick†**

## Nutrition Facts

Serving Size 1 Tbsp (14 g)
Servings Per Container 32

Amount Per Serving

**Calories**  100   Calories from Fat  100

% Daily Value\*

**Total Fat**  11 g                                         17%

  Saturated Fat  2 g  ←                          10%

  *Trans* Fat  3 g  ←

**Cholesterol**  0 mg              →           0%

### Saturated Fat : 2 g
### + *Trans* Fat   : 3 g
### Combined Amt.: 5 g
### Cholesterol: 0% DV

**Margarine, tub†**

## Nutrition Facts

Serving Size 1 Tbsp (14 g)
Servings Per Container 32

Amount Per Serving

**Calories**  100   Calories from Fat  60

% Daily Value\*

**Total Fat**  7 g                                         11%

  Saturated Fat  1 g  ←                          5%

  *Trans* Fat  0.5 g  ←

**Cholesterol**  0 mg              →           0%

### Saturated Fat : 1 g
### + *Trans* Fat   : 0.5 g
### Combined Amt.: 1.5 g
### Cholesterol: 0% DV

\* Nutrient values rounded based on FDA's nutrition labeling regulations. Calorie and cholesterol content estimated.

\*\* Butter values from FDA Table of *Trans* Values, January 30, 1995.

† Values derived from 2002 USDA National Nutrient Database for Standard Reference, Release 15.

**Figure 12.3**  Examples of *trans* fat labels.

configuration, rather than the *cis* form commonly found in nature. Because of this undesirable type of double bond, alternatives to using partially hydrogenated fats are needed. Completely hydrogenated oils do not contain double bonds and thus do not contain *trans* fatty acids. They can be modified to obtain the desired characteristics by blending with an oil that has not been hydrogenated.

> 0.5 g /serving

Evidence is accumulating that *trans* fatty acids elevate the levels of low density lipoprotein (LDLs), which may increase the risk of coronary heart disease. As a result, the Food and Drug Administration issued a ruling in 2003 that the amount of *trans* fat in a serving must be listed on a separate line under saturated fat on the Nutrition Facts panel (unless the total fat in a serving is less than 0.5 g per serving and no claims are made about fat, fatty acid, or cholesterol content). If *trans* fat is not listed, a footnote must be included saying the food is "not a significant source of *trans* fat." The ruling on labeling *trans* fats became effective from January 2006 (Figure 12.3).

The levels of *trans* fats are higher in stick margarines than in tub and liquid margarines because more hydrogenation is involved in producing the firmer sticks (Table 12.1), but today's levels are about 25 percent lower than they used to be. The type of *trans* fatty acid also is of importance. **Elaidic acid** (a *trans* isomer of oleic acid), which is produced when fat is hydrogenated, elevates LDL levels, while **vaccenic acid**, a different *trans* fatty acid isomer of oleic acid occurring naturally in butterfat, does not cause elevated LDLs.

**Elaidic acid**
*Trans* isomer (t9-18:1) of oleic acid produced during hydrogenation; raises LDLs.

**Vaccenic acid**
*Trans* isomer (t11-18:1) of oleic acid occurring naturally in butterfat; does not raise LDLs.

**Table 12.1**  Fat Content in Selected Foods (1 Tablespoon)

| Product | Total Fat (g) | Saturated Fat (g) | *Trans* Fat (g) | Combined Saturated and *Trans* Fats (g) | Cholesterol (mg) |
|---|---|---|---|---|---|
| **Butter**[a] | 11.52 | 7.29 | 0.1 | 7.30 | 31.1 |
| **Margarine, stick**[a] | 11.30 | 2.13 | 3.4 | 5.53 | 0.0 |
| **Margarine, tub**[a] | 9.65 | 1.23 | 0.6 | 1.83 | 0.0 |
| **Smart Balance** | 9.05 | 2.34 | 0.1 | 2.35 | 0.0 |

[a] Values derived from 2002 USDA National Nutrient Database for Standard Reference, Release 15.

---

### FOOD FOR THOUGHT: Fat Facts

Consumers are targets of so many nutrition messages and information that they may be more bewildered than helped in knowing how to eat for good health. However, Federal Register Final Rule: *Trans Fatty Acids in Nutrition Labeling, Nutrient Content Claims, and Health Claims*, approved in 2003 and mandatory in 2006, requires a statement of the *trans*-fat content on food labels. This label provides consumers with a clear statement of the amount of this undesirable type of fat. Packages loudly boast 0 g of *trans* fat when such a statement is possible. Actually, the legislation allows a food containing less than 0.5 g *trans* fat to state that it has 0 g *trans* fat, a statement that is legal, but not completely accurate.

Unfortunately, many consumers read only the 0 g *trans* fat statement on the package. They may fail to see that the total fat content in a serving actually is quite high. Instead, they may interpret the food as being low in calories and proceed to eat a large serving without realizing the caloric impact of the food. Certainly, *trans*-fat labeling can be helpful in making food choices, but consumers must become aware of the need to read the entire nutrition label, paying particular attention to the statements of total fat and saturated fat content. Labels provide the information people need to make healthy food choices, but consumers have the responsibility of reading all of the nutrition information provided. Then it is possible to eat a diet consistent with achieving and maintaining a healthy weight.

---

The negative effects of *trans* fatty acids in the diet have spurred efforts to make more healthful spreads with reduced amounts of *trans* fatty acids. Chill fractionation in which the higher-melting *trans* fatty acids are removed is one viable approach. Consumers have been reducing their intake of hydrogenated vegetable oils, which is an effective means of reducing *trans* fatty acid intake, too.

Industry efforts to reduce *trans* fatty acids range from changing the formulations of shortenings and spreads to the development of plant oil sources with a modified fatty acid profile. An effective reduction to <10 percent *trans* fatty acids has been accomplished using carbon dioxide in the hydrogenation process. Biotechnologists are working to change the composition of plant oils in soybeans to reduce levels of saturated fat and linolenic acid and increase oleic acid content.

## Inter- and Intraesterification

**Interesterification**
Treatment of a fat, usually lard, with sodium methoxide or another agent to split fatty acids from glycerol and then to reorganize them on glycerol to form different fat molecules with less tendency to form coarse crystals.

Metal salts and/or lipases can be used to remove fatty acids from glycerol. When this is done to oils or fats that are composed of quite an array of triglycerides with differing fatty acids, these freed fatty acids can then be joined back onto glycerol to form new triglycerides. The catalyst used for this is usually sodium methoxide; the process is termed **interesterification** or chemical and/or enzymatic interesterification. This technique results in the formation of fats with altered characteristics, such as a higher melting point or a difference in crystallization tendencies. However, no *trans* fatty acids are created.

It is possible to use interesterification to produce shortenings and margarines with a higher melting point and with good spreading characteristics while avoiding the *trans* fatty acids that are part of the product when hydrogenation is used to raise the melting point of vegetable oils. NovaLipid™ is a fat that is essentially free of *trans* fat; it is based on interesterification using a mix of approximately 1 part fully saturated soybean oil [high in stearic acid with a melting point of ~70°C (~158°F)] and 3 parts liquid soybean oil. The result is a fat that is appropriately firm at room temperature for use in baking but that meets the concern of lowering the content of *trans* fats. Stearic acid is useful in this interesterified product because it not only contributes a high melting point, but it also appears to not influence LDL blood levels.

If the fat is melted for the interesterification process, all of the fat molecules are able to participate, and the procedure is termed **randomized interesterification**. Only the lipid molecules that are in the fluid state are altered by sodium methoxide when fat is kept below its melting point,. This type of reaction is called **directed interesterification**. Directed interesterification is useful when the goal of the manufacturer is to raise the melting point of the fat. Either method is useful for improving the textural characteristics of a fat. In the United States, lard commonly undergoes interesterification in its processing for the market.

Fatty acids that have been removed from glycerol as a consequence of the presence of another substance such as sodium methoxide have the potential to recombine with this glycerol, thus maintaining the same fatty acids of the triglyceride, but in different positions on the glycerol. This process of reorganizing the molecule is called **intraesterification**. Intraesterification, like interesterification, alters the textural characteristics of a fat by modifying the ease of crystallization and crystal aggregation.

**Randomized interesterification** Interesterification accomplished using melted fat.

**Directed interesterification** Process of interesterification in which the fat is kept below its melting temperature.

**Intraesterification** Catalyzed reaction in which the fatty acids split from glycerol and rejoin in a different configuration, but with the same fatty acids being retained in the molecule.

## FATS AND OILS IN THE MARKETPLACE

### Fats and Today's Health Challenges

Consumers today are bombarded by messages regarding two health risks—obesity and *trans* fats (Figure 12.4). Now that information about the *trans* fat content of a food appears on nutrition labels, the food industry and consumers are keenly aware of the importance of reducing the amount of it in products. In addition, overweight and obese consumers also are

**Figure 12.4**    In Saudi Arabia, corn oil carries a visual message suggesting it is good for heart health, and olive oil can be identified by the picture on the container even by those who cannot read the words.

*efforts to reduce trans*

seeking food products with overall reduced fat levels. Needless to say, the food industry has been spending considerable effort and money developing products that meet both of these concerns.

Efforts to reduce *trans* fats include changes in oils used, extent of hydrogenation, interesterification, alterations in diacylglyceride levels, adding short- and medium-chain fatty acids, and changing the fatty acids in plants through selective breeding and genetic alterations.

## Sources

Today's marketplace provides a testimonial to the industriousness of "fat chemists," as consumers and food manufacturers confront a most impressive array of fats and oils from which to choose the product best suited to their specific food preparation problems. One choice is between animal fats and plant lipid products. Animal fats are confined primarily to butters and lards, but the additional choices among butters include salted or unsalted and whipped or unwhipped, as well as reduced-fat. The options in plant lipids range from fluid oils from various sources to solid fats, some of which are the result of chemical modification of the plant oils.

The composition of fats varies with the animal or plant source (Figure 12.5). Generally, animal fats are higher in saturated fatty acids and lower in unsaturated fatty acids than are plant lipid sources. In particular, plant sources as a group are higher in the polyunsaturates than are animal sources, although fish oils provide an exception to this generalization. Table 12.2 presents a comparison of the fatty acid composition and cholesterol content of selected animal and plant sources of lipids.

***Animal Fats.***    Fats in the food supply that are obtained from animal sources include beef tallow and butterfat (milk fat) from cattle and lard from pigs (Figure 12.6). Of nutritional interest is the fact that cholesterol occurs only in animal fats and never in plant lipids.

Butter is high in saturated fatty acids (myristic, palmitic, and stearic), and oleic acid is the major monounsaturated fatty acid. Butyric acid contributes its distinctive quality to the *flavor*

| DIETARY FAT | CHOLESTEROL mg/tbsp. | FATTY ACID CONTENT NORMALIZED TO 100 PERCENT | | |
|---|---|---|---|---|
| Safflower oil | 0 | 77% | 13% | 10% |
| Sunflower oil | 0 | 69% | 20% | 11% |
| Corn oil | 0 | 62% | 25% | 13% |
| Soybean oil | 0 | 61% | 24% | 15% |
| Cottonseed oil | 0 | 55% | 18% | 27% |
| Peanut oil | 0 | 33% | 49% | 18% |
| Canola oil | 0 | 32% | 62% | 6% |
| Lard | 12 | 12% | 47% | 41% |
| Palm oil | 0 | 10% | 39% | 51% |
| High-oleic sunflower oil | 0 | 9% | 82% | 9% |
| Olive oil | 0 | 9% | 77% | 14% |
| Beef fat | 14 | 4% | 44% | 52% |
| Butter fat | 33 | 4% | 30% | 62% |

POLYUNSATURATED FAT    MONOUNSATURATED FAT    SATURATED FAT

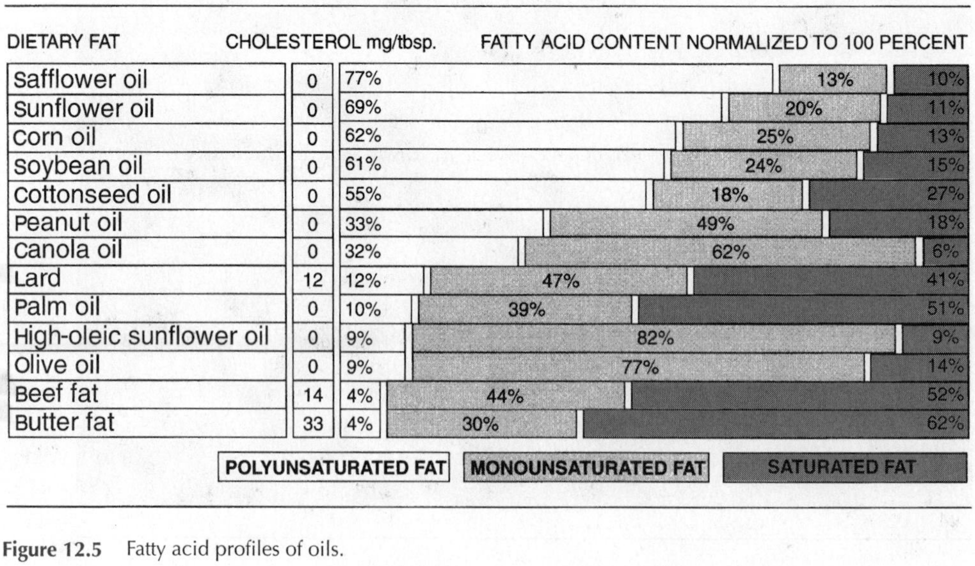

**Figure 12.5**    Fatty acid profiles of oils.

**Table 12.2**    Percent Fat, Fatty Acids, and Cholesterol in Selected Fats

| Fat Source | Fat (%) | Fatty Acids (%)[a] | | | Cholesterol (mg/100 g) |
| | | Saturated | Monounsaturated | Polyunsaturated | |
|---|---|---|---|---|---|
| *Animal sources* | | | | | |
| Beef tallow[a] | 100 | 49.8 | 41.8 | 4.0 | 109 |
| Butter | 81 | 50.5 | 23.4 | 3.0 | 219 |
| Lard | 100 | 39.2 | 45.1 | 11.2 | 95 |
| *Plant sources* | | | | | |
| Cocoa butter | 100 | 59.7 | 32.9 | 3.0 | 0 |
| Coconut oil | 100 | 86.5 | 5.8 | 1.8 | 0 |
| Corn oil | 100 | 12.7 | 24.2 | 58.7 | 0 |
| Cottonseed oil | 100 | 25.9 | 17.8 | 51.9 | 0 |
| Olive oil | 100 | 13.5 | 73.7 | 8.4 | 0 |
| Palm oil | 100 | 49.3 | 37.0 | 9.3 | 0 |
| Palm kernel oil | 100 | 81.4 | 11.4 | 1.6 | 0 |
| Peanut oil | 100 | 16.9 | 46.2 | 32.0 | 0 |
| Rapeseed oil[b] | 100 | 5.0 | 68.1 | 22.5 | 0 |
| Safflower oil | 100 | 9.1 | 12.1 | 74.5 | 0 |
| Sesame oil | 100 | 14.2 | 39.7 | 41.7 | 0 |
| Soybean oil | 100 | 14.4 | 23.3 | 57.9 | 0 |
| Sunflower oil | 100 | 10.1 | 45.4 | 40.1 | 0 |
| *Margarine, stick* | | | | | |
| Corn oil | 80.5 | 13.2 | 45.8 | 18.0 | 0 |
| Safflower, soybean | 80.5 | 3.8 | 31.7 | 31.4 | 0 |
| Soybean | 80.5 | 16.7 | 39.3 | 20.9 | 0 |
| Sunflower, soybean cottonseed | 80.5 | 11.9 | 28.5 | 36.6 | 0 |
| *Margarine, soft (tub)* | | | | | |
| Corn oil | 80.4 | 12.1 | 31.6 | 31.2 | 0 |
| Safflower oil | 80.4 | 9.2 | 23.2 | 44.5 | 0 |
| Soybean oil | 80.3 | 13.5 | 36.4 | 26.8 | 0 |
| Sunflower, peanut oils | 80.4 | 16.1 | 30.7 | 30.1 | 0 |

[a] For specific information on the content of saturated fatty acids (4–18 carbon atoms), monounsaturated fatty acids (16–22 carbon atoms), and polyunsaturated fatty acids (18–22 carbon atoms), refer to the source for this table.
[b] Erucic acid (22 carbon atoms, 1 double bond) content is 45 percent and higher.
Adapted from *Composition of Foods: Fats and Oils Raw, Processed, Prepared.* Consumer and Food Economics Institute. Agriculture Handbook No. 8–4. *Science and Education Administration.* U.S. Department of Agriculture. Washington, DC, 1979.

flavor profile of butter, although it is present in a small amount. Lard has more oleic acid, but somewhat less saturated fatty acid content (primarily palmitic and some stearic) than does butter.    *lard = more oleic, less saturated than butter*

***Plants.***    Plant sources include olives, palm fruit and palm kernels, cottonseeds, soybeans, rapeseed (canola oil), corn, sunflower seeds, safflower seeds, grape seeds, coconuts, peanuts, cacao beans, walnuts, macadamia nuts, and rice bran.

Plant oils have become prominent in the American diet because they are high in polyunsaturated fatty acids, which are valued for their potential role in reducing serum cholesterol levels. Almost three-fourths of the fatty acids in safflower oil are polyunsaturated. About 60 percent of the fatty acids in soybean oil are polyunsaturated. In contrast, cocoa butter, coconut oil, and palm kernel oils have a high saturated fatty acid content (as much or even more saturated fatty acids than in animal fats).

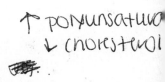
*↑ polyunsaturated
↓ cholesterol*

**Figure 12.6**    Some markets in Egypt stock ghee (clarified butter), a traditional form of fat in the diet in India and also among some Egyptians.

ive oil+ rapeseed

m mono

Monounsaturated fatty acids may be significant in decreasing serum cholesterol. Olive oil is unique in its high content of monounsaturated fatty acids, although rapeseed (canola) oil is a very close second.

***Modifying Plant Lipids.***    The fatty acid profiles of the lipids in plants have some excellent features, but other aspects are not ideal for promoting human health. Consequently, two approaches have been taken to develop better products. Selective plant breeding is a technique for enhancing the levels of certain fatty acids. The newer approach is genetic engineering, which involves modification of the genetic information in the cells. Safflower has been bred to almost eliminate erucic acid (a fatty acid), because of its possible impact on the heart.

### FOOD FOR THOUGHT: Profiting from Grape Seeds?

Grape seed oil is a relatively new arrival in the consumer marketplace. The rise in wine production is a serendipitous event because grape seed oil production is an economically productive answer to disposal of this part of the waste from the grapes used to make wine (Figure 12.7). Linoleic, an omega-6 fatty acid, is the primary fatty acid in grape seed oil; next in abundance is oleic, then palmitic and stearic. Linolenic content is low, which promotes shelf life.

Grape seed oil is made by first removing the seeds from the pulp and skins after they have been partially dried in a rotating drum. Then the seeds are dried fully in preparation for the key step—extraction. Cold pressing of the dried seeds is the method used to produce high-quality grape seed oil. Not surprisingly, it takes a large quantity of dried grape seeds to produce a rather small amount of oil—25 kilos (55 pounds) of seeds yield about half a liter of oil.

With a high smoke point [216°C (420°F)], grape seed oil is well suited for frying foods. It has a mild flavor, which makes it a suitable choice for use in salad dressings and other cooking applications. Its high price tends to limit its use and adds to its image as a gourmet oil.

**Figure 12.7**    Grape seed oil is a specialty oil that can be savored with balsamic vinegar as dressing on a salad or for dipping a warm piece of artisanal bread.

Genetic and plant-breeding research created several oilseeds (sunflower, soybean, and canola) that provide oils with fatty acid profiles of value in formulating food products low in *trans* fats. In Canada, canola was bred from the rape plant to create a seed oil containing less than 2 percent erucic acid and less than 30 μmoles glucosinolates, changes that significantly improved performance and healthiness. Sunflower oil with a high oleic acid level has been developed through plant breeding; its value is in its greater stability during frying. Both genetic and plant-breeding efforts are directed primarily to raising oleic acid and lowering linolenic acid levels.

**Asoyia™ ultra low linolenic acid soybean oil**
Soybean oil produced through plant breeding for reduced linolenic acid and no *trans* fat.

Soybeans have been used for many years as a source of vegetable oil, but the high content of linolenic acid limits shelf life and frying because it oxidizes comparatively quickly. **Asoyia™ Ultra low linolenic acid soybean oil** has no *trans* fat and its linolenic acid content has been reduced from 7 percent to 1 percent. These changes were created through plant breeding for the desired characteristics, and the result is an oil that can be used successfully for an extended period, is well suited to frying, and requires no hydrogenation.

**Docosahexanoic acid (DHA)**
Omega-3 fatty acid containing 22 carbon atoms and 6 double bonds.

**Eicosapentanoic acid (EPA)**
Omega-3 fatty acid containing 20 carbon atoms and 5 double bonds.

*Fish Oils.*   Fish oils are gaining prominence as attention is directed increasingly to the roles that certain foods may play in health in addition to being a source of nutrients. Omega-3 long-chain polyunsaturated fatty acids (PUFA) include **docosahexanoic acid (DHA)** and **eicosapentanoic acid (EPA)**, both of which are found in fish oils. DHA is a fatty acid containing 22 carbon atoms and 6 double bonds, the first of which occurs at the third carbon from the methyl end (hence the designation as an omega-3 long-chain polyunsaturated fatty acid). EPA has 20 carbon atoms and 5 double bonds, beginning at the third carbon from the omega end of the molecule. Their structures follow:

Docosahexanoic Acid (DHA)

Eicosapentanoic Acid (EPA)

**Omega-3 fatty acid**
Polyunsaturated fatty acid with the first double bond on the third carbon from the methyl end of the molecule.

Food technologists considering ways of incorporating **omega-3 fatty acids** into new products are confronted with a somewhat daunting challenge because of possible fishy flavors associated with the fish oils. One approach with good potential is the production of DHA by *Crypthecodinium cohnii*, a strain of marine algae. This source produces DHA that is somewhat more resistant to developing off flavors from oxidation. The impetus for using omega-3 fatty acids is their possible role in reducing the incidence of coronary heart disease and strokes.

## Products

*Oils.*   Various oil products are available to consumers and the food industry. Peanut, corn, cottonseed, safflower, canola, sunflower, olive, and soybean oils are the ones used most often, either as a single type of oil or as a blend of two or more oils. Some special oils also can be found in some markets, including various nut oils, such as macadamia nut, and also rice bran oil. Rice bran oil is of some interest because it contains two types of sterols—(**oryzanols** and **tocotrienols**)—that are being studied intensively for possible health benefits. Oryzanols and tocotrienols are of significance for their antioxidant properties.

**Oryzanols**
Class of sterols in rice bran oil of significance for their antioxidant properties.

**Tocotrienols**
Class of sterols related to vitamin E valued for their antioxidant properties; found in rice bran and palm oils.

*Spreads.*   Butter, the water-in-oil emulsion formed when milk fat is churned sufficiently to reverse the emulsion, was long the unchallenged fat of choice as a spread on breads. The focus on obesity and on fats in the diet has resulted in availability of reduced-fat butter, a product containing at least 40 percent milk fat and usually skim milk, water, and gelatin.

Decades ago, technological advances in the ability to hydrogenate plant oils led to the development of margarines, which have been very competitive with butter (Figure 12.3). Present margarine products afford a choice of regular or **stick margarine** (Figure 12.4), engineered to approximate butter in most characteristics, and **soft or tub margarine**, which still can be spread because of their formulations (may include a higher content of water and/or polyunsaturates with low melting points). **Whipped margarine** and whipped butter are still other options. These whipped products have increased volume; a lighter, more airy texture; and fewer calories per given volume than the original spreads because of the air that has been whipped in. Manufacturers of margarines also have the option of reducing the amount of fat in margarines so claims can be made on the label:

- Reduced-fat or reduced-calorie diet margarine—no more than 60 percent oil (25 percent calorie reduction)
- Light/lower fat margarine—no more than 40 percent oil (50 percent or more calorie reduction)
- Fat free margarine—less than 0.5 g per serving

Health concerns have prompted the food industry to develop alternative margarines designed to reduce cholesterol when used regularly in the diet. Stanol and sterol esters have emerged as the effective compounds in alternative spreads. Phytosterols (e.g., β-sitosterol) have structures similar to that of cholesterol but with different side chains that alter their action in the body. It is necessary to esterify phytosterols with a fatty acid and also to hydrogenate them to incorporate them into a margarine. Phytostanols (β-sitostanol or campestanol) are called "tall oils" because they are derived from pine tree pulp. They also have to be esterified with a fatty acid, but hydrogenation is not required. The structures for β-sitostanol and β-sitosterol are shown here:

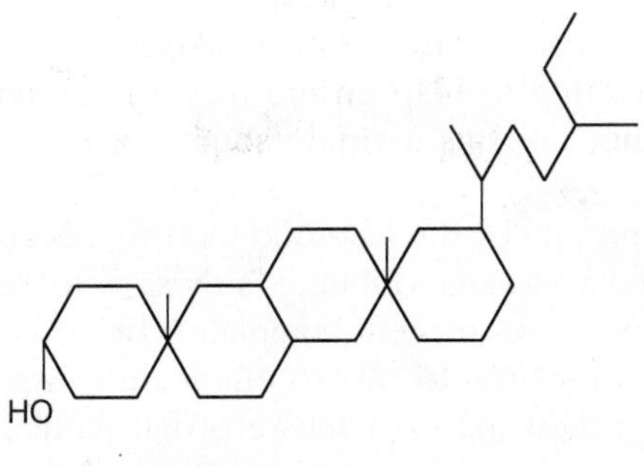

β-sitosterol (in Take Control®) 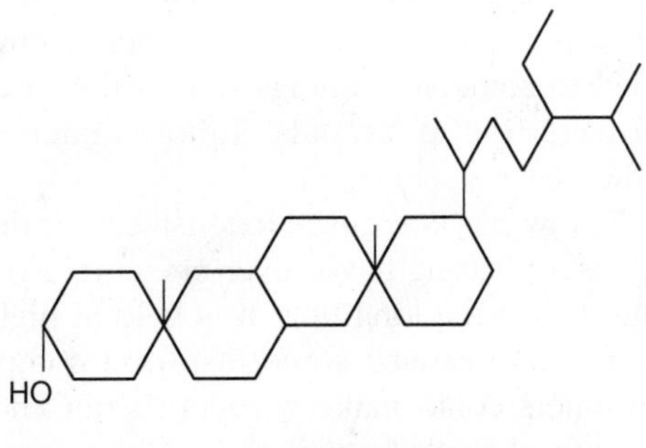          β-sitostanol (in Benecol®)

Take Control®, the spread developed by Lipton, Unilever's subsidiary, contains a plant sterol ester derived from soybeans; the effective agent in Benecol® (marketed by McNeil Consumer Healthcare, a division of Johnson & Johnson) is a plant stanol ester. These products work by preventing absorption of LDL cholesterol, thereby effecting a lowering of cholesterol levels, which helps to reduce the risk of coronary heart disease.

These spreads are effective even when the intake is only two servings each day, which is fortunate because they are both very expensive compared with regular margarines. Despite the fact that sales of both of these products have been relatively small, other competitors (e.g., Reducol™) are being developed for this same functional foods ingredient market.

Some margarines are made from a single oil source, such as corn oil margarines. One of the motivations for this is to appeal to consumers seeking margarines high in polyunsaturates as a possible aid in reducing their serum cholesterol. Many margarines are made by blending oil from several sources. Package ingredient labels often indicate that one or more of the listed oils is used in the product. This allows manufacturers to change their

**Stick margarines**
Spreads made by hydrogenating plant oils and adding water, milk solids, flavoring, and coloring to achieve a product similar to butter.

**Soft or tub margarines**
Spreads with melting points lower than those of stick margarines because of a higher content of polyunsaturated fatty acids.

**Whipped margarines**
Stick margarines that have been whipped mechanically into fat foam; increased volume results in fewer calories per given volume.

formulations at various times, depending on the relative cost of the oils identified in the ingredient label.

Peanut butter and the various nut butters are other examples of spreads made with plant oils. However, these differ from butter and margarine in that they contain not only oil but also some protein and other components of the nuts, which are ground after shelling to make a paste.

***Shortenings and Lard.***    Shortenings and lard, like oils, are essentially all fat. However, they are solids with considerable **plasticity** (ability to be spread or whipped to a heavy fat foam). This plasticity is the result of the physical nature of these solid fats, which actually are composed of large numbers of fat crystals with oil interspersed throughout the system.

Lard, the fat product rendered from pigs, naturally has a somewhat grainy texture because the fat molecules resulting from the rendering of lard have spatial configurations that associate readily into organized crystalline aggregates. Such a texture is not optimal for preparing fine-grained cakes. Interesterification during the processing of lard overcomes this crystalline disadvantage. Other modifications, particularly in fatty acid composition, can be accomplished by altering the diet of the pigs and even by some genetic manipulations.

Shortenings are quite sophisticated products produced through the efforts of technology. The main part of their manufacturing has been the hydrogenation of vegetable oils to achieve the desired consistency. In response to the demand to reduce or eliminate *trans* fats, shortenings have been modified in their formulation, often by changing oils and using intra- and interesterification and fractionation to replace the former hydrogenated, *trans* fat-containing products. Mono- and diglycerides are added to improve the ability of the shortening to form an emulsion in batters and doughs. To simulate the advantages of butter as an ingredient in baked products, shortenings that have β-carotene added to achieve a yellow color and butter-like flavoring compounds are available.

Beef tallow is yet another solid fat that has some applications within the food industry. Tallow is rendered from cattle to yield a characteristically hard fat. Small flakes of tallow are added to some shortenings during their manufacturing to help ensure that the shortening will have fine, β′ crystals. Tallow generally is not used as a single source of fat in the marketplace.

Tallow has been used extensively for deep-fat frying by the fast-food industry because it imparts a pleasing flavor overtone and also has the advantage of being fairly slow to deteriorate during the long time it is held at high temperatures. Recent complaints by consumer groups have caused some fast-food companies to switch to oils to eliminate *trans* fats. Consumers could make an even better choice by reducing their intake of fat, particularly that in fried foods.

***Substituting Fats.***    Sometimes necessity or health reasons require that a fat or oil in a recipe or formula be replaced with another fat or oil. Any substitution will influence the outcome to some extent even when the quantity has been substituted appropriately. The changes are due to the difference in the composition of the fats. Each type of fat has its unique properties as a consequence of its actual fatty acid composition and possibly the amount of water. Ideally, the type of fat used for a particular product should optimize the desired characteristics. However, that choice is not always possible.

Butter may need to be replaced with another type of fat. The problem becomes determining how much of this fat should be used as a substitute. It is important to remember that butter is only about 80 percent fat, and it has a water content of about 16.5 percent. Since stick margarines are similar in composition, an equal amount of margarine can be substituted for butter. However, subtle differences in flavor and texture can be noted by careful comparison.

Tub margarines may have the same ratio of fat and water as is found in stick margarines and butter. This suggests that equal amounts of tub margarine could be substituted for either stick margarine or butter. This is true on the basis of tenderness, but the lower melting point

**Plasticity**
Ability of a fat to be spread or creamed.

of the tub margarines causes cookies to spread far more than is the case when stick margarine or butter is used. Also, some tub margarines have a higher water content.

Shortening and lard can be used in place of butter or margarine, but not in the same quantities. Only about 90 percent as much shortening should be used when replacing butter. This amount ($\frac{7}{8}$ cup shortening for 1 cup of butter) compensates for the fact that shortening is entirely fat. Lard can be substituted in the same amount as the value given for shortening because lard also is 100 percent fat. A very small amount of salt may need to be added when using shortening or lard as a replacement for butter to improve the flavor.

Oil, like shortening and lard, is 100 percent fat. For an amount of fat that equals that of butter, the same substitution values as are suggested for shortening and lard are appropriate. However, some products can be made with considerably reduced fat levels when oil is substituted for butter. Because of the difference in mouthfeel and spreading characteristics that oil brings to the different products, the exact amount to use needs to be determined experimentally for various products

## FUNCTIONAL ROLES OF FATS

### Color

Butter contributes a yellow to creamy color to products. The importance of this pleasing color is evidenced by the fact that all margarines are colored to simulate the color of butter. Even some vegetable shortenings now have β-carotene added to provide the desired yellow color. Fats also aid in developing a pleasing color on the surface of fried and baked products.

### Flavor

Fats contribute a richness of flavor when used in a variety of food products. In addition, specific fats provide unique flavor qualities. Butter, for example, has a complex flavor profile contributed by butyric and other fatty acids, as well as by lactones, aldehydes, and ketones. The flavor of butter is so popular that most margarines and some shortenings have synthetic butter flavoring added to them to simulate the natural flavor of butter. Olive oil and lard are examples of other types of fats that contain distinctive flavor components. Most other fats have a pleasing richness of flavor, yet limited unique overtones in their flavor profiles.

### Texture

Fats influence textural characteristics in several different ways, depending on the type of food considered. In pastry, the distribution of fat in small pieces contributes flakiness to the baked product. Butter or shortening can be creamed with sugar to obtain a very fine cell structure of great uniformity in a shortened cake. In this case, sharp sugar crystals create numerous tiny spaces in the fat where steam and carbon dioxide collect and expand during baking to produce a fine-textured cake. Fat in a bread dough keeps the crumb and crust soft in comparison with a similar bread made without any fat.

Fried foods develop a crisp texture on their surfaces, the result of being heated at the very high temperatures that can be reached when fats or oils are the cooking medium. Hash browns and French fries can provide excellent examples of the crisp textures that result when the frying fat is very hot. If the fat is too cool, the result unfortunately is a soggy, greasy texture.

### Tenderness

Although tenderness actually is an aspect of textural properties, fat is so vitally important in baked products that it is appropriate to consider this function of fat separately. The unique composition of each type of fat that might be used in preparing a baked product will determine the specific capability that a particular fat may have in tenderizing.

Different qualities are desired for particular baked products; selection of the best fat for a specific application requires consideration of several qualities. One of the most important qualities is the ability of a fat to aid in creating a tender baked product. The ability of various fats and oils to tenderize a product is determined by their ability to interfere with the development of gluten, the structural protein complex in wheat flour products.

Formation of the gluten complex in a batter or dough requires water, wheat flour, and manipulation of these two key ingredients. When flour proteins become moistened with water and are stirred or beaten, the gluten complex begins to develop, resulting ultimately in a somewhat elastic network which, when placed in the oven and baked, stretches and then is set permanently into the structure of the baked product.

One of the ways that fats and oils interfere with gluten development is by physically preventing or inhibiting contact between water and flour proteins. This obstruction is accomplished by mixing the fat or oil with the flour so that the lipid gradually coats the surface of the gluten complex that is starting to form. Water is unable to penetrate a layer of lipids because lipids are hydrophobic and repel water. A soft fat or an oil can physically be spread over a much larger surface area than can a firm fat. Consequently, such lipids are effective tenderizing agents.

A closer look at the chemical structures of the fatty acids that may be found in fats helps clarify the mechanism by which fats are able to cover surface area and partially block water from gluten. The ability of a lipid to accomplish this tenderizing action is called its **shortening power**.

Much of the fatty acid molecule is a carbon chain to which hydrogen is attached. Saturated carbon chains are hydrophobic and are repelled by water. However, the double bonds in unsaturated fatty acids are hydrophilic and have an attraction for water. Furthermore, the carboxyl group of free fatty acids is attracted to water. In essence, the majority of the molecule does not attract water, but the double bonds and carboxyl group are drawn toward water. This contradictory behavior is theorized to cause unsaturated fatty acid molecules to align themselves along the interface and block the passage of water by the presence of the carbon chains.

The amount of surface area covered by a single molecule is determined by the amount of unsaturation in the fatty acid. A fatty acid with a single double bond is able to cover more surface than can a saturated fatty acid but is no more effective than a fatty acid with two double bonds. However, a fatty acid with three double bonds is able to cover more surface area than one with only one or two double bonds. Apparently, the chain length between the first and third double bond in a fatty acid with three double bonds is sufficiently long to enable both of these double bonds to reach the interface, as shown in the diagram (Figure 12.8). Because of their content of unsaturated fatty acids, oils have great shortening power.

**Shortening power**
Ability of a fat to cover a large surface area to minimize the contact between water and gluten during the mixing of batters and doughs.

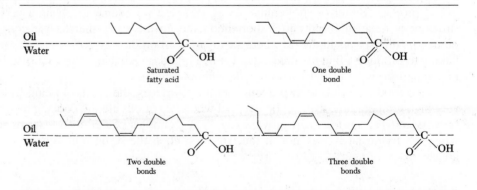

**Figure 12.8** Schematics illustrating the shortening power of fatty acids with different numbers of double bonds. Note that fatty acids with one or two double bonds are more effective in covering surface area than is a saturated fatty acid, but far more area is covered by fatty acids with three double bonds.

In addition to the ability of fat molecules with unsaturation to collect at the interface and block some of the water from the gluten, unsaturation also influences the physical nature of the fat. As noted earlier, lipids with a high degree of unsaturation are fluid oils at mixing temperatures and, therefore, are able to flow readily throughout any batter or dough in which they are included.

In contrast, fats with more saturated fatty acids and less unsaturation are solids, which restricts their movement throughout the mixing process and limits their ability to coat the gluten. Solid fats do vary considerably in their spreadability because of differences in melting points and variations in the extent of saturation. Solid fats that are soft enough to be manipulated and spread quite easily are said to be plastic.

Plastic fats have appreciably more shortening ability than do those that are quite hard and difficult to spread. The hard fats are high in saturated fatty acids; consequently, they have high melting points and only a moderate amount of shortening power. This physical nature and the fact that the fatty acid molecules are primarily hydrophobic mean that fat covers only a limited amount of surface area when it is used in mixing batters and doughs. Therefore, such fats have limited shortening power.

Shortened cakes are made with a solid fat as one of the major ingredients. In these cakes, the fat is incorporated as a heavy foam, the result of **creaming** the fat and sugar together. For creaming to be effective, the fat needs to be able to retain the air pockets that form. Obviously, oils do not meet this criterion. Shortenings have been tailored to meet this requirement, for they have an excellent plastic range. At room temperature, shortenings can be creamed into comparatively light fat-sugar mixtures, entrapping air.

**Creaming**
Vigorous blending of fat and sugar to incorporate air to promote fine, light texture in cakes.

Butter and margarines also can be creamed, but with more difficulty than is encountered in creaming shortenings. Butter is quite hard and almost brittle at refrigerator temperatures, yet it can become so soft that it is almost fluid when it is creamed with sugar on a warm day. Margarine has a broader temperature range of plasticity than does butter because of the varied fatty acid content. This makes it easier to cream margarine than butter without causing the fat to become so fluid that the emulsion in the batter breaks during later mixing.

Broken emulsions should be avoided in cake batters; a stable emulsion helps promote the fine cell size desired in shortened cakes. Lard, because of its unique crystallinity, tends to clump as it is creamed into a batter. This quality is better suited to use in pastries than in cakes.

Chiffon cakes rely on an egg white foam for their cell structure because they are made with oil rather than a solid fat. This emphasizes the need for some form of foam as the basis for the cells and illustrates the inability of oil to perform this role. However, the fluidity of the oil promotes tenderness in the cake. This tenderness is a definite contrast to the textural characteristics of sponge and angel cakes, neither of which contains any fat.

Fat is a major ingredient in pastry and often is included in weights that are about half of the weight of the flour. However, the cutting in of the fat until it is in moderately coarse particles results in its very inefficient use as a tenderizing ingredient. All of the fat on the interior of each piece is unavailable to interfere with gluten development. This explains why pastry often is fairly tough despite the large amount of fat it contains.

**Flakiness** is a highly desirable textural characteristic in pastry. When fat is left in pieces in a pastry dough, it melts during the baking period and flows, leaving a hole where steam collects and pushes upward against the upper surface of the resulting cell. The gluten in the pastry is denatured during baking, and the cell is locked into the extended position. This results in a flaky pastry. Solid fats facilitate formation of a flaky pastry, but the flow properties of oil interfere with formation of the cell pockets needed for flakiness.

**Flakiness**
Quality of thin layers of pastry that shatter when cut or chewed.

Quick breads and most yeast breads contain fat or oil. Biscuits utilize a hard fat that can be cut into pieces in a fashion similar to the preparation of pastry. This promotes flakiness in biscuits. In muffins and yeast breads, the fat is melted to obtain maximum tenderizing. Flakiness is not a possibility in breads and muffins when the fat is in the liquid state when incorporated into the mixture. The fat used in breads does contribute some flavor, tenderness, and color, but it is much less important in breads than in cakes, which contain a far higher proportion of fat than is found in breads.

From the preceding discussion, it is evident that oils are particularly effective tenderizing agents and that shortenings and lards also are appropriate to use in making tender batter and dough products. Margarines and butters are other possible choices, but it is important to recognize that both of these types of spreadable fats are not pure fat. In fact, they are about 16 percent water and 80 percent fat. Their substitution in a recipe specifying shortening means that too little fat and too much water are being used unless adjustments are made. Unless more butter or margarine and less liquid are incorporated, the baked product will be less tender than anticipated.

## Emulsification

Virtually all food systems have an aqueous liquid of some type in their formula, and many also contain fats or oils. The result is that these ingredients may be uniformly dispersed as an emulsion (see Chapter 6) or curdled. For optimal textural qualities, stable emulsions are desired. In most food systems, the type of emulsion formed is an oil-in-water emulsion, in which the oil droplets are dispersed in the aqueous phase. An emulsifying agent of some type is needed to facilitate the formation of stable emulsions.

Shortenings are formulated today with added mono- and diglycerides so that the fat and milk in cake batters are emulsified, resulting in a fine-textured cake. These mono- and diglycerides are effective emulsifiers because they have one or two hydroxyl groups in place of one or two fatty acids, leaving at least one fatty acid carbon chain, which is largely hydrophobic. The hydroxyl groups draw the molecule toward the aqueous phase. The result is that a monomolecular layer of mono- and diglycerides tends to collect around the surface of the fat droplets, serving as a protective coating to keep the fat suspended in the liquid batter (Figure 12.9).

Other ingredients also may serve as emulsifying agents. The best of the emulsifying agents available in the home is the lecithin in egg yolks. Lecithin is a phospholipid that is unusually effective in covering the surfaces of droplets in an emulsion to keep them from coalescing.

Even in the presence of emulsifying agents, some oil-in-water emulsions in foods break because of excessive evaporation of water. This is seen as a particular problem in thick

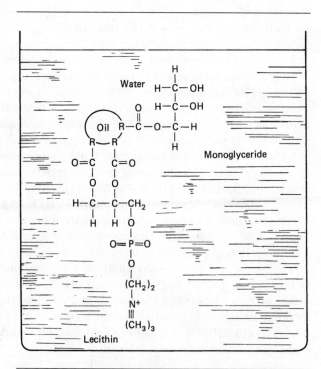

**Figure 12.9**   Orientation of monoglyceride and lecithin molecules at the interface between oil droplets and water provide a protective layer that stabilizes the emulsion by interfering with the union of oil droplets.

sauces for soufflés, especially in chocolate and cheese soufflés that contain fat from the flavoring ingredients, as well as the fat in the sauce. A similar problem may be found in some gravies. In such instances, the emulsion can be re-established by replacing the water lost through evaporation during an extended heating and/or holding period. A broken emulsion vividly reinforces the fact that fats and oils do not evaporate when they are heated in normal cookery procedures.

## Cooking Medium

Fats and oils are used as a cooking medium because of their ability to be heated to high frying temperatures well above the boiling temperature of water. Such high temperatures not only cook foods more quickly than can be done in water, but they also cause changes in texture, flavor, and color that are enjoyed by many people. The type of fat selected for frying influences the actual characteristics of fried foods.

Butter and margarines are popular because of their pleasing flavor and color and occasionally are used in shallow frying because of these qualities. Unfortunately, they deteriorate quite rapidly at the high temperatures used in frying. Part of the problem is that the water in these fats causes excessive splattering and increased aeration of the fat during frying. This hastens the development of rancidity if these fats were to be stored for later reuse. To compound the problem, the milk solids they contain begin to brown quickly and soon burn during the frying period.

Shortenings, because of their added mono- and diglycerides, are not well suited to frying. These molecules lose their fatty acids rather quickly, leaving free glycerol. In very short order, the glycerol begins to break down to form acrolein (see Chapter 11), and the smoke point rapidly drops. When the fat is smoking at frying temperatures, the acrolein not only irritates the eyes of the chef but also begins to be detectable in the flavor of the food.

For these reasons, oils usually are the type of lipid selected for frying foods, although olive oil is not an acceptable choice because of its low smoke point. The smoke points of high-quality oils generally are well above frying temperatures; however, the water from food being fried in oils, the splattering that occurs when foods are placed into the hot oil, and the extended use of the oil all combine to reduce the smoke point gradually until it drops to the point where the oil is smoking at frying temperatures. The flavor changes and the irritation from the acrolein formed at this stage make continued use of the oil undesirable.

The process of cooking food in deep fat is quite complex when studied from the perspectives of chemistry and physics. Surprisingly, deep-fat frying can be viewed as a dehydrating process, for the intense heat of the fat on the surface of a food such as a French fry evaporates water quickly into the cooking medium. This water loss causes mass transfer to occur in the French fry. This mass transfer of water from the central portion toward the surfaces to replace the lost water is of considerable importance, for soluble materials are carried with the water toward the surface, and some eventually escape into the fat.

Loss of water from the food into the hot frying oil helps avoid burning on the surface of the food because the water carries thermal energy from the food surface. The temperature on the surface of a frying food is about 100°C (212°F) despite the fact that the oil itself may be 180°C (356°F). This temperature gradient is due in large measure to the heat of vaporization that is required for converting the escaping water to steam.

Water beneath the surface of a French fry plays a key role in conducting heat into the interior of the piece so that gelatinization of the starch can occur. Enough heat must reach the starch to gelatinize it, but not so much that water is drawn back out of the starch gel, creating a rather spongy texture. If the fat is much too hot, cross-linkages cause hardening of the surface of the fry and considerable polymerization of the fatty acids in the oil.

Fat absorption during frying is a matter of interest from the perspectives of both food quality and health. As little fat as possible should be absorbed. Maintenance of a frying temperature of at least 175°C (347°F) and preferably 190–195°C (374–383°F) is an important factor in minimizing fat absorption during frying. The upper part of this range is more effective than 175°C (347°F) in keeping fat absorption as low as possible. At temperatures above 195°C (383°F), foods tend to brown too much before they are heated adequately in the center.

The key to frying is to heat the fat to the correct temperature before adding food and then cooking only a small amount of food at a time. Otherwise, the temperature of the fat drops too low, and the food is fried at a lower temperature than intended. The result is a rather greasy product because of the increased absorption of fat. Use of a fresh oil rather than a frying oil that has begun to deteriorate and become somewhat more viscous from previous use is another means of reducing the amount of oil absorbed (and adsorbed) during frying.

The formulation of a batter or dough to be fried has an effect on fat absorption. Increasing levels of sugar and fat in the mixture will result in more fat absorption than will occur in less sweet and/or rich batters and doughs. All-purpose and bread flours in batters and doughs absorb less fat during frying than does cake flour.

## FAT REPLACEMENTS

One of the most active areas of food research recently has been the development of fat replacements that can mimic the desired qualities contributed by fat without adding such a large number of calories. Various approaches to developing replacement products have been used. These range from carbohydrate-based products to protein-based replacements and even lipid-based ingredients; caloric contributions differ with the specific replacement (Table 12.3). Their suitability in food products differs because of the physical properties inherent in the individual replacement products.

### Protein-Based Replacement

Simplesse®, a product made by Kelco, is used as a fat replacement in frozen dessert. This substitute contains milk proteins and egg white in small particulates (0.1–0.3 microns, which is one-tenth the size of powdered sugar) that are able to convey the mouthfeel of fat because of the way the particles move as the product is eaten. The protein content of Simplesse® prevents its use in products requiring heat. Approval by the FDA was comparatively simple because Simplesse® is made of proteins from food and water (in a ratio of 1 part protein to 2 parts water), all of which are considered safe. The high percentage of water

**Table 12.3**    Overview of Some Fat Replacements

| Trade Name | Type of Base | Calories/g | Where Used | Comments |
|---|---|---|---|---|
| Simplesse® | Protein | 1.3 | Ice cream | Cannot be heated |
| N-Lite | Carbohydrate | 4 | Many foods | Modified food starch |
| Stellar™ | Carbohydrate | 4 | Many foods | Cornstarch |
| Slendid™ | Carbohydrate | 0 | Many foods | Pectin |
| Oatrim | Carbohydrate | <1 | Many foods | Maltodextrins from oat flour |
| Rice* Trin® 3 Complete | Carbohydrate | <1 | Many foods | Maltodextrins from rice flour |
| Avicel® | Carbohydrate | 0 | Many foods | Cellulose gel |
| Litesse® | Carbohydrate | 1 | Many foods | Polydextrose |
| Paselli Excell | Carbohydrate | 4 | Many foods | Maltodextrins |
| Kelcogel® | Carbohydrate | 0 | Many foods | Gums |
| Olean® | Fat | 0 | Baked goods | Olestra |
| Caprenin | Fat | 5 | Chocolates | Palm kernel/coconut oils |
| Benefat® | Fat | 5 | Baked goods | Salatrim (2-C fatty acid(s) |
| Olestra | Fat/carbohydrate | 0 | Potato chips | Sucrose polyester |

means that Simplesse® provides 1.3 calories per gram, which is a dramatic improvement over 9 calories from the fat that would have been needed to provide the smooth texture desired in a quality ice cream.

## Carbohydrate-Based Replacements

N-Lite products are fat substitutes based on starch and containing other natural food materials, including gums and nonfat dried milk. National Starch and Chemical Company is the developer and producer of this line of fat substitutes, which includes five different patterns for use in diverse types of foods.

Stellar™ is another starch-based fat substitute. This product, made by A. E. Staley Manufacturing Company, is made from cornstarch. Present applications include use in cheese spreads and frostings. Slendid™, a fat substitute made by Hercules, Inc., is yet another product based on a carbohydrate. However, the carbohydrate utilized for Slendid™ is pectin.

Oatrim is a fat substitute made from oats by a process patented by the U.S. Department of Agriculture. This fat substitute is made using alpha-amylase to catalyze the formation of maltodextrins from amylose and amylopectin in oat flour. Beta-glucan, also present in oatrim, is the main soluble in oats. When oatrim is heated, the gel it forms provides less than 1 calorie per gram. Probable uses for oatrim include milk-containing beverages, salad dressings, meats, cheese spreads, and high-fiber breads. Rice* Complete® is a similar product derived from rice.

Avicel® cellulose gel is a microcrystalline carbohydrate derivative that can be used as a fat substitute in salad dressings. Other fat substitutes are made using cellulose gel in combination with guar gum or with maltodextrins and xanthan gum.

Polydextrose is a starch polymer plus a little sorbitol and citric acid. It is used as a bulking agent, texturizer, and humectant in a range of products including salad dressings, puddings, candies, and other products in which it can replace some of the sugar as well as the fat. Although polydextrose is not well utilized in the body, it does provide about 1 kilocalorie per gram. Its somewhat limited digestibility contributes to its laxative effect when consumed at levels above 90 g per day.

Modified food starches made from various plant sources are other carbohydrate-based fat replacements. The sources for these starches include corn and potato. Paselli SA2 is a potato starch maltodextrin that has many applications in dips, bakery products, dressings, ice creams, and fillings.

## Fat-Based Replacements

Surprisingly, fat-based products have been developed for use as replacements for the usual fats and oils. Salatrim is an example of this approach. Marketed under the name Benefat®, **Salatrim** actually is a family of structured triglycerides that result from interesterification of lipids with long- and medium-chain (6–12 carbons) fatty acids. The introduction of acids with only two carbon atoms on one or two of the possible binding sites on glycerol impacts how readily the body utilizes Salatrim. The result is that Salatrim contributes 5 kcal/g.

**Salatrim**
Acronym for short and long acyltriglyceride molecules.

Another structured triglyceride, caprenin, uses capric and caprylic acids (10 and 8 carbon fatty acids, respectively) in combination with behenic acid, a saturated fatty acid with 22 carbon atoms. The source of the capric and caprylic acids in caprenin is coconut and palm kernel oil. Behenin, the saturated long-chain fatty acid, is available from peanuts, fish oils, and hydrogenated rapeseed oil. The hydrogenated rapeseed oil is the chosen source for manufacturing caprenin. Behenic acid is absorbed inefficiently in the small intestine, which means that much of it is excreted. The medium-chain acids are metabolized to a limited extent. The net result is that caprenin provides only about 5 calories per gram, rather than the usual 9 calories. This Procter & Gamble product is used in chocolate coatings for candy and doubtless will find numerous other applications in the future.

The structure of caprenin is as follows:

caprenin

Olestra is a prominent fat replacement that is a bit of a hybrid between a carbohydrate and a fat. It is actually a compound classified as a sucrose polyester. This esterification between the hydroxyl groups on six or eight of the carbon atoms in sucrose and the corresponding number of medium-chain (8–12 carbon atoms) fatty acids results in quite a bulky molecule that can be used as a fat replacement. Development by Procter & Gamble and the approval process required 25 years. In 1996, olestra was finally approved by the FDA. Although sucrose polyester is composed of digestible carbohydrate and metabolizable fatty acids, this compound cannot be digested and therefore cannot be absorbed. Consequently, olestra (marketed as Olean®) does not provide calories to the body.

During the developmental tests, olestra was found to cause diarrhea and flatulence in some subjects fed high levels. Therefore, FDA approval was accompanied by the requirement that consumers be informed of possible side effects. Consequently, a warning that consumption of olestra could cause these symptoms was required on product information labels if olestra (or Olean®) was an ingredient. In 2003, the FDA dropped the warning requirement in the label; research during the intervening seven-year period did not demonstrate that olestra caused significantly more gastrointestinal problems than did the placebo.

## SUMMARY

Fats undergo numerous processing steps, including extraction, refining, fractionation, crystallization, and quality checks before they are ready for the marketplace. Chemical modifications that may be a part of their production include hydrogenation, interesterification, and intraesterification.

The large array of fats and oils in the market are derived from animal, marine, and plant sources. Emphasis on the amount of *trans* fats in food products has resulted in labeling of that information on the nutrition facts label. Agricultural researchers have worked to alter the fatty acid content of plant oils through selective plant breeding and genetic engineering. The food industry has responded to this consumer concern by making adjustments in the sources of oils and changes in hydrogenation and interesterification processes. These new fats and oils are now being transformed into the diverse products valued in food preparation.

Among the functional roles of fat are contributing color, flavor, texture, tenderness, emulsification, and a cooking medium.

Fat replacements are used widely in commercial food products to reduce calories and/or enhance health. These replacements may be protein based, carbohydrate based, or fat based.

## STUDY QUESTIONS

1. Explain the winterizing of oils. Why is this an important step in processing salad oils?

2. What is the purpose of tempering a fat? How is this accomplished?

3. Describe the processes of (a) hydrogenation and (b) interesterification.

4. Why may interesterification be preferred to hydrogenation in the manufacturing of margarine?

5. Name and write the chemical structures of two $\Omega$-3 fatty acids.

6. Why may oryzanols and tocotrienols be of potential nutritional interest? Identify a food source of these compounds.

7. What is a possible mechanism for a fat or oil in tenderizing a baked product?

8. What can be done to minimize the absorption of fat in frying foods?

9. Identify an example of each of the following types of fat replacements: protein based, carbohydrate based, and fat based.

10. True or false. Equal weights of butter and shortening contain the same amount of fat.

11. True or false. Olive oil is pressed from the seed of olives.

12. True or false. Winterizing is done to remove fatty acids with high melting points.

13. True or false. Tempering is done to control crystal size in shortening.

14. True or false. Shortening power is a term used to describe the role of fats in cakes.

15. True or false. Lecithin is effective as a shortening agent in making pastry.

## BIBLIOGRAPHY

Albers, M.J., et al. 2008. 2006 marketplace survey of *trans* fatty acid content of margarines and butters, cookies and snack cakes, and savory snacks. *J. Am. Dietet. Assoc. 108* (2): 367.

Allison, D. B., et al. 1999. Estimated intakes of *trans* fatty and other fatty acids in the U.S. population. *J. Am. Diet. Assoc. 99* (2): 166.

Becker, C. C. and Kyle, D. J. 1998. Developing functional foods containing algal docoshexanoic acid. *Food Technol. 52* (7): 68.

Belitz, H. D. and Grosch, W. 1999. *Food Chemistry*. 2nd ed. Springer. New York.

Bell, S. J., et al. 1997. The new dietary fats in health and disease. *J. Am. Diet. Assoc. 97* (3): 280.

Berry, D. 2005. Designer lipids. *Food Product Design 14* (12): 118.

Blumenthal, M. M. 1991. New look at the chemistry and physics of deep-fat frying. *Food Technol. 45* (2): 68.

Boyle, E. 1997. Monoglycerides in food systems: Current and future uses. *Food Technol. 51* (8): 52.

Clark, J. P. 2005. Fats and oils processors adapt to changing needs. *Food Technol. 59* (5): 74.

Clydesdale, F. 1997. Olestra: The approval process in letter and spirit. *Food Technol. 51* (2): 104.

Decker, K. J. 2004. ABCs of omega-3s. *Food Product Design 14* (11): 81.

deRoos, K. B. 1997. How lipids influence food flavor. *Food Technol. 51* (1): 60.

Dreher, M., et al. 1998. Salatrim: Triglyceride-based fat replacer. *Nutr. Today 33* (4): 164.

Duxbury, D. 2005. Omega-3s offer solutions to *trans* fat substitution problems. *Food Technol. 59* (4): 34.

Fortin, N. D. 2005. Fats in the fast lane. *Food Product Design 14* (12): 148.

Foster, R. J. 2009. Checking the oil for snacks. *Food Product Design 19* (11): 54.

Hazen, C. 2005. *Trans*-formulation alternatives. *Food Product Design 15* (5): 71.

Hernandez, E. and Lucas, E. W. 1997. Trends in transesterification of cottonseed oil. *Food Technol. 51* (5): 72.

Hicks, K. B. and Moreau, R. A. 2001. Phytosterols and phytostanols: Functional food cholesterol busters. *Food Technol. 55* (1): 63.

Hollingsworth, P. 2001. Margarine: Over-the-top functional food. *Food Technol. 55* (1): 59.

Kuntz, L. A. 2001. Fatty acid basics. *Food Product Design 11* (8): 93.

Kuntz, L. A. 2002. Designer fats for bakery. *Food Product Design 12* (8): 55.

Kuntz, L. A. 2005. *Trans*-lating formulas. *Food Product Design 15* (7): 14.

Leake, L. L. 2007. *Trans* fat to go. *Food Technol. 61* (2): 66.

Liu, K. S. and Brown, E. A. 1996. Enhancing vegetable oil quality through plant breeding and genetic engineering. *Food Technol. 50* (11): 67.

List, G.R. 2004. Decreasing *trans* and saturated fatty acid content in food oils. *Food Technol. 58* (1): 23.

Luff, S. 2004. Ascendancy of omega-3s. *Food Product Design: Functional Foods Annual* (September): 67.

Luff, S. 2005. Better ingredients through biotechnology. *Food Product Design 14* (10): 91.

Marangoni, A. G. and Hartel, R. W. 1998. Visualization and structural analysis of fat crystal networks. *Food Technol. 52* (9): 46.

McCaskill, D. R. and Zhang, F. 1999. Use of rice bran oil in foods. *Food Technol. 53* (2): 50.

Miraglio, A. M. 2002. The low-down on *trans* fatty acids. *Food Product Design 12* (1): 31.

O'Brien, R. D. 2003. *Fats and Oils: Formulating and Processing for Applications.* 2nd ed. CRC Press. Boca Raton, FL.

Ohr, L. M. 2005. Functional fatty acids. *Food Technol. 59* (4): 63.

Ohr, L. M. 2009. Functional fat fighters. *Food Technol. 63* (10): 59.

Palmer, S. 2005. New current for tropical oils. *Food Product Design 15* (2): 87.

Pszczola, D. E. 2004. Fats: In *trans*-ition. *Food Technol. 58* (4): 52.

Remig, V., et al. 2010. *Trans* fats in America: Review of their use, consumption health implications, and regulation. *J. Amer, Dietet. Assoc. 110* (4): 585.

Saguy, I. S. and Pinthus, E. J. 1995. Oil uptake during deep-fat frying: Factors and mechanism. *Food Technol. 49* (4): 142.

Sloan, A. E. 2005a. Top 10 global food trends. *Food Technol. 59* (4): 20.

Sloan, A. E. 2005b. Time to change the oil? *Food Technol. 59* (5): 17.

Spano, M. 2010. Heart health and fats. *Food Product Design 20* (3): 22.

Tarrago-Trani, M. T., et al. 2006. New and existing oils and fats used in products with reduced *trans*-fatty acid content. *J. Am. Diet. Assoc. 106* (6): 867.

Tiffany, T. 2007. Oil options for deep-fat frying. *Food Technol. 61* (7): 46.

## INTO THE WEB

*http://www.theepicentre.com/tip/nutandseedoils.html*—Brief descriptions of nut and seed oils.

*http://www.innvista.com/health/nutrition/fats/process.htm*—Overview of steps in processing fats and oils.

*http://oliveoilsource.com/making_olive_oil.htm*—Information on making olive oil.

*http://www.americanpalmoil.com/*—Extensive information on palm oil.

*http://www.mpoc.org.my/*—Official web site of Malaysian Palm Oil Council.

*http://www.ghirardelli.com/bake/chocolate_tempering.aspx*—Instructions and video on tempering chocolate.

*http://www.cdr-mediared.com/food-diagnostics/foodlabfat/control-quality-edible-oils-fats?gclid=CO7y9tuGkZ8CFQZfagod0jpAxQ*—Instrument for measuring quality of fats and oils.

*http://www.aocs.org/*—Site for American Oil Chemists' Society (AOCS).

*http://www.aoac.org/*—Site for Association of Official Analytical Chemists (AOAC)

*http://www.codexalimentarius.net/web/index_en.jsp*—Site for FAO and WHO Codex Alimentarius Commission (CAC)

*http://www.iupac.org/*—Site for International Union for Pure and Applied Chemistry (IUPAC).

*http://www4.agr.gc.ca/AAFC-AAC/display-afficher.do?id=1171307588659&lang=eng*—Ways of reducing *trans* fats.

*http://www.foodprocessing.com/articles/2005/459.html*—Discussion on ways to reduce *trans* fats in food products.

*http://www4.agr.gc.ca/AAFC-AAC/display-afficher.do?id=1172579557273&lang=eng*—Techniques for reducing *trans* fats.

*http://www.webexhibits.org/butter/working.html*—Description of making butter.

*http://www.preparedfoods.com/HTML/Application_Videos/BNP_GUID_9-5-2006_A_100000000000000386871?id=TC1&id2=RDEast08*—Videocast on long-chain omega-3 fatty acids in functional foods.

*http://www.reducol.com/index2.htm*—Site about GRAS status for Reducol™.

*http://www.margarine.org/howtousemargarine.html*—Discussion of reduced-fat spreads and substitutions in cooking and baking.

*http://www.caloriecontrol.org/sweeteners-and-lite/fat-replacers*—Overview of fat replacements.

*http://www.cwu.edu/~geed/440project044.html*—Poster sessions of research projects done by students enrolled in an experimental foods course.

*http://www.cpkelco.com/products-mwpc.html*—Information on Simplesse®.

# 5

# Proteins

Legumes are a significant source of protein for many people around the world.

# CHAPTER 13

# Overview of Proteins

## Chapter Outline

## OBJECTIVES

After studying this chapter, you will be able to:

1. Describe the structure of proteins.
2. Identify the types of proteins that comprise various foods.
3. Interpret how electrical charges and hydrolysis influence food preparation of some foods high in protein.
4. Discuss the significance of denaturation and coagulation in preparing protein-rich foods.

## INTRODUCTION

Proteins are complex in their composition and behavior in food products. Careful attention needs to be paid to the preparation of protein-rich foods because the treatment of the protein greatly influences the quality of the final product. The proteins in food are unforgiving of abuse during preparation; heating for too long or to too high a temperature can cause some highly detrimental, irreversible changes in the proteins of a food. The significance of heat treatment will become apparent with the presentation of the structure and behavior of proteins. The importance of pH in dealing with protein-rich foods also will be evident in the succeeding chapters.

## COMPOSITION

Proteins are molecules composed of many **amino acids** joined together by peptide linkages. A protein may contain several hundred amino acid moieties linked together into a complex molecule. Amino acids are organic substances containing two characteristic functional groups: the amino ($-NH_2$) group and the **carboxyl group** ($-COOH$). The amino and the acid groups and various organic components are attached to a single carbon (the $\alpha$ carbon). These various organic components (designated as R groups) range from the single hydrogen atom of glycine to dual-ring structures, such as that in tryptophan. Table 13.1

**Amino acids**
Organic compounds containing an amino ($-NH_2$) group and an organic acid ($-COOH$) group.

**Carboxyl group**
Organic acid group:

$$(-C\overset{O}{\underset{\phantom{x}}{\diagup}}OH)$$

**Table 13.1**   Amino Acids in Foods

| Name | Formula | Isoelectric Point | Type |
|---|---|---|---|
| Alanine | | 6.0 | Neutral—aliphatic |
| Glycine | | 6.0 | Neutral—aliphatic |
| Isoleucine* | | 6.0 | Neutral—aliphatic |
| Leucine* | | 6.0 | Neutral—aliphatic |
| Valine* | | 6.0 | Neutral—aliphatic |
| Serine | | 5.7 | Neutral—hydroxy |
| Threonine* | | 6.2 | Neutral—hydroxy |
| Cysteine | | 5.1 | Neutral—sulfur-containing |
| Cystine | | 4.6 | Neutral—sulfur-containing |
| Methionine* | | 5.7 | Neutral—sulfur-containing |
| Asparagine | | 5.4 | Neutral—amide |
| Glutamine | | 5.7 | Neutral—amide |

(continued)

**Table 13.1** Continued

| Name | Formula | Isoelectric Point | Type |
|------|---------|-------------------|------|
| Phenylalanine* | | 5.5 | Neutral—aromatic |
| Tryptophan* | | 5.9 | Neutral—aromatic |
| Tyrosine | | 5.7 | Neutral—aromatic |
| Aspartic acid | | 2.8 | Acidic |
| Glutamic acid | | 3.2 | Acidic |
| Arginine | | 11.2 | Basic |
| Histidine* | | 7.6 | Basic |
| Lysine* | | 9.7 | Basic |
| Hydroxyproline | | 5.8 | Imino acid |
| Proline | | 6.3 | Imino acid |

*Essential amino acid.

provides the formulas for the amino acids of importance in foods, all of which are L-amino acids.

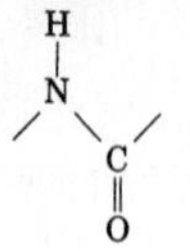

The **peptide linkage** is a covalent bond formed between the nitrogen of one amino acid and the carbon of the carboxyl group of another amino acid; a molecule of water is eliminated in the reaction:

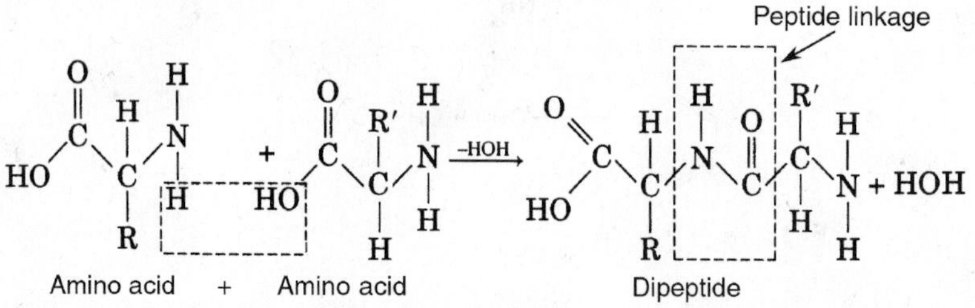

**Peptide linkage**
Linkage from the nitrogen of one amino acid to the carbon of the carboxyl group of another amino acid:

Amino acid    +    Amino acid                                    Dipeptide

Proteins are large molecules containing 100 or more amino acid residues, which are linked together covalently by peptide bonds. The net result of this arrangement is that a backbone chain with a repeating pattern (—N—C—C—N—C—C, etc.) forms, and the R group and the =O extend outward from this backbone. This level of protein organization is the rudimentary molecule and is termed the **primary structure** of proteins, as shown:

**Primary structure**
Covalently bonded backbone chain of a protein: —C—C—N— C—C—N—C—C—N—.

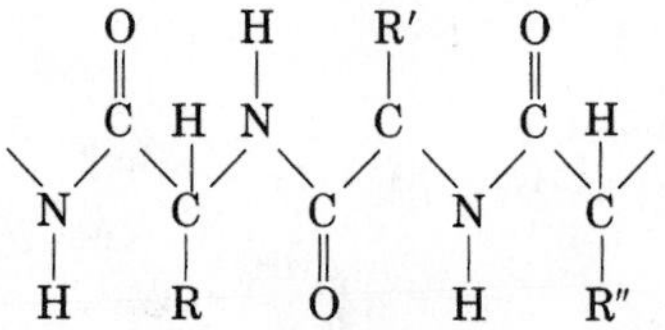

Although the primary structure can be visualized as a linear chain, it is important to remember that R represents different structures, depending on which amino acid moiety is there. For example, the R for glycine is simply a hydrogen atom. However, for many of the more complex amino acids, such as tryptophan, the R is actually a bulky structure extending outward from the backbone chain, as can be seen in Table 13.1. Although each protein has a different amino acid sequence, the configuration of the R for each of these must be fitted along the chain.

The primary structure is the foundation of the protein molecule, but the extended linear molecule is stressed somewhat. The energy level required to maintain the extended primary structure of a protein can be reduced if the molecule is coiled, as occurs in many **native protein** molecules. In this coiled arrangement, called an α-helix, the long chain of the primary structure is formed by **hydrogen bonds** into a right-handed, spring-like configuration, the **secondary structure** [Figure 13.1(a)]. This structure has a comparatively low energy state. There are 3.7 amino acid residues in each turn of the helix; the turn is determined by the hydrogen bonding between the nitrogen (—NH) and the carbonyl (C=O) of the residues above or below each other in the helical form.

The helical secondary structure in native protein is convoluted and folded into various shapes, which are held in their native configurations by secondary bonding forces between the R groups that extend from the backbone chain. The bonding forces may be hydrogen bonds, salt bridges, disulfide (covalent bonding force) linkages, and hydrophobic interactions. This shape, characteristic of many native proteins, is the **tertiary structure** [see Figure 13.1(b)].

The hydrophobic nature of some parts of the protein molecule is an important aspect of the shape of specific proteins. Some of the amino acid structures are quite hydrophobic; these include methionine, tyrosine, tryptophan, leucine, isoleucine, valine, and alanine. The hydrophobicity of the R groups causes the protein molecule to be drawn together somewhat tightly in places, thus facilitating the formation of hydrogen bonds in the native protein.

**Native protein**
Protein molecule as it occurs naturally without external influences such as heat or changes in pH.

**Hydrogen bond**
Secondary bond formed between a hydrogen atom (covalently linked to an electronegative atom such as nitrogen) and an electronegative atom (such as oxygen in the carbonyl group of a protein).

**Secondary structure**
Typically the α-helical configuration of the backbone chain of many proteins; held by secondary bonding forces, notably hydrogen bonds; also may be in other forms (e.g., β-pleated sheet).

**Tertiary structure**
Distorted convolutions of the helical configuration of a protein; the form in which many proteins occur in nature and which is held by secondary bonding forces.

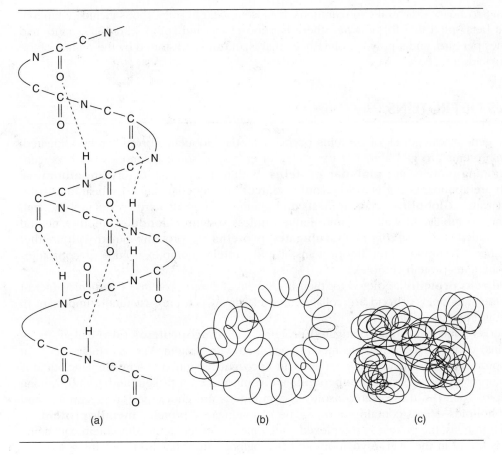

**Figure 13.1**   Organizational levels of many proteins: (a) α-helix (secondary structure of a protein), (b) possible configuration of the tertiary structure of a globular protein, and (c) quaternary structure of some proteins.

In most proteins, the tertiary structure is the final level of organization for the protein molecule, but in some instances (e.g., hemoglobin), two or more peptide chains may be held tightly together in a quaternary structure [Figure 13.1(c)]. The quaternary structure represents a close aggregation of protein segments.

Although many food proteins exist in the right-handed α-helix at the secondary structural level, some have a highly hydrophobic nature, which can force the molecule into a beta (β) turn that changes direction at every fourth amino acid residue. This can create a spatial arrangement that is referred to descriptively as a β-**pleated sheet** (Figure 13.2) when such molecules are linked in a parallel or antiparallel structure by interchain hydrogen

**β-Pleated sheet**
Secondary structure of some hydrophobic proteins that form a pleated left-handed turn that creates a pleated arrangement.

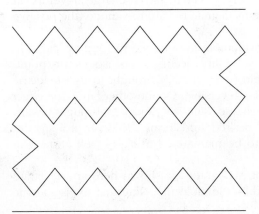

**Figure 13.2**   Diagram of part of a protein in a β-pleated sheet configuration.

bonds. Secondary structural configurations that also appear in some foods include a random coil, a beta spiral that forms when there is a loosely turned spiral with 13.5 amino acid residues per turn, and a poly-L-proline helix that is an α-helix distorted by the rigidity of the amino acids in it.

## TYPES OF PROTEINS

**Globular proteins**
Native proteins with a tertiary structure that is rather spherical.

**Albumins**
Water-soluble globular proteins that can be coagulated by heat.

**Globulins**
Globular proteins that can be coagulated by heat, but with limited solubility in water.

**Conjugated proteins**
Compounds containing protein attached to another substance.

**Fibrous proteins**
Insoluble, elongated protein molecules.

**Mucoproteins**
Conjugated protein that includes a carbohydrate moiety.

**Lipoprotein**
Conjugated protein that includes a lipid moiety.

**Metalloprotein**
Conjugated protein that includes a metal.

**Nucleoprotein**
Conjugated protein that includes a nucleic acid.

**Phosphoprotein**
Conjugated protein that includes an inorganic phosphate.

**Amphoteric**
Capable of functioning as either an acid or a base, depending on the pH of the medium in which the compound occurs.

**Isoelectric point**
The pH at which a protein molecule is electrically neutral; the specific pH differs for various proteins.

Three general categories of proteins occur in foods: globular, fibrous, and conjugated. Many proteins are globular. For example, all enzymes, some hormones, and oxygen-transporting proteins are **globular proteins**. Within this large category are **albumins**, which are abundant (egg is a noteworthy source), readily coagulated by heat, and soluble in water. **Globulins** occur in meats and legumes. They also are coagulated by heat, but their solubility in water is quite limited unless sodium chloride or another neutral salt is added (called *salting in*). **Conjugated proteins** are proteins joined with another substance. Examples are lipoproteins (lipid–protein complex) and glycoproteins (carbohydrate–protein complex).

**Fibrous proteins** are noted for their insolubility. Fibrous proteins of particular interest from the standpoint of food are collagen and elastin, which are structurally important in meats and poultry.

Foods contain various conjugated proteins. The **mucoproteins** (also called glycoproteins) contain a carbohydrate moiety combined with protein. Various sugars occur in mucoproteins. Ovomucoid in egg white is an uncoagulable mucoprotein. Hemagglutinin is a mucoprotein in soybeans. **Lipoproteins** are composed of a protein and a lipid. Among the lipids found in these water-insoluble compounds are cholesterol, triglycerides, and phospholipids. Meats contain another type of conjugated protein, **metalloprotein**. In metalloproteins, the protein is complexed with a metal. Ferritin, a metalloprotein containing iron, is found in the liver. Myoglobin and hemoglobin are other iron-containing metalloproteins. **Nucleoproteins**, proteins combined with nucleic acids, also are found in foods. **Phosphoproteins** include the casein in milk. In these conjugated proteins, inorganic phosphates are linked with the protein.

## ELECTRICAL CHARGES

Some individual amino acid residues in a protein have the potential to be charged electrically in opposite ways, depending on the pH of the medium in which the protein is found. Individual protein molecules are **amphoteric**, because they have the potential to function as either an acid or a base, depending on the pH.

When the number of positive and negative charges on a protein is equal, the protein is at its **isoelectric point**. At pH values below the isoelectric point (more acidic), protein molecules are drawn toward the cathode (the negative electrode), as a consequence of the net positive charge on the molecules. Conversely, a net negative charge develops at pH values above the isoelectric point, and proteins migrate toward the anode (the positive electrode).

The isoelectric point of proteins is of considerable importance in food preparation (Figure 13.3) because a protein's minimum solubility occurs at its isoelectric point. Therefore, when a fluid food containing protein (e.g., milk) is brought to its isoelectric point, curdling is likely to occur, a change that is irreversible. On the other hand, the effect of pH can be used to advantage when making cheese. Milk is deliberately brought toward pH 4.6, the isoelectric point of casein, and the desired curd forms. However, if a smooth milk product is the goal, the mixture needs to be maintained at a pH well above the isoelectric point.

The ionization states of glycine at different pH values are shown here and illustrate the effect of pH on the electrical charge carried by the molecule. Note that the isoelectric

**Figure 13.3**   Cheese is made by bringing milk closer to the isoelectric point of casein to coagulate some of the protein and form a curd.

point of glycine is reached at pH 6.0. See Table 13.1 for the isoelectric points of other amino acids.

$$^+H_3NCH_2COOH \underset{H^+}{\overset{OH^-}{\rightleftharpoons}} \, ^+H_3NCH_2COO^- \underset{H^+}{\overset{OH^-}{\rightleftharpoons}} H_2NCH_2COO^-$$
$$\text{pH} \sim 1 \qquad\qquad\qquad \text{pH } 6 \qquad\qquad\qquad \text{pH} \sim 11$$

## HYDROLYSIS

Protein molecules may undergo hydrolysis to form shorter chains. This chemical change occurs at the peptide linkage between amino acids in the primary structure when the hydrogen of a molecule of water joins with the nitrogen in one amino acid moiety to form an amino group and the OH joins with the carbonyl of the adjoining amino acid to form a carboxyl group. This reaction splits the primary chain into two shorter peptides, as shown here.

The reaction usually is the result of enzymatic action by peptidases, but sometimes collagen is cleaved by acid hydrolysis. The shorter chains resulting from hydrolysis show increased solubility and decreased ability to thicken food products.

## DENATURATION AND COAGULATION

When proteins are subjected to stresses, particularly heat, agitation, and ultraviolet light, they may undergo modifications that result in decreased solubility or loss of ability to catalyze reactions (if the protein is an enzyme). These physical changes are caused by alteration of the shape of the protein molecule, a process called **denaturation**. Different R groups appear on the surface of the molecule, causing different behavioral properties, such as net electrical charge on the surface.

During denaturation, protein molecules relax from their tertiary state and begin to resemble more closely their secondary helical structure without distortion. As the somewhat twisted, spherical shape of molecules gradually relaxes into a rather elongated helical form, it is possible for other elongated protein molecules to align themselves in clumps joined primarily by hydrogen bonding. These coagulated protein aggregates increase the viscosity of the mixture perceptibly, even when the mixture is hot. In fact, in instances where the concentration of protein is sufficiently high, the fluid mixture may coagulate into a solid, as is true in the cooking of eggs (Figure 13.4). This second stage in which the denatured molecules clump together is referred to as **coagulation** of protein.

Denaturation and coagulation are physical rather than chemical changes in the protein molecule. They are brought about by the introduction of energy into the protein-containing system, energy that is ordinarily provided by heating or beating. The beating of egg whites provides a good illustration of the effectiveness of agitation in bringing about denaturation and coagulation of proteins. The cell walls of the foam result from denaturation and coagulation of part of the protein in the whites.

As is seen in egg cookery, heating also enables permanent physical changes to take place in the protein. Most of the physical changes of denaturation and coagulation are irreversible. Therefore, care must be exercised in the preparation of food products to ensure the use of optimal techniques whenever protein-rich foods are the ingredients.

**Denaturation**
Relaxation of the tertiary to the secondary structure of a protein accompanied by decreasing solubility of a protein.

**Coagulation**
Clumping together of denatured proteins (often as a result of energy input, such as heating or beating).

**Figure 13.4**   Heat causes denaturation and coagulation to transform liquid egg into a solid, as can be seen with this hard-cooked egg.

The use of **enzymes** in food work also is related to this discussion on denaturation. The fact that enzymes are proteins must always be remembered when foods containing enzymes are prepared. If enzyme action is desired, as it is when invertase is added to a fondant to aid in liquefying the center of a chocolate-dipped candy, the enzyme cannot be added until the heating and cooling of the fondant are complete. Otherwise, the enzyme may be denatured and its catalytic ability lost because of the change in its physical shape.

Conversely, there are instances when enzyme activity should be avoided, as in the preparation of a gelatin salad with pineapple. Bromelain in fresh pineapple is a proteolytic enzyme that hydrolyzes gelatin, resulting in a loss of gel strength. Cooked or canned pineapple can be added to gelatin, however, because heating the pineapple inactivates the enzyme.

The action of enzymes occurs at certain active site(s) on the surface of the enzyme molecule. In the native protein, the surface character or shape enables the enzyme to lock temporarily with another compound, the **substrate**, in a food. This intimate arrangement facilities a specific chemical reaction, which results in a change in the food. Subsequently, the enzyme and the resulting compounds unlock, and the enzyme is free to interact with another substrate molecule.

There are many different enzymes in various foods, and their effects need to be considered when preparing foods. For example, vegetables to be frozen are blanched to inactivate (denature) the enzymes that could continue to cause some deterioration during storage. In other instances, adding an enzyme may enhance the usefulness of a food. An example of this is the addition of lactase to milk to produce a milk product that is low in lactose, making it suitable for people with lactose intolerance.

Whenever enzymes are used in food preparation, both the pH and the temperature of the food must be considered to achieve the desired result. Alteration of the pH can greatly retard or even block catalysis by the enzyme because of changes in the electrical profile on its surface. Heat can alter the surface shape of the enzyme, making it impossible for it to lock with the substrate to catalyze a reaction.

**Enzyme**
Protein capable of catalyzing a specific chemical reaction.

**Substrate**
A general term for the compound that an enzyme alters.

## FUNCTIONAL ROLES

Among commonly eaten foods, milk and milk products, meats, poultry, fish, eggs, and legumes are particularly important sources of protein. Cereal grains and gelatin also contain useful amounts. Actually, each food source contains many proteins that are unique to that food. The behavioral properties of the different proteins offer a variety of opportunities to utilize their specific functions in food preparation.

### Foam Formation

An important functional property of gelatin and egg proteins is their ability to form stable foams. These proteins in their native state can be whipped, spreading them into thin foams encasing air. The energy available from the beating action causes denaturation of some of the protein, and it is this denatured protein that gives rigidity and some stability to the cell walls of the foam. These stable foams provide a light and airy product with a large volume (Figure 13.5).

Gelatin foam is essential to provide the light, airy texture of marshmallows, prune whip, chiffon pie fillings, and other whipped desserts. Angel food, sponge, and chiffon cakes and also meringues are familiar examples of the foaming power of egg whites and their importance in preparing light products with a large volume. The volume is generated in large measure as a result of the extensibility of the native protein, but denaturation imparts rigidity and permanency during baking.

### Thickening Agent

Egg proteins are effective as thickening agents when they are denatured and coagulated by heat. When custards are baked, the proteins denature by gradually unwinding from their tertiary structure to their secondary structure. Then the denatured egg proteins coagulate to form a gel structure by cross-linking as hydrogen and other secondary bonds form to establish a continuous network of protein molecules. Hollandaise, some other sauces, and

**Figure 13.5**   The proper amount of beating and some cream of tartar create a stable egg white foam which is stabilized still more by baking this soufflé.

cooked salad dressings provide additional examples of the use of egg yolk and egg white proteins to serve as thickening and emulsifying agents when they are denatured and coagulated. Milk proteins undergo denaturation and coagulation to precipitate and form the curd from which cheese is made. Soybean proteins are transformed to tofu similarly.

---

### FOOD FOR THOUGHT: The Clot Thickens

Tofu provides a useful illustration of a food product that is made by taking advantage of the unique properties of proteins. First, soybeans are soaked to soften them, and these are washed before being whirled in a blender with some water to make a smooth, rather light paste. This protein-containing paste is boiled briefly with a small amount of water until it foams. The heat causes physical changes (including some gelatinization of starch and denaturation of proteins) so that the mixture can be pressed through a colander to separate the soymilk from the pulp.

Soymilk at this point contains trypsin inhibitor, an enzyme that interferes with digestion of proteins. Fortunately, this enzyme in the soymilk is inactivated by heating the liquid enough to denature and change the shape of the trypsin inhibitor molecules.

The next step in making tofu is to dilute the soymilk with water and heat to boiling before adding either a dispersion of a salt, such as calcium sulfate, or enough lemon juice or other acid to adjust the pH to 4.2 (the isoelectric point of soy proteins). If a salt is used, precipitation of the soy proteins will occur because the electrical charges provided by the ionized salt cancel the electrical charges on the soy protein molecules. An acid accomplishes precipitation by altering the pH so the electrical charges on the soy proteins are minimal, which allows clotting to occur. Subsequently, whey is pressed from the tofu curd.

## Structural Component

Gluten, the protein complex that forms when wheat flour is manipulated with water, provides the protein network that is responsible for much of the textural characteristics of baked products, including both cell structure and volume. The stretching capability or the elasticity provided by the gluten during mixing defines the potential structure of the product. During baking, the cell walls stretch under the pressure from gases within the cells. When oven heat coagulates the stretched gluten, the walls become rigid enough to maintain the extended cellular structure.

Edible films can be made from milk, wheat, and corn proteins using special commercial processing techniques. The wheat films are made with gluten and a heated alkaline alcohol–water mixture. Such material can be used as coatings for dry-roasted peanuts and similar coating applications. Zein is the mixture of proteins from corn that is used for making edible films and coatings. Nuts, dried fruits, and jelly beans may be coated with zein films. Casein films and whey films are other examples of edible protein films.

## SUMMARY

Proteins are composed of amino acids, which are compounds distinguished by the presence of an amino group and an organic acid radical. These amino acids join through peptide linkages between the nitrogen of one amino acid and the carbon of the carboxyl group in the next amino acid to form the primary structure or backbone chain of a protein. Specific R groups distinguish the various amino acids and contribute to the behavioral properties of the protein molecule.

This primary structure then is coiled into an α-helical configuration, the secondary structure, which is held by secondary bonding forces in this relaxed shape or other forms. Superimposed on the secondary structure is the distorted or twisted tertiary structure often found in a spherical or globular shape in native food proteins. This tertiary structure also is held by secondary bonding forces. Only rarely are the proteins in food found in a quaternary structure. A few proteins in foods are fibrous, rather than globular, and some are conjugated proteins.

When heat or other energy (e.g., beating) is applied to a food containing protein, the protein begins to denature or gradually relax from the tertiary to the secondary, low-energy structure. With continued energy input, molecules may coagulate by cross-linking and precipitating, which is evidenced by thickening and loss of solubility. Coagulation also occurs very readily when proteins are at their isoelectric point, because the reduction in electrical charge on the molecules permits them to clump together rather than repel each other.

Proteins function in several ways in food preparation, including foam formation, thickening agent, and structural component. A particularly important use is the formation of foams. Egg whites and gelatin are especially effective in forming foams. Egg white foams can be denatured and coagulated to create stable products. Gelatin foams give stability when cooled. Egg proteins often are used as thickening agents, and milk proteins can be precipitated to form the curd used in making cheese. Gluten provides the structural network for baked products. Gluten, zein, and milk proteins (casein and whey) can be the key ingredients in making edible films and coatings.

## STUDY QUESTIONS

1. Write the basic structure of an amino acid and then unite it with a peptide linkage to a second amino acid. Be sure to include the actual configuration of the R group for both of the amino acids. Did the amino acids you used seem crowded when the R groups were written out?

2. Select four amino acids and join them with peptide linkages to make part of a molecule. Now write that molecule in the form of an α-helix according to the information provided in Figure 13.1(a).

3. When proteins are subjected to heat and undergo denaturation so that the tertiary structure is altered,

what happens to the R groups in the interior and on the surface of the molecule? Why is this important?

4. What influence does the clumping together of protein molecules during coagulation have on the solubility of the protein?

5. What effect does heat have on an enzyme's activity? Explain why this is so.

6. Draw structures showing the change in electrical charge on an amino acid as the pH decreases from the isoelectric point to a lower pH.

7. Find a recipe for each of the following and explain the function of any protein-rich ingredients: (a) scrambled egg, (b) mayonnaise, (c) angel food cake, (d) bread.

8. True or false. The primary structure of a protein is broken during denaturation.

9. True or false. For egg yolk to coagulate, at least some amino acids must be split from the primary structure.

10. True or false. Many of the proteins in foods are globular proteins.

11. True or false. During denaturation and coagulation, the tertiary structure of proteins usually is modified.

12. True or false. Because enzymes are proteins, heat must be controlled to prevent denaturing the protein and halting its enzymatic function.

13. True or false. Tofu is made by precipitating the proteins in egg whites.

14. True or false. Proteins can undergo a shortening of their primary structure due to acid hydrolysis.

15. True or false. A peptide linkage is the same as a hydrogen bond.

## BIBLIOGRAPHY

Belitz, H. D. et al. 2009. *Food Chemistry*. 4th ed. Springer. New York.

Birschbach, P., et al. 2004. Enzymes: Tools for creating healthier and safer foods. *Food Technol. 58* (4): 20.

Decker, K. J. 2009. Protein: A functional powerhouse. *Food Product Design 19* (6): 34.

Damodaran, S., et al. 2007. *Fennema's Food Chemistry*. 4th ed. CRC Press. Boca Raton, FL.

Foster, R. J. 2005. pHood phenomena. *Food Product Design 14* (11): 61.

Giese, J. 1994. Proteins as ingredients: Types, functions, applications. *Food Technol. 48* (10): 49.

Haard, N. F. 1998. Specialty enzymes from marine organisms. *Food Technol. 52* (7): 64.

Hazen, C. 2004. Proteins: Nutrition and function. *Food Product Design 14* (5): 34.

Holsinger, V. H. and Kligerman, A. E. 1991. Applications of lactase in dairy foods and other foods containing lactose. *Food Technol. 45* (1): 92.

Penet, C. S. 1991. New applications of industrial food enzymology: Economics and processes. *Food Technol. 45* (1): 98.

Pszczola, D. E. 1999. Enzymes: Making things happen. *Food Technol. 53* (2): 74.

Pszczola, D. E. 2004. Dawning of the age of proteins. *Food Technol. 58* (2): 56.

Wade, L. G. 2009. *Organic Chemistry*. 7th ed. Prentice Hall. Upper Saddle River. NJ.

## INTO THE WEB

*http://www.elmhurst.edu/~chm/vchembook/566secprotein. html*—Discussion of secondary structure of proteins.

*http://academic.brooklyn.cuny.edu/biology/bio4fv/page/ beta_pl.htm*—Diagram of β-pleated sheet.

*http://class.fst.ohio-state.edu/FST822/lectures/Denat.htm*— Discussion of denaturation.

*http://highered.mcgraw-hill.com/sites/0072943696/student_ view0/chapter2/animation__protein_denaturation. html*—Animation of protein denaturation.

*http://www.foodproductdesign.com/articles/2004/04/food- product-design-applications—april-2004—t.aspx*— Discussion of egg and other protein foams.

*http://practicalaction.org/docs/technical_information_ service/tofu_soymilk.pdf*—Information on making tofu.

*http://www.ars.usda.gov/is/AR/archive/nov05/milk1105. htm*—Discussion of edible casein films.

*http://news.ucanr.org/newsstorymain.cfm?story=508*— Discussion of whey films.

Whey Cheese from goat's milk is being made by this artisanal cheese maker in a mountain village on Crete.

# CHAPTER 14

# Milk and Milk Products

## Chapter Outline

## OBJECTIVES

After studying this chapter, you will be able to:

1. Describe the composition and nutritional contributions of milk.
2. Explain the various ways and the rationale for processing milk.
3. Identify the range of dairy products.
4. Discuss chemical and physical effects on milk products caused by heat, enzymes, acid, and salts.
5. Explain formation of milk and cream foams.

## COMPONENTS

Milk is a complex fluid that contains a remarkable array of chemical compounds dispersed in an aqueous medium. Whole cow's milk is approximately 88 percent water, 5 percent carbohydrate, 3.5 percent protein, and 3.3 percent fat. Its nutrient content varies from species to species, from breed to breed, seasonally, and even from the beginning of the milking to the end of the process. In the United States, cow's milk is used most commonly as a beverage and in food preparation. Therefore, this chapter discusses cow's milk throughout.

In 1994, **recombinant bovine somatotropin** (usually abbreviated as **rbST** or **bST**) was approved for injection in dairy cattle to increase milk production. The process to produce this genetically engineered hormone begins when the gene for bovine somatotropin (a natural bovine hormone) is inserted into a special bacterium and the result is harvested and processed into the recombinant bST that is ready for use. Estimates are that use of bST increases milk production averages per cow by about 1,800 pounds (from approximately 14,841 per cow to 16,641) annually (Burrington, 2002). The milk itself is essentially identical with milk produced without this hormone.

## Lipids

Fat content in cows varies considerably. Guernsey and Jersey cows are noted for their comparatively high fat content, actually in excess of 5 percent. In contrast, federal tables of food composition report milk nutrient content on the basis of milk containing only 3.3–3.7 percent fat; some of the fat has been removed to achieve this accepted standard.

Although milk contains phospholipids, carotenoid pigments, sterols, and fat-soluble vitamins, it is the milk fat (containing triglycerides in abundance) that generates particular interest in food products. Milk fat is a notable type of fat because of the array of fatty acids found in its triglyceride molecules. The length of the fatty acid carbon chains ranges from 4 to 26, and these chains contain primarily an even number of carbon atoms.

Altogether, 64 different fatty acids have been identified in milk fat. Butyric acid (4 carbons) adds to its flavor. By far the most common of the saturated fatty acids is palmitic acid (16 carbon atoms); oleic acid (18 carbons) is even more abundant than palmitic acid and is the most plentiful monounsaturated fatty acid in milk fat, as Table 14.1 shows. Linoleic acid is the most abundant of the polyunsaturated fatty acids, but it is found in rather small quantities compared with both palmitic and oleic acids.

The triglycerides containing these diverse fatty acids are dispersed in milk in the form of tiny fat globules, each of which is surrounded by a **fat-globule membrane** containing phospholipids and proteins, including enzymes and lipoproteins (Figure 14.1). Lipoproteins are effective emulsifying agents because they contain hydrophilic and hydrophobic components, enabling them to form a protective layer around the triglyceride molecules. Even with this

### FOOD FOR THOUGHT: Cloning and Casein

Cloning of cows and other domestic farm animals has been a subject of great interest and argument for several years. Scientists hope to achieve higher-quality animals by a reproductive technique called *somatic cell nuclear transfer*. Such experiments clone cells of animals that have the desired characteristics. The eventual goals are consistently healthy and well-formed clones that yield meat and other products of very high quality.

Meat and milk from cloned animals are not available in markets at the present time. Research is still underway to determine the health risks to animals and the safety of products from these animals. The Food and Drug Administration will regulate food products from cloned sources when their safety has been established. Researchers in New Zealand have been introducing genetic alterations in dairy cows to increase the content of β- and κ-casein in their milk, and a few of the cows have been cloned successfully. The level of β-casein in their milk increased by as much as 20 percent, the κ-casein content was doubled, and total protein also increased in the milk produced by the experimental cows. The next question to be answered is how the milk with modified levels of caseins and total protein will affect clotting and production of yogurt and cheese.

**Table 14.1**   Fatty Acid Composition of Milk Fat

| Type of Fatty Acid | Fatty Acid | Composition[a] | Percentage of Total Fat |
|---|---|---|---|
| *Saturated* | | | 62.3 |
| | Butyric | 4:0 | 3.3 |
| | Caproic | 6:0 | 1.7 |
| | Caprylic | 8:0 | 1.2 |
| | Capric | 10:0 | 2.3 |
| | Lauric | 12:0 | 2.6 |
| | Myristic | 14:0 | 10.2 |
| | Palmitic | 16:0 | 26.3 |
| *Monounsaturated* | | | 28.7 |
| | Stearic | 18:0 | 12.0 |
| *Polyunsaturated* | | | 3.6 |
| | Linoleic | 18:2 | 2.4 |

[a]Number of carbon atoms: number of double bonds.
Adapted from Composition of Foods: Dairy and Egg Products. *Agr. Handbook* 8–1. ARS. Washington, DC: U.S. Department of Agriculture, 1976.

protection, the cream separates slowly from the aqueous portion of fresh, unprocessed milk. Before the era of widespread homogenization of milk, people often poured the cream from the top of the milk on their cereal and in their coffee.

## Carbohydrate

A cup of milk contains 11–12 g of carbohydrate, almost all of which is in the form of lactose. This disaccharide, which is only about a fifth as sweet as sucrose, is quite an uncommon carbohydrate and is not found in significant amounts in foods other than milk and milk-containing products.

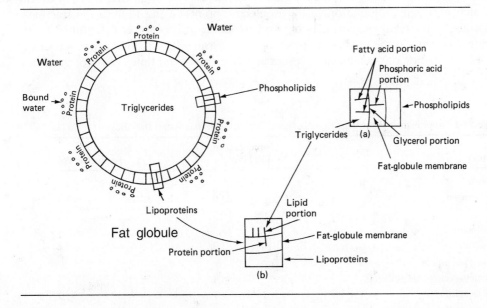

**Figure 14.1**   Schematic of a fat globule in milk; enlarged areas show (a) arrangement of a phospholipid at the fat-globule membrane and (b) lipoprotein at the fat-globule membrane.

Among the familiar sugars found in foods, lactose stands out as being particularly diffi-cult to dissolve and to keep in solution. Instead, it tends to precipitate fairly easily, especially at cool temperatures (see Chapter 8). These lactose crystals are irregular in shape and are noticeable on the tongue, sometimes creating a significant textural problem because of the gritty quality they can impart to ice cream. The majority of the problem with sandy-textured lactose crystals is the result of precipitation of crystals of α-**lactose**, which are distinctly less soluble than β-lactose crystals. Unfortunately, as α-lactose crystals precipitate in cool or frozen milk products, the β-lactose in the milk gradually shifts to the less soluble α-lactose. The process of crystallization, isomerization, and crystallization may continue until the texture becomes quite sandy.

Crystals of α-lactose form during the processing of dried milk solids. These highly hygroscopic crystals attract and hold water to form a crystalline hydrate, causing powdered milk to begin to lump. Lumping is a large problem in powdered milks unless moisture is blocked from coming into contact with the dried milk.

In addition to the lactose dissolved in milk, there are very small amounts of galactose and glucose, the two component sugars of lactose. Minor quantities of oligosaccharides also are found in milk. These other carbohydrates are present in such small amounts that they have little apparent effect on the characteristics of milk. In contrast, the level of lactose is sufficient to give a distinctly sweet taste to milk and to contribute to the browning of heated milk products.

## Proteins

***Casein.***    The two basic categorizations of milk proteins are casein and the proteins in **whey**. The reason for this distinction is evident when the pH of milk is adjusted to 4.6, for at this acidity, casein is quite insoluble and precipitates readily to form a soft **curd**, which can be separated from the remaining liquid (whey) by cutting and draining the casein curd. The whey proteins are but a part of the resulting whey, which is a watery mixture that also contains lactose, some minerals, and water-soluble vitamins.

The proteins in the curd are called **casein**, but they actually represent a group of three primary forms of casein and very small amounts of other proteins. The three pre-dominant forms of casein are $\alpha_s$-casein (actually $\alpha_{s1}$ and $\alpha_{s2}$), β-casein, and κ-casein. A fourth form, γ-casein, is not as abundant as the other forms. The isoelectric point of casein is 4.6, which represents integration of the isoelectric points of the four casein fractions. The most abundant form of casein is $\alpha_s$, and its isoelectric point is 5.1 (Table 14.2). Countering this figure is the isoelectric point of the κ form, which is lower than 4.6 (actu-ally 3.7–4.2). Together, the caseins are present at levels about four times those of the whey proteins.

**α-lactose**
Less soluble form of lactose, the disaccharide prominent in milk; form of lactose largely responsible for the sandy texture of some ice creams.

**Whey**
Liquid that drains from the curd of clotted milk; contains lactose, proteins, water-soluble vitamins, and some minerals.

**Curd**
Milk precipitate that contains casein and forms readily in an acidic medium.

**Casein**
Collective name for milk proteins precipitated at pH 4.6.

**Table 14.2**    Approximate Percentage Composition of Major Milk Proteins and Their Isoelectric Points

| Protein | Percentage of Total Protein | Isoelectric Point |
|---|---|---|
| *Caseins* | 78 | 4.6 |
| $\alpha_{s1}$ casein | 42.9 | 5.1 |
| β-casein | 19.5 | 5.3 |
| κ-casein | 11.7 | 3.7–4.2 |
| γ-casein | 3.9 | 5.8 |
| *Whey Proteins* | 17 | |
| β-lactoglobulin (an albumin) | 8.5 | 5.3 |
| α-lactalbumin | 5.1 | 5.1 |
| Immunoglobulins | 1.7 | 4.0–6.0 |
| Serum albumin | 1.7 | 4.7 |

Adapted from Zapsalis, C. and Beck, R. A. 1985. *Food Chemistry and Nutritional Biochemistry*. Wiley. New York, p. 112.

The various types of casein molecules are joined into raspberry-like organized aggregates called *micelles*. These micelles also contain many phosphate groups bridged by calcium. The surface of **casein micelles** is composed of κ-casein molecules that effectively block their aggregation. The hydrophilic portion of these molecules promotes solvation of micelles and reduces interactions. The other forms of casein are shielded within the micelles, with the result that their hydrophobic nature is not of significance. Casein micelles, although they are much smaller than fat globules, effectively block light transmission in milk to cause its characteristic opacity.

Two different substances—acid or rennin—often are used in milk processing to precipitate casein when curd formation is desired. Acid can be added to milk to shift the pH closer to the isoelectric point of κ-casein, the protein that prevents the micelles from precipitating. The negative electrical charges that are normally on the surface of casein micelles are counteracted by the positive charges of the hydrogen ions from the acid, and the repulsive forces no longer keep the micelles apart.

**Rennin**, the proteolytic enzyme in the stomach lining of calves, destabilizes casein micelles in quite a different way. This enzyme (also called chymosin) splits off the hydrophilic portion of κ-casein that was primarily responsible for the stabilizing effect of κ-casein on the surface of the casein micelles. In the presence of calcium, this para-κ-casein becomes insoluble. Consequently, the micelles then can aggregate easily to form a gel (Figure 14.2). Clearly, the two mechanisms—the alteration of the pH to approach the isoelectric point of casein and the use of rennin—are quite different, but both are effective in precipitating casein.

**Casein micelle**
Casein aggregate that is comparatively stable and remains colloidally dispersed unless a change such as a shift toward the isoelectric point or the use of rennin destabilizes and precipitates casein.

**Rennin**
Enzyme from the stomach lining of calves that eliminates the protective function of κ-casein in micelles and results in curd formation.

***Whey Proteins.***   The various caseins account for a little less than 80 percent of the total protein in milk (see Table 14.2); the various whey proteins contribute the remainder. These whey proteins sometimes are categorized as either lactalbumins or lactoglobulins. These designations are confused a bit by the finding that one of the lactalbumins is called β-lactoglobulin. Whey proteins are sensitive to heat.

***Enzymes.***   Milk contains many enzymes, including alkaline phosphatase, lipase, protease, and xanthine oxidase. The resistance to denaturation by heat varies with the enzyme. Protease is quite resistant, a fact that can create problems in some milk that is subjected to high-temperature, short-time heat treatment methods. Conveniently, alkaline phosphatase is inactivated when milk is heat-treated adequately to destroy potentially harmful microorganisms; thus, testing for the presence of active alkaline phosphatase can determine the adequacy of pasteurization. Lipase is responsible for catalyzing lipolysis of the fats in milk, a chemical change that can be avoided by heat denaturation. Xanthine oxidase is a useful enzyme in milk because of its ability to catalyze the breakdown of flavin-adenine dinucleotide (FAD) to yield riboflavin.

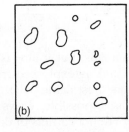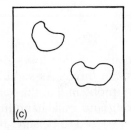

**Figure 14.2**   Sketch of the gradual clumping of casein molecules to form a soft gel when rennin is present: (a) 1 minute; (b) after 4 minutes; (c) after 8 minutes.

---

**FOOD FOR THOUGHT: Encasing with Casein**

Researchers at the Eastern Research Center of the USDA's Agricultural Research Service developed an edible film with casein as its main ingredient. In this patented process, milk is subjected to carbon dioxide under high pressure to isolate casein. The pure protein then mixes with glycerol, is spread out, and solidifies into a thin film that is flexible and water resistant. The entire process required to make casein films is about 3 hours. A less expensive version of this casein film can be made by substituting up to 20 percent of the casein with dried, nonfat milk.

Casein films are of merit because they are biodegradable, as well as edible. Fruits and vegetables can be coated with them to minimize moisture loss during marketing, and the film does not need to be removed before the food is eaten. The fact that these films are water resistant makes them particularly suitable for use as the protective covering in packages of dairy foods such as yogurt and cottage cheese.

---

## Vitamins and Minerals

Milk naturally is an excellent source of riboflavin, calcium, and phosphorus. It also provides valuable amounts of thiamin, niacin, and vitamin A. In fact, milk is an outstanding source of most nutrients except iron and vitamin C, which are present in very small amounts. Vitamin D is added to almost all milk sold today (as the label indicates); this product is the best source of vitamin D in the foods commonly consumed. Vitamins A and D are fat soluble and are not found in the whey that is separated from the casein curd in cheese making. However, whey is so high in riboflavin that it even has a greenish-yellow color.

## Flavor Components

The compounds that contribute to the overall flavor profile of milk are varied and complex. They include volatile organic compounds, notably aldehydes, ketones, and acids. The actual compounds in a particular milk sample are determined by previous treatment of the milk— for example, heating, fermentation, and storage. During these processes, chemical changes alter the flavor components. Heat-treated milk often is described as having a "cooked" flavor, which is the result of heat-assisted chemical reactions promoting degradation of the lactose and interaction with proteins.

Fermentation by microorganisms results in the formation of acid from lactose, a change that alters both flavor and texture. Lipase action on lipids, as well as oxidative changes during storage, releases butyric acid and other fatty acids that ultimately have a strong influence on flavor. Sunlight also can alter the flavor of milk by triggering formation of different sulfur-containing compounds, such as hydrogen sulfide. Even the feed the lactating cattle eat influences the flavor of the milk they produce.

## PROCESSING

### Pasteurization

Milk not only is a remarkably fine food for humans, but it also is a medium in which many microorganisms can thrive. As soon as milk is expressed from the cow's udder, contamination can be a problem. For this reason, most dairies try to maintain as clean an environment for milking and milk handling as is possible. Cows' udders are washed carefully, handlers are checked to be sure they are not carriers of diseases such as tuberculosis and typhoid fever, and the equipment coming into contact with milk is sanitized. Even with all of these precautions and prompt refrigeration, some microorganisms are present in milk.

Milk that comes directly from the animal is designated as raw milk, meaning that no heat treatment has been used on it. In a few places in this country, **certified raw milk** is marketed. This milk is so-named because it has been produced in an environment clean enough to keep the microorganism count low enough to meet the criteria required for this designation. Milk marketed as certified raw milk may not be safe to drink, however. Any microorganisms that are in the milk are alive, and some of them can cause serious illnesses, such as tuberculosis and undulant fever.

By far, the majority of the milk on the market today has undergone **pasteurization** (a heat process) and is marketed as pasteurized milk. Pasteurized milk (named after Louis Pasteur, the inventor of this important sanitizing process) has been heated to kill potentially harmful microorganisms. As is true with the sanitization of any food, the higher the temperature, the shorter the time required to kill the dangerous microorganisms. To pasteurize milk, it is necessary to kill harmful microorganisms without seriously reducing the quality of the milk. Some benign microorganisms may still be viable in pasteurized milk, but they do not cause illness.

The **hold method** of pasteurizing requires that milk be heated to only 63°C (145°F) and held there for 30 minutes, followed by a quick cooling to 7°C (45°F). A common pasteurization method is the **HTST method** (high-temperature short-time). In this method, the milk is heated to 72°C (161°F) and held for at least 15 seconds before cooling to 10°C (50°F).

Milk pasteurized by the preceding methods or even slightly more rigorous ones is safe to consume when kept refrigerated, but some deteriorative changes do occur gradually, ultimately making it unsatisfactory for food use.

Because refrigeration is minimal or lacking in much of the world, researchers have tried to develop a processing technique that produces milk that can be stored safely without refrigeration. The current answer is **UHT pasteurization** (UHT stands for ultrahigh temperature). In this rigorous process, milk is quickly heated to 138°C (280°F), held there for at least 2 seconds, and then stored in a sterile container.

Such a severe treatment kills all the microorganisms in the milk, making it possible to store UHT-treated milk at room temperature until the carton is opened. As soon as it is opened, live microorganisms can enter the milk, making refrigeration necessary, just as it is for any other type of milk. The high temperature used in UHT pasteurization causes the development of a slightly "cooked" flavor, but this product is finding a market niche in the world.

The importance of pasteurization of milk and milk products springs into the news periodically. Certified raw milk has been the source of salmonellosis outbreaks. *Salmonella dublin*, a particularly hazardous species of *Salmonella*, has been traced to raw milk in the diets of several people in each outbreak in spite of the fact that the milk was certified. The headlines about soft Mexican cheeses manufactured in California in 1985 drew attention to another microorganism, *Listeria monocytogenes*, which also entered the food supply from raw milk and ultimately resulted in the deaths of a number of people. Such examples underline the importance of using only pasteurized milk and milk products.

## Homogenization

When allowed to stand, the fat globules in milk tend to aggregate into clusters and rise to the top of the milk. The clustering of fat globules is facilitated by the presence of a protein called *agglutinin*, which apparently serves as an adhesive to form collections of fat globules of fairly large diameter that rise to the surface. This process of separation of cream from the aqueous portion of the milk is called **creaming** Although creaming is useful if the cream is to be separated from milk, it is a nuisance to shake milk thoroughly each time it is used just to disperse the fat uniformly throughout. This inconvenience led to the development of the process of homogenization.

**Homogenization** is a mechanical process in which milk is forced through tiny apertures under a pressure ranging from 2,000 to 2,500 psi. These apertures are so small that the

**Certified raw milk**
Milk that has a small microorganism population but has not been heat treated and, therefore, may cause serious illnesses.

**Pasteurization**
Heat treatment of milk adequate to kill microorganisms that can cause illness in people.

**Hold method**
Pasteurization in which milk is heated to 63°C (145°F) and held there 30 minutes before it is cooled to 7°C (45°F).

**HTST method**
High-temperature short-time pasteurization in which milk is heated to 72°C (161°F) and held there at least 15 seconds before it is cooled to 10°C (50°F).

**UHT pasteurization**
Extreme pasteurization [138°C (280°F) for at least 2 seconds] that kills all microorganisms and makes possible the storage of milk in a closed, sterile container at room temperature.

**Creaming**
Separation of fat from the aqueous portion of milk that takes place when fat globules cluster into larger aggregates and rise to the surface of the milk.

**Homogenization**
Mechanical process in which milk is forced through tiny apertures under a pressure of 2,000–2,500 psi, which breaks up the fat globules (3–10 microns in diameter) into smaller units (less than 2 microns in diameter) that do not separate from the milk.

fat globules split into units (less than 2 microns in diameter) that are no longer capable of coalescing and rising to the surface. Instead, these tiny units remain dispersed uniformly throughout the milk, eliminating the formation of a layer of cream on the surface and thus the need to shake the milk before use.

Homogenization causes milk to lose its ability to cream not only because it yields tiny fat globules but also because these globules now have a reduced ability to bind together, apparently as a result of the formation of an adsorbed layer of protein around the individual globules. Increased viscosity and a whiter appearance are other physical changes in homogenized milk.

Homogenized milk also is less stable to heat, is more sensitive to oxidation caused by light, and foams more readily than it did before processing. In addition, curds formed from homogenized milk are softer than they would be if the milk had not been homogenized. This may result from the binding of many casein molecules to the surface of the numerous tiny fat globules in the homogenized product, leaving less casein free in the plasma phase of the milk.

Milk flavor is less distinctive after homogenization, which may result, at least in part, from the casein coating of the fat globules in homogenized milk. It also is possible that the fat, with its increased surface area, may begin to develop a slightly rancid flavor in homogenized milk exposed to light.

## Evaporation

**Evaporated milk**
Sterilized, canned milk that has been concentrated to about half its original volume by evaporation under a partial vacuum.

The large percentage of water in milk (just under 90 percent) contributes greatly to the bulk of milk that is to be stored. Consequently, various canned milks are produced by evaporation of enough of the water to about double the concentration of protein and fat. The milk is evaporated under a partial vacuum so that water can be removed at a temperature well below that required at normal atmospheric pressure. This process is of some help in minimizing the flavor and color changes that would occur if the temperature of evaporation were higher.

Homogenization is a key step in the preparation of **evaporated milk** products. Without this vital step, the fat in the milk would separate and cause significant textural difficulties. Fortunately, homogenization results in an emulsified fat that is quite stable during evaporation.

Although milk can be evaporated at moderate temperatures, canned evaporated milk has to be sterilized at 116°C (241°F) for 15 minutes to ensure destruction of any microorganisms that might be present. This intense heat causes some of the lactose and milk protein to undergo the Maillard reaction (see Chapter 8).

**Sweetened condensed milk**
Canned milk to which sugar is added (contains more than 54 percent carbohydrate because of milk sugar and added sugar); evaporation of about half the water and heat treatment to kill harmful microorganisms precede the canning process.

A related product, **sweetened condensed milk**, is particularly susceptible to nonenzymatic browning as a consequence of the large amount of sugar added to evaporated milk (1.8 pounds of sucrose per 10 pounds of milk) before evaporation. This level of sugar (42 percent sucrose or glucose, plus about 12 percent lactose) promotes browning during storage and/or heating and also is an effective antimicrobial agent. Sweetened condensed milk is also available as low fat and fat free versions.

## Drying

Milk is dried to produce a food that can be stored for an extended period of time without refrigeration and/or to reduce the problems of transporting fluid milks, which are subject both to spoilage and to high shipping costs because of the large amount of water in them. When dried, the milk powder consists of lactose in either an amorphous or a crystalline state, fat in globules or free, and protein in the form of casein micelles and precipitated whey proteins, with air interspersed throughout.

The components of dried milk tend to lump together when they are rehydrated with water. Various creative solutions have overcome this problem. An effective technique is to instantize the dried milk by adding moisture to the dried product to make it sticky and then drying it a second time to obtain rather spongy particles. When water is added to reconstitute the **instantized dried milk**, the particles absorb water and sink downward, dispersing as they fall.

Lumping is but one of the problems with dried whole milk. Spray-dried milk causes lactose to solidify in an amorphous or glass-like state. The highly hygroscopic nature of **glassy lactose** increases the absorption of moisture during storage; effective airtight packaging is important to good shelf life.

## Fermentation

Various microorganisms are used commercially to ferment lactose in milk and milk products. The goal is the production of lactic acid by one or more types of microorganisms. Frequently, *Streptococcus lactis* initiates the fermentation process, with lactobacilli of various types (e.g., *Lactobacillus casei, L. bulgaricus, L. lactis,* and *L. helveticus*) continuing the fermentation as the pH drops into their effective ranges.

The thickening associated with fermented products is the result of the association of casein micelles, often accompanied by β-lactoglobulin. When a comparatively large amount of β-lactoglobulin is bound to the casein micelles, a fairly stable gel forms and syneresis is minimal. Buttermilk is somewhat thickened as a result of fermentation, whereas yogurt is acidified to the point at which a gel forms as a result of fermentation and controlled heat. Whey can be fermented to produce comparatively sweet cheeses such as Mysost and Gjetost.

Unfortunately, undesirable microorganisms can grow under essentially the same conditions as are needed to achieve controlled, desirable fermentation. Products may become ropy in consistency if a microorganism such as *Bacillus subtilis* is present. *Lactobacillus tardus* can generate unpleasant aromas or flavors by suppressing lactic acid production and promoting citric acid formation. *Pseudomonas putrefaciens* is an example of a microorganism that forms compounds with a putrid odor. Molds may grow in butter fairly easily because of the low water content. *Alcaligenes* and *Aerobacter* are but two of the types of bacteria that can cause a slimy curd to develop. *Pseudomonas nigrificans* can result in the formation of a black color in butter. These are just some examples of undesirable fermentation in dairy products. Clearly, the problem is to control the microorganisms that are present by promoting the growth of desirable ones and eliminating the undesirable ones through careful sanitation and attention to quality control.

**Instantized dried milk** Milk that has been dried, moistened until sticky, and then re-dried into spongy aggregates of solids that rehydrate readily without lumping.

**Glassy lactose** Amorphous (noncrystalline) milk sugar.

---

## INGREDIENT HIGHLIGHT: Labneh

Labneh is a traditional food and ingredient in the Middle East where people centuries ago learned to ferment milk to make yogurt and then strained it to separate the curd from the whey, creating this acidic, flavorful soft cheese. Appropriately, its name is derived from an Arabic word *labni*, which means milk or white. In earlier times, the acidity of labneh was particularly important because it helped to retard spoilage in the heat of the region where it originated. Labneh may be made with added butterfat or dried milk solids to enhance flavor and ability to tolerate high temperatures in cooking. It is popular in South Asia and the Middle East as an ingredient in a variety of savory and sweet recipes. In India, water buffalo milk is fermented instead of cow's milk to make *dahi*. Disposable clay pots sometimes are used as the container when making *dahi* (Figure 14.3).

**Labneh** Soft cheese made by separating the curd from yogurt.

**Figure 14.3**   A pot of curd from the milk of water buffalo and honey is a dessert savored in Sri Lanka.

## PRODUCTS

### Milks

**Cultured buttermilk**
Low fat or nonfat milk containing *S. lactis* and *L. bulgaricus* that has been incubated to produce some lactic acid.

**Sweet acidophilus milk**
Unfermented milk to which *L. acidophilus* has been added.

**Kefir**
Fermented milk that is about 3 percent alcohol because of fermentation by *Lactobacillus kefir*, which also adds $CO_2$.

**Lactaid®**
Milk in which lactose content has been reduced by an enzyme (lactase) that splits lactose into glucose and galactose.

**Yogurt**
Milk clotted by inoculating with *S. thermophilus* and *L. bulgaricus* and fermenting to pH ~5.5.

Milk and milk products are available to consumers and food manufacturers in several different forms. The fluid form of milk can be categorized on the basis of fat content [nonfat (fat free), low fat, reduced fat, and whole], as Table 14.3 shows. The rheological properties of nonfat milk are quite different from the other three milks. All of these are excellent sources of most nutrients except vitamin C and iron; the lower the fat content of these milks, the higher the content of the other nutrients.

Fermented milks include cultured buttermilk, sweet acidophilus milk, and kefir. **Cultured buttermilk** is milk that has been inoculated with *S. lactis* and *Leuconostoc bulgaricus* and incubated for 20 hours to produce 0.9 percent lactic acid. The texture of buttermilk often is stabilized with gums (carrageenan or locust bean). **Sweet acidophilus milk** is inoculated with *Lactobacillus acidophilus*, but it is not allowed to ferment and therefore is sweet rather than acidic. Many people choose fermented milks, such as **kefir**, because of the pleasing qualities and the uniqueness of each type. Another potential benefit is that lactose-intolerant people often are able to consume these products without discomfort.

Lactase-containing milks, such as **Lactaid®**, are particularly effective for people who are lactose intolerant (Figure 14.4). Technically, lactase-containing milk is not fermented because acid is not made. Instead, lactase reduces the amount of lactose by digesting a considerable portion of lactose into its component monosaccharides (galactose and glucose).

**Yogurt** is clotted milk. *Streptococcus thermophilus* and *Lactobacillus bulgaricus* are the microorganisms that ferment milk to yogurt. Fermentation produces lactic acid, but yogurt is not allowed to continue to ferment once an acidity of pH 5.5 is reached because the taste becomes too acidic.

**Table 14.3**   Distinguishing Characteristics and Uses of Milk and Milk Products

| Name of Product | Fat Percentage | Characteristics | Uses |
|---|---|---|---|
| *Fluid Milks* | | | |
| Whole | 3.25+ | Rich flavor | Beverage, cooking |
| Reduced fat | 2.0 | Some richness | Beverage, cooking |
| Low fat (light) | 1.0 | Slight richness | Beverage, cooking (calorie and cholesterol control) |
| Fat free (skim) | 0.1 | Not rich flavor, somewhat thin | Beverage, cooking (calorie and cholesterol control) |
| Cultured buttermilk | 0.1 | Tangy flavor (lactic acid bacteria), somewhat thick | Beverage, baking |
| Kefir | | 3% alcohol, $CO_2$ | Beverage |
| Sweet acidophilus milk | 3.25+ | Somewhat sweet (*L. acidophilus*) | Beverage, cooking (for lactose intolerance) |
| Yogurt | 0.1–3.25+ | Tangy coagulum (*S. thermophilus*, *L. bulgaricus*) | Dressing, dessert, frozen dessert (tofutti) |
| Lactaid® | 0.1–3.25+ | Slightly sweet (lactase) | Beverage, cooking (for lactose intolerance) |
| *Creams* | | | |
| Sour | 18 | Tangy | Dips, toppings, baking |
| Half-and-half | 10.5–18.0 | Slightly viscous | Added to cereals and beverages |
| Coffee (light) | 18–30 | Somewhat thick | Added to coffee, sauces |
| Light whipping (whipping) | 30–35 | Whips to fairly stable foam | Whipped topping |
| Heavy cream (heavy whipping) | 36 | Whips easily to stable foam | Whipped topping |
| Butter | 80 | Yellow water-in-oil emulsion | Spread, baking, flavoring agent |
| *Cheese* | | | |
| Natural | Varies | Firmness and flavor vary with manufacturing and ripening; acid and rennet used to form clot | Sliced, grated; eaten alone or as an ingredient |
| Process | Varies | Emulsifier added to natural cheeses; 41% moisture | Casseroles, sandwiches |
| Process cheese food | Varies | Process cheese with 45% moisture | Casseroles, warm dip |
| Process cheese spread | Varies | Process cheese with 50% moisture | Spread |
| *Ice Creams* | | | |
| Ice cream | 10 or more | 20% milk solids, sweeteners, gums, or other stabilizers | Dessert |
| Low-fat ice cream | 2–7 | 11% milk solids | Dessert |
| Sherbet | 0 | 2–5% milk solids | Dessert |
| Mellorine | 10 | Butterfat is replaced by less saturated fat | Dessert |
| Parevine | 10 | No butterfat or milk solids | Dessert |

Yogurt and other fermented milk products are gaining some momentum in the marketplace because they are sources of probiotics and prebiotics (see Chapter 1). **Inulin** and other fructooligosaccharides are prebiotics that may be added to yogurt during manufacturing. Benefits of inulin include improving flavor in conjunction with sugar alcohols, serving as a prebiotic fiber, and promoting calcium absorption. In the large intestine, beneficial *acidophilus* and *bifidus* bacteria digest inulin to produce a more acidic environment, which is unfavorable to pathogenic bacterial growth and survival.

**Inulin**
Fructose polymer linked by β-2,1-linkages and with a glucose unit at the end; obtained from chicory root.

Yogurt with reduced fat content lacks the textural properties of yogurt made with whole milk. Gelatin and a modified food starch often are added to stabilize the product. The amount of gelatin has a strong impact on the final texture. Whey protein concentrate (WPC), agar, and pectin are other stabilizing agents that may be used.

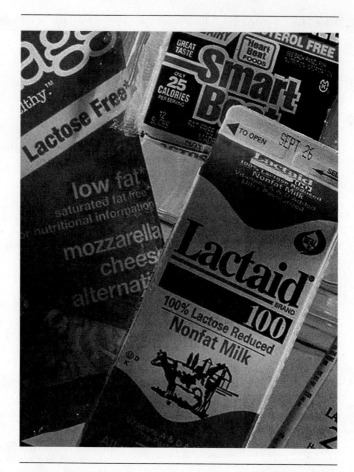

**Figure 14.4** Lactase hydrolyzes lactose to galactose and glucose in these milk products, which enables people who are lactose intolerant to drink milk without discomfort. (Courtesy of Agricultural Research Service.)

**Soymilk**
Beverage made from whole, finely ground defatted soybeans; a beverage designed to compete with milk.

   **Soymilk** is not a dairy product, but it is used as a beverage by some people who are allergic to milk or who are seeking ways of increasing their intake of the isoflavones available from soybeans. The protein content (3.2 percent) of soymilk is comparable to cow's milk; calcium is added to many soymilks to enhance the content of this key mineral. Whole, defatted soybeans are ground to a fine particle size in hot water to begin production of soymilk. To mask or offset the beany flavor that is naturally present in soymilk, flavors such as chocolate or even coffee are added.

## Butter

Butter technically is a dairy product; it actually is a water-in-oil emulsion containing about 15 percent water and at least 80 percent fat. Its manufacture is accomplished through churning to reverse the colloidal dispersion, transforming the water from the continuous to the dispersed phase by establishing tiny water droplets.

   The fat is sufficiently warm at churning temperatures to form a continuous network. Subsequent chilling produces fat crystals in the $\beta'$ form. Fat crystals give rigidity to the butter-like mixture, which then can be squeezed sufficiently to force out enough water to bring the level to about 15 percent. Usually, sweet cream is the milk product from which butter is churned. If the color is not sufficiently yellow at certain times of year, annatto or another coloring agent is added to provide the expected yellow color. Usually salt is added for flavor, although there also is a market for unsalted butter.

## Creams

Creams vary in fat content from 10.5 percent (half-and-half) to 36 percent (heavy whipping cream). They are produced by the centrifugation of milk to separate varying amounts of the lighter cream from the aqueous portion of the milk, depending on the type of cream desired.

Production of cultured sour cream begins with a 30-minute holding period at 74–82°C (165–180°F) to pasteurize the cream and destabilize some of the protein. Next is the addition of lactic acid-forming bacteria and controlled incubation to generate the acid needed to form the clot. Gums may be added to contribute some thickening; carrageenan is effective at reducing syneresis in sour cream and dips made with it. Reduced-fat sour cream requires considerably more stabilizer than the regular product because the fat content is too low to give much body.

## Cheeses

Cheese production requires the formation of a curd (Figure 14.5) and removal of a considerable amount of water (whey). Clotting is facilitated with the addition of various bacterial cultures, such as *Lactococcus lactis* and *Streptococcus cremoris,* to generate lactic acid. Acid causes calcium phosphocaseinate in the casein micelles to eliminate calcium. Acid precipitation is used to make cottage cheese and cream cheese. Mozzarella and provolone often are made using *S. thermophilus* to produce the acid for clotting and *L. casei* to enhance flavor development during ripening. Other bacteria that sometimes are used to make different cheeses and give them their distinctive qualities include *L. helveticus* and *Leuconostoc.*

For making most other cheeses, rennin, the active enzyme from the stomach lining (fourth stomach) of calves, traditionally has been used to convert κ-casein into para-κ-casein. Para-κ-casein then participates in curd formation by uniting with calcium, which forms an insoluble material. Researchers are looking for alternative sources of enzymes capable of precipitating casein.

Potential microbiological sources of enzymes include *Escherichia coli* and *Saccharomyces cerevisiae*. The enzymes they produce appear to achieve better results than the enzymes from *Mucor milhei, M. pusillus,* and *Endothia parasitica.*

**Figure 14.5**   Curd formation, the first step in making cheese. (Courtesy of Dr. Phil Andrews.)

## A Matter of State

States have state flowers that they proudly promote, but Wisconsin has hit upon a designation to top all others. It has identified *Lactococcus lactis* as the state microbe (Figure 14.6). Certainly this is appropriate because Wisconsin prides itself on its cheese-making prowess, and *L. lactis* plays a key role in triggering lactic acid production from lactose to precipitate proteins and form the curd needed in cheese production.

It is too soon to know if other envious states may leap on the bandwagon and pick out their own special microorganism. One can't help but wonder if Wisconsin would have been so bold if *L. lactis* had not undergone a name change from *S. lactis,* a name with a somewhat more sinister sound.

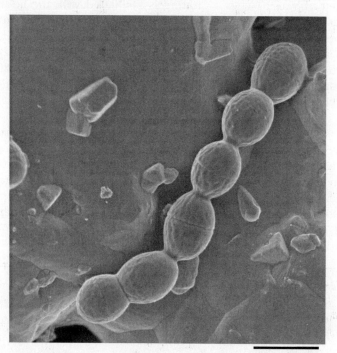

**1μm 18000X**

**Figure 14.6** *Lactococcus lactis* is the nominee to be the state of Wisconsin's official microbe. (Courtesy of Joseph Heintz at the UW-Madison BBPIC lab and UW-Madison Department of Bacteriology.)

The curd is cut to release the whey (Figure 14.7). Gentle heating [35°C (95°F) for hard cheeses and 32–33°C (90–91°F) for soft cheeses] and squeezing cause the curd to draw tighter so that the whey can drain, leaving the highly concentrated protein and fat-containing curd. The whey contains the lactose, water-soluble vitamins (notably riboflavin), and serum (whey) proteins. Usually the moisture content of cheeses is around 37–40 percent, which is quite a reduction from the 87 percent water content of whole milk.

***Natural Cheeses.*** **Natural cheeses** are categorized according to their firmness or moisture content. Some of these natural cheeses undergo a ripening or aging period in a controlled environment at a cool temperature. Microorganisms that could be harmful to humans are no longer viable after 60 days at a storage temperature of around 1.7°C (35°F), which makes it possible to use raw milk in making some well-aged cheeses.

**Natural cheese**
Any cheese made by clotting milk to form a curd and then concentrating the curd by draining the whey; variations are produced by varying the curd concentration and by ripening with or without the addition of selected microorganisms or other ingredients.

**Figure 14.7**  Following curd formation, the next step in making cheese is cutting the curd in preparation for draining away the whey. (Courtesy of Agricultural Research Service.)

Even when pasteurized milk is used for making cheese, ripening is beneficial because of the numerous chemical changes that occur. These changes result in a variety of end products, ranging from some individual amino acids via some minor degradation of the original proteins to large polypeptides. Distinctive flavor and textural changes are associated with these proteolytic reactions. Casein proteolysis is responsible, in part, for the rather soft texture of semi-soft and soft cheeses.

A textural comparison of green (unripened) and well-ripened Cheddar cheeses reveals a change from the rubbery, tough green product to a more tender and soft or even crumbly texture as ripening progresses. These proteolytic reactions during ripening result in a cheese that can be incorporated readily into a wide range of hot foods; the unripened cheese blends with considerable difficulty.

Chemical changes in the fats in cheeses during aging cause some of the distinctive flavors associated with ripened cheeses. Fatty acids and other organic compounds such as lactones, alcohols, methyl ketones, various other ketones, aldehydes, and esters combine to delineate the complex flavors of ripened cheeses (Figure 14.8).

---

### FOOD FOR THOUGHT: Burrata

Mozzarella is a cheese that immediately brings to mind Italian recipes such as lasagna. This familiar cheese from southern Italy may be made using milk from water buffalo, cows, or goats. In Apulia, a town in the heel of southeastern part of Italy's "boot," cheese makers developed a special ball of cheese called *burrata*. Manufacturing begins with the making of mozzarella. The soft curd of hot mozzarella is worked manually from a stringy character into a small ball within seconds and then blown up gently like a balloon. Soft burrata curd and cream then are added to fill the "cheese balloon" before it is twisted to seal the top.

Unlike mozzarella, burrata is a unique cheese that must be eaten within two or three days of its manufacture to be at its best because of the high moisture content of the filling in the burrata ball. This soft, creamy interior makes burrata a particularly pleasing cheese to serve without heating—for example, a sliced tomato salad topped with some burrata and a drizzle of olive oil, or some burrata on a bed of mixed greens.

**Figure 14.8** Swiss cheese and cream cheese are featured in the array of dairy products that include milk and whipping cream.

Microorganisms often are added deliberately to cheeses that are to be aged, and it is the work of these microorganisms that causes many of the chemical degradative reactions in the fats. A familiar example is the various blue cheeses. Incorporation of dried spores of the blue mold *Penicillium roqueforti* allows the mold to break the fat-globule membrane so that the lipase present in the cheese can split the fat molecules into their components and release free fatty acids. It is these free fatty acids that are the key to the distinctive flavor of Roquefort cheese.

Part of the flavor of well-aged Swiss cheeses, such as Emmentaler, is the result of lipolytic breakdown to such short-chain fatty acids as butyric (4 carbons), caproic (6 carbons), caprylic (8 carbons), and capric (10 carbons). However, the sweet flavor detected in Swiss cheeses is due primarily to free amino acids derived from the proteolytic reactions that occur during aging.

The appearance of some cheeses is unique because of the changes during ripening. One readily apparent result of ripening in the blue cheeses is the marbled or streaked appearance where the blue mold has penetrated the product. The distinctive eyes in the various Swiss-type cheeses also are the result of microorganisms that reproduce during the ripening period. The holes or eyes are caused by the formation of carbon dioxide from lactic acid, the product of various *Propionibacterium* species or other lactic acid–producing bacteria, including several *Lactobacillus* species and *Streptococcus thermophilis*.

Camembert cheese undergoes textural change during ripening as the mold *Penicillium camemberti* grows on the surface of the green cheese. This mold is the source of the proteolytic enzymes that migrate slowly into the Camembert and catalyze proteolysis to produce a very soft to almost fluid cheese. This reaction limits the shelf life of Camembert because the continued action of these enzymes ultimately can produce so much ammonia that the flavor becomes unpleasant. This problem can be reduced by the presence of other bacteria, notably *Candida, Geotrichum,* or *Mucor* strains. Table 14.4 presents some of the key aspects of the production of the unique flavors and textures of aged cheeses.

**Table 14.4** Unique Aspects of Selected Natural Cheeses

| Variety | Milk | Content (Percentage) | Approximate Moisture Description | Aging (Months) | Microorganisms Added |
|---------|------|----------------------|----------------------------------|----------------|----------------------|
| Parmesan | Low-fat cow's | 30 | Very hard, used grated | >12 | *L. bulgaricus, S. thermophilus* |
| Sapsago | Nonfat cow's | 38 | Green, pungent due to powdered clover; sour whey is added | 5 | |
| Cheddar | Whole cow's | 37 | White or yellow, firm to crumbly | 1–12+ | *S. lactis, S. cremoris; L. casei* and *L. plantarum* appear during aging |
| Swiss (Emmentaler) | Whole cow's | 39 | Holes (eyes), white, smooth | 3–10 | *Propionibacterium, L. helveticus. L. bulgaricus, L. lactis, S. thermophilus, S. lactis* |
| Mozzarella | Whole or low-fat cow's | 45–55 | Mild, soft | 0 | *L. bulgaricus, S. thermophilus* |
| Camembert | Whole cow's | 53 | Very soft; mold-coated rind | 1–2 | *P. camemberti; Brevibacterium linens, Candida, Geotrichum,* and *Mucor* may develop |
| Blue (bleu) | Whole cow's or ewe's [a] | 40 | Firm, heavily veined with blue mold | 2–6 | *S. lactis, S. cremoris, P. roqueforti* |

[a]Roquefort cheese is made from ewe's milk and aged in caves in the southwest of France at the town of Roquefort. Similar cheeses include Stilton, which is made in England from cow's milk, and Gorgonzola, the Italian blue-veined cheese made from cow's and/or goat's milk. European blue-veined cheeses often are designated as *bleu* cheeses, in contrast to the American spelling—*blue* cheese.

## What's in a Name?

The problems of nomenclature are enough to make cheese makers in Stilton, England, as blue as their cheese. Their problem is that they cannot legally label their cheese as Stilton cheese even if it is made within the city limits of their town. Under the European Union Protected Food Name Scheme, Stilton cheese can only be so-named if it is made by any of the six creameries in the Stilton Cheesemakers Association, and none of these is in Stilton. It seems that this professional association obtained Product Designation of Origin status for its Stilton cheese without including the town of Stilton or any product made there in its petition. That leaves the town celebrating its heritage with a Stilton Cheese Rolling Festival based on a product that is sold but not made there. Perhaps that balances out all right, because competitors roll chunks of wood in place of cylinders of the actual cheese at this annual event.

Stilton is a very famous English cheese that has a tangled history, but it appears that it got its name at least two centuries ago when the blue-veined cheese was brought from Leicestershire, Nottinghamshire, and Derbyshire to the market in Stilton. Production is a lengthy process that begins with pasteurization of cow's milk prior to adding a controlled culture of *Penicillium roqueforti*, clotting with rennet, and subsequently cutting and draining the curd. The curds are molded into cylinders of cheese that are monitored and stored under carefully controlled conditions as they undergo some moisture loss, development of blue veins of the mold, and ripening. At the end of 12 weeks, the cylinders are ready to market to cheese connoisseurs who fancy the creamy-textured, well-veined cheese that is not made in Stilton.

**Stilton cheese**
Blue-veined cheese made by an approved process in any of six creameries in Leicestershire, Nottinghamshire, and Derbyshire, England.

**Product Designation of Origin**
Legal designation that the European Union can grant to food products from specific locations.

**Pasteurized process cheese**
Cheese product made by heating natural cheeses with an emulsifier and then cooling in a brick form; moisture level is about 41 percent.

**Process cheese food**
Process cheese product with a moisture content of about 45 percent, which causes the food to be comparatively soft, yet firm.

**Process cheese spread**
Spreadable process cheese product with a moisture content of about 50 percent.

**Coldpack (club) cheese**
Cheese product made by adding an emulsifier to a mixture of natural cheeses.

*Process Cheeses.* **Pasteurized process cheeses** (Table 14.5) represent a variation from the natural cheeses. These products are made from natural cheeses by heating them and then adding an emulsifying agent (e.g., disodium hydrogen phosphate or sodium citrate). The heat and emulsifying agent alter the cooking characteristics of the original natural cheeses. Microorganisms are killed by the heating, which reduces the potential for spoilage during storage. Addition of the emulsifying agent is effective in controlling the separation of fat from the cheese during cooking, thus eliminating a significant problem associated with natural cheeses.

The combination of heat [65–71°C (149–160°F)] and citrate and phosphate salts alters the protein molecules by reducing their molecular size, a change that increases the protein's solubility and water-binding capacity. The freshly made natural cheeses (less than seven days old) used in the manufacturing of process cheeses contributes a rather rubbery consistency and a bland flavor; to compensate for these limitations, some ripened or aged cheeses usually are included in the formula to enhance the flavor and textural characteristics of process cheese.

**Process cheese food** is similar, but the moisture level usually is about 45 percent, around 4 percent higher than in process cheese. **Process cheese spread** has a comparatively high moisture level (about 50 percent), which permits easy spreading on bread and crackers.

**Coldpack (club) cheeses** are also manufactured from natural cheeses with added emulsifiers. They differ from process cheeses because they are not subjected to the heat treatment that causes proteolysis and increased solubility in the process cheeses.

**Table 14.5**  Composition of Various Pasteurized Process Cheese Items

| Type | Percentage Water | Percentage Milk Fat |
|---|---|---|
| Process cheese | maximum 43 | 47 |
| Process cheese food | maximum 44 | minimum 23 |
| Process cheese spread | 44–60 | 20 |

Nevertheless, the added emulsifier does enhance the spreadability of coldpack cheeses over that of the natural cheeses used in their manufacture.

***Reduced-Fat Cheeses.***   Reduced-fat cheeses first appeared on the market in 1986 in response to consumer interest in reducing fat and cholesterol in the diet. Various approaches have been taken in an attempt to develop cheeses that are low in fat and high in palatability. The most straightforward way is simply to remove the fat, which can be done comparatively successfully up to a critical level.

Fat performs several functions in cheese, including contributing significantly to mouth-feel, firmness, and adhesiveness, as well as to flavor. Results are satisfactory if the fat level is not reduced more than 25 percent in a Cheddar-type cheese, but a 50-percent reduction produces a cheese of definitely inferior quality, albeit one with fewer calories.

Another approach to making a reduced-fat cheese is to remove all of the butterfat from the milk before beginning manufacturing and then adding some oil. The fluidity of the oil overcomes some of the palatability problems at a lower calorie level than can be achieved with whole milk as the starting point. Another solution is to add various gums to help achieve the desired textural characteristics of the product. Considerable research is still being conducted to develop acceptable reduced-fat cheeses because the market for these cheeses seemingly is a strong one, due to both health concerns and the problem of over-weight and obesity in the United States. Imitation cheeses produced from soy protein also are available.

## Whey

Whey protein concentrate is a by-product of cheese manufacturing that is available in large quantities as a result of all the cheese that is manufactured. In whey, two proteins, α-lactalbumin and β-lactoglobulin, constitute approximately 80 percent of the proteins, with β-lactoglobulin contributing 55 percent of the total. These proteins contribute to the viscosity and stability of food products.

Whey, in the form of whey protein concentrate and whey protein isolate, contributes desirable textural properties. For example, Simplesse® (CP Kelco) utilizes a mixture of whey protein solution and egg white that has been heated and sheared to make extremely tiny spheres (0.1–3.0 μm). Surprisingly, these tiny spheres provide a mouthfeel similar to that contributed by fat, but with far fewer calories.

Whey protein concentrate is useful in promoting water retention in some meat products; this action results from its tendency to promote gel formation. The abundance of whey in the food industry spurs research efforts to find ever more uses of such a healthful and economical ingredient.

Whey protein concentrate can be texturized into powders, crumbs, ribbons, and chunks to meet the needs of the food industry in formulating a variety of products. Whey's use as a textural ingredient to replace some of the carbohydrate gums and starch is nutritionally desirable because of its protein content.

## Ice Creams and Frozen Desserts

**Plain ice cream** is a frozen mixture of cream (which contributes considerable milk fat), milk solids, and flavorings into which some air has been stirred. A federal standard of identity has been established for this food: at least 10 percent milk fat, 20 percent total milk solids (which may be reduced as a result of added flavorings), and optional additives (0.5 percent stabilizer and 0.2 percent emulsifier).

The comparatively high fat content of ice creams contributes the smooth texture and rich flavor that are prized by consumers. The added milk solids further enhance the texture and flavor. Sweeteners (sugar or various corn sweeteners) lower the melting and freezing points of ice cream, a physical phenomenon that aids in production of a smooth ice cream with a pleasingly light texture. Carboxymethyl cellulose, gelatin, or gums such as guar gum, alginates, or carrageenan are useful stabilizers in ice creams because of their ability to bind

**Plain ice cream**
Frozen dessert containing at least 10 percent milk fat and 20 percent total milk solids and no more than 0.5 percent edible stabilizer; flavoring particles must not show.

**Composite ice cream**
Frozen dessert containing at least 8 percent milk fat and 18 percent total milk solids and no more than 0.5 percent edible stabilizer; flavoring particles are not to exceed 5 percent by volume.

**Frozen custard**
Ice cream–like product that is a frozen, egg yolk custard.

**Low-fat ice cream**
Frozen dairy product containing 2–7 percent milk fat and 11 percent total milk solids.

**Sherbet**
Somewhat acidic frozen dessert containing from 2 to 5 percent milk solids, and no milk fat.

**Mellorine**
Imitation ice cream in which the milk fat has been removed and replaced by a different fat.

**Parevine**
Imitation ice cream in which both the milk fat and the milk solids have been replaced by nondairy ingredients.

water and increase the viscosity of the ice cream mixture prior to freezing. Monoglycerides are added to serve as emulsifying agents.

Similar to plain ice cream, **composite ice creams** are similar high-fat (at least 8 percent milk fat) ice creams with at least 18 percent total milk solids. Actually, the difference between plain and composite ice creams is caused by the bulkiness of the flavoring components in the composite ice creams; these components cannot exceed 5 percent of the volume of ingredients.

Several other related products also are on the market. These range in composition from **frozen custards** to ices and even imitation ice creams. Frozen custard differs from ice cream in that it is made by cooking added egg (usually yolks) to form a custard before freezing the mixture. Egg is added at the rate of 1.4 percent for plain ice creams and 1.12 percent for composite ice creams containing visible pieces of flavoring substances.

**Low-fat ice cream** contains between 2 and 7 percent milk fat and at least 11 percent total milk solids. It is definitely less rich than ice cream, which has least 10 percent milk fat and a minimum of 20 percent total milk solids. This compositional difference results in a reduced level of calories, a distinction of interest to dieters. The textural problems in formulating an acceptable product often are solved by adding a microcrystalline-cellulose-based stabilizer/emulsifier system, sugar alcohols, and polydextrose. Consumers need to be aware of the potential laxative effect of the sugar alcohols.

**Sherbet** may be even lower in fat than some ice milks; it contains between 2 and 5 percent total milk solids and essentially no fat. The acidity level—at least 0.35 percent—is specified in sherbets, but not in ice milk or ice cream.

In some markets, imitation ice cream products are available. One type is **mellorine**. Mellorine is a frozen dessert with basically the same ingredients as ice cream, except that a different type of fat replaces the milk fat. This dessert may be of interest to people who are trying to eliminate butterfat from their diets and replace that fat with less saturated fats. The other type of imitation ice cream is **parevine**, a frozen dessert containing no milk fat or milk solids. Parevine fits within the guidelines of Jewish food laws, making this a popular frozen dessert for people of the Jewish faith.

A frozen ice-cream product is a colloidal dispersion classified as a foam, with the ice crystals forming the solid continuous phase and the air incorporated from agitation during freezing being the discontinuous or dispersed phase (Figure 14.9). Preparation of ice cream begins with the blending of ingredients, followed by rapid cooling and freezing, preferably with some agitation. Rapid freezing is required for the formation of fine ice crystals, which are essential to achieving the desired smooth texture.

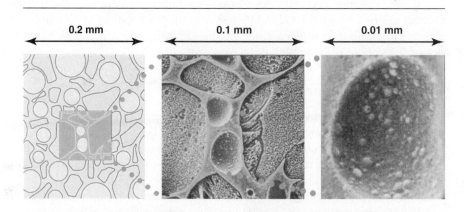

**Figure 14.9**    The structure of ice cream, progressively enlarged: (left) thick sugar syrup bubbles containing air bubbles, fat globules, and ice crystals; (middle) syrup layers between air bubbles and ice crystals; (right) fat globules on surface of air bubble. (Courtesy of Unilever Research. Viaardingen, The Netherlands.)

The necessary fast rate of freezing can be accomplished by surrounding the ice cream container with a mixture of eight parts ice to one part salt (by volume). This ratio of ice to salt is sufficiently cold to enable small ice crystals to form quickly, yet it still allows time for adequate agitation to incorporate sufficient air during the freezing process.

The addition of salt to ice to facilitate freezing utilizes the effect of salt on the freezing point of water. Substances that dissolve in water depress the freezing point of the solution by 1.86 C° (3.35 F°) for each gram molecular weight dissolved in 1,000 g of water (Table 14.6). Because salt ionizes, it depresses the freezing point of water 3.72 C° (6.7 F°) for each gram molecular weight (58 g) dissolved in 1,000 g of water (1 liter).

When the salt is sprinkled on the ice, it begins to melt, and a salt solution forms. It has a different vapor pressure than does the ice, a difference that causes the ice to continue to melt in an attempt to establish equilibrium between the two media. The melting ice absorbs the energy from the surrounding salt solution as its physical state changes to water. This endothermic reaction requires 80 calories per gram of ice melted. The maximum effect of salt in lowering the freezing point of ice occurs when the ratio of ice to salt is about 3:1. At this high concentration of salt, the freezing point is depressed to about –22°C (–7.6°F) (see Chapter 5).

The process of freezing ice cream requires the reduction of the temperature of the ice cream mixture below the freezing point of water because the sugar in the formula depresses the temperature required for freezing to occur. However, the maximum effect of salt is not utilized in making ice cream. Such a high ratio of salt to ice not only increases the cost, but also causes the texture of the ice cream to become too solid and compact. Freezing occurs so rapidly that there is little opportunity for the agitation to incorporate air or promote coalescence of some of the fat globules, both of which aid in achieving the optimal texture.

Stirring during freezing helps control various aspects of texture. A key function of agitation is to achieve a comparatively uniform temperature throughout the freezing mixture (Figure 14.10). Without agitation, the mixture touching the edges of the container is much colder than in the interior. This lack of uniformity promotes the formation of comparatively large ice crystals because the small ones that form first quickly melt and then recrystallize to make larger aggregates of ice crystals that feel coarse on the tongue.

Another physical effect of agitation is the coalescence of some of the fat globules. Some clumping is desirable to present a smooth mouthfeel. However, excessive agitation prior to freezing can cause rich ice cream mixtures to churn into a reversal of the emulsion to form the water-in-oil emulsion found in butter. The rate of agitation and the time and temperature at which freezing occurs need to be coordinated to avoid breaking the oil-in-water emulsion, and yet to allow some melding of fat globules.

When the ice cream mixture is chilled and begins to freeze, agitation results in the incorporation of some air into the system and increased volume. This increased volume of the frozen mixture is termed **overrun**. A desirable level of overrun in commercial ice creams generally is between 70 and 80 percent. At this level, the texture is pleasingly light, yet neither frothy nor compact. Frothiness is quite detectable when overrun approaches 100 percent. Sherbets usually have only about 30–40 percent overrun. In home-cranked ice cream, the overrun level rarely exceeds 50 percent.

**Overrun**
Increase in volume (expressed as a percentage) that occurs when ice cream is frozen with agitation. Calculated as: (volume of dispersion-original volume/original volume) x 100.

**Table 14.6**   Effect of Salt on Freezing

| Salt/Liter of Water (g) | Freezing (°C) | (°F) |
|---|---|---|
| 58 (3 tbsp) | −3.72 | 25.3 |
| 116 (6 tbsp) | −7.44 | 18.6 |
| 174 (9 tbsp) | −11.16 | 11.9 |
| 232 (3/4 c) | −14.88 | 5.2 |
| 290 (15 tbsp) | −18.6 | −1.5 |
| 348 (18 tbsp) | −22.32 | −8.2 |
| 406 (21 tbsp) | −26.04 | −14.9 |

**Figure 14.10**   Agitation during freezing of ice cream maintains some uniformity of temperature throughout the mixture, promotes coalescence of some fat globules, and includes of air to promote a smooth, light texture.

Ice creams frozen in the freezer unit of a refrigerator frequently have poor textural characteristics compared with similar products made in a crank-type home or commercial freezer. The lack of agitation in these still-frozen ice creams results in rather coarse ice crystals and extremely limited overrun. Some air can be included if whipped cream, egg white, or evaporated milk foams are components of the mixture. Crystal growth also can be influenced by choice of ingredients. Gelatin, chocolate, egg white, and starch-thickened pastes are effective in stabilizing the textural character of the ice crystals because they interfere somewhat with the growth of crystal aggregates, even when agitation is lacking.

In freshly frozen ice creams, the process is quite dynamic at first. Small ice crystals will melt in the sweet, unfrozen portion of the ice cream, and other crystals will form as heat is removed from the freezing mixture. Larger crystal aggregates tend to form gradually, replacing the labile fine crystals formed initially.

As long as agitation is continued in the freezing mixture, the dasher in the center of the freezer helps prevent this aggregation. However, the undisturbed hardening that occurs after agitation has ceased does allow some opportunity for aggregation to occur. This is when stabilizing agents are important, for they make it difficult for ice crystals to clump together into coarse, large aggregates. For example, an ice cream made with a large amount of fat tends to be noticeably smoother in texture than one made with only the minimal amount. The fat content of chocolate makes this a useful ingredient in promoting a fine-textured ice cream. Protein provided by egg white or added milk solids also performs a stabilizing function.

## PHYSICAL AND CHEMICAL EFFECTS ON MILK PRODUCTS

### Heat

Whey or serum proteins, notably α-lactoglobulins and β-lactalbumins, precipitate gradually with prolonged heating; the higher the temperature, the faster the denaturation of the whey proteins. After half an hour of heating at 70°C (158°F), a little less than a third of the α-lactoglobulin and β-lactalbumin is denatured, whereas at 80°C (176°F) more than 80 percent is denatured. This gradual relaxation of these susceptible proteins from their native tertiary state to the secondary state exposes an increased number of sulfhydryl (–SH) groups, a transition that contributes to the typical "cooked" flavor of heated milk.

Visual evidence of the denaturation of these whey proteins is the thin layer of precipitated protein that collects on the bottom of the pan in which the milk is heated. This precipitated protein becomes quite hot because of its concentration on the hot bottom surface of the pan, and scorching soon becomes evident. The color and flavor changes that occur as the precipitated whey proteins interact with the lactose in milk are gradual as nonenzymatic browning (the Maillard reaction) proceeds.

Scorching is one of the chief problems in milk cookery, because whey proteins precipitate to the bottom of the pan of milk and begin to interact with lactose to produce undesirable browning and flavor changes. A heavy pan, such as one made of heavy aluminum, distributes heat quite uniformly and reduces scorching. However, thorough and frequent stirring still needs to be done to prevent the denatured whey proteins from collecting on the bottom of the pan and becoming overheated.

Denaturation of the whey proteins also leads to scum formation in milk-containing products heated to and held above 60°C (140°F), because the denatured protein molecules gradually join together. Evaporation of water from the surface of heating milk compounds the problem of scum formation by increasing the concentration of casein and the salts in milk, particularly calcium phosphate. This causes decreased solubility of the protein salts, and they coalesce gradually as heating continues. Even when the scum is removed, newly precipitated milk proteins quickly form a new layer on the surface. Steam pressure builds under the scum, and the milk may boil over easily.

Casein micelles are distinctly resistant to heat treatment when the pH of the medium is essentially neutral, but this stability gradually changes if casein is heated in an acidic environment. Excessive heating at a pH of approximately 7, however, can cause casein to form a coagulum. This change may result, in part, from the splitting of some of the phosphate that had been esterified to the casein and also from some proteolysis of peptide bonds in the casein molecules. The time and temperature usually involved in milk cookery are not ordinarily sufficiently severe to cause denaturation and coagulation of casein within the milk product.

### Enzyme Action

Occasionally, rennet is used in making clabbered (clotted) desserts. This is the same process used in the manufacture of cheese when rennet is added to hasten curd formation. Maintenance of the temperature between 15 and 60°C (59 and 140°F) is essential to the action of the enzyme. Outside these extremes, the proteinaceous enzyme is inactivated, and clotting is blocked. The clot that forms is stronger when the pH of the milk is about 5.8 than when it is either more acidic or more neutral.

Milk used in making rennet-clabbered desserts should not be heated above 60°C (140°F) at any time, either before or during the making of the dessert. Although the reason that previous heat treatment influences the rate of clabbering is not clear, there does appear to be an interaction between the casein micelles and the β-lactoglobulin in the whey; α-lactalbumin and β-lactoglobulin complexed with κ-casein are not coagulated by the action of rennet.

## Acid

Casein, the most abundant protein in milk, is least soluble and most easily precipitated at a pH of about 4.6, which is its isoelectric point. Although the pH of milk normally is above the isoelectric point of casein, adding fruits and certain acidic vegetables can reduce the pH of the milk enough to approach the isoelectric point. The result is decreased solubility of casein and increased likelihood of curdling.

Curdled milk-containing products are not aesthetically satisfying. When fruits or quite acidic vegetables, such as tomatoes, are included in with milk in a recipe, the acidic ingredient should be stirred in carefully and heated just as short a time as possible. This technique keeps the likelihood of curdling to an absolute minimum.

Curdling is more likely to occur when asparagus, peas, string beans, or carrots are cooked in milk than when cabbage, cauliflower, or spinach is the vegetable used. The different behavior of these various vegetables probably results from the difference in polyphenol (tannin) content, as well as in pH.

## Salts

Salts of various types also influence the stability of the proteins in milk-containing recipes. The ability of salt ions to interact with electrical charges on the surface of milk proteins enables the conflicting electrical charges to be reduced to an absolute minimum, which favors the denaturation and coagulation of milk proteins. The valence of the salt ions and the concentration play critical roles in determining the influence of a specific salt on milk proteins in a food product. A concentration of between 0.2 and 0.6 percent calcium chloride causes milk to coagulate if the temperature is maintained between 45 and 65°C (113 and 149°F).

Salts can complicate the problem of heating meats in milk, as meats naturally contain some salts. Cured meats have a considerable amount of added salt, which is likely to result in formation of milk curds during cooking. The problem is accentuated if salty meats and acid-containing ingredients are combined in a recipe (e.g., ham and scalloped potatoes).

## EFFECTS OF HEAT ON CHEESE PRODUCTS

The moisture and fat content of natural cheeses, as well as the pH, influence the ease with which cheeses can be utilized in heated products. Its high moisture content enables cream cheese to blend readily with other ingredients during heating. In comparison, Cheddar cheese, which is lower in moisture, is more difficult to blend; however, Cheddar that has been aged for several months blends readily. It appears that the solubility of some of the protein increases as a result of the chemical changes that occur during ripening of cheeses with a comparatively high fat content. However, aging cheeses with less fat does not significantly improve their cooking characteristics; instead, they are likely to become increasingly stringy and tough as aging proceeds.

Stringiness can be a problem in products containing heated natural cheeses. The stringy nature of these heated cheeses is accentuated at a pH between 5 and 5.6, but the textural difficulties are reduced detectably if the pH of the cheese and other ingredients is at least 5.8 or higher. This difference may result from the interaction of calcium with phosphates at the higher pH, a change promoting easier blending of the casein.

Stringiness, toughness, and fat separation are problems when products made with natural cheeses are heated excessively. Either an extended period of heating or a high cooking temperature is likely to cause a cheese-containing food to develop some or all of these characteristics. The longer the heating period and the higher the temperature, the greater are the detrimental changes.

With such high energies in the food, the protein molecules draw ever tighter together, forcing fat and water out of the system and increasing the density of the proteins. Such changes are not reversible and can be controlled best by using aged cheese, moderate temperatures, and keeping heating periods to a minimum. The presence of an emulsifying agent in process cheeses helps avoid the loss of fat and the consequent increased concentration of protein when using these cheeses for cooking.

## FOAMS

### Milk Foams

The proteins and water in milk can be extended into thin films by agitation. These thin films enclose small air bubbles, making a foam in which the protein and water provide the continuous network of the colloidal dispersion, and the air is the discontinuous or dispersed phase. This arrangement is possible because the native proteins in milk have a low surface tension and low vapor pressure. The low surface tension makes it possible to spread the liquid proteins into thin films, and the low vapor pressure reduces the likelihood that evaporation will occur.

In fluid milks, the concentration of protein is too low to permit production of a foam with any stability. However, evaporated milk can be whipped into a foam with a very large volume. The increased protein and fat concentrations of undiluted evaporated milk make it possible for a foam to form, and it will even have some limited stability. Foam formation and stability are enhanced if the undiluted evaporated milk is chilled until ice crystals start to form in it. This causes the fat to be rather firm, which concentrates the protein in the remaining unfrozen water and also helps give some rigidity to the cell walls in the foam. Lemon juice or other acid can be added to help precipitate the milk proteins and strengthen the cell walls.

Although evaporated milk foams can be formed, they are of limited usefulness unless a stabilizer, such as gelatin, is added. Even with gelatin, evaporated milk foams may be unacceptable to consumers unless they are used in a product with sufficiently strong flavors to mask the cooked flavor imparted by the evaporated milk.

Nonfat dried milk solids also can be used to make milk foams. To do this, the solids are combined with an equal volume of water. This level of dilution results in a concentrated protein mixture that readily forms a foam with large volume and limited stability. As is true with evaporated milk foams, gelatin is needed to provide suitable stability for this type of foam.

Foams made with nonfat dried milk solids gain their limited stability from the comparatively high concentration of denatured milk proteins. Essentially no fat is present to enhance stability. Although this type of foam has little flavor because of the absence of fat, it does have the advantage of being quite low in calories.

### Cream Foams

Whipping cream, with its fat content of at least 30 percent, can be beaten to a foam, but cream with 36 percent fat can be beaten quickly to a more stable foam. At a fat level of at least 30 percent, the foam forms fairly readily if the cream is chilled (but not frozen). The fat is quite firm and contributes considerable rigidity to the cell walls in a whipped cream foam that is kept chilled. Overbeating of whipped cream causes reversal to a water-in-oil emulsion, and butter results.

Since fat is the principal component contributing to the strength of the cell walls, it is essential that whipped cream be stored under refrigeration until the time it is served. If the cream is allowed to begin to warm to room temperature, the fat will start to soften, and the rigid cell walls containing the warming fat will weaken. If warm enough, the whipped cream foam may melt into a liquid system.

Whipping cream ordinarily is pasteurized to ensure the absence of harmful microorganisms. This heat treatment has a slightly detrimental effect on foam formation, but the benefit of safety outweighs this small disadvantage. On the other hand, homogenization usually is not done because the disruption of the fat globules into smaller spheres interferes significantly with foam formation and stability and offers no advantage.

Cream that has aged for about three days whips somewhat better than very fresh cream. This effect is thought to result from the slight drop in pH during this period.

Sugar often is added to whipped cream for flavor despite the slight negative effect it has on stiffness. This effect is noted whether sugar is added before, during, or after the cream is whipped.

## INGREDIENTS FROM DAIRY FOODS

Whey is available in large quantities as a by-product of cheese manufacturing. In addition to lactose, whey also carries water-soluble B vitamins and proteins as it drains from the precipitated curd. A particularly interesting ingredient produced from whey is whey protein concentrate. By the process of ultrafiltration, whey protein can be concentrated to a level from 34 to 80 percent protein. Whey protein concentrate can be tailored for use in baked products, meats, sauces, dips, and related products to give structure to gels, bind water, and promote browning.

Another product is whey protein hydrolysates (WPH), which are shorter chains of whey protein produced by the hydrolytic action of appropriate enzymes. WPH is used in some products for people needing protein that is fairly easy to absorb and digest.

Lactoferrin is available in small amounts from cow's milk, but it is of interest as an ingredient. It binds free iron, making it unavailable to bacteria and parasites needing iron for growth. It also may inhibit growth of cancers, viruses, and fungi. Another whey protein product that may become important in the future is glycomacropeptide (GMP). This protein results when chymosin (rennin) interacts with κ-casein while the curd in cheese forms. GMP may be effective in promoting dental and intestinal health. Its possible role in regulating appetite is being explored.

## SUMMARY

Milk contains less than 4 percent fat, more than half of which is saturated and about a third of which is monounsaturated. Lactose is the type of carbohydrate in milk. Casein is the protein complex comprising the curd formed when milk reaches the isoelectric point of casein (pH 4.6) or is treated with rennin, a milk-clotting enzyme. The whey proteins include β-lactoglobulin and α-lactalbumin.

Pasteurization is a heat treatment that kills potentially harmful microorganisms that might be present in milk. Homogenization, which forces milk through tiny apertures, modifies the size of fat globules and alters the protein slightly. Evaporation and drying are techniques used to preserve milk for later use. Fermentation alters the physical and chemical properties of some milk products, such as buttermilk.

Numerous products are available in the dairy department, ranging from homogenized, pasteurized fluid milks (whole, reduced fat, low fat, and nonfat) and chocolate milk, through fermented milks (cultured buttermilk, acidophilus milk, yogurt, and Lactaid), creams of varying fat content, butter, canned milks (evaporated milks of varying fat levels and sweetened condensed milk), and dried milks.

Cheese is made from milk by forming a curd (with the use of acid, rennin, or both), draining much of the whey, and then heating and pressing gently to achieve the desired moisture level, usually around 40 percent. These natural cheeses often are aged to modify flavor and texture. Natural cheeses are heated with an emulsifier to make process cheeses.

Ice cream (either plain or composite), frozen custard, low-fat ice creams, and sherbet are frozen dairy desserts. Mellorine is an imitation ice cream in which the milk fat has been replaced, and parevine is an imitation product in which both the milk fat and milk solids have been substituted.

Ice creams are frozen while being agitated in a container surrounded by a mixture of ice and salt in the ratio of about 8:1. Freezing needs to occur rapidly enough that small ice crystals can be formed while the mixture is agitated. Agitation incorporates some air, producing overrun and a pleasingly light texture with fine crystals. Fat, egg white, various sweeteners, and other interfering substances can be included in an ice cream formula to help prevent the formation of large ice crystals and lactose crystals, thus enhancing texture.

Natural cheeses blend with other foods, depending on moisture content, aging, and pH. Blending is best when using cheeses with high moisture content. Aging of high-fat natural cheeses increases solubility of the protein, causing them to blend well with other ingredients. Stringiness, toughness, and fat separation can be problems when cooking with natural cheeses unless the heating period is short and the temperature comparatively low. Process cheeses are less likely to show these detrimental changes because of the emulsifier they contain. Reduced-fat cheeses are being developed and marketed.

Rennet can be used to make clabbered desserts. The pH should be about 5.8, and the temperature should be within the range from 10 to 65°C (50 to 149°F).

When milk is heated, some β-lactoglobulin and α-lactalbumin denature and precipitate to form a thin layer of protein on the bottom of the pan. This protein gradually undergoes nonenzymatic browning (the Maillard reaction) with lactose, leading to scorching. In addition, a scum can form when milk is heated. This rough protein scum coats the surface and can easily hold in steam until enough pressure builds up, forcing it to rise, and the milk boils over. Casein is quite resistant to precipitation when heat is applied, but severe heating can cause casein to form a curd at a pH of 7.

When preparing products with milk, it is important to be aware of the isoelectric point of casein and the acidity of fruits and some vegetables. If the acid is added to the milk and the heating time is kept short, curdling can be avoided or kept to a minimum. Salts can increase the likelihood of milk protein curdling by providing ions that combine with the electrical charges on the protein molecules to reduce the electrical repulsion between the molecules. This change facilitates aggregation and precipitation of the milk proteins.

Fluid milks do not form stable foams. Undiluted, chilled evaporated milk and nonfat dried milk solids diluted with an equal volume of water can be whipped into foams that can be stabilized with gelatin or gums so they can be used in food products. The cooked flavor of an evaporated milk foam is detectable unless it is masked with other strong-flavored ingredients. The foam made with nonfat dried milk solids is mild in flavor and low in calories because it lacks fat. Protein is the stabilizing agent in foams made with nonfat dried milk solids; protein and fat help stabilize evaporated milk foams. Whipped cream gets its stability primarily from its high fat content (preferably 36 percent); it must be kept chilled so that the fat will be firm.

## STUDY QUESTIONS

1. Identify the principal proteins in milk. Describe the form in which casein occurs.

2. How is milk pasteurized? Why is this treatment important?

3. What changes occur in milk as a result of homogenization?

4. Name three fermented milk products and describe the production of each.

5. Compare the advantages and disadvantages of fluid milks, evaporated milks, sweetened condensed milk, and nonfat dried milk.

6. Describe the production of cheese. How does the production of natural cheese differ from that of process cheese?

7. What are the differences between ice cream, sherbet, frozen custard, mellorine, and parevine?

8.  What changes occur in milk when it is heated? How does the addition of acid influence the product?

9.  Describe the effect of salts on milk that is used in a heated cream soup.

10. Identify three ways of achieving a fine texture in ice cream. Explain the action of each.

11. True or false. Raw milk has been proven to be free of bacteria.

12. True or false. UHT milk does not have to be refrigerated until it has been opened.

13. True or false. Homogenizing increases the amount of fat in milk.

14. True or false. Ripened cheeses blend into sauces more readily than do unripened natural cheeses.

15. True or false. Oven temperature has little effect on ripened cheeses.

## BIBLIOGRAPHY

Aguilera, J. M. 1995. Gelation of whey proteins. *Food Technol.* 49 (10): 83.

Baggs, C. 2002. Saying more than just cheese. *Food Product Design 11* (12): 72–79.

Beardmore, G. 2006. Cheese, the right stuff(ing). *Food Technol. 60* (5): 21.

Bren, L. 2003. Cloning: Revolution or evolution in animal production? *FDA Consumer 37* (3): 28.

Brody, J. 2010. Say Mozzarella. *Food Product Design 20* (4): 22.

Burrington, K. J. 2002. New dairy ingredients "moove" to enhance products. *Food Product Design 12* (1): 63.

Burrington, K. J. 2002. More than just milk. *Food Product Design 12* (7): 37.

Feder, D. 2009. Smile and say *formaggio*. *Food Product Design 19* (10): 30.

Foster, R. J. 2005. pHood phenomena. *Food Product Design 14* (11): 61.

Hazen, C. 2004. Proteins: Nutrition and function. *Food Product Design 14* (5): 34.

Hegenbart, S. 2002. Soy: The beneficial bean. *Food Product Design 11* (10): 83.

Hollingsworth, P. 2003a. Culture revolution. *Food Technol. 57* (3): 20.

Hollingsworth, P. 2003b. Frozen desserts: Formulating, manufacturing, and marketing. *Food Technol. 57* (5): 26.

Klahorst, S. 2002. Nutrients from dairy sources. *Food Product Design: Functional Foods Annual* (September): 89.

Knehr, E. 2004. Whey protein gives beverages a boost. *Food Product Design 14* (5): 31.

Kuntz, L. A. 2010. Concentrating on whey protein isolate. *Food Product Design 20* (3): 18.

Marshall, R. T. and Goff, D. 2003. Formulating and manufacturing ice cream and other frozen desserts. *Food Technol. 57* (5): 32.

Mermelstein, N. H. 2002. Concentrating milk. *Food Technol. 56* (3): 72.

Miraglio, A. M. 2004. Wheying the positives. *Food Product Design 13* (11): 33.

Ohr, L. M. 2003. Dairy derivations. *Food Technol. 57* (5): 81.

Ohr, L. M. 2004. Powerhouse proteins soy and whey. *Food Technol. 58* (8): 71.

Roos, Y. H. 2002. Importance of glass transition and water activity to spray drying and stability of dairy powders. *Lait 82* (4): 475.

Shapiro, L. S. 2001. *Introduction to Animal Science.* Prentice Hall. Upper Saddle River, NJ.

Sloan, A. E. 2002. Got milk? Get cultured. *Food Technol. 56* (2): 16.

Thomsen, M. K., et al. 2005. Two types of radicals in whole milk powder. Effect of lactose crystallization, lipid oxidation, and browning reactions. *J. Agri. Food Chem. 53* (5): 1805.

Van Hekken, D. L. and Farkye, N. Y. 2003. Hispanic cheeses: The quest for queso. *Food Technol. 57* (1): 32.

Zammer, C. M. 2002. Soymilk: Coming of age? *Food Product Design FFA* (September): 23.

## INTO THE WEB

*http://www.rbstfacts.org/*—Information about recombinant bST.

*http://www.foodsci.uoguelph.ca/dairyedu/chem.html#lipids3*—Description of milk fat globules.

*http://www.foodsci.uoguelph.ca/deicon/casein*—Examination of the casein micelle.

*http://www.arserrc.gov/CaseinModels/*—Molecular modeled structures of various caseins.

*http://www.foodsci.uoguelph.ca/deicon/casein.html*—Discussion of casein micelles.

*http://www.foodsci.uoguelph.ca/dairyedu/chem.html#protein3*—Review of casein micelles and their characteristics.

*http://www.vivo.colostate.edu/hbooks/pathphys/digestion/stomach/rennin.html*—Description of rennin and its role in clotting milk.

*http://www.ars.usda.gov/is/AR/archive/nov05/milk1105. htm*—Information on casein films.

*http://www.fda.gov/Food/FoodSafety/Product-Specific-Information/MilkSafety/ucm122062.htm#pasteur*—FDA information about raw milk and its potential risks.

*http://findarticles.com/p/articles/mi_m3301/is_3_106/ ai_n27867236/?tag=content;col1*—Effect of time and temperature of pasteurization.

*http://www.foodsci.uoguelph.ca/dairyedu/pasteurization. html*—Discussion of methods of pasteurizing milk in Canada.

*http://www.milkfacts.info/Milk%20Processing/Heat%20 Treatments%20and%20Pasteurization.htm*—Information on conditions for pasteurization and UHT treatment of milk.

*http://www.foodsci.uoguelph.ca/dairyedu/homogenization. html*—Discussion of homogenization of milk.

*http://www.foodsci.uoguelph.ca/dairyedu/yogurt.html*—Examines various aspects of fermented milk products.

*http://www.foodsci.uoguelph.ca/dairyedu/yogurt*—Overview of production of yogurt.

*http://www.stiltoncheese.com/*—Information on Stilton cheese, a type of blue cheese.

*http://www.euprotectedfoodnames.org.uk/*—Background on E.U. Product Designation of Origin.

*http://www.stiltoncheese.com/making_stilton*—Video on making of Stilton cheese.

*http://www.cpkelco.com/market_food/prod-simplesse.html*—Information about Simplesse®.

*http://www.pacode.com/secure/data/007/chapter39/s39.22. html*—Legal requirements for identifying a product as mellorine.

*http://www.recipetips.com/kitchen-tips/t—1385/homemade-ice-cream-freezing-methods.asp*—Information on freezing ice cream.

*http://www.milkfacts.info/Milk%20Composition/protein. htm*—Information on milk proteins and heat.

*http://www.adpi.org/DairyProducts/Whey/tabid/63/Default. aspx*—Some information on various whey products.

*http://www.daviscofoods.com/fractions/gmp.cfm*—Information about glycomacropeptide (GMP).

Fish are important in the diet as a source of protein and omega-3 fatty acids.

# CHAPTER 15

# Meats, Fish, and Poultry

## Chapter Outline

## OBJECTIVES

After studying this chapter, you will be able to:

1. Describe the structure and composition of muscle, connective tissue, and fat.
2. Identify the pigments in meat and the changes that occur during heating and cutting.
3. Discuss the factors determining meat quality.
4. Identify meat cuts, grades, and the marketing process.
5. Explain the methods of meat cookery and the rationale for using each.
6. Identify various modified meat products.
7. Explain the composition of gelatin and its properties.

## CLASSIFICATION

Flesh foods usually are categorized as *meat, poultry,* or *fish*. **Meat** includes all red meats from animal sources, although the only ones commonly available are beef, veal, pork, and lamb (or mutton in some countries). **Poultry** is the inclusive term for turkey, chicken, and

**Meat**
Red meats, including beef, veal, pork, and lamb.

**Poultry**
Fowl, notably turkey, chicken, and duck.

**Fish**
Broadly defined as aquatic animals, but more narrowly defined to designate those with fins, gills, a backbone, and a skull.

**Shellfish**
Subclassification of fish; includes mollusks and crustaceans.

**Mollusks**
Shellfish with a protective shell.

**Crustaceans**
Shrimp, lobsters, crabs, and other shellfish with a horny covering.

duck, as well as pheasants and other less available fowl. **Fish**, in the broad sense, designates aquatic animals, but frequently it is the more narrow classification that includes only fish with fins, gills, a backbone, and a skull. **Shellfish**, the other classification of aquatic animals, is subdivided into **mollusks** and **crustaceans**, the former having a shell and the latter a horny covering.

---

## FOOD FOR THOUGHT: Fish and the Environment

The public has heard the message of the health benefits of omega-3 fatty acids, which are found in abundance in seafood. Increased consumption of these fish has created concern about overfishing of various popular fish, and limits have been placed on the take for both commercial and sport fishermen. Nature also plays a significant role in determining when and where seafood will thrive. Water temperatures in the oceans have been changing. Measurements taken by the National Oceanic and Atmospheric Administration (NOAA) along the Pacific Coast indicate that water temperature recently has been rising faster than it did at the beginning and during much of the past century.

Fishermen are reporting changes in the types and numbers of seafood in their locales. Humboldt squid and marlin, normally found in waters off Baja California, are now being caught as far north as Washington; the squid population found near San Francisco is increasing rapidly. San Diego fishermen are reporting reduced numbers of albacore, but more yellowfin tuna (the kind valued for making sushi). Fishermen off the coast of Oregon and Washington have been dealing with significant limits on the salmon they can take home; the good news is that more albacore tuna are found in this region as warmer waters have promoted their movement northward.

In the East, some people blame the warming water in Chesapeake Bay for causing an illness (mycobacteriosis) in bass living there. Algae are flourishing and taking some oxygen from the water, which may be a factor in the health of the bass population. Reduced oxygen content in the water is a potential problem in coastal waters and can cause a drop in the number of fish on both coasts in the future.

Predictions on the extent of glacial melt and the long-term effects on sea level suggest that present marshlands along the coast may disappear as the water gets deeper, and the area is transformed into open sea. When this happens, the population of crabs, shrimp, and menhaden, as well as some other types of sea life that reproduce in the sheltered marshy setting, will be reduced. These are some of the problems that need to be monitored and alleviated as much as possible to avoid losing valuable variety and quantities of seafood to feed the nation.

Quite a different environmental problem is developing for fish in the Great Lakes. The current situation has evolved from importing of some Asian carp by a fish farmer in Arkansas in 1972. The imported carp were effective in keeping holding pens clean and caused no concern until flooding in the 1990s released them into local streams. Unlike the common carp, that were already there and weighed around 25 pounds, the Asian carp ate voraciously and displaced most of the native fish as they worked their way north in streams and rivers. These giant fish may weigh as much as 100 pounds, and they are so strong that they can injure people in small boats when they soar as high as 8 feet out of the water. Not only are these giant carp badly behaved, they don't even taste good and are full of bones, all of which removes them from the risk of appearing on dining tables. In other words, they are a menace to the edible fish supply and to hapless boaters. Now giant carp have arrived in Chicago, where vigorous and costly efforts are being made to keep them from escaping into Lake Michigan. This effort amounts to environmentalists' last chance to protect all of the Great Lakes from giant Asian carp. Failure to stop them will have a devastating impact on the fishing industry and sport fishing in all of these lakes.

# STRUCTURE

## Muscle Tissue

***Components.***  Water is the primary constituent in muscle; actually, it is about 75 percent water. The next most abundant substance, protein, is a distant second, constituting only about 18 percent of the total. The amount of fat is highly variable, but commonly ranges from 4 to 10 percent. Carbohydrate, primarily in the form of glycogen plus a small amount of glucose and glucose 6-phosphate, accounts for a little more than 1 percent of the total. Vitamins, minerals, and trace amounts of various organic compounds complete the picture.

The specific composition of muscle tissue varies from muscle to muscle and even from one spot to another within a muscle. This variation makes the testing of meats a challenging task. Additional complications occur because of the differences between carcasses.

***Proteins.***  The three most abundant muscle proteins are myosin, actin, and tropomyosin. **Myosin** is a comparatively long, thin protein molecule. **Actin** is found in two forms—G-actin and F-actin. F-actin is very long, and its molecular weight is extremely large, ranging into the millions. G-actin is much smaller; it can aggregate to form F-actin. **Tropomyosin** is considerably smaller than myosin, actually only about one-fourth as long.

Actin and myosin unite to form an important myofibrillar protein called **actomyosin**. The formation of actomyosin is a reversible reaction catalyzed by **adenosine triphosphate (ATP)** and the presence of calcium and magnesium ions. Five lesser muscle proteins (tropomyosin, troponin, M-protein, α-actinin, and β-actinin) may influence the formation of actomyosin or its degradation to actin and myosin. The transformation of myosin and actin into actomyosin and the reversal releasing these two protein components accompanies the contraction and relaxation of muscle tissue.

Enzymes contribute to the total protein found in muscle tissue. Some soluble **adenosine triphosphatase (ATPase)** occurs in the sarcoplasm, the jelly-like protein in muscle fibers. The action of this enzyme postmortem ultimately leads to lactic acid formation, causing a drop in the pH of tissues that may be largely responsible for the onset of rigor mortis. **Neutral pyrophosphatases (PPase)** may influence the water-holding capacity of meat. **Cathepsins** and **calcium-activated factor (CAF)** are other enzymes found in muscle tissue. These are important because of their probable role in proteolytic reactions leading to the softening of tissue that marks the passing of rigor mortis.

**Myosin**
Principal myofibrillar protein; long, thin molecule.

**Actin**
Myofibrillar protein existing primarily in two forms [F (very long) and G (smaller)].

**Tropomyosin**
Very small, least abundant of the three principal myofibrillar proteins.

**Actomyosin**
Muscle protein formed from the union of actin and myosin during muscle contraction.

**Adenosine triphosphate (ATP)**
Key compound in metabolism that contains energy-rich bonds.

**Adenosine triphosphatase (ATPase)**
Enzyme in muscle tissue involved in glycolytic reactions leading to lactic acid formation.

**Neutral pyrophosphatase (PPase)**
Group of enzymes in muscle tissue influencing the water-holding capacity of meat.

**Cathepsins**
Group of proteolytic enzymes that can catalyze hydrolytic reactions leading to the passing of rigor mortis.

**Calcium-activated factor (CAF)**
Proteolytic enzyme activated by calcium; contributes to tenderizing of aging meat.

---

**Muscle**
*Myofilaments*
    Thick (myosin)
    Thin (actin, plus proteins such as troponin, tropomyosin, and α-actin)
*Sarcomeres* (organized myofilaments) that form bands (Z lines, I band, H band, A band)
*Myofibrils*
    Sarcomeres
    Myofilaments
*Fibers*
    Myofibrils
    Sarcoplasm
    *Bundles of fibers*
    *Muscle (bundles of bundles)*

**Binding Components**
Sarcolemma (encases fibers)
Endomysium (between fibers)
Perimysium (surrounds bundles of fibers)
Epimysium (surrounds many bundles of fibers to encase muscle)

**Chart 15.1**  Structural components and organization of meat

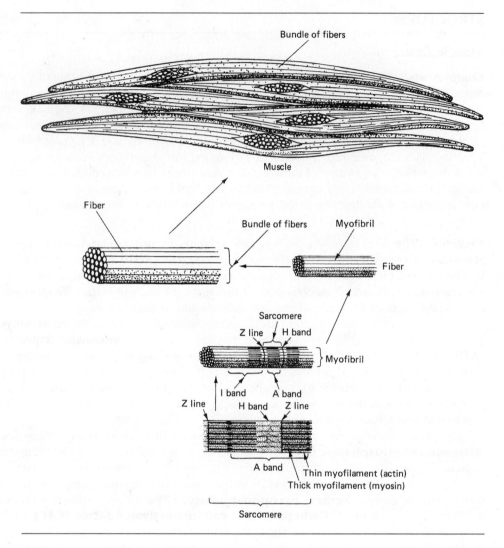

**Figure 15.1**    Schematic diagrams of thin (actin) and thick (myosin) myofilaments organized into higher levels to form a muscle.

**Myofilament**
Simplest level of organization in muscle; classified as thick or thin myofilaments.

**Thick myofilament**
The thicker, longer type of myofilament; composed of myosin molecules joined together to form a screw-like, thick, and elongated filament.

**Thin myofilament**
Thin filament formed by the helical twisting of two strands of polymerized actin.

**Sarcomere**
Portion of a myofibril.

**Myofibril**
Linear bundle of several myofilaments that contain a number of sarcomeres.

***Organization.***    The structure of muscle tissue is quite complex in its organization; it begins with the association of protein molecules (including myosin and actin) into **myofilaments** and continues as these initial structures associate with increasingly larger structures until finally the complete muscle is defined (Figures 15.1 and 15.2). In **thick myofilaments**, myosin molecules are arranged in a linear, head-to-foot orientation. **Thin myofilaments** contain spherical actin molecules polymerized into strands, with two of these strands twisted in a helical configuration to form a single, thin myofilament. The actin of the thin myofilaments associates with other muscle proteins (including troponin, tropomyosin, and α-actinin), which adds to the complexity of the basic organization of muscle.

These myofilaments are clustered in units called **sarcomeres**, within which they are arranged quite regularly in distinct patterns of Z lines and A and H bands. These sarcomeres are organized linearly into units called **myofibrils**.

The organization of the myofilaments of myosin and actin is important to muscle contraction and relaxation. Contraction occurs when thin actin filaments bond briefly with a portion of the myosin filaments to form actomyosin and then slowly migrate toward the middle of the sarcomere, causing the sarcomere to shorten or contract. Relaxation and lengthening to the original dimensions of the sarcomere occur when the actomyosin dissociates into the component actin and myosin myofilaments (Figure 15.3).

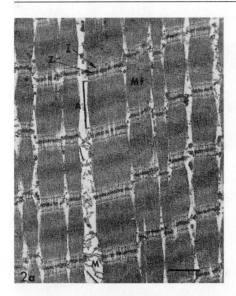
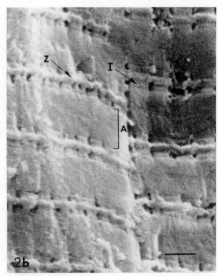

**Figure 15.2**    Muscle cell structure in aged bovine semitendinosus muscle viewed by (left) transmission electron microscopy and (right) scanning electron microscopy. Myofibrils (Mf) run vertically. Bandings on the sarcomeres are the Z, I, and A bands. Mitochondria (M) are degraded. Bars = 1μ. (Courtesy of S. B. Jones, U.S.D.A. Agricultural Research Service, North Atlantic Area Eastern Regional Research Center. Reprinted from Food Technology. 1977. Vol. 41 (4): 83. Copyright © Institute of Food Technologists.)

Myofibrils are held in long bundles called **fibers**. Each fiber consists of many myofibrils oriented together in a linear fashion and surrounded by a distinctly viscous protein sol, the **sarcoplasm**. A thin, transparent membrane called the **sarcolemma** holds this assembly together. The diameter and length of these fibers in meat vary considerably with the muscle and the animal (Figure 15.4).

**Fiber**
Bundle of myofibrils and sarcoplasm encased in the sarcolemma.

**Sarcoplasm**
Jelly-like protein surrounding the myofibrils in muscle fibers.

**Sarcolemma**
Thin, transparent membrane surrounding the bundle of myofibrils that constitute a fiber.

**Figure 15.3**    Severely contracted fibers of cold-shortened longissimus dorsi muscle (SEM micrograph). (Courtesy of S. B. Jones, U.S.D.A., Agricultural Research Service, North Atlantic Area Eastern Regional Research Center. Reprinted from Food Technology. 1977. Vol. 31 (4): 83. Copyright © by Institute of Food Technologists.)

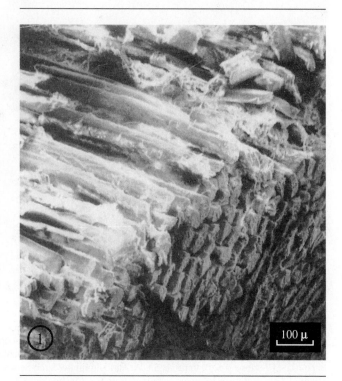

**Figure 15.4** Muscle fibers of bovine semitendinosus muscle imaged at low magnification by scanning electron microscopy. (Courtesy of S. B. Jones, U.S.D.A., Agricultural Research Service, North Atlantic Area Eastern Regional Research Center. Reprinted from Food Technology. 1977. Vol. 31 (4): 3. Copyright © by Institute of Food Technologists.)

## Connective Tissue

***Proteins.*** Four categories of materials are abundant in the connective tissue of meats. All of these are either pure protein or conjugated protein compounds. Of the four, collagen is perhaps the most important. The others are elastin, reticulin, and ground substance.

**Collagen** is of importance in meats because it is the fibrous protein found in the structural sheaths both within and between muscles. It is a rather complex protein that contains strands of tropocollagen, which are produced by the cells and then transferred to the ground substance for actual integration into molecules of collagen. **Tropocollagen** is a fibrous, coiled molecule consisting of three strands linked together to make a long, thin unit (Figure 15.5).

The presence of an abundance of hydroxyproline and proline (about 25 percent) accounts for the fibrous nature of tropocollagen, because the linkage of this amide through the **pyrrolidine ring** sterically hinders the molecule from assuming the usual helical configuration that leads to a spherical protein. In other words, the nitrogen in either proline or hydroxyproline is involved in the primary structure (the backbone chain) of the strands of tropocollagen, and the planar rigidity of the pyrrolidine ring prevents the bonding angles that lead to the usual α-helix and ultimate spherical nature of most food proteins. This rigidity is significant because of the unusually large quantity of the two pyrrolidine-ring amino acids, proline and hydroxyproline, as shown here.

**Collagen**
Fibrous protein composed of three strands of tropocollagen.

**Tropocollagen**
Fibrous protein consisting of three strands twisted together and containing large amounts of glycine, proline, and hydroxyproline.

**Pyrrolidine ring**
Organic ring structure containing one atom of nitrogen; linkage to another amino acid through this nitrogen favors formation of a linear, fibrous protein molecule.

Collagen

Tropocollagen

**Figure 15.5**   Schematic diagram of a portion of a collagen molecule.

The other unique aspect of the chemical composition of tropocollagen is that it contains a large amount of the extremely simple amino acid glycine. Actually, a third of the molecule is glycine. A variety of amino acids constitute the remaining approximately 42 percent of the amino acids in tropocollagen.

The formation of hydrogen bonds causes the association of the three strands to form the tropocollagen molecule. In turn, these molecules are held in the larger collagen molecule as a result of a combination of bonding forces, including hydrogen bonding. The stability of the native collagen molecule is due in large measure to the formation of covalent bonds that cross-link the three strands of tropocollagen to form a molecule of collagen. The number of covalent bonds formed between the three tropocollagen constituents increases gradually over time, which helps explain the increasing toughness of the meat from animals as they grow older.

**Elastin**, in contrast to the relative abundance of collagen, is found in limited amounts intramuscularly. The yellow color of elastin distinguishes it from collagen, which is white. Unlike collagen, which can be converted to gelatin during cooking, elastin is resistant to chemical change. The rubbery character of elastin accounts for its name. Two unusual amino acids, desmocine and isodesmocine, provide important structural contributions to

**Elastin**
Yellow connective tissue occurring in limited amounts intramuscularly and in somewhat greater concentrations in deposits outside the muscles.

elastin because their tetracarboxylic–tetraamino acid functional groups permit them to cross-link with as many as four chains of amino acid residues. Only a somewhat limited amount (less than 3 percent of the total protein) of hydroxyproline is found in elastin.

**Ground substance** is a protein-containing material in meat; its main constituents are plasma proteins and glycoprotein, including amino acids with excess carboxylic acid groups. In contrast to collagen, ground substance proteins are low in glycine and void of proline. Interestingly, it is apparently in ground substance that three tropocollagen strands are cross-linked securely so that collagen molecules form. The proteins in the ground substance frequently are bound tightly to mucopolysaccharides, and these glycoproteins form an amorphous matrix in which collagen and elastin can be held to form connective tissue.

**Reticulin** is the fourth category of protein in connective tissue. Although it is a fibrous protein similar to collagen, the linkage with myristic acid (a 14-carbon fatty acid) clearly distinguishes it from collagen.

***Organization.*** Connective tissue, designated the **endomysium** (innermost), consists primarily of ground substance and collagen and is found between each fiber. The next level of organization in muscle tissue, the **perimysium**, is the bundling together of several fibers to create a thicker fibrous bundle, which then is encased in still more connective tissue. Finally, many of these bundles of fibers, each surrounded by its perimysium, are gathered into large collections of fibers and surrounded by yet another layer of connective tissue called the **epimysium** (outer). These structures are the **muscles** contained in meats (Figure 15.6).

Fish flesh is structurally similar to that of red meats. Myofibrils in fish are much like those found in red meats, except that some (but not all) of the myofibrils in fish muscle are flat, rather than cylindrical. The fibers containing fish myofibrils are only about 3 centimeters long, yet they are comparatively thick. The ultimate level of organization in fish tissue is parallel layers of these fibers, **myotomes**, which then are attached to sheets of the connective tissue, **mycomatta**. The actin and myosin levels (and consequently the actomyosin level) are not only higher in fish than in red meats, but the actomyosin and collagen in fish also are more sensitive to heat. These characteristics dictate the need for careful temperature control when preparing fish.

## Fat

Lipids occur in muscle tissue and also in fatty deposits or fat depots. The fatty acids found most abundantly in the triglycerides in the fat depots are oleic (18 carbon atoms, 1 double bond), palmitic (16 carbon atoms, no double bonds), and stearic (18 carbon atoms, no

---

**Ground substance**
Undifferentiated matrix of plasma proteins and glycoproteins in which fibrous molecules of collagen and/or elastin are bound.

**Reticulin**
A type of connective tissue protein associated with a fatty acid (myristic acid).

**Endomysium**
Delicate connective tissue found between fibers.

**Perimysium**
Connective tissue surrounding a bundle of several fibers.

**Epimysium**
Connective tissue surrounding an entire muscle (many bundles of bundles of fibers).

**Muscle**
Aggregation of bundles of bundles of fibers surrounded by the epimysium.

**Myotomes**
Fibers in fish; these are thick and about 3 centimeters long.

**Mycomatta**
Sheet-like connective tissue in fish.

---

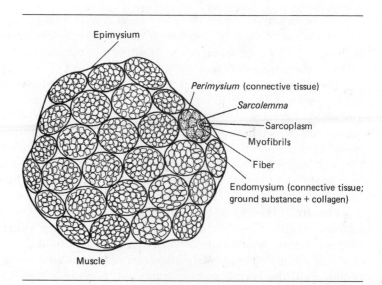

**Figure 15.6** Diagram of a cross section of muscle.

**Table 15.1**    Fat Components of Selected Foods (g/100 g edible portion)

| Food | Total Fat | Fatty Acids | | | | | | Cholesterol (mg) |
| | | Total Saturated | Total Mono unsaturated | Total Poly unsaturated | 18:3[a] | 20:5[b] | 22:6[c] | |
|---|---|---|---|---|---|---|---|---|
| Beef, chuck, raw | 23.6 | 10.0 | 10.8 | 0.9 | 0.3 | | | 73 |
| Lamb, leg, raw | 17.6 | 8.1 | 7.1 | 1.0 | 0.3 | | | 71 |
| Chicken, light, no skin, raw | 1.7 | 0.4 | 0.4 | 0.4 | d | d | d | 58 |
| Chicken, dark, no skin, raw | 4.3 | 1.1 | 1.3 | 1.0 | d | d | d | 80 |
| Turkey with skin, raw | 9.7 | 2.8 | 3.2 | 2.5 | 0.1 | d | d | 82 |
| Ocean perch | 1.6 | 0.3 | 0.6 | 0.5 | d | 0.1 | 0.1 | 42 |
| Salmon, sockeye | 8.6 | 1.5 | 4.1 | 1.9 | 0.1 | 0.5 | 0.7 | — |
| Crab, Alaska king | 0.8 | 0.1 | 0.1 | 0.3 | d | 0.2 | 0.1 | — |
| Scallop, Atlantic | 0.8 | 0.1 | 0.1 | 1.1 | 0.2 | 0.5 | d | 101 |

From Hepburn, F. N., Exler, J., and Weihrauch, J. L. 1986, Provisional tables on content of omega-3 fatty acids and other fat components of selected foods. *J. Am. Dietet. Assoc. 86:* 788.
[a] Linolenic acid
[b] EPA (eicosapentanoic acid)
[c] DHA (hexanoic acid)
[d] trace

double bonds). In the cells, the lipid and lipid-related compounds include cholesterol, **glycolipids**, **phospholipids**, plasmalogens, and sphingomyelin. These lipid components are deposited in fat cells in a matrix of connective tissue, primarily collagen. The presence of fat contributes to the perception of juiciness and flavor of meats, and it also is of interest from the perspective of nutrition.

Beef fat, with its comparatively high content of saturated, long-chain fatty acids (Table 15.1) is quite different from pork, lamb, or various types of poultry and fish. Nutritional concern regarding the composition of fats in flesh foods has generated interest in the content of omega-3 fatty acids. Omega-3 fatty acids are particularly abundant in fish oils.

Changes in fats during storage or cooking can affect the flavor and aroma of the food (see Chapter 11). Phospholipids are particularly susceptible to chemical changes. Oxidative rancidity may increase with the presence of phospholipids. Chemical changes may occur as a result of high temperatures in deep-fat frying and broiling, particularly if the time is extended.

**Glycolipid**
Molecule with a sugar moiety and a lipid portion.

**Phospholipid**
Complex phosphoric ester of a lipid.

# PIGMENTS

## Myoglobin and Related Compounds

The two key pigments responsible for the color of meats are **hemoglobin** and, more prominently, **myoglobin**. Actually, myoglobin and hemoglobin are closely related chemically, for they both are iron-containing pigments with heme as the common component. They differ only in their protein component. In contrast to hemoglobin with its four polypeptides, myoglobin has only one strand of protein polymer and a molecular weight of about 17,000.

Hemoglobin is about four times as large as myoglobin. It consists of four heme–polypeptide polymers joined together. **Heme** contains four pyrrole rings linked covalently to form a large complex that is joined to a central atom of iron by attachment to the nitrogen atoms in each of the pyrrole rings (Figure 15.7). In turn, four heme–polypeptide polymers also are linked to make the large molecule designated as hemoglobin. Its molecular weight is approximately 68,000.

The iron atom in the center of heme can complex with other atoms or compounds to form new compounds, resulting in alterations in the color of the meat. Other color changes result from changes in the valence of the iron atom itself.

Myoglobin is of particular interest in the study of meat color because it is the predominant pigment and contributes about three times as much color as does hemoglobin.

**Hemoglobin**
Large, iron-containing compound consisting of four heme–polypeptide polymers linked together; contributes to meat color.

**Myoglobin**
Purplish-red pigment consisting of heme-containing ferrous iron and a polypeptide polymer (globin).

**Heme**
Compound composed of four adjoining pyrrole rings linked to an atom of iron.

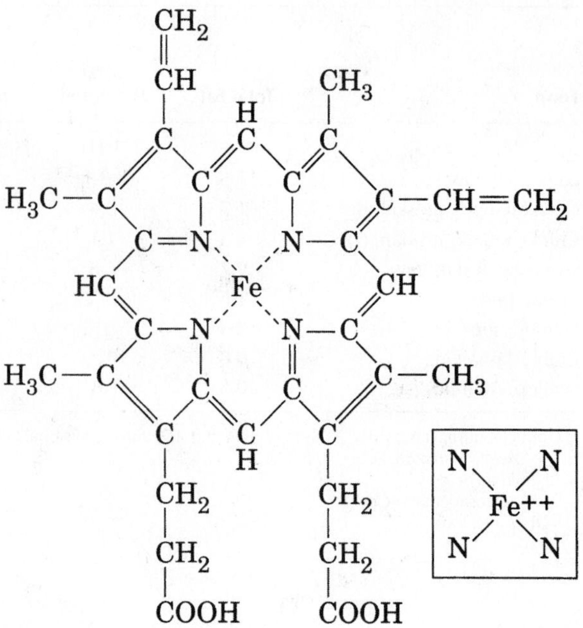

**Figure 15.7**   Structure of heme (and the abbreviated representation).

The purplish-red color of myoglobin is seen because iron is in the ferrous (2+) state and free of additional atoms or compounds. When meat is fresh and protected from contact with air, the dominant color is the purplish-red color of myoglobin.

In the presence of air, myoglobin readily adds two atoms of oxygen to form a new compound, **oxymyoglobin**, which is responsible for the rather intense red seen on the cut surface of meat that has been exposed to air for a while. Availability of an abundance of oxygen favors the formation of oxymyoglobin and ensures that meat will have a pleasingly bright red color.

The use of a plastic wrap permeable to oxygen for packaging pre-cut meats helps ensure that the cuts will have the bright red color of oxymyoglobin that appeals to consumers. Tetrasodium pyrophosphate, sodium erythorbate, and citric acid are an effective combination to maintain oxymyoglobin longer when meat is marketed in modified-atmosphere packaging.

If the oxygen supply available to myoglobin is rather limited or meat is exposed to fluorescent or incandescent light too long, a brownish-red pigment, **metmyoglobin**, forms. This less desirable color results from oxidation of the iron atom to the ferric (3+) state and complexing of a molecule of water. Metmyoglobin can be reduced back to myoglobin. Depending on the environment, myoglobin pigments may be converted between oxymyoglobin, myoglobin, and metmyoglobin according to the following scheme:

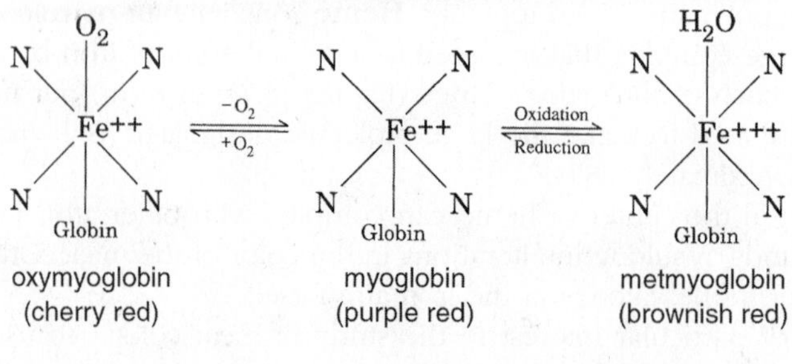

**Oxymyoglobin**
Cherry red form of myoglobin formed by the addition of two oxygen atoms.

**Metmyoglobin**
Brownish-red form of myoglobin formed when the ferrous iron is oxidized to the ferric form and water is complexed to the oxidized iron.

In contrast to the red color of various animal meats, fish and poultry generally are pigmented quite lightly. Hemoglobin contributes to the light coloration in poultry. Quite a few vertebrate fish have two muscles—a light-colored, large lateral muscle and a less desirable dark muscle deeply pigmented with myoglobin. The lateral muscle in salmon derives its unique color from the presence of **astaxanthin**, which is classified as a carotenoid pigment.

**Astaxanthin**
Reddish-orange carotenoid pigment in salmon and in cooked crustaceans.

## Changes Effected by Heating

While red meat cooks, heat changes the pigments. First, the myoglobin present in the interior of muscles changes to oxymyoglobin. Continued heating converts the oxymyoglobin into **denatured globin hemichrome**, the grayish brown associated with well-done meats. Denatured globin hemichrome is the counterpart of metmyoglobin, but heat has denatured the protein (globin) component. This reaction is illustrated here.

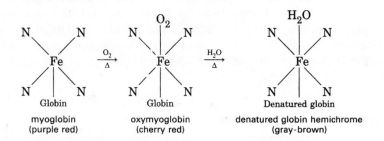

myoglobin (purple red) → oxymyoglobin (cherry red) → denatured globin hemichrome (gray-brown)

**Denatured globin hemichrome**
Myoglobin derivative formed when heat triggers the oxidation of iron to the ferric (+3) state and denatures the globin portion of the compound while the oxygen of oxymyoglobin is replaced with water complexed to the iron atom, resulting in a gray-brown color.

Heating enhances the light color of fish by increasing opacity, but this is not a dramatic change. Quite a different situation occurs with crustaceans, for the drab, blackish-green color of uncooked crab, lobsters, and shrimp changes as astaxanthin becomes dominant and the former predominant pigment loses its effect due to denaturation of the protein with which it is complexed.

As is true with fin fish, poultry ordinarily is essentially colorless when cooked. If young poultry has been frozen and some hemoglobin has leaked from the marrow, there may be some hemoglobin in the flesh close to the bones, which becomes dark when cooked. Sometimes poultry that has been subjected to intense heat during preparation develops a reddish-pink color. This is the result of hemoglobin reacting with carbon monoxide and nitric oxide generated by an electric heating element or flames when barbecuing.

## Changes Effected by Curing

Meats sometimes are cured; this process usually involves treatment with either nitrates or nitrites to preserve meats for long-term storage. One of the important functions of nitrite is to prevent botulism in cured meats (see Chapter 19).

Of particular interest in a discussion of pigments is the effect of nitrites on meat color. The nitric oxide, which forms from the nitrates and nitrites in meat curing, combines with myoglobin to form nitric oxide myoglobin. This compound changes to nitric oxide myochrome when a second nitroso group replaces the globin during the slow heating involved in curing. Nitric oxide myochrome is a key pigment in cured meats and contributes to the stability of their familiar pinkish-red color. Exposure to light and air causes oxidation of the ferrous iron to the ferric (3+) state, which results in development of a brownish color.

Nitrosyl-hemochrome forms during the curing of meats when a nitroso group joins with myoglobin, and heat denatures the globin portion of the molecule. This compound also is a pink pigment abundant in cured meats. On oxidation of the iron in nitrosyl-hemochrome, the pigment structure changes to that of denatured globin nitrosyl-hemochrome, which is brownish. The reactions during curing are as follows:

myoglobin (purple red) — nitric oxide myoglobin (red) — nitrosyl-hemochrome (pink) — denatured nitrosyl-hemochrome (brown)

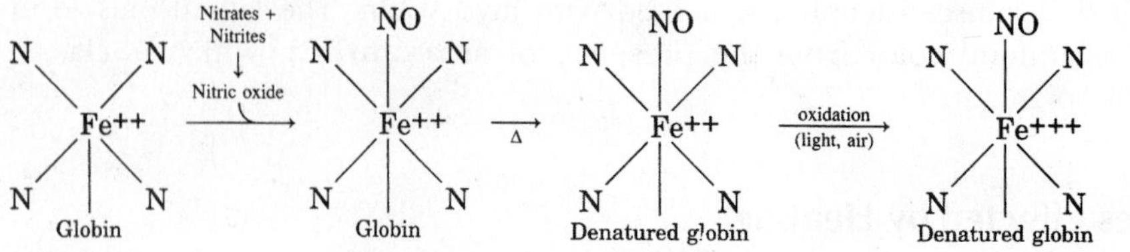

Exposure to light and additional oxygen hastens the breakdown of pigments in cured meats. In particular, light promotes removal of the nitroso group from pigments. This sets the stage for oxidation of ferrous pigments to the ferric state, with the resulting discoloration. Occasionally, the porphyrin ring is oxidized, which leads to fading of pigments and sometimes development of a fluorescent green or yellow color. The almost rainbow-like appearance occasionally noted on the surface of packaged cured meats may result from the way light is refracted from the pigments.

## MEAT QUALITY

### Factors Affecting Quality

***Maturity.*** The physical changes that occur from the time an animal is born until it is slaughtered affect the characteristics of the resulting meat. Young animals have a comparatively low ratio of lean to bone. They also have a relatively large amount of connective tissue and little fat. These characteristics can be seen in veal, which comes from animals not more than three months old. Somewhat greater range is seen in lamb, which comes from animals up to the age of 14 months. However, mutton is more than two years old. Unlike the distinctions made for different ages of both cattle and sheep to facilitate marketing, pork generally is from animals that are six months or just slightly older.

The increased fat content of mature animals influences the flavor of the meat and contributes to apparent juiciness, an important contribution because moisture content decreases. Connective tissue within the lean tissue increases in total amount as an animal matures, but it is present in somewhat smaller percentages than those when the animal was very young. Despite this small shift, meat from mature animals may be less tender than a comparable cut from a young animal. This decrease in tenderness may result from increased formation of cross-linkages between the fibers of collagen within the lean muscle as the animal matures and grows older.

As cattle age, beef flavor undergoes change. Fat content influences this, but other changes also may contribute to the stronger, characteristic flavor of mature beef. The color of the muscles gradually becomes redder and sometimes darker. The pH of muscle also may decrease. Differences between carcasses make information regarding changes caused by maturation difficult to verify.

***Postmortem Changes.*** Biochemical processes in a carcass continue several hours after slaughter and influence the quality of meat. The level of glycogen stores in the animal at the time of slaughter is paramount in determining onset of **rigor mortis** and key palatability factors in the meat when it is ready to be marketed. Glycogen is important because this complex carbohydrate undergoes biochemical degradation to produce lactic acid after slaughter.

**Rigor mortis**
Temporary rigidity of muscles that develops after death of an animal.

Desirably, the pH in the flesh of cows and other mammals drops from approximately neutral (commonly, pH 7.0 to as high as 7.2) to a pH of about 5.5 (only pH 6.2–6.5 in fish).

This level of postmortem lactic acid production occurs if the animal is in a rested, comparatively calm state at the time of slaughter. However, stress and exercise just prior to slaughter reduce glycogen levels, which limits the amount of lactic acid that is formed postmortem.

The pH reached during rigor is about at the isoelectric point of the proteins, which causes the fibers to tend to pack together and force out some of the water. Moisture level in meat is important because of its influence on the juiciness of the cooked product. Phosphate ions can be added to loosen the fibers enough for some water to be retained and to promote juiciness. Solutions can be injected to enhance juiciness of roasts and other cuts. Other additives that may promote juiciness in poultry and pork include carrageenans and starches (e.g., corn and potato).

The chemical changes associated with the onset of rigor mortis begin with the loss of available oxygen from blood when circulation ceases. Anaerobic reactions occur as a result of this change. ATP (adenosine triphosphate) can continue to be formed from ADP (adenosine diphosphate) until there is no more creatine phosphate available. Then the ATP level falls, and lactic acid forms anaerobically from glycogen. Muscles lose extensibility, because the lack of ATP blocks the unlocking of actin–myosin links. Accumulation of calcium ions resulting from the lack of ATP causes contraction and rigidity of the muscles. Some fluid also is forced from tissues.

The time of onset of rigor mortis differs among species and even a bit among carcasses of the same species. Fish may begin to develop rigor mortis an hour after being killed (Table 15.2), although onset may be delayed by as much as 7 hours.

Rigor mortis is extended in fish if they are iced as soon as they are killed and maintained in a chilled storage environment. This extends the time that fish remain fresh, because bacterial spoilage commences only after rigor mortis has passed. However, even careful icing during storage cannot extend the storage time of fresh fish more than about a week from the time of death.

Onset of rigor mortis usually is somewhat slower in poultry than in fish. For chickens, rigor mortis should begin and end after at least 4 hours have elapsed, and for turkeys at least 12 hours after slaughter is normal. Cooking or freezing should not begin until rigor mortis has passed because the flesh will be tough. Prompt chilling of poultry carcasses by immersion in ice water as soon as possible after slaughter is important in retarding rigor mortis and achieving tenderness.

Animal carcasses emerge from rigor mortis somewhat more slowly than do poultry or fish carcasses. Pork needs to be aged at least a day for it to pass through rigor. The various muscles in beef exhibit different behavior; some need to be aged 11 days before achieving maximum tenderness. During this aging period, desirable storage conditions include ultraviolet light to control growth of microorganisms, a controlled humidity of 70 percent, and a temperature just above freezing. These measures are important to keep the meat from becoming microbiologically unsafe or too dry. However, beef may be held at a temperature of 16°C (61°F) for between 16 and 20 hours after slaughter before being aged at 2°C (36°F) to promote tenderness.

Electrical stimulation (between 100 and 600 volts) of carcasses for 1 or 2 minutes within 45 minutes of slaughter promotes tenderness in poultry and meats. The electricity stimulates fast muscle contractions, promoting both physical and biochemical changes that favor increasing tenderness.

When meat is chilled too rapidly after slaughter, the muscles contract drastically, a phenomenon termed **cold shortening**. This shortening of muscle length varies from one muscle to another and also is influenced by whether or not the muscle is attached to a bone,

**Cold shortening**
Severe contraction of muscles in carcasses that have been chilled too quickly and severely after slaughter.

**Table 15.2**   Usual Elapsed Time Between Slaughter and Passage of Rigor Mortis

| Species | Elapsed Time from Slaughter to Passage of Rigor Mortis |
| --- | --- |
| Fish | 1–7 hours (unless stored in ice) |
| Chicken | 4 hours or more |
| Turkey | 12 hours or more |
| Pork | 1 day or more |
| Beef | 11 days |

**Table 15.3**   USDA Meat Grades

| Beef | Veal/Calf | Lamb/Yearling | Barrows/Gilts | Sows |
|------|-----------|---------------|---------------|------|
| Prime | Prime | Prime | U.S. No. 1 | U.S. No. 1 |
| Choice | Choice | Choice | U.S. No. 2 | U.S. No. 2 |
| Good | Good | Good | U.S. No. 3 | U.S. No. 3 |
| Utility | Standard | Utility | U.S. No. 4 | Medium |
| Cull | Utility | Cull | Utility | Cull |

**Aging (of meat)**
Holding of meat while it passes through rigor mortis and sometimes for a period extending several days or even two weeks, depending on the quality and type of carcass; storage of meat to enhance tenderness.

**Dark-cutting beef**
Very dark, sticky beef from carcasses in which the pH dropped to only about 6.6.

**PSE pork**
Pale, soft, and exudative pork from carcasses with a low pH, usually ranging between pH 5.1 and 5.4.

**Figure 15.8**  Federal inspection stamp is required on primal cuts of all meat crossing state lines.

**Figure 15.9**  Federal grading stamp rolled continuously along each primal cut exhibits that a federal meat grader has determined the quality of the carcass.

which restricts contraction somewhat. The very tough quality evidenced when a muscle is deeply in rigor mortis is the result of considerable formation of cross-linkages between myosin and actin in the contracted muscle.

Even in carcasses held under optimum temperature controls, rigor mortis does develop. Fortunately, the contraction that occurs during rigor is reversed gradually as the carcass is **aging**. The Z lines (see Figure 15.1) begin to become indistinct, myofibrils begin to break up, and the meat gradually becomes tender. This increased tenderness is theorized to be the result of both some action of cathepsins and calcium-activated factor (proteolytic enzymes) on the muscle proteins to produce simpler proteins and the breaking of linkages between actin and myosin. Clearly, aging promotes the development of tenderness in beef and other types of meat from animals.

Tenderness also is enhanced by carefully hanging carcasses so that the muscles are stretched beyond their normal, resting state prior to the onset of rigor. Ordinarily, the carcass is hung from the Achilles tendon, a position that stretches the tenderloin muscle to more than half again its resting length. Another option is to suspend the carcass from the aitch-bone (hip bone); this position favorably influences the tenderness of several other muscles.

The color of beef and pork can be influenced by the pH reached in the carcass (a reflection of glycogen stores at the time of slaughter) during the aging period as the carcass passes through rigor mortis. In the case of beef, a final pH of about 6.6 (well above the normal value of around 5.5) is accompanied by the presence of a very deep red, almost black color in the flesh and a sticky, rather slimy feeling. Such beef is called **dark-cutting beef**.

The detrimental variation seen in pork as a consequence of reaching an abnormally low pH during rigor mortis is termed **PSE pork**—pale, soft, and exudative pork—and occurs when the pH drops as low as 5.1, or even 5.4. This excessively acidic condition results in light-colored, mushy pork with considerable drip loss during cooking. Both of these pH problems cause the carcasses to bring a low market price.

## Inspection and Grading

Federal inspection is mandatory for slaughter and interstate marketing of all meat, although state laws apply if meat will not be crossing state lines. This is done to assure that the animal was not diseased and that adequate sanitation standards and temperature controls have been maintained so that the meat is safe when it is shipped to market. A federal inspection stamp on the area of each primal cut is evidence of the appropriate inspection for meat safety (Figure 15.8).

Quality of meat usually is determined by federal graders, who evaluate carcasses according to established criteria based on such physical characteristics as marbling, texture, and yield (Figure 15.9). Thus, grade designation is a guide to palatability, but does not indicate safety. Table 15.3 lists the grades for various types of meats.

## IDENTIFYING MEAT CUTS

In the marketplace, meats are cut and named according to industry standards for beef, veal, pork, and lamb. Carcasses are first carved into large primal cuts, each of which will bear the federal inspection stamp if entering interstate markets. Subsequently, the primal cuts are

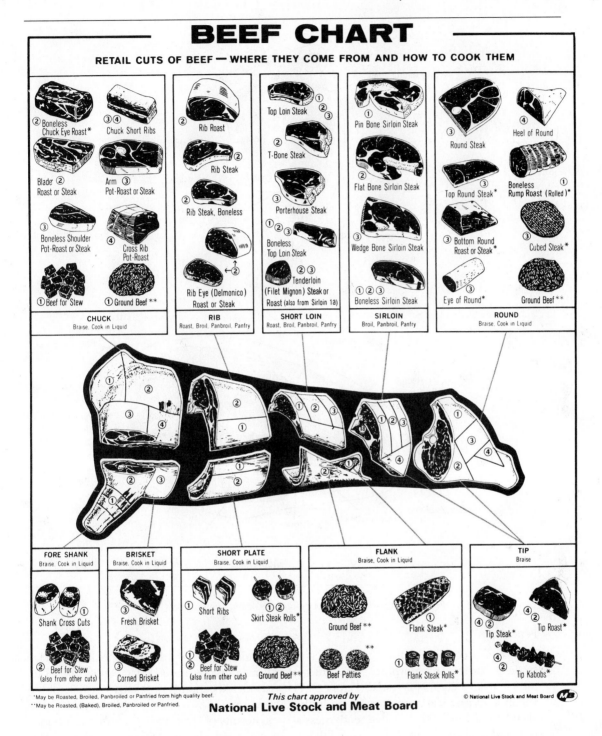

**Figure 15.10**   Primal and retail cuts of beef.

subdivided into retail cuts and labeled according to established standards. Consumers are able to anticipate the characteristics they can expect in a particular cut if they read the identifying label and know grading standards and the anticipated tenderness from the specific primal cut (Figure 15.10).

Researchers conducting studies on meat need to describe the meat cuts used by identifying the species, approximate age (if known), presence (or removal) of bone, cut (if a standard

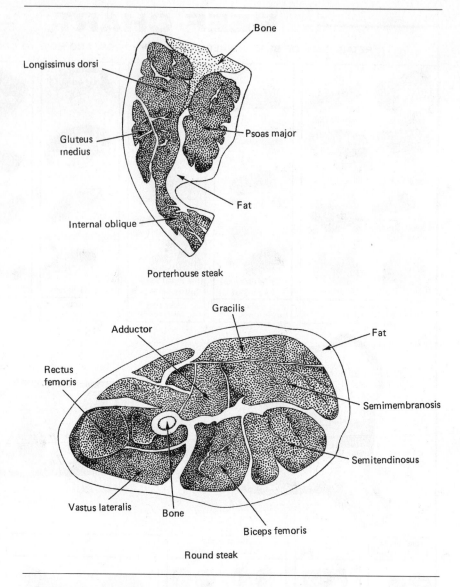

**Figure 15.11**    Muscles in porterhouse steak and round steak.

retail cut, according to the Uniform Retail Meat Industry Standard), and specific muscles studied. For research purposes, the names of various muscles are used to identify the exact muscle from a cut that is tested (Figure 15.11). This identification is necessary because of the variation in physical properties found in the different muscles within a single cut, as can be seen by comparing the relative tenderness scores. Results of shear tests (reported in pounds required to shear the muscle) showed that psoas major was the most tender and the internal oblique was the least tender of the muscles tested: the smaller the value, the more tender the muscle (Table 15.4).

The psoas major, commonly referred to as the tenderloin, and the longissimus dorsi are two muscles often used in meat research, but the T-bone steak that contains these two muscles also may include the gluteus medius and the internal oblique muscles. The longissimus dorsi is such a long muscle that it also is the principal muscle in standing rib roast. In contrast to the limited number of muscles in cuts from the rib and short loin, round steak has several muscles: rectus femoris, adductor, semimembranosus, semitendinosus, biceps femoris, and vastus lateralis (Figure 15.11).

**Table 15.4**   Relative Tenderness of Selected Muscles as Measured by Shearing

| Muscle | Force Needed[en] To Shear (pounds) |
| --- | --- |
| *Porterhouse steak* | |
| Psoas major | 7.2 |
| Longissimus dorsi | 7.9 |
| Gluteus medius | 7.5 |
| Internal oblique | 12.6 |
| *Round steak* | |
| Adductor | 10.0 |
| Semimembranosus | 12.0 |
| Semitendinosus | 11.0 |
| Biceps femoris | 10.7 |
| Vastus lateralis | 11.4 |
| Rectus femoris | 9.0 |

Ramsbottom, J. M. and Strandine, E. J. 1948. Comparative tenderness and identification of muscles in wholesale cuts of beef. *Food Res. 13*: 315.

# MARKETING

## Fresh/Frozen

Meat, poultry, fish, and shellfish that are available in retail markets may have been produced in the United States or may have come from distant places such as Thailand, Vietnam, and New Zealand (Figures 15.12 and C37). Products from distant ports often are frozen so they can be shipped as economically as possible; some domestic products also are frozen prior to the marketing process. Freezing retards deterioration of these protein-rich foods and extends their shelf life.

Labeling regulations require manufacturers to inform consumers about the previous temperature treatment of meat, poultry, fish, and shellfish displayed for sale. Products marketed as **fresh** must never have reached a temperature below –3.3°C (26°F). Poultry and other flesh foods chilled to –3.3°C (23°F) are still pliable but are very close to being frozen.

At retail markets some poultry, fish, and meat items are identified as **previously frozen**. These items have been frozen at –17.8°C (0°F) and then allowed to thaw to at least –3.3°C (26°F).

**Fresh**
Designation for meat, poultry, fish, and shellfish that have not been chilled below –3.3°C (26°F).

**Previously frozen**
Statement required on meat, poultry, fish, and shellfish if freezing occurred, but the item is thawed to at least –3.3°C (26°F) before it is purchased.

---

### FOOD FOR THOUGHT: Kosher/Halal

Niche markets provide the opportunity for developing food to meet the specific requirements of a particular group of consumers. However, the goal of the company making the food may be to sell to others beyond that relatively small market, as well. For many years, kosher meats, dairy, and a variety of baked and other food products have been produced according to kashruth (Jewish dietary laws) and certified as kosher, a designation shown on each package. Buyers of kosher foods include Jews and many other people who are seeking the oversight in production and/or the pleasure of eating a food they enjoy. Approximately 86,000 food products bearing the kosher designation now appear in markets in the United States.

The recent increase in the Muslim population in the United States is beginning to create a niche market for foods labeled as halal (meats and food products that have been prepared for market meeting Muslim dietary laws). Both kashrut and halal certifications require that meats (pork is prohibited for both) and poultry be butchered and bled according to the ritual specified, which differs with the religion. Predictions are that food products labeled as halal will begin to make their way into supermarkets across the nation soon.

**Figure 15.12**   In Vietnam, fish are unloaded from the fishing boat to the shore and iced, the first step in their trip to distant markets.

This tells consumers that the textural changes that occur in frozen products will be present. It also signals the need to prepare the item fairly soon and to avoid refreezing it.

## Domestic/Foreign

Meat, poultry, fish, and shellfish available to consumers come from domestic and foreign sources. Entry of imported products is of interest not only to U.S. customs officials but also to the various other federal agencies (e.g., USDA and FDA) responsible for safeguarding the nation's food. Regulations that permit entry into the national food supply are of great importance because of the potential for imported foods to be unsafe for human consumption. The American public expects all foods in the marketplace to be safe to eat. Standards for safety and sanitation of food vary widely in various countries, thus creating a large challenge to those involved in assuring that imported meats and other foods are safe.

Sodium and potassium lactate may be added to meats to inhibit growth of *Clostridium botulinum, E. coli 0157:H7, Salmonella,* and *L. monocytogenes* and help extend the shelf life

---

**INGREDIENT INSIGHTS: Apples and an Antimicrobial Edible Film**

An edible film made using apple peels, carvacrol (from oregano), and cinnamaldehyde (from cinnamon) is being tested for possible use in commercial meat packaging and in homes. The purpose is to retard growth of harmful microorganisms (notably, *E. coli* and *0157:H7 Listeria monocytogenes*) on meat during storage. The benefits of using such a film as a wrap for meats could be twofold: improved food safety and a commercial product that generates income from material that would become waste.

of meats. Diacetate may be used in conjunction with lactate salts because the two together are particularly effective against *Listeria* in meats that are marketed as ready to eat. Lactates also help retain the red color of beef during marketing.

Contamination of fish because of environmental pollution has triggered extensive efforts to clean up the environment. Particular concerns regarding industrial and human waste contamination resulted in cleaner water in streams, rivers, and even parts of the ocean. Although these efforts need to continue, levels of contamination in wild and farmed fish that are sent to market are monitored and are deemed to be safe for human consumption. The levels of polychlorinated biphenyls (PCBs) in farmed salmon are well within the recommended range for safety and are not considered to be a risk.

A particularly graphic example of international food safety concerns is the continuing effort to stamp out **mad cow disease**. Only a very few cases of Creutzfeldt–Jakob disease in humans have occurred in Europeans who had eaten beef linked to mad cow disease. Attempts to control bovine spongiform encephalopathy (BSE), or mad cow disease, have been ongoing in Europe and many other parts of the world where cattle are raised.

**Mad cow disease**
Common name for bovine spongiform encephalopathy (BSE).

Control efforts in Canada in 1993 included destroying all cows that had been imported from the United Kingdom. Both the United States and Canada banned ruminant-derived protein feed as a measure to prevent the spread of BSE from contaminated feed (the recognized mode of transmission). Despite its efforts, Canada has had seven cases of BSE in cattle since 2003, the most recent one in July 2006.

Closure of the U.S. border for Canadian cattle and cattle products has occurred as a means of reducing the possibility of spreading BSE to cattle in the United States. Another proactive measure is removal of brain and spinal-cord material before carcasses are processed because these tissues are the most likely to transmit BSE infection. USDA regulations prevent beef from nonambulatory, or "downer," animals from entering the human food supply.

## Carcass/Case-Ready

Traditionally, beef, veal, pork, and lamb carcasses or sides of beef have been shipped to markets where they are hung in a chilled room until the store's butcher(s) cut and package them for customers. Today, large quantities of **case-ready meats** are available in markets throughout the nation. Case-ready meats are prepared and packaged in large facilities in which safety is monitored carefully according to the Hazard Analysis and Critical Control Points (HACCP) procedures approved for each facility.

**Case-ready meats**
Meats that are processed and packaged in retail packaging at a centralized site for distribution to retail markets.

One of the compelling reasons for the production of case-ready meats is the quality of review and enforcement of sanitation standards that are possible. Oversight of these facilities

---

### What's in a Name?

Particularly in sushi bars and in offices monitoring populations of endangered species of fish, there is a need to know the specific kind of fish. Endangered species numbers are a concern to the planet and need to be monitored to avoid overfishing and extinction. The matter closer at hand is what kind of fish is a consumer actually buying?

Truthfulness in menu descriptions presented to customers ordering sushi is an issue in a restaurant or market, where prices vary according to the type of fish that is supposed to be the central ingredient. Profit is doubtless the motive when restaurateurs substitute a less expensive fish, but this action might be detected by a very knowledgeable customer, who would not eat there again. The truth is that many people probably would not realize that they were being cheated.

In the future, there may be a means of checking the kind of fish. A short piece of the DNA of a fish has been found that can serve as a type of bar code to identify the species. Although a 90-minute laboratory test is required to do the testing presently, the goal is to develop technology to the point where consumers could do the test while in a restaurant. False fish names might quickly depart from menus then.

> **FOOD FOR THOUGHT: Shrimp Ahoy**
>
> Shrimp from the Gulf of Mexico are a familiar commodity in fish markets, but a combination of a "dead zone" in the Gulf and imports from Latin America and Asia (from shrimp farms and the sea) is challenging shrimp fishermen in the Gulf. Shrimp have been dying in a huge area (dubbed the "dead zone") of the Gulf along the coast of Texas and Louisiana. Shrimp deaths in this area are due to lack of oxygen in the water, a condition that has varied over the past two decades.
>
> The problem can be traced to the fertilizers used on farms in the Midwest. Fertilizer residues and other pollutants ultimately drain into the Mississippi River and finally into the Gulf of Mexico, where algae thrive on these pollutants. The formation of huge blooms of algae depletes the ocean water of the oxygen that shrimp and other ocean species need to live, which creates the dead zone. The size of the dead zone varies from season to season. Surprisingly, the dead zone in the summer of 2003 was found to be much smaller, and fishermen were catching shrimp in surprising unusually large numbers.
>
> Credit for this environmental improvement was given to the powerful storms that whipped up the waters so vigorously that waters in the dead zone were oxygenated sufficiently for shrimp to flourish once again. Although these severe storms wreaked havoc on other sectors of the economy along the Gulf, they clearly sparked the shrimp market. Nevertheless, improvements in preventing pollution of the waterways also are necessary to protect the Gulf fishing industry.

can be done efficiently because the number of sites is quite small compared with the number of smaller meat markets throughout the nation.

Developments in preparing case-ready meats produce products that can be put on sale in display cases without additional handling and packaging by butchers in retail stores. Packaging materials and techniques are constantly studied as the industry works to extend the retail shelf life beyond the few days that are assured presently. Extended shelf life helps keep meat costs down. Labor costs also are kept to a minimum with case-ready meats because of the efficiencies in use of labor at the large processing and packaging centers.

## PREPARATION

### Changes Effected by Heat

**Water-binding capacity**
Amount of water held by muscle protein as bound water; heating reduces capacity.

Fat melts and proteins are denatured when meats are heated. The overall effect on palatability depends on the conditions used in heating the meats. Water is lost while cuts heat. Initially during heating, some bound water converts to free water as the **water-binding capacity** of the meat is reduced. This newly available free water offsets the water lost in the early period of cookery, and the meat remains juicy. When meat reaches temperatures between 74° and 80°C (165° and 176°F), the well-done stage, bound water is converted to free water very rapidly. However, the water loss exceeds the water available from this conversion, resulting in reduced juiciness.

Muscle fibers undergo changes in dimensions as a result of heating (Figure 15.13). Shrinkage in their width begins to occur soon after heating begins and is completed at a temperature of 62°C (144°F); water-binding capacity also is reduced. The narrowing width of fibers appears to result from the unwinding of the tertiary structure of the proteins, a change that is followed by cross-linkage of the coagulating proteins, causing shrinkage of the length. Fibers begin to shrink lengthwise at about 55°C (131°F) and continue until about 80°C (176°F).

Other changes in muscles ensue. One of the changes that can be noted under magnification is cracking in the I bands. These cracks widen to leave distinct gaps in the actin

**Figure 15.13**   Some shrinkage occurs during roasting, causing the ends of the bones to be more visible than they were before being heated.

filaments of the I band. Particularly in the biceps femoris and semimembranosus of round steak, cracks in the I band lead to a granular or mealy character because of the large amount of structural disintegration. In contrast, the longissimus dorsi of the rib and short loin tends to form an increasingly solid and less tender muscular mass as the temperature rises.

Muscle proteins in meat become less tender when heated. This toughening is a two-step process. When meat fibers are heated to between 40° and 50°C (104° and 122°F), myosin becomes less soluble, and hydration of myosin and other muscle proteins decreases. The second phase affecting the tenderness of muscle proteins occurs when they reach temperatures between 65° and 75°C (149° and 167°F).

Connective tissue in meats also needs to be considered in a discussion of the tenderness of cooked meats. Elastin is not modified, but collagen molecules slowly change when subjected to moist heat, because hydrogen bonds begin to break between the component tropocollagen strands. This permits some movement within the collagen molecules, and the gelatin components of collagen begin to move away from each other. Evidence for this formation of gelatin can be seen when drippings from a pot roast are refrigerated, causing the gelatin to form a gel.

Conversion of collagen to gelatin has a considerable effect on the tenderness of meat cuts heated for an extended period, which occurs when preparing less tender cuts. The length of time that meat is held above 65°C (149°F) is important in promoting collagen conversion to gelatin.

Clearly, the tenderizing effect of heating collagen for an extended period is opposed by the toughening effect of heat on muscle proteins. Optimal preparation of a particular cut requires a choice of cookery compatible with the meat's composition. If collagen content is high, as is true in less tender cuts, extended heating is desirable to permit considerable conversion of collagen to gelatin. This tenderizing action will more than compensate for the toughening of the muscle proteins that occurs at the same time.

Tender cuts of meat will become less tender with extended heating if the meat reaches temperatures above 60°C (140°F). This effect is the result of the toughening of the muscle proteins, a change that cannot be offset by the conversion of limited amounts of collagen to gelatin.

The importance of the opposing effects of collagen and muscle proteins during heating is seen particularly clearly in the preparation of fish. Only a small amount of collagen occurs in fish, which means that the major effect of heating is change in muscle proteins. By heating fish just until it flakes, some softening of collagen occurs to permit easy separation of fibers while some denaturation of the muscle proteins also occurs. At this point, the fish flesh is still tender. Continued heating increases toughness, the result of continuing detrimental changes in the muscle proteins.

Overcooking of fish causes considerable loss of palatability. The temperatures used for cooking fish are quite unimportant in determining the quality and overall palatability of the finished product as long as heating ceases as soon as the fish muscle can be separated into flakes. However, overcooking is more likely to occur when high temperatures are used because of the rapid rise in temperature of the fish.

***Dry Heat.***    Dry heat cookery methods (roasting, broiling, pan broiling, pan frying, and deep-fat frying) are designed to maximize the quality of muscle proteins, rather than considering the changes in collagen if heating were extended longer (Figure 15.14). Tender cuts of meat, because of their relatively high proportion of muscle protein and reduced quantity of collagen, are well suited to dry heat cookery. Only in roasting is there much opportunity to convert collagen to gelatin; cooking times for other dry heat methods are too short for effective conversion to occur. Recommended final interior temperatures are given in Table 15.5.

The usual temperature for roasting is 163°C (325°F). This temperature produces meats that are pleasingly juicy, tender, and flavorful, and the cooking losses are less than occur with an oven temperature of 218°C (425°F) or higher. Less tender roasts, such as top round, can be cooked at 93°C (200°F) to achieve a pleasing roast. The very long roasting period

**Figure 15.14**    Turkey placed breast down in a V-shaped rack is ready to be roasted (uncovered), a dry heat method.

**Table 15.5** Recommended Interior Temperatures for Selected Meat, Poultry, and Fish to Assure Safety

| Food | Final Temperature (°F) |
|---|---|
| *Meats* | |
| Beef, veal, lamb: | |
| Ground | 160 |
| Roasts, steaks | 145 (medium rare) |
| | 160 (medium) |
| | 170 (well done) |
| *Pork* | |
| Ground | 160 (medium) |
| | 170 (well done) |
| Roasts, chops | 160 (medium) |
| | 170 (well done) |
| *Ham* | |
| Fresh | 160 |
| Fully cooked (reheat) | 140 |
| Poultry (Chicken, Turkey) | |
| Ground | 165 |
| Whole, unstuffed | 170 (medium) |
| Whole, stuffed | 180 (well done)[a] |
| Breasts | 170 |
| Thighs, wings | Juices run clear |
| *Fish* | |
| All types[b] | 145 |

Food Safety and Inspection Service. USDA rule mandates safe handling statements for raw meat and poultry products. *FSIS Backgrounder,* May 1994: 3; National Fisheries Institute. News about seafood safety and consumer tips. 1994. Arlington, VA.
[a]Stuffing must be at least 74°C (165°F).
[b]Time often must be used because fish is likely to be too thin for accurate thermometry. Suggested times are baking at 232°C (450°F), broiling, steaming, grilling, or poaching—10 min/inch; boiled lobster—4–6 min/lb; shucked shellfish—until plump and opaque; raw shrimp—3–5 minutes (until pink); scallops—3–4 minutes.

required for the meat to reach an internal temperature of 67°C (152°F) enables the collagen to convert effectively into gelatin.

Broiling is a direct heat method in which meat is subjected to an intense heat until the desired degree of doneness is reached. Not surprisingly, the higher the internal temperature reached in the meat, the greater are the cooking losses. When thick pork chops are broiled, they are less juicy than comparable chops about half as thick. For broiled ground beef patties, juiciness and tenderness decrease, while flavor improves, as the interior temperature rises. Cooking losses are greater in broiling than in microwaving, but less than in roasting.

Pan broiling is similar to pan frying except that fat is drained frequently, rather than remaining in the pan as is done in pan frying. In deep-fat frying, meats are immersed in hot fat, usually at 190°C (375°F). Meats cook rapidly by these methods; deep-fat frying causes rapid browning on the exterior, but the interior may not have reached a safe temperature. It is essential to check the interior temperature with a thermometer.

Fast-food operations and some institutional settings use restructured meats, which often are prepared by either frying or grilling. Breading is especially important for palatability if deep-fat frying is the method chosen. Broiling, grilling, or roasting produces meats that are juicier and have a better texture than results from deep-fat frying.

Microwave cooking of meats results in greater cooking losses and less juiciness than when other meat cookery methods are used. This may be due to extreme tightening of protein molecules, which could force water from the meat. A potential advantage of microwave cookery of meat is reduced fat content, possibly the result of comparatively rapid heating of fat in the inner portions, which may facilitate drainage of fat from the cut. Fat, particularly fatty acids with double bonds, is agitated even faster than water when subjected to microwaving. The flavor of the outer areas of meat prepared in a microwave oven is not as appealing as comparable cuts that have been oven roasted.

Although microwave cookery of meat is essentially a dry heat method, heating of the meat occurs in a somewhat different manner than in other methods of dry heat meat cookery. Microwave heating is accomplished when microwaves cause oscillation of water and fat molecules in the meat (see Chapter 6). This heat is distributed farther into the interior of the cuts by conduction, a process that requires time. Ordinarily, meats heated in a microwave oven are prepared using a moderate setting, which automatically alternates periods of microwaving with standing time to provide the opportunity for conduction and to enhance the equalization of heat in the cuts.

One of the shortcomings of meats prepared in a microwave oven is the unattractive gray surface. Tenderness also is affected adversely. Surprisingly, microwaved meat cuts usually are less tender than comparable cuts prepared by roasting and broiling even though microwaving reduces the cooking time. Bacon can be microwaved successfully, perhaps because of its high fat content.

The particular benefit of a microwave oven in meat cookery is for reheating meats that have been cooked previously. The flavor of meats reheated in a microwave oven is more appealing than that of meats reheated by traditional methods.

***Moist Heat.***   Moist heat, either braising or stewing, is designed to provide sufficient time for collagen to be converted to gelatin without toughening the muscle proteins unduly. The liquid in which the meat is braised or stewed prevents the surface of the meat from becoming hot enough to dry and brown excessively (Figure 15.15). Sufficient heat input is needed to maintain the liquid at a simmering temperature or even at a gentle boil, so the meat will be hot enough for collagen to begin to unwind and separate slowly into molecules of gelatin. The likelihood of evaporating all of the cooking liquid with this controlled rate of heating is minimal. As long as water is present, the meat cannot get hot enough to burn and toughen extensively.

Less tender cuts of meat that are braised or stewed until they reach and maintain an interior temperature of about 98°C (208°F) for about 25 minutes will be fork tender. Cuts that

**Figure 15.15**   Less tender cuts of meats become tender when prepared using moist heat stew, as was done to make this reindeer stew.

are high in connective tissue are well suited to this type of meat preparation because of their comparatively high collagen content. The increased tenderness resulting from the conversion of collagen more than offsets the toughening in the muscle proteins. Tender cuts of meat are not well suited to prolonged moist heat cookery; the comparatively high amount of muscle protein is toughened by the heat and tends to counteract the effect of collagen conversion.

Poaching or steaming can be used to prepare fish, but the preparation time needs to be just long enough to coagulate the muscle proteins. There is so little connective tissue in fish that the muscle proteins clearly are the dominant type of protein to consider.

Crock pots or slow cookers are small electric appliances designed specifically for moist heat cookery. The temperature reached in different models varies somewhat but ordinarily is less than 107°C (225°F). Several hours are required for the interior of the meat to be heated sufficiently to tenderize the connective tissue and kill microorganisms that may be present. Meats prepared in a slow cooker are palatable, and tenderness scores are better than when a pressure saucepan is used.

In contrast to the crock pot, the pressure saucepan speeds moist heat meat cookery because the pressurization results in a hotter temperature for braising than can be achieved in a regular Dutch oven or other nonpressurized, covered pan. When braising is done in a pressure saucepan, the meat is less juicy than when a nonpressurized pan is used. Flavor and overall acceptability are not altered by use of a pressure saucepan for braising.

Sometimes meats are roasted in aluminum foil or in special roasting bags. These devices trap moisture around the meat, changing roasting from a dry heat cookery method to a moist heat one. At the oven temperatures ordinarily used for roasting, 150–163°C (302–325°F), wrapping the meat in aluminum foil increases cooking time significantly. Compared with roasts prepared without foil, foil-wrapped roasts prepared in an oven at 150°C (302°F) are less pleasing in flavor and are also less juicy and tender. Roasting bags produce results similar to those obtained using a foil wrapping.

*Cooking Losses.*   **Cooking losses**, although frequently discussed as a single entity, actually are the combination of evaporative and drip losses. **Drip losses** include both juices and melted fat. **Evaporative losses** are calculated as the difference between the weight of the uncooked meat and the weight of the cooked meat plus drippings.

Because of both the relatively high cost of meats and the reduced juiciness associated with losses during cooking, the effect of the method of meat cookery on cooking losses and the resulting impact on yield is important. Use of aluminum foil or film wrapping when heating meats causes greater cooking losses than occur when the wrapping is omitted. Cooking losses are greater for meats prepared in a slow cooker than when a faster cooking method is used.

Predictably, the higher the final temperature of meat, the greater is the cooking loss. Cooking losses for sirloin roasts heated in a convection oven are less when the oven temperature is 93°C (200°F) than when it is 149°C (300°F), and the yield also is greater, as would be anticipated. In a conventional oven, cooking losses are lower at 125°C (257°F) than at 163°C (325°F), but the long roasting period required for the lower temperature makes the higher oven temperature the one ordinarily used.

## Effects of Altering pH

Hydration of meat is important to evaluation of juiciness in the cooked product. Adding an alkaline ingredient darkens the color of the meat by increasing pH and minimally influencing hydration. Increased tenderness does not develop, so the addition of soda is not recommended. Adding an acid is another possibility; marinating a less tender cut in undiluted vinegar for two days can result in increased juiciness and tenderness when the meat is braised. Possible negative effects on aroma, flavor, and acceptability may offset the improved tenderness and juiciness.

**Cooking losses**
Total losses from meat by evaporation and dripping during cooking.

**Drip losses**
Combination of juices and fat that drip from meat during cooking.

**Evaporative losses**
Losses of weight from meat during cooking as a result of evaporation.

## Effect of Salt

The main effect of salt on meats during preparation is enhanced water retention. This ability to hold water in the meat improves juiciness. It may have a minor role in promoting tenderness, but the overall impact of salt on palatability of meat is much too minor to outweigh the health advantage of avoiding use of excess salt.

## Meat Tenderizers

**Papain**
Single proteolytic enzyme from papaya or a blend of three enzymes from this fruit.

*Enzymes.*    Certain proteolytic enzymes can increase the tenderness of less tender cuts of meat. The most common of these is a commercial blend of enzymes from papaya and salt, a blend that is referred to simply as **papain**. The three enzymes in this substance are chymopapain, papain, and a peptidase. Papain is applied to the surface of the meat, and then the meat is pierced repeatedly with a fork to help carry the enzymes into the interior. Unless piercing is done, the enzymes will tenderize only the surface of the meat and a very short distance (no more than 2 millimeters) into the muscle because of the limited penetrating capability of the enzymes.

Papain has little effect at room temperature, but it becomes active when the temperature of the meat reaches 55°C (131°F) and increases in activity with additional heating to even 80°C (176°F). Activity ceases when the enzyme is denatured by heat; it is definitely inactive at 85°C (185°F).

Much of the tenderizing effect is the result of the enzyme destroying the sarcolemma surrounding the myofibrils in the fibers, hydrolyzing actomyosin, and then continuing hydrolytic breakdown of various proteins in the fiber. Collagen also may be hydrolyzed to contribute still further to the tenderizing effect. The result of this enzymatic action often is the development of a somewhat mushy texture in regions where the enzyme has acted. This is true whether or not the enzyme has been allowed to stand on the meat for a period before cooking, because the enzyme exhibits its major action in the hot meat.

**Bromelain**
Proteolytic enzyme in pineapple.

**Ficin**
Proteolytic enzyme in figs.

Although papain is the principal enzyme used for tenderizing meats, other proteolytic enzymes also can be utilized for this purpose. For example, **bromelain** is an enzyme found in fresh pineapple. Its action sometimes occurs when the fresh fruit is an ingredient in recipes such as those for kabobs or stir-fried chicken. Bromelain is inactivated between 77° and 82°C (170° and 180°F). **Ficin**, a proteolytic enzyme in figs, is another possible enzyme for tenderizing meat.

*Mechanical Tenderizing.*    Commercially, meat can be run through a tenderizer equipped with needles or blades to cut some of the connective tissue and increase tenderness. This makes it possible for some less tender cuts to be used in the same way as tender cuts. Mechanical tenderization changes the texture somewhat, but it does not produce the mushy character sometimes found in cuts tenderized by enzymes.

Another means of mechanical tenderization is pounding with a meat hammer to break some of the muscle fibers and connective tissue. Yet another technique, one a bit more rigorous than pounding, is cubing. In cubing, the meat is passed through a machine that cuts through a fair portion of the muscle, a process that may be repeated to increase tenderness still more.

The ultimate means of mechanical tenderizing is a meat grinder, which is used to make ground meats. This intensive shearing of the fibers and connective tissue results in very tender meat from less tender cuts.

# MODIFIED MEAT PRODUCTS

## Reduced-Fat Meats

Consumer interest in weight control and cholesterol has created a demand for meats and meat products with reduced fat levels. An early approach toward satisfying this demand was modification of the beef grading criteria to enable beef with a comparatively low fat content to be graded as USDA Select. Now, the meat mixtures themselves are modified to formulate ground meats with a fat content as low as 10 percent.

## Restructured Meats

Portion control and quality control are two major problems of purchasing meat for institutional use. **Restructured meats** provide the answer to these problems and a challenge to food technologists. Aspects of production that require careful attention include the size of the protein flakes, the amount of fat, and the content of connective tissue.

Flakes no larger than 6 millimeters produce a more desirable restructured meat than flakes a little more than twice that size. Large pieces of connective tissue are detrimental to the quality of the product. Although the resulting products are not identical to regular meat cuts, restructured meats are sufficiently pleasing to have gained a strong entry into the food-service industry and even into the home.

**Restructured meats**
Meats made from meat cuts that are somewhat less expensive; made by creating small particles, adding fat and other ingredients, and shaping into uniform portions.

## Comminuted Meats

Meats can be chopped into tiny pieces, mixed with water, and heated to denature the proteins. Phosphate, sodium chloride, and other salts may be added to improve the physical characteristics of the meat mixture and increase sensory qualities such as juiciness and texture. Modifications in fat type and amount are emerging to expand the acceptability of these **comminuted meats**, which include hot dogs and sausages.

**Comminuted meats**
Products made by almost pulverizing meats and adding the desired fat and salts before heating the resulting mixture.

## Structured Seafood Products

Public interest in nutrition is reflected by increased consumption of fish, a change that has stimulated efforts to provide new types and products in the marketplace. One of the products of fish processing is minced fish meat, which is washed thoroughly to eliminate fat, pigments, and other compounds that would present flavor and storage problems during frozen storage. A preservative of some type, commonly sorbitol or sugar, is added before minced fish is frozen in preparation for its use in structured seafood products. This intermediate seafood product is called **surimi** (Figure 15.16).

**Surimi**
Purified and frozen minced fish containing a preservative; intermediate seafood product used in making structured seafood products.

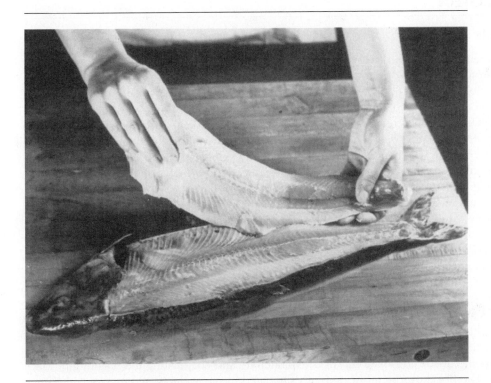

**Figure 15.16**   Fish being filleted. Filleted white fish, usually pollack, is used to make surimi. (Courtesy of the U.S. Department of Agriculture.)

Technically, surimi could contain any type of minced fish. However, pollack is the predominant choice; other types of fish (especially the less popular fish that are caught along with more costly types) are included as available. Preparation of surimi requires deboning of either filleted or headed and gutted fish; filleting (Figure 15.16) results in a higher grade surimi with lighter color and better gel-forming ability.

Mincing of the deboned fish with a drum that has small perforations (1–2 millimeters) produces a particularly high-quality product. The minced fish requires careful, thorough washing, rinsing, straining, and some dehydration to achieve a moisture level of around 82 percent. The last step prior to freezing is incorporation of the preservative or cryoprotectant.

The principal use of surimi for human food has been in fabricating structured seafood products, notably crab and shrimp analogs. As minced fish alone does not have the textural characteristics associated with crab and shrimp, egg white and a starch (the type used depends on the product being made) are added to the surimi to achieve the desired firmness without rubberiness. Each of these two ingredients appears to regulate the influence of the other, with starch promoting rubberiness and egg white interfering with the structural matrix that contributes a rubbery texture to the original surimi. A limited amount of oil (not more than 4 percent) is added to improve the freeze-thaw characteristics of these structured seafood products.

The mixture undergoes fiberization to achieve an end product that rather closely approximates the texture of crabmeat or shrimp. This mixture is extruded as a sheet and heated briefly so that the sheet is elastic and can be folded and molded. After the extruded mixture is in the desired final form, it is heated sufficiently to gelatinize the starch and completely denature the protein without causing excessive toughness and rubberiness. The surface is colored to simulate crab and shrimp products.

Considerable research effort has been directed toward the development of surimi-based structured seafood products, and these products have become an accepted choice by many consumers. The distinct economy of the shrimp and crab analogs and their present level of quality have made them viable fish products.

## SOY PROTEIN

### Products

Plant proteins have been of continuing research interest for many years as alternatives for animal proteins. So far, soy protein has proven to be by far the most versatile and economically feasible of the plant proteins that have been studied. At the present time, several soy protein products are available to consumers to add variety in flavor and texture to many main dishes and other recipes (Figure 15.17). Dried soybeans are remarkably high in protein, actually about 34 percent. The blend of essential amino acids is unusually complete for a plant protein, which is an important reason why soy protein is recommended by nutritionists and dietitians as an excellent means of meeting part of the day's protein needs. Methionine is the limiting essential amino acid in soy protein.

Soybeans undergo several steps in their processing to obtain the products seen in the marketplace. After the beans are dehulled, the oil is extracted, and the remaining portion is soy flour, with a protein content of about 50 percent. Additional steps utilizing alkali and acid at various points to alter pH result in a variety of **soy protein concentrates** and **soy protein isolates**, with protein levels ranging from 70 percent in the concentrates to as high as 95 percent in the isolates. The concentrates are important in formulating vegetarian meat alternatives, often in conjunction with cereals; isolates are particularly well suited for use in beverages and energy bars.

From the isolates, texturized soy protein products are made by various processes including the extrusion of cooked soy flour and spinning fibers; the latter is accomplished by forcing the dissolved isolate through a spinnerette and coagulating the resulting product as fibers. The fibers can be fabricated into simulated meat products, sometimes with rather good success. NutriSoy Next™ is a textured extruded meat alternative combining soy and vegetable protein that is suitable for use in a variety of vegetarian products. Imitation bacon

**Soy protein concentrate** Defatted soy product usually containing about 70 percent protein.

**Soy protein isolate** Defatted, highly concentrated (up to 95 percent) soy protein; used to make many textured soy products.

**Figure 15.17**   Dried soybeans are made into a wide array of food products. (Courtesy of Agricultural Research Service.)

bits of textured soy protein are a familiar example. **Textured soy protein (TSP)** is also called **textured vegetable protein (TVP)**.

    Although textured vegetable protein is the form of soy used most frequently, soy flour and soy grits also have some specific uses. **Soy grits** are made by grinding defatted flakes of soy to a distinctly coarse particle size. These are incorporated into some commercial food products to alter textural characteristics of some ground meat products and cereals.

    If grinding is continued so that a fairly fine powder is formed from the soy flakes, the product is **soy flour**, and its protein content is 50 percent. For optimal shelf life, defatted soy flour is suggested. This type of flour is used as an ingredient in some bakery products because of its nutritional merits and favorable influence on tenderness and crust color. However, it needs to be used at levels no greater than 3 percent of the weight of wheat flour in some baked products and certainly no higher than about 10 percent in doughnuts. The lack of the structural protein complex (gluten) limits the usefulness of soy flour in batter and dough mixtures, making it necessary that enough wheat flour is incorporated to ensure adequate strength of structure.

    A different type of soy product is made by forming a curd from soy milk, a process rather similar to making cheese from animal milks. The result is a bland, slightly spongy precipitated soy protein food called **tofu**. Tofu is made in three different forms (Table 15.6) commonly available to provide the texture needed for easy use in a wide variety of recipes. Firm tofu is the most concentrated, dense form and is well suited to recipes that use tofu cubes. Soft tofu has a high enough moisture content to assure easy blending into soups or

**Textured soy protein (TSP) or textured vegetable protein (TVP)**
The end product of a series of steps producing fibers from soybeans.

**Soy grits**
Coarsely ground soy flakes.

**Soy flour**
Finely ground soy flakes.

**Tofu**
Soybean curd.

**Table 15.6**   Tofu Products and Their Uses

| Product | Calories from Fat (percentage) | Uses |
|---|---|---|
| Firm tofu | 45 | Cubes |
| Soft tofu | 52 | Soups, sauces |
| Silken | 30 | Toppings, custard-like desserts |

sauces. Silken tofu, with its custard-like texture, also blends well or can be served with a simple topping. Nutritionally, silken tofu is interesting because it has a fat content providing only 30 percent of the calories, compared with fat calories of 45 percent from firm tofu or 52 percent from soft tofu. Tofu needs to be kept covered with water during refrigerated storage, and the water needs to be changed every day.

Tofu is a common ingredient in many Asian recipes and is included to add variety in texture and flavor, as well as to increase the protein content of the dish. This meat alternative may be used effectively as large, distinct cubes in main dishes, soups, and salads. It also may be puréed so that it blends smoothly with other ingredients in sauces or even in desserts such as cheesecake. Yet another illustration of the versatility of tofu in food preparation is a frozen dessert that is the soybean counterpart of ice cream.

A few types of soy cheeses are available—a mozzarella simulation is one. These products contain approximately 6 g of protein per ounce, part of it caseinate from cow's milk. The isoflavone content from soybeans is 6–7 milligram per 100 g of soy cheese.

Fermented soy products are popular in Asian cuisines. **Tempeh** is a chewy, cake-like product made by cooking soybeans and then adding a culture to ferment the beans. **Natto** also is fermented after cooking whole soybeans. Fermentation results in a smelly, viscous (almost cheese-like) product for use in soups or as a spread.

**Miso** is a pasty condiment made by allowing soybeans, a mold, and salt (often with rice or other grain) to age together for as long as three years. This salty fermented soy product is popular in Japan to season soups, sauces, and other dishes.

**Tempeh**
Fermented cooked soybean product resembling cake.

**Natto**
Fermented cooked soybean product useful as a spread or in soups.

**Miso**
Paste made by fermenting a mold culture with soybeans, salt, and often rice for up to three years.

## Using Textured Soy Protein

Although textured soy protein products can be made into meat analogs, a different application as a meat extender has gained reasonable acceptance. By mixing some textured soy protein with ground meats, a given quantity of meat can serve an increased number of people, thus reducing food costs. The amount that can be used has been studied to determine just how much soy can be added successfully. An important contribution made by the added soy protein is improved juiciness in comparison with all-beef patties. However, flavor and texture are influenced negatively when the level of soy protein is close to 20 percent.

The comparatively low cost of soy protein as contrasted with that of beef makes the use of textured soy protein as an extender attractive when beef prices are high. Federal school lunch programs are permitted to extend meat by using up to 25 percent textured soy protein in ground meat mixtures. Many sausages, chicken nuggets and other formed poultry, and ground meat products are made with soy protein isolates and concentrates to enhance juiciness and increase yield.

The water-binding capability of textured soy protein and other soy products enhances juiciness, in part because of water content. However, fat that may be present in a beef patty or meat loaf made with soy protein will be bound by the soy protein and retained in the final product. This retentive characteristic limits the usefulness of TVP as a meat extender; retained fat cannot be removed, which may mean the final product has a higher fat content than if the fat were carefully removed from a 100 percent beef patty or loaf.

## Quorn™

**Mycoprotein**
Protein produced by a fungus.

**Quorn™**
Trade name for products based on hyphae harvested from *Fusarium venenatum* mixed with a binder and then heated and frozen.

An innovative **mycoprotein** suitable for making meat analogs was developed in the United Kingdom and marketed with the name **Quorn™**. Researchers there harvested the hyphae from *Fusarium venenatum,* a fungus growing in a field. Production includes controlled fermentation with glucose as the medium, followed by harvesting of the fibrillar hyphae. The hyphae tubules are mixed with a binding agent. Finally, the mixture is heated and frozen in a patented process that develops a texture similar to that of meat. A rather wide range of entrées can be found in some markets, particularly those catering to vegetarians and others seeking alternative protein sources.

---

### FOOD FOR THOUGHT: Petri Burgers?

Imagine a burger that started in a Petri dish, a product of bioengineering by scientists merging medical research frontiers with food science. Will it ever happen? The answer to that question is far from certain at the present time, but research efforts in laboratories in Australia, Europe, and North America are developing techniques for growing muscle cells into tissues.

Researchers at the University of Western Australia and Harvard University produced a thin monolayer of cells that are nourished in a bioreactor that maintains a sterile environment. These cells then need to be developed into a thicker cluster. Some researchers at the International Institute of Biophysics in the Netherlands have explored developing a scaffold or matrix approach to facilitate development of clusters to give dimension to the cell cluster beyond the monolayer. The ultimate goal of this type of research for medicine is to synthetically produce organs.

Another possible future development could be muscle tissue that could serve as a food source, thus providing a protein source that would eliminate current concerns such as mad cow disease. Far into the future, space travelers may benefit by being able to carry a regenerating meat source without having to solve the problems of transporting animals. It all sounds like science fiction today, but basic research undertaken today may provide answers to people in the future.

---

## GELATIN

### Composition

**Gelatin** is a stabilizing and gelling agent that has a variety of applications in the food industry as well as at home. Although it must compete with an array of carbohydrate gums in the food industry, this unique protein remains unchallenged in home food preparation. The origin of gelatin is the tropocollagen strands of collagen, a protein in the connective tissue of meat that undergoes a change during heating (see p. 331).

**Gelatin**
Protein derived from collagen via tropocollagen when heated for an extended period; useful in gel formation.

### Properties

*Hydration.*    When gelatin is used in food preparation, the first step is to convert it from a firm, friable substance to one that is soft and pliable. The elongated gelatin molecules have many polar groups exposed, making it possible for water to be bound by hydrogen bonds at many points along each molecule. As water is bound, the gelatin swells noticeably. This can be observed easily by placing a packet (tablespoon) of gelatin in one-fourth cup of water. The remarkable ability of gelatin to bind water and also to trap free water in the interstices between molecules enables a small amount of gelatin to hold a far greater amount of water in a very short period. Evidence of this is the swollen volume of the hydrated gelatin and the lack of water that can be poured off.

The swelling of gelatin when it adsorbs water is influenced by the pH and the presence of salts. Swelling is bimodal in that it is less at its isoelectric point than it is at either a more acidic or a more alkaline pH. For many gelatins, the pH at which hydration is at a maximum is around pH 3.2–3.5 or 9. The effect of different salts on hydration is variable, but salts do promote swelling at the isoelectric point of gelatin. Unfortunately, no specific isoelectric point can be given for gelatin because of its heterogeneous nature and the fact that different production methods influence the pH of the isoelectric point. Usually, the isoelectric point for alkaline-processed gelatins is between 4.75 and 5.2 and that for acid-processed gelatins is between 5.5 and 6.5.

***Sol Formation.***    Unless gelatin has been hydrated first in cold water, a gelatin sol can be derived only with considerable difficulty when the hot liquid is added. After hydration in cold water, the swollen gelatin can be dispersed readily in hot water or other hot liquid. Apparently this ability to be dispersed as a sol is aided by the physical dissociation of the gelatin molecules from their closely packed, dehydrated state.

The distancing that occurs between gelatin molecules during hydration in cold water weakens intermolecular association and enables hot water to complete the dissociation, allowing the long gelatin molecules to move freely as the discontinuous phase in the sol. The high temperature facilitates the breaking of weak hydrogen bonds that exist in the hydrated gelatin.

For gelatin to be an acceptable ingredient in any food product, the individual molecules must be dispersed in a very hot liquid. When molecules adhere to each other rather than dispersing, their fibrous and compact nature makes them quite rubbery and tenacious, characteristics that definitely are undesirable in foods. For this reason, hot gelatin sols are stirred until they appear to be absolutely transparent and totally homogeneous. If this state is not reached with a reasonable amount of stirring, it may be necessary to heat the sol longer to provide the energy needed for dissociation of the molecules of gelatin.

***Enzymatic Hydrolysis.***    Various proteolytic enzymes are able to cleave the long gelatin molecules into shorter polypeptides. This change in molecular length quickly eliminates the usual ability of gelatin to form gels. If ingredients containing enzymes capable of catalyzing this proteolysis are incorporated into gelatin mixtures, a satisfactory gel cannot form. Among the foods containing these enzymes are papaya, which contains papain; pineapple, which contains bromelain; figs, which contains ficin; and kiwi fruit, which contains **actinidin**. These enzymes lose their catalytic capability if they are heated until denatured, which explains why canned pineapple can be used in gelatin salads whereas frozen and fresh pineapple cannot.

**Actinidin**
Proteolytic enzyme in kiwi fruit.

***Gel Formation.***    Under appropriate conditions, a gelatin sol can convert to a gel as the sol is cooled. The formation of a gelatin gel is only one familiar example of gel formation. In the case of a gelatin gel, the gelatin molecules cross-link to form a continuous network of solid protein to which some water is bound and in whose interstices additional water is trapped as the discontinuous phase.

The formation of a gelatin gel is endothermic and occurs gradually as the energy of the system dissipates. A surface film forms as some of the gelatin molecules cross-link in a comparatively compact configuration. When the interior begins to gel, the molecules of gelatin are organized quite randomly.

There is considerable formation and subsequent disruption of secondary bonds and reformation of new bonds with the molecules in slightly different positions as the gel structure ages. Gradually, a somewhat more organized arrangement evolves in a gelatin gel that has been stored many hours. In other words, a gelatin gel actually is a somewhat dynamic colloidal dispersion and is subject to gradual change, which is evidenced by decreasing tenderness during storage.

Conversion from a gelatin sol to a gel depends on several factors. First, enough gelatin molecules must be present to cross-link through the entire system to form the continuous network. If the concentration is too dilute, the sol will remain in that state. Even when there is enough gelatin to form a gel, the concentration influences gel formation. As the concentration increases, the rate of gel formation also increases—that is, the gel forms more quickly with an increasing concentration of gelatin.

It is necessary to avoid too high a concentration of gelatin because the concentration influences the texture of the gel formed as well as the rate of its formation. Increasing the concentration of gelatin causes the gel to become increasingly firm and less tender. Too high a concentration is undesirable because of the rubbery consistency that develops. A satisfactory gel often can be formed with a concentration of only 1.5 percent gelatin by weight, but in some other systems, the level of gelatin may need to be about 3 or even 4 percent. An average figure of about 2 percent is appropriate for many applications.

The temperature to which a gelatin sol must be cooled for gelation to occur is influenced by the rate of cooling. The temperature of gelation will be lower if the sol is cooled rapidly by adding ice to the dispersed gelatin or by packing the bowl containing the sol in ice than if the gel is cooled slowly at refrigerator temperature. Despite the fact that the gel sets at a lower temperature when ice is used, the time required for gelation is less than is necessary when cooling occurs in the refrigerator because the temperature drops quickly as the ice melts.

A gelatin sol with a high enough concentration of gelatin can be gelled in about three hours by allowing it to cool at room temperature, a rate that is obviously slower than that occurring in the refrigerator. Even when the gelatin concentration is very high, a gelatin sol must be cooled to at least 35°C (95°F) before a gel can form. The energy of the system is too great to permit the necessary bonding between molecules if the sol is above that temperature, and there is inadequate stability to any bonds that might be able to form.

Stability of the gel is influenced by the rate at which the gel forms. Gelatin gels that are set in the refrigerator require considerable patience because of the long time required for them to set compared with those using ice to cool. They do have the advantage that they are much more resistant to melting back to a fluid sol when they are served on a warm day. This method of gel formation may be appealing if the gelatin gel must be held at room temperature or warmer for a long period, as often is true at buffets or picnics.

Fortunately, for people with severe restrictions on their preparation time, addition of ice cubes to the dispersed gelatin or chilling of the gelatin in a bowl placed in ice water makes a gel that remains sufficiently firm for easy service at most meals if the product is made far enough in advance to allow time for the gel to become stronger by cross-linking during storage.

Use of milk as the dispersing liquid in place of water produces a stronger gelatin gel than does a comparable amount of water. The increased strength may result from the interactions of the gelatin with the milk proteins and salts. This is of particular interest when gelatin is added as a stabilizing agent in ice cream, and it also has application when cream cheese, cottage cheese, and other dairy products are incorporated into a gelatin salad.

To a limited extent, variations in the pH of the gelatin system influence the strength of the gelatin gel formed. Greatest strength is found in systems between pH 5 and 10, but the flavor generally is particularly pleasing at pH 3–3.5, significantly more acidic than the isoelectric point of gelatin. Fortunately, gel strength is adequate at this range.

Sugar frequently is added to temper the acidic taste of tart gelatin mixtures. The effect of sugar on gelation time and gel strength depends on the concentration of sugar. In concentrations up to 0.02–0.03 M, sugar delays gelling. At levels above 0.1 M (68.4 g or about $\frac{3}{8}$ cup/liter), sugar has the opposite effect and actually speeds gelation. This level is less than half the amount of sugar that frequently is added to sweetened and acidified gelatin sols. Therefore, it is reasonable to anticipate that gelatin sols containing sugar ordinarily set more rapidly than they would without the sugar. However, these large amounts of sugar in gelatin do increase the tenderness of the resulting gel.

Gelatin gels are thixotropic—that is, they can revert to a sol when agitated. This reversal is due presumably to the breaking of hydrogen bonds between gelatin molecules in the gel. This is seen when fruit or another ingredient is stirred into a gelatin mixture that has gelled enough to pile; it softens and becomes quite smooth with the agitation necessary to blend the fruit uniformly into the entire mixture.

Gels also can be reversed to sols if the temperature rises sufficiently. The reversibility of gelatin gels to sols and back to gels can be demonstrated repeatedly, for the hydrogen bonds and other secondary bonds that may be responsible for establishing the gel structure can be broken by providing sufficient energy and re-formed by removing energy, specifically by cooling. Interestingly, the temperature required for gelation to occur is somewhat lower than the temperature required for reversal to form a sol. One of the curious aspects of gelatin is that gelation occurs more rapidly the second time the product is gelled.

Gelatin can be beaten to form a foam if handled properly. The excellent foaming properties of this protein make it possible for gelatin to increase as much as threefold in volume.

The optimal time to beat gelatin into a foam is when the dispersed gelatin sol has cooled so that the mixture has the consistency of a thick syrup. The surface tension is low enough so that the gelatin can be spread into extensive thin films surrounding bubbles of air, and the cooled gelatin will congeal quickly to add stability to the cell walls in the foam. This capability of gelatin is used in preparing some stabilized foam desserts and whipped gelatin salads. Beating must be done before the gelatin actually congeals because the gelatin will then be so brittle that it will break into pieces and will not be capable of being spread to form the cell walls needed for foam formation. Fortunately, gelatin can be warmed a bit when this happens so that the problem can be rectified.

Gelatin is used in numerous recipes, most frequently as the background matrix for a congealed salad with various other ingredients added to give variety. In some cases, plain granulated gelatin is used; other recipes utilize pulverized, sweetened gelatin with coloring and flavoring added. Plain gelatin requires hydration with cold water prior to dispersal in a hot liquid. Otherwise, it clumps badly.

The finer consistency of the sweetened gelatin product and the dilution of the gelatin by the sugar enable this type of gelatin to be dispersed directly in a hot liquid without preliminary hydration. Yet another gelatin product is sweetened with a sugar substitute to reduce the calories. This product also does not require preliminary hydration.

## SUMMARY

Flesh foods are categorized as meat, poultry, or fish; the latter is divided into fish and shellfish (mollusks and crustaceans). These foods contain both muscle and connective tissue. The proteins in muscle consist of enzymes and the abundant myofibrillar proteins. The principal myofibrillar protein is myosin, a long and thin molecule. Actin, another myofibrillar protein, is found as F-actin (heavy and long) and G-actin (small subunits that can aggregate to make F-actin). The other prominent myofibrillar protein is tropomyosin, a comparatively short and small molecule. In the presence of ATP, calcium, and magnesium ions, actin and myosin unite to form actomyosin in a reaction that is reversible and is accompanied by contraction and relaxation of muscles (when actin and myosin are released from actomyosin).

Connective tissue also is composed of proteins, including elastin, reticulin, ground substance, and collagen, the most important in meat cookery. Tropocollagen, the basic component of collagen, is made of three fibrous strands. Three strands of tropocollagen, in turn, are twisted together to form long and fibrous strands of collagen. Hydroxyproline and proline, amino acids present in unusual abundance in tropocollagen and collagen, account for the fibrous nature of these proteins. Elastin is the yellow, tough connective tissue that is found to a small extent intramuscularly and also in some large deposits intermuscularly. Ground substance provides the protein matrix in which collagen and elastin are deposited to form connective tissue. Reticulin is another fibrous protein associated with a fatty acid in connective tissue.

Triglycerides containing a variety of fatty acids (particularly palmitic, oleic, and stearic) are the predominant form of lipids in meats. Others are cholesterol, glycolipids, phosphoglycerides, plasmalogens, and sphingomyelin. These various components constitute the fat depots in which they are embedded in a matrix of connective tissue.

Organization of muscle tissue is complex, beginning with thick and thin myofilaments consisting of myosin and actin, respectively. Myofibrils are formed by the orderly alignment of myofilaments into sarcomeres, which then are organized into the larger myofibrils. Z, I, H, and A bands in muscle tissue are the result of the overlap of thick and thin myofilaments in a rather organized fashion in the sarcomeres. A thin membrane (sarcolemma) encases many myofibrils that are surrounded by sarcoplasm to form a fiber. Connective tissue, the endomysium, encases the fibers. Several fibers are held together in bundles by more connective tissue, the perimysium. Bundles of these bundles are finally encased in still more connective tissue, the epimysium, to form the completed muscle.

Hemoglobin and myoglobin are the two principal iron-containing pigments in red meats. Myoglobin, in its various forms, is particularly important. In the ferrous (2) form, it is the purple-red color of fresh meat. Exposure to air adds two atoms of oxygen to make oxymyoglobin, a bright red pigment. Metmyoglobin, a brownish-red pigment, forms if the iron is oxidized to the ferric (3) state. Heating causes gradual changes in pigment to the grayish-brown compound, denatured globin hemichrome. With the addition of nitrites, nitric oxide myoglobin forms during curing of meats and undergoes conversion to nitric oxide myochrome when heated, resulting in the stable reddish color of cured meats. Another pigment found in cured meats is nitrosylhemochrome. Poultry and fish generally have little pigmentation, although poultry may have some reddish color from hemoglobin, and fish may have some dark muscles colored by myoglobin. Salmon red is the result of astaxanthin, a carotenoid pigment.

Prominent among changes after slaughter is development and passage of rigor mortis. The rate of onset and passage of rigor is the result of the species and the physical condition at the time of slaughter. Quality of the various meats is judged on the basis of texture, marbling, and overall palatability; yield is based on the amount of muscle in relation to bone and fatty deposits. Meat cuts can be identified on the basis of the size of the cut, color, muscles present, and bone shape.

When heated, muscle fibers shrink a bit lengthwise and lose some of their water-binding capacity. Concurrently, connective tissue (specifically collagen) begins to be converted slowly to gelatin. An acid marinade is of limited benefit in promoting palatability of less tender cuts of meat.

Meat tenderizers are somewhat effective in destroying the sarcolemma surrounding fibers and in hydrolyzing actomyosin, as well as possibly hydrolyzing some of the collagen. The overall effect may be to create a somewhat mushy texture in some areas. This action occurs during the heating of the meat. Mechanical devices can tenderize less tender cuts of meat prior to heating.

Dry heat methods for tender cuts include roasting, broiling, pan broiling, pan frying, deep-fat frying, and microwave cookery. Less tender cuts may be prepared effectively by braising and stewing; fish may be poached. Cooking losses vary with various factors such as temperature and method of heating, as well as final temperature of the meat.

Textured soy protein can be made into meat analogs such as imitation bacon bits and can also be used to extend ground meats. Soy grits and soy flour are used to some extent in the baking industry to enhance the protein content of baked products. However, they can replace only about 3 percent of the wheat flour in most baked products, whereas a substitution of from 20 to 25 percent can be made when textured soy protein is used to extend meat. Other analogs use surimi, a minced fish intermediate product that usually contains pollack and other available fish. Tofu, tempeh, miso, natto, and cheese are other soy products.

Gelatin is the protein derived by extraction from the collagen obtained from animal skins and bones. The molecules are somewhat varied but are fibrous as a result of the high content of proline and hydroxyproline. Commercial gelatin usually is obtained by alkaline extraction and finally is dried and marketed as granular or pulverized gelatin. Plain gelatin requires hydration in cold water to disperse and swell. Water is bound and trapped in the interstices between gelatin molecules. When hydrated gelatin is dispersed in hot water, it forms a sol, which on cooling forms a gel. The presence of proteolytic enzymes in a gelatin gel results in liquefaction as the gelatin molecules are cleaved to shorter, more soluble molecules by the enzymes.

A gel forms as the gelatin molecules begin to establish a continuous, solid network by forming hydrogen bonds and other secondary bonds. Water is bound to these molecules and also is trapped within the framework. The strength of the gel depends on the concentration of gelatin, the rate of cooling, the temperature of the gel, the age of the gel, the presence of electrolytes, the pH, and the amount of sugar present. Mechanical agitation causes thixotrophy in gelatin gels. Gelatin gels and sols are reversible, depending on the temperature. When gelatin is just beginning to congeal, it can be beaten into a light foam for use in desserts and salads.

## STUDY QUESTIONS

1. What are the two major categories of proteins in muscles? Identify and briefly describe at least three proteins in each category.

2. Name and describe four types of proteins in connective tissue.

3. Describe the chemistry of collagen in some detail, being sure to discuss its unique amino acid composition.

4. Explain the organization of a muscle, beginning with the composition of myofilaments.

5. Describe the chemical changes and the resulting color shifts that occur when myoglobin is subjected to different conditions.

6. Why is cured meat a different color than uncured meat?

7. What changes occur in the carcass after slaughter?

8. Describe the changes that occur in meats when heated.

9. What are the effects of extending ground meats with textured soy protein?

10. What is surimi and how is it used?

11. How does the chemical nature of gelatin influence its physical behavior in food preparation?

12. What factors influence the temperature at which gelatin mixtures form gels?

13. What factors determine the tenderness of a gelatin gel?

14. True or false. Dry heat meat cookery tenderizes meat.

15. True or false. Beef cuts from the loin are best when prepared using a moist heat method.

## BIBLIOGRAPHY

Bernard, D. T. and Scott, V. N. 1999. Listeria monocytogenes in meats: New strategies are needed. *Food Technol. 53* (3): 124.

Bolton, D. J., et al. 1999. Integrating HACCP and TQM reduces pork carcass contamination. *Food Technol. 53* (4): 40.

Brody, A. L. 2002. Case-ready fresh red meat: Is it here or not? *Food Technol. 56* (1): 77.

Brody, A. L. 2004. Case for case-ready red meat packaging. *Food Technol. 58* (8): 84.

Cassens, R. G. 1995. Use of sodium nitrite in cured meats today. *Food Technol. 49* (7): 72.

Cole, B. and Kuecker, B. 2002. Packaging up case-ready profits. *Food Product Design 11* (11): 113.

Decker, E. A. and Xu, Z. 1998. Minimizing rancidity in muscle foods. *Food Technol. 52* (10): 54.

Decker, K. J. 2004. Meat analogues enter the Digital Age. *Food Product Design 14* (1): 14.

DeLoia, J. 2005. "Shrimply" irresistible. *Food Product Design 15* (1): 70.

Dubberley, M. 2002. Fowl play: Creating delicious chicken dishes. *Food Product Design 12* (5): 78.

DuBois, C. M., et al. 2008. *World of Soy.* University of Illinois Press. Urbana, IL.

Egbert, R. and C. Borders. 2006. Achieving success with meat analogs. *Food Technol. 60* (1): 28.

Foster, R. J. 2004a. A horse of another flavor. *Food Product Design 14* (6): 15.

Foster, R. J. 2004b. "Meating" consumer expectations. *Food Product Design 14* (9): 38.

Foster, R. J. 2005. pHood phenomena. *Food Product Design 14* (11): 61.

Giese, J. 2001. It's a mad, mad, mad, mad cow test. *Food Technol. 55* (6): 60.

Giese, J., 2004. Testing for BSE. *Food Technol. 58* (3): 58.

Grün, I. U., et al. 2006. Reducing oxidation of meat. *Food Technol. 60* (1): 36.

Hazen, C. 2004. Mainstreaming soy protein. *Food Product Design: Functional Foods Annual* (September): 87.

Hazen, C. 2005. Antioxidants "meat" needs. *Food Product Design 15* (1): 61.

Hazen, C. 2006. Adding soy ingredients for health. *Food Product Design 16* (2): 63.

Hazen, C. 2010. Richer meat flavors. *Food Product Design 20* (3): 48.

Hegenbart, S. 2002. Soy: The beneficial bean. *Food Product Design 11* (10): 82.

Hogan, B. 2002. Putting punch in meat flavor profiles. *Food Product Design 12* (4): 94.

Hogan, B. 2004. Beef—beyond the burger. *Food Product Design 14* (1): 14.

Katz, F. 1998. That's using the old bean. *Food Technol. 52* (6): 42.

King, J. W., Turner, N. J., and Whyte, R. 2006. Does it look cooked? Review of factors that influence cooked meat color. *J. Food Sci. 71* (4): R31.

Kolettis, H. 2004. It's a mad, mad world. *Food Product Design 13* (11): 21.

Kolettis, H. 2005. Study indicates meaty results. *Food Product Design 15* (1): 22.

Lanier, T. C. 1986. Functional properties of surimi. *Food Technol. 40* (3): 107.

Lillard, H. S. 1994. Decontamination of poultry skin by sonication. *Food Technol. 48* (12): 72.

Liston, J. 1990. Microbial hazards of seafood consumption. *Food Technol. 44* (12): 56.

Luff, S. 2005. Better ingredients through biotechnology. *Food Product Design 14* (10): 91.

Lusk, J. L., et al. 1999. Consumer acceptance of irradiated meat. *Food Technol. 53* (3): 56.

Lynch, B. 2002. From sea to shining sea. *Food Product Design: Food Service Annual* (November): 56.

Mandigo, R. W. 2002. Ingredient opportunities in case-ready meats. *Food Product Design 11* (11): 96.

Mermelstein, N. H. 2010. Analyzing for histamine in seafood. *Food Technol. 64* (2): 66.

Murphy, P. A., et al. 1997. Soybean protein composition and tofu quality. *Food Technol. 51* (3): 86.

Ohr, L. M. 2004. Powerhouse proteins soy and whey. *Food Technol. 58* (8): 71.

Pszczola, D. E. 1999. Ingredients that get to the meat of the matter. *Food Technol. 53* (4): 62.

Ravishankar, S., et al. 2009. Edible apple film wraps containing plant antimicrobials inactivate food borne pathogens on meat and poultry products. *J. Food Sci. 74* (10): M440.

Regenstein, J. M. 2004. Total utilization of fish. *Food Technol. 58* (3): 28.

Resurreccion, A. V. A. and Galvez, F. C. F. 1999. Will consumers buy irradiated beef? *Food Technol. 53* (3): 52.

Sablani, S. S., et al. 2009. Apple-peel-based edible film development using a high pressure homogenization. *J. Food Sci. 74* (7): E372.

Santerre, C. R. 2004. Farmed salmon: Caught in a numbers game. *58* (2): 108.

Sebranek, J. G., et al. 2006. Carbon monoxide packaging of fresh meat. *Food Technol. 60* (5): 184.

Shapiro, L. S. 2001. *Introduction to Animal Science*. Prentice Hall. Upper Saddle River, NJ.

Sierengowski, R. 2004. Ethnic sausages. *Food Product Design 13* (11): 76.

Silver, D. 2003. Oceans of opinions. *Food Product Design 13* (4): 120.

Sloan, A. E. 2006. Prime time for meats and poultry. *Food Technol. 60* (3): 19.

Warner, K., et al. 2001. Use of starch-lipid composites in low-fat ground beef patties. *Food Technol. 55* (2): 36.

Zino, D. 2005. Taking a closer look at beef. *Food Product Design 15* (9): 62. 106.

## INTO THE WEB

*http://www.aamp.com*—American Association of Meat Processors; summaries of current concerns and news about meat.

*http://www.mbayaq.org/cr/cr_seafoodwatch/sfw_ac.asp*—Monterey Bay Aquarium site providing background information on aquaculture and other topics on seafood.

*http://minnesota.publicradio.org/display/web/2009/12/02/carp-kill-inadequate/*—Report on the problem of blocking the giant carp from entering Lake Michigan.

*www.beefretail.org/*—Information about beef cuts and cookery.

*http://www.theotherwhitemeat.com/*—Information about pork.

*http://www.americanlamb.com/*—Information about lamb.

*http://www.fsis.usda.gov/Fact_Sheets/Beef_from_Farm_to_Table/index.asp*—Overview of beef and its preparation.

*http://www.foodproductdesign.com/news/2009/09/edible-film-protects-meat-poultry.aspx*—Article on protective edible film.

*http://www.nytimes.com/2009/12/31/us/31meat.html?_r=1&emc=eta1*—Ammoniated beef trimmings and safety.

*http://barcoding.si.edu/*—Information on the Consortium for the Barcode of Life.

*http://www.fishbol.org/*—Fish Barcode of Life Initiative site.

*http://www.bseinfo.org/*—Current information about bovine spongiform encephalopathy.

*http://www.surimiseafood.com/*—Information on various surimi products.

*http://www.soyfoods.org/locate/retail-soy-products*—Information about various soy products.

*http://www.soyfoods.com/soyfoodsdescriptions/tofu.html*—Background information on tofu.

*http://www.lifescientists.de/members/vanwijk.htm*—Background information regarding research related to Food for Thought: Petri Burgers?

*http://www.nsf.gov/pubs/2004/nsf0450/app_2b.pdf*—Provides information regarding federal support of tissue-engineering research.

*http://www.quorn.us/*—Information about Quorn™.

Food technologist Deana Jones (foreground) and hyperspectral imaging specialist Jerry Heitschmidt examine eggs for shell quality defects as biological science aide Vicky Broussard uses a pressure device to test eggs for cracks. (Courtesy of Agricultural Research Service.)

# CHAPTER 16

# Eggs

## Chapter Outline

## OBJECTIVES

After studying this chapter, you will be able to:

1. Describe egg formation, structure, and composition.
2. Discuss egg quality, safety, and preservation.
3. Explain the functions of eggs in preparation of various food products.

## FORMATION

Egg formation begins in the ovaries of poultry with the development of a yolk that contains the germ cell (ovum). From the ovaries, the yolk moves into the oviduct where layers of egg white are secreted to surround the yolk. If sperm pass through the oviduct and reach the yolk before it is encased by this coating of egg white proteins, they fertilize the yolk and create a fertile egg.

Encasement of the yolk and egg white proceeds gradually as the developing egg continues its passage down the oviduct. Two membranes—an inner, somewhat fragile membrane and a tough outer membrane—confine the yolk and white. Finally, a shell develops around the outer membrane as the egg moves through the region of the oviduct that secretes minerals to provide this strong outer protection for the egg. Calcium is the mineral of particular importance in the shell.

During egg formation, various defects may develop. If a blood vessel happens to rupture in the ovary or along the oviduct, blood spots can occur in the yolk or in the white, depending on the location of the lesion in relation to the stage of the egg's development.

**Table 16.1**    U.S. Weight Classes for Consumer Grades for Shell Eggs

| Size | Minimum Weight/Dozen (oz) |
| --- | --- |
| Jumbo | 30 |
| Extra large | 27 |
| Large | 24 |
| Medium | 21 |
| Small | 18 |
| Peewee | 15 |

Although uncommon, chickens may have an infection or even a parasite in the oviduct, which then becomes a part of the developing egg and is enclosed by the shell. When hens first begin laying, their eggs are quite small, but the size increases as they mature. Federal classification of size is based on the weight of eggs (Table 16.1).

## STRUCTURE

**White yolk**
Small sphere of light-colored yolk at the center of the yolk.

**Germinal disk**
Blastoderm of the yolk, which is located at the edge of the yolk and is connected to the white yolk.

**Latebra**
Tube connecting the white yolk to the germinal disk in the yolk.

**Vitelline membrane**
Sac enclosing the yolk.

**Chalazae**
Thick, rope-like extensions of the chalaziferous layer that aid in centering the yolk in the egg.

**Chalaziferous layer**
Membranous layer surrounding the vitelline membrane of the yolk.

**Albumen**
White of an egg; consists of three layers.

The yolk is more complex structurally than may be thought at first glance. At the very center is a light-colored structure called the **white yolk**. Leading from this white yolk to the **germinal disk** or blastoderm is a connecting tube called the **latebra**. These structures, plus concentric rings of light and dark yolk layers, constitute the contents of the yolk. The **vitelline membrane** serves as the enclosure or sac for the yolk. The strength of this membrane is important when yolks and whites need to be separated in cookery.

Two important, rather twisted membranous structures extend from opposite sides of the yolk into the white. These **chalazae** are extensions of the **chalaziferous layer** (membranous layer surrounding the yolk) and help keep the yolk centered in the white by impeding movement through the thick white (Figure 16.1).

The **albumen**, often called the white, surrounds the yolk to provide a protective buffer. Although all of the albumen appears to be transparent and somewhat fluid, the white

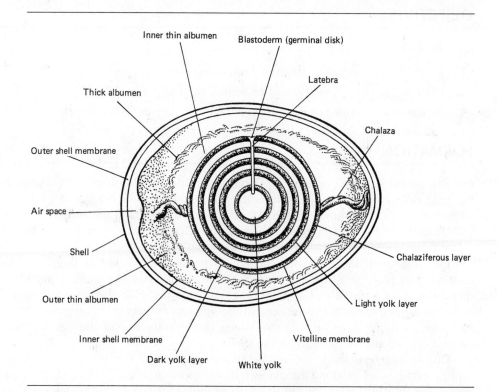

**Figure 16.1**    Cross-sectional diagram of an egg in the shell.

**Figure 16.2**   A specialized camera captures images of illuminated eggs inside this see-through case. Inside, the eggs are subjected to a slight vacuum (negative pressure) that enhances existing shell microcracks, making them more visible. (Courtesy of Agricultural Research Service.)

actually contains three layers of material. The innermost layer is a fairly fluid material called the inner **thin albumen**. A distinctly viscous layer, the **thick albumen** or albuminous sac, encompasses the inner thin albumen. Another layer of thin albumen, the outer thin albumen, is the final layer of white.

Two **shell membranes** enclose the entire contents of the white and yolk of an egg. The inner shell membrane is immediately adjacent to the outer thin albumen and serves to confine this fluid outer layer of white. An outer shell membrane is right next to the shell. At the large end of the egg, an air space or **air cell** develops between the inner and outer shell membranes as the freshly laid egg cools.

The final layer is the shell, which surrounds the entire egg and is quite a rigid packaging material for the total egg (Figure 16.2). Although at first glance the shell appears to be a solid, it actually has many tiny pores that permit passage of gases or tiny microorganisms into and out of the shell. The egg's natural protection against such a microorganism invasion is a natural, thin coating called **bloom**, which is already on the egg when it is laid and effectively seals the pores in the calcium carbonate-rich shell.

**Thin albumen**
Rather fluid egg white adjacent to the yolk and to the inner membrane.

**Thick albumen**
Viscous white forming the middle layer of albumen.

**Shell membranes**
Inner membrane encasing the white and an outer membrane adjacent to the shell of the egg.

**Air cell**
Space between the inner and outer shell membranes at the large end of the egg.

**Bloom**
Natural protective coating sealing the shell pores when an egg is laid.

## COMPOSITION

Gross differences in composition occur between the white and the yolk, differences that are somewhat evident even from casual observation. First, there is considerably more white than yolk. In fact, the white weighs about twice as much as the yolk. On the basis of volume,

**Table 16.2**   Approximate Composition of Chicken Egg Yolk and White (in percentage)

| Portion | Percentage | Constituents | | | |
|---|---|---|---|---|---|
| | | Water | Protein | Fat | Ash |
| Whole egg | 100 | 65.5 | 11.8 | 11.0 | 11.7 |
| White | 58 | 88.0 | 11.0 | 0.2 | 0.8 |
| Yolk | 31 | 48.0 | 17.5 | 32.5 | 2.0 |

| | | Carbonates | | | |
|---|---|---|---|---|---|
| | | Calcium | Magnesium | Calcium Phosphate | Organic Matter |
| Shell | 11 | 94.0 | 1.0 | 1.0 | 4.0 |

U.S. Department of Agriculture.

approximately 20 milliliters of yolk and 30 milliliters of white ($1\frac{1}{3}$ tablespoons of yolk and 2 tablespoons of white) equal one egg. This ratio remains fairly constant despite the actual size of the egg.

Pure yolk contains quite a different ratio of fluids than the white. As Table 16.2 shows, the white is fluid because it contains close to 90 percent water and almost no fat. On the other hand, the yolk is not quite half water and almost one-third lipids. The protein concentration of the yolk is about 1.5 times that of the white.

When the actual content of one egg white is compared with that of one egg yolk, a somewhat different picture emerges because of the difference in the relative amounts of yolk and white. On this basis, an egg white provides about $1\frac{1}{3}$ times as much protein as the yolk (Table 16.3). One egg white has only a trace of fat, while one yolk contains 4.6 g; thus, one yolk provides about 3.5 times as many calories as one white. The yolk contains some iron (0.8 milligram) and a bit of vitamin A, neither of which occurs in the white.

The yolk contains several different lipid compounds. About half of the lipids in the yolk are triglycerides. Phospholipids, including lecithin and cephalin, account for about a fifth of the egg yolk solids. Lecithin is of special interest because of its remarkable emulsifying ability, the result of the hydrophobic nature of the fatty acids in the molecule and the hydrophilic character of the phosphoric acid component (see Chapter 6).

**Cholesterol** is a sterol of nutritional concern found in the lipid component of egg yolks. The yolk of an egg contains about 186 milligrams of cholesterol, which may eliminate egg yolks from the diets of people who need to limit their cholesterol intake. Whites do not contain lipids or cholesterol and can be used freely in the diet.

Other key lipids are the lipoproteins, which combine a protein with a lipid. Lecithin, for example, combines with a protein to make a specific type of lipoprotein designated as a lecithoprotein. The protein components of the lipoproteins are discussed in the next section.

**Cholesterol**
A sterol found in abundance in egg yolk; the structure is

**Table 16.3**   Composition of One Chicken Egg Yolk versus One White

| Component | White | Yolk |
|---|---|---|
| Weight | 29 g | 15 g |
| Calories | 15 | 52 |
| Protein | 3.2 g | 2.4 g |
| Fat | Trace | 4.6 g |
| Iron | Trace | 0.8 mg |
| Vitamin A | 0 IU | 510 IU |

Adapted from U.S. Department of Agriculture Handbook No. 456: *Nutritive Value of American Foods in Common Units*. Washington, D.C: U.S. Department of Agriculture, ARS, 1975.

# PROTEINS

## Albumen

The albumen proteins, 12 of which have been identified, contribute various characteristics to the behavior of eggs in food preparation. The most abundant of these proteins is **ovalbumin**, which accounts for more than half of the protein in egg white (Table 16.4). Ovalbumin is denatured comparatively easily by heat.

**Conalbumin** is significant because of its ability to bind metals, forming undesirable colors if ions of iron, aluminum, copper, or zinc are present. Complexes of iron and conalbumin cause a red color in egg whites, whereas copper ions (2+) form a yellow complex with conalbumin, as can be noted when beating egg whites in a copper bowl. Although these metal-conalbumin compounds are stable to heat, conalbumin itself is heat sensitive. As these ions usually are not available to bind it, conalbumin is considered susceptible to denaturation when heated. However, egg whites to be pasteurized can be treated with metals to give conalbumin heat stability.

Ovomucoid is a glycoprotein that may contain a variety of carbohydrates, including mannose, galactose, and deoxyglucose. This protein is quite resistant to denaturation by heating unless it is in an alkaline medium.

**Lysozyme** is an unusual albumen protein, because it has a fraction ($G_1$ globulin) that has an isoelectric point of 10.7. Only avidin, among the other albumen proteins, has an isoelectric point in the alkaline range, and the isoelectric point of avidin is only 9.5. Another characteristic of lysozyme is that it has bactericidal action because of its ability to hydrolyze a polysaccharide in the cell wall of specific bacteria to help prevent bacterial spoilage in eggs.

Two ovoglobulins important for their foaming ability are $G_2$ globulin and $G_3$ globulin (isoelectric points are pH 5.5 and pH 5.8, respectively). These globulins, along with ovomucin and conalbumin, are of great merit in the foaming of egg whites when they are beaten.

**Ovomucin** has a somewhat fibrous character. This rigidity of structure accounts for the difference in the flow properties of thin versus thick albumen, for ovomucin is four times more abundant in thick albumen than it is in thin albumen. These fibers of ovomucin may be cut by the action of a rapidly moving eggbeater during the whipping of whites to shorter lengths that enhance the stability of egg white foams. Although ovomucin contributes to the stability of egg white foams, it is quite resistant to denaturation by heating.

**Ovalbumin**
By far the most abundant protein in egg albumen; readily denatured by heat.

**Conalbumin**
Protein in egg albumen capable of complexing with iron ($Fe^{3+}$) and copper ($Cu^{2+}$) ions to form red and yellow colors, respectively.

**Lysozyme**
Albumen protein with an isoelectric point of pH 10.7; notable for its ability to hydrolyze a polysaccharide in the cell wall of some bacteria, thus protecting against contamination by these bacteria.

**Ovomucin**
Rather fibrous protein occurring in thick white at about four times the concentration in which it occurs in thin white; protein contributing significantly to the viscous, gel-like texture of thick white.

**Table 16.4**    Proteins in Egg Albumen

| Protein | Relative Amount in Albumen (percentage) | Isoelectric Point | Molecular Weight | Characteristics |
|---|---|---|---|---|
| Ovalbumin | 54 | 4.6 | 45,000 | Phosphoglycoprotein |
| Conalbumin | 13 | 6.6 | 80,000 | Binds metals |
| Ovomucoid | 11 | 3.9–4.3 | 28,000 | Inhibits trypsin |
| Lysozyme ($G_1$ globulin) | 3.5 | 10.7 | 14,600 | Lyses some bacteria |
| $G_2$ globulin | 4.0? | 5.5 | 30,000–40,000 | — |
| $G_3$ globulin | 4.0? | 5.8 | ? | — |
| Ovomucin | 1.5 | ? | ? | Sialoprotein |
| Flavoprotein | 0.8 | 4.1 | 35,000 | Binds riboflavin |
| Ovoglycoprotein | 0.5? | 3.9 | 24,000 | Sialoprotein |
| Ovomacroglobulin | 0.5 | 4.5–4.7 | 760,000–900,000 | — |
| Ovoinhibitor | 0.1 | 5.2 | 44,000 | Inhibits some proteases |
| Avidin | 0.05 | 9.5 | 53,000 | Binds biotin |

Powrie, W. D. Characteristics of edible fluids of animal origin: Eggs. In *Principles of Food Science I*. Fennema, O. R., ed. New York: Dekker, 1976, p. 665.

Among the other albumen proteins that have been studied fairly extensively is **avidin**. This protein is of interest nutritionally because of its ability to bind biotin, thereby preventing absorption of this vitamin. This fact was used in structuring an experimental diet to determine the role of biotin in humans. This ability to bind biotin is eliminated when avidin is denatured by heating and hence does not present a realistic dietary problem for people.

## Yolk

Egg yolk consists of four types of particles—yolk spheres, granules, low-density lipoproteins, and myelin figures—dispersed in a plasma containing livetin, a globular protein, and low-density lipoproteins. Most of the yolk spheres are found in the white yolk, yet a few are located in the yellow yolk, which constitutes almost all of the yolk. Granules account for almost a fourth of the total solids in yolks, and about 60 percent of the content of these granules is protein.

**Livetin**, a prominent protein in the plasma of egg yolk, actually can be separated into three fractions: α-, β-, and γ-livetin. These three forms of livetin are thought to originate in the blood of the hen developing the egg.

**Low-density lipoproteins (LDLs)** constitute the remainder of the proteins found in yolk plasma. Actually, the LDL fraction consists largely of lipids, with protein representing somewhat more than 10 percent.

The granules in egg yolk contain three types of protein—**lipovitellins** (classified as high-density lipoproteins or HDLs), phosvitin, and low-density lipoprotein. They also contain phosphorus primarily in the form of phosphatidylcholine.

In addition to the lipovitellins, yolk granules contain about 16 percent phosvitin and somewhat less (about 12 percent) low-density lipoprotein. **Phosvitin** is a comparatively small protein (molecular weight approaching 40,000) with a phosphorus content of about 10 percent and a high content (almost one-third of the amino acid residues) of serine. Of particular interest is that phosvitin, with its ability to bind ferric ions in a soluble complex, is the means by which iron is incorporated into the yolk.

## EGGS BY DESIGN

Some chickens today are getting assistance with their egg production from the farmers who care for them. **Omega-3 eggs** are available for health-concerned consumers willing to pay their comparatively high price to obtain eggs with an elevated content of omega-3 fatty acids (0.4 g compared with 0.1 g in an ordinary egg). The increased level of omega-3 fatty acids is achieved by feeding laying hens a diet that contains between 10 and 20 percent flaxseeds (a particularly rich source of these healthful fatty acids).

The diets of chickens can be manipulated to alter other components of the eggs that they lay. A diet high in canola oil produces eggs that are marketed as lower in saturated fat and cholesterol. Under FDA regulations, this claim can be made only if the content of both saturated fat and cholesterol is at least 25 percent lower than it is in a standard egg. Altering the diet of laying hens also can increase the vitamin E content in eggs.

Sometimes eggs are identified as **cage-free** or **free-range eggs** (Figure 16.3). Such designations are of interest to consumers who are concerned about the treatment of the hens. Although cage-free chickens are not held in cages, they still may be raised on the floor of a building. Free-range eggs are laid by chickens that are in outdoor pens during the day, but generally are sheltered in a barn at night. Despite these special living conditions, cage-free and free-range eggs are comparable nutritionally to eggs laid by caged hens.

The price of eggs that are modified nutritionally or from cage-free and free-range chickens usually is higher than that of standard eggs. For some people, the added cost is not a deterrent, and they believe the health benefits of the nutrient-modified eggs may be important enough to justify the added expense. However, cage-free and free-range eggs do not provide any nutritional benefit for their added cost.

**Figure 16.3**   Choices in eggs sometimes include such variations as omega-3, organic, cage-free and more.

## EGG QUALITY

The quality of eggs can be graded in or out of the shell. Table 16.5 shows the grade designation assigned to the federal standard descriptions of the shell, air cell, white, and yolk for the three federal grades. The date on which eggs are graded appears on the carton and can be helpful information when buying eggs. The date is indicated as a 3-digit number, beginning with January 1 as 001 and ending with December 31 as 365.

Candling, the technique used for grading eggs in the shell, is based on observing the eggs in silhouette while they are rotated (Figure 16.4). This process reveals mobility of the yolk and the size and location of the air cell in silhouette. In an egg of high quality, the yolk is centered and shows limited movement when the egg rotates.

Egg quality can also be determined by breaking the egg out of the shell. A special micrometer measures the height in **Haugh units** of the thick albumen in the white in relation to the weight of the egg. Other measures of quality out of the shell are **albumen index** and **yolk index**.

**Haugh unit**
Units used to denote the quality of albumen; correlates thick albumen height with egg weight.

**Albumen index**
Grading measurement of albumen to determine quality on the basis of the amount of thick white.

**Yolk index**
Measurement of egg quality based on the ratio of yolk height to yolk width.

**Table 16.5**   Standards for Grading Eggs in the Shell

| Grade | Shell | Air Cell | White | Yolk |
|---|---|---|---|---|
| AA | Clean, unbroken; practically normal | $\frac{1}{8}$ inch or less in depth | Clear; firm (72 Haugh units or higher) | Outline slightly defined; practically free from defects |
| A | Clean, unbroken; practically normal | $\frac{3}{16}$ inch or less in depth | Clear; may be reasonably firm (60–72 Haugh units) | Outline may be fairly well defined; practically free from defects |
| B | Clean to very slightly stained, unbroken; may be slightly abnormal | $\frac{3}{8}$ inch or less in depth | Clear; may be slightly weak (31–60 Haugh units) | Outline may be well defined; may be slightly enlarged and flattened; may show definite but not serious defects |

U.S. Department of Agriculture, 1968.

**Figure 16.4**   As eggs are candled by passing over a bright light, they are checked for shell cleanliness and integrity and also yolk position and size of air cell. (Courtesy of Ontario Farm Animal Council photo library.)

Particularly in shell eggs, the quality indicated by the grade designation may not be what actually is found at the time of use because egg quality declines continuously after laying. The extent of the changes depends on the conditions under which eggs are stored and the length of storage. However, the changes occur in a predictable fashion and are clearly visible, whether viewed from above or sideways when eggs have been broken from the shell (Figure 16.5).

The air cell at the large end of the intact shell forms between the inner and outer membranes as the warm, freshly laid egg cools to ambient temperature and contracts. This air cell continues to grow as the egg loses both moisture and carbon dioxide through the pores during storage. Loss of carbon dioxide also elevates the pH of the white. The egg albumen gradually increases in alkalinity, from its original pH of about 7.6 to as high as 9.4.

## AA Quality

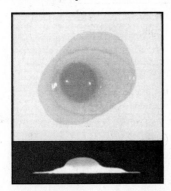

1. Egg covers small area; much thick white surrounds yolk; has small amount of thin white; yolk round and upstanding.
2. White—firm—72 Haugh units minimum.

## A Quality

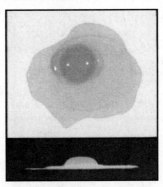

1. Egg covers moderate area; has considerable thick white; medium amount of thin white; yolk round and upstanding.
2. White—reasonably firm—60 Haugh units minimum.

## B Quality

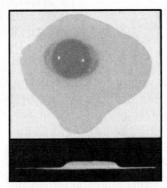

1. Egg covers very wide area; has no thick white; large amount of thin white thinly spread; yolk very flat and enlarged.
2. White—weak and watery—less than 60 Haugh units.

**Figure 16.5**    Quality guide to eggs broken out of the shell (top and side views). (Courtesy of the U.S. Department of Agriculture.)

The increasing alkalinity of the albumen of low-quality eggs is unique among foods, for virtually all other foods become increasingly acidic as they deteriorate.

A gradual transition in the ratio of thick white to thin white is another change that occurs as eggs lose quality. In a high-quality egg, the thick albumen is held fairly firmly around the yolk and does not spread very far. A gradual change to albumen that is flatter and spread more widely occurs as the ovomucin in the thick albumen undergoes degradation and the thin white increases. The appearance of the egg changes from one that is relatively thick and occupying a comparatively small space to one that spreads thinly over a large area. The thinner albumen allows the yolk to move away from the center and toward the edge of the egg.

The yolk also changes in appearance as the quality of an egg declines. A strong vitelline membrane in an egg of high quality results in a well-rounded yolk. However, the change in the pH of the albumen appears to have an effect on the chalaziferous layer and the vitelline membrane, with the result that the vitelline membrane stretches and becomes weaker. These changes cause the yolk to flatten gradually and to spread farther. Weakening of the vitelline membrane also makes it more difficult to separate yolks from the whites without breaking them. The migration of a small amount of water from the white into the yolk contributes to the stretching and weakening of the vitelline membrane.

**Table 16.6**   Storage Recommendations for Eggs and Egg Products

| Product | Refrigerator | Freezer |
|---|---|---|
| Raw eggs in shell | 3–5 weeks | Do not freeze |
| Raw egg whites | 2–4 days | 12 months |
| Raw egg yolks | 2–4 days | Do not freeze well |
| Hard-cooked egg | 1 week | Do not freeze |
| Egg substitutes, liquid | | |
| Unopened | 10 days | Do not freeze |
| Opened | 3 days | Do not freeze |
| Egg substitutes, frozen | | |
| Unopened | After thawing, 7 days, or "use-by" date on carton | 12 months |
| Opened | After thawing, 3 days, or "use-by" date on carton | |
| Casseroles with eggs | 3–4 days | After baking, 2–3 months |
| Eggnog, commercial | 3–5 days | 6 months |
| Eggnog, homemade | 2–4 days | Do not freeze |
| Pies, pumpkin or pecan | 3–4 days | After baking, 1–2 months |
| Pies, custard or chiffon | 3–4 days | Do not freeze |
| Quiche, any filling | 3–4 days | After baking, 1–2 months |

Adapted from Food Safety and Inspection Service information dated May 2006.

Although the pH of the albumen changes much more than that of the yolk during storage, both parts of the egg reflect the loss of carbon dioxide through the pores. Accompanying this increase in pH is a subtle modification of flavor. The bright flavor of a fresh egg slowly changes to reveal increasing overtones of sulfur as the egg ages and loses quality. Oil-dipping can seal the pores and block loss of water and carbon dioxide from the egg during storage. However, this oil traps volatile flavoring compounds that are generated within the egg, which can lead to off flavors if the egg is held in storage for a prolonged period.

Eggs can be held in cold storage for six months and retain satisfactory quality, but eggs held at room temperature will drop at least a grade level within a week. Storage of eggs in closed containers during refrigeration slows the rate of deterioration (Table 16.6). Egg cartons are suitable for this purpose.

## SAFETY

Eggs can provide an excellent medium for the growth of microorganisms. Therefore, care must be taken to avoid contamination from external sources. Egg producers employ strict sanitation standards to maintain clean facilities and to wash and sanitize eggs. Such measures are effective in eliminating numerous potential hazards, but they do not address the contamination of eggs by ***Salmonella enteritidis*** (see Chapter 19). Unfortunately, this bacterium infects the yolk of the egg while the egg is forming and before the shell is made. Therefore, sanitary handling of the eggs has no effect on the presence and viability of *Salmonella enteritidis*.

Control of this potential source of food-borne illness must focus on sanitary procedures at all stages of handling of the laying flock as a means of helping to prevent the hens from being infected with this dangerous form of *Salmonella*. At the present time, there probably is only a 1 in 10,000 chance of an egg being contaminated with *Salmonella enteritidis*, but even this risk level necessitates careful attention to adequate heat treatment to kill any bacteria that might be present. To assure a safe product when preparing a recipe in which egg will not be cooked, the egg and a liquid from the recipe need to be blended together and heated gently to a temperature of 71°C (160°F) before being mixed with the remaining ingredients.

***Salmonella enteritidis***
Form of bacteria that can be incorporated into the yolk of an egg as the hen forms the egg; potential source of food-borne illness if egg is not heated sufficiently during preparation.

The use of pasteurized egg products avoids this source of infection. Good sanitary measures also are necessary. Refrigeration of shell eggs at 7.2°C (45°F) or colder by warehouses, transport trucks, and retailers is now required (FSIS and FDA regulations) to promote the safety of eggs during storage and marketing.

Even if eggs are not contaminated with *Salmonella enteritidis,* they still are a perfect medium for the growth of most microorganisms with which they might become infected during food handling. To assure that they are safe when eaten, eggs that have not been pasteurized should be cooked to a temperature of at least 72°C (162°F). A thermometer can be used to check the interior temperature of egg-containing dishes.

Yolks and whites should be heated until firm; scrambled eggs should not be runny. Prompt refrigeration of eggs and egg-containing dishes or holding of hot egg dishes at or above 60°C (140°F) should be the rule at all times to help avoid food-borne illnesses that could be caused by improperly handled eggs. Fresh eggs may be stored in the refrigerator for up to five weeks.

## PRESERVATION

### Pasteurization

The potential for shell eggs to be contaminated with *Salmonella enteritidis* stimulated efforts to eliminate this hazard, and the industry responded by developing a technique to pasteurize eggs in the shell by heating enough to kill the microorganisms without altering the egg proteins. In stores marketing eggs that have undergone this treatment, the cartons are labeled as pasteurized.

When food is prepared in institutional quantities, the labor involved in breaking eggs can become quite costly. This has led to a market for eggs sold out of the shell. Unfortunately, eggs can spoil extremely rapidly once they are removed from their protective shells. Therefore, pasteurization of eggs marketed outside their shells is necessary to ensure the safety of these products.

Various methods of pasteurization are suitable; a common technique is to hold liquid whole eggs at 61°C (142°F) for 3.5 minutes, a temperature sufficient to kill harmful microorganisms without causing undesirable denaturation in the eggs. Yolks can be pasteurized in a similar fashion, although those to which sugar or salt is added require a higher temperature to kill *Salmonella* and other potentially harmful microorganisms. Even though egg whites are more sensitive to pasteurization than are either yolks or whole eggs, they still are satisfactory for producing foams after pasteurization.

### Drying

Preservation of eggs by drying is done for convenience and safety, but these products are available primarily to the food industry for use in manufacturing other food items. Spray drying is the commonly used technique to dry eggs. Dried eggs are of interest because of the convenience of the long-term storage that is possible without refrigerated or frozen storage facilities. Among the dried egg products available are whole-egg solids, yolk solids, fortified whole-egg solids with varying proportions of whites and yolks, and whites.

Production of spray-dried egg whites presents a particular challenge because of the potential changes that can impair whipping characteristics, color, and flavor. During a fermentation period by bacteria or yeast, glucose can be eliminated from the whites prior to drying, thus preventing the negative impact of this sugar on the dried whites. When glucose remains in dried egg whites, browning occurs as the Maillard reaction proceeds during storage.

Dried whites that have undergone bacterial fermentation prior to drying can be used effectively in making products containing foams. Adding a surface-active agent to dried egg whites in angel food cake mixes enhances foaming to help overcome the slightly detrimental effect from small amounts of fat that may be incorporated as a contaminant from broken yolks during the initial separation process.

Glucose also complicates the production of dried yolks and dried whole eggs. During storage, glucose can react with cephalin in the yolk to produce an off flavor. The fat in the yolks makes it necessary to eliminate oxygen from the packaging to prevent oxidation during storage. If the glucose is eliminated by fermentation prior to drying and the adequately dried yolk or whole-egg mixture is packaged tightly with inert gas replacing any oxygen, these egg products have excellent storage life and are useful in many food products.

## Freezing

In freezing, whites perform well and do not require special treatment. Unlike when whites are dried, glucose does not present a problem when whites are frozen. With normal freezer packaging precautions, frozen egg whites can be stored in the freezer and then thawed for use in any way that fresh whites would be used. Foaming power and flavor are excellent after freezing and thawing.

Unfortunately, egg yolks form gels as a consequence of freezing, making thawed yolks or whole eggs too viscous and unsatisfactory to use unless either sugar or salt is added prior to freezing. For commercial purposes, 10 percent salt or sugar (by weight of the yolks) can be added to either whole eggs or yolks before freezing. If yolks or whole eggs are to be frozen in small quantities, the yolks or whole eggs are blended, and then either half a teaspoon of salt or two tablespoons of sugar per cup are added to reduce the problems of yolk gelation.

Of course, addition of salt or sugar alters the flavor of the product, making it necessary to choose the appropriate flavor of frozen egg yolk or whole egg. For example, frozen whole eggs containing sugar would not be acceptable for preparing scrambled eggs.

Unlike many other frozen foods, eggs stored at colder freezing temperatures are not better than those stored at somewhat more moderate freezing temperatures, because of the impact of extreme cold on the gelling properties of protein. Storage at about −17°C (0°F) causes fewer problems with gelation than storage at −23°C (−10°F).

Consumer demand for pre-prepared products has provided an incentive for developing frozen products containing cooked egg. One of the key problems to overcome is syneresis. Several approaches may help improve the quality of cooked egg products that ultimately will be frozen, thawed, and reheated. The addition of gums helps bind the water within the egg mixture. Citric acid can alter the pH from essentially neutral to a somewhat acidic pH of around 6.0. Salt also may be added, although consumer concerns about salt in processed foods may be a factor in promoting other solutions to syneresis.

---

### FOOD FOR THOUGHT: Thousand-Year Eggs

In some markets in China and other Asian countries, shoppers may find three types of thousand-year eggs: Hulidan, Dsaudan, and Pidan. These uncooked eggs may be between one month and several years old. The preservation of each of these types of thousand-year eggs is done a bit differently, so it is not surprising that their characteristics vary.

Eggs can be coated with a mixture of salt and wet clay before storing them for at least a month to preserve them. Eggs preserved in this way have a darker white, a partly hardened yolk, and a salty taste. They are called Hulidan eggs.

Dsaudan eggs are made by packing them in cooked rice and salt for a storage period of six months, conditions that produce a softened shell and coagulated interior with a wine-like flavor.

Pidan eggs are aged in a mixture of wood ash, lime, salt, and a tea infusion for at least five months. The resulting egg has a grayish-green yolk and a white that has changed to a deep brown color and a jellied texture. The flavor is reminiscent of lime.

## FUNCTIONAL PROPERTIES AND APPLICATIONS

Eggs, either as yolks, whites, or whole eggs, are unique among food ingredients because of their versatility in the functional roles they fulfill in a variety of products. These functions can be categorized as (1) coloring agent, (2) emulsifier, (3) thickening agent, and (4) texturizing agent (Figure 16.6). These are contributions above and beyond the excellent nutrient content eggs provide. Their only negative role is as a source of cholesterol (approximately 213 milligrams per yolk) and protein, to which a few people are allergic.

### Coloring Agent

Four bright carotenoid pigments in the yolk add color to light-colored food products in which they are used. The two carotenoid pigments contributing the majority of the color are lutein and zeaxanthin; the former is a xanthophyll with a closed ring and two hydroxyl groups, and the latter possesses a single hydroxyl group and two closed rings (see Chapter 10). Cryptoxanthin (another xanthophyll) and carotene also contribute to the overall color of the

**Figure 16.6**  Lemon meringue pie demonstrates three roles of egg: coloring agent and thickening due to yolk proteins in the filling and texture resulting from the foaming of the egg white proteins in the meringue.

yolk. The bright color of the yolk is particularly important in deviled eggs, where it contrasts sharply when stuffed into the white.

Yolk color varies somewhat with the season of the year in the United States, but it ranges from a fairly pale yellow through a slightly subtle orange tone. The yolks in Guatemala are an amazing nasturtium-orangish red color. The hen's diet has a strong impact on yolk color.

## Emulsifying Agent

Egg yolk itself is an example of a naturally occurring oil-in-water emulsion. Lecithin and lysolecithin are key compounds responsible for the remarkable ability of egg yolk to act as an emulsifying agent in food systems in which it is incorporated. An oil-in-water emulsion forms when egg yolk is an ingredient in food mixtures containing some type of fluid fat and water or other aqueous liquid. Yolk particles (including low-density lipoproteins, myelin figures, and high-density lipoproteins) contribute to the stability of an emulsion by interacting at the surface of the oil droplets to form a layer. Mayonnaise is a classic example of a food emulsion stabilized by egg yolk (see Chapter 5).

Hollandaise sauce is a classic sauce noted for its capricious behavior. Its high fat content (approaching 50 percent) needs to be formed into an emulsion, with lemon juice and water serving as the liquid portion. The emulsion is traditionally prepared by gently heating the yolks and liquid to thicken the mixture by partial denaturation, at which point the yolk mixture is removed from the heat and cooled slightly by the addition of some butter to reduce the risk of overheating and curdling the yolk mixture. Then, melted butter is added very gradually with agitation to emulsify the fat as it is incorporated into the sauce.

Temperature control is essential to the retention of this delicate oil-in-water emulsion, for excess heat will cause the yolk proteins to draw more tightly together, reducing their ability to cover a large amount of surface area at the interface between the oil droplets and the aqueous phase. Loss of too much liquid from evaporation also can cause curdling. If this happens, a small amount of liquid must be added to replace the water. The key to success with a hollandaise sauce is formation of a somewhat viscous, stable emulsion, with egg yolks providing the essential emulsifying agent, lecithin.

Baked products containing egg yolk also benefit from the emulsifying action of lecithin and lysolecithin. Cake batters sometimes have a tendency to curdle, particularly in warm weather when the fat becomes quite soft, but egg yolk in the recipe can provide a definite deterrent. Soufflés also use egg yolk to form a stable emulsion.

Undoubtedly the most dramatic example of egg yolk as an emulsifying agent in baked products is cream puffs. The paste formed for cream puffs is made by melting butter in boiling water and making a thick paste with added flour. At this point, egg yolks are beaten in, one at a time, to form an oil-in-water emulsion. This process alters the texture of the paste to a velvety smoothness and a viscosity capable of being dropped as small balls onto a baking sheet. The remarkable increase in volume that occurs when cream puffs are baked in a very hot oven depends on the formation of a stable oil-in-water emulsion, which is possible because of the generous amount of egg yolk available to serve as the emulsifying agent.

## Thickening Agent

Eggs are used as thickening agents in a wide array of food products because of the effect of heat on their proteins, resulting in denaturation and coagulation (see Chapter 13). One of the confounding problems of using egg as a thickening agent is that coagulation occurs over a rather wide temperature range, depending on the specific conditions applied. Unlike the preparation of candies in which security is provided by carefully boiling the candy to the correct final temperature, thickening with eggs requires personal judgment to determine when the product is done. Failure to judge this point correctly can lead to such problems as curdled sauces, weeping custards, or tough fried eggs.

The proteins in egg whites begin to coagulate at about 60°C (140°F) and lose their ability to flow when the temperature reaches approximately 65°C (149°F) if no other

- Source of protein
  White only (lowest)
  Whole egg (in between)
  Yolk (highest)
- Rate of heating (fast rate raises it)
- Concentration of protein (added ingredients raise it)

**Chart 16.1**    Factors affecting coagulation temperature of eggs

ingredients are added (Chart 16.1). Yolk proteins are somewhat more resistant to the impact of heat and begin coagulation around 65°C (149°F); they cease to flow when the temperature rises to about 70°C (158°F) when no liquid or other ingredient dilutes them. Coagulation occurs at a higher temperature when the rate of heating is rapid than when it is slow; the slower rate of heating provides more time for molecules to unwind to the secondary structure and clump as they denature and coagulate before the temperature rises too far into the range within which coagulation occurs.

When liquids or other ingredients (e.g., sugar) are added to eggs, the coagulation temperature of the mixture rises. Undiluted egg products, such as poached or fried eggs, coagulate at temperatures lower than those needed to coagulate custards or egg-containing dessert sauces. Custards (Figures 16.7) and dessert sauces are examples of egg-thickened products in which liquid (usually milk) and sugar dilute the protein and raise the coagulation temperature.

If a thermometer is inserted in a custard during baking and the temperature is observed carefully as the egg proteins in the custard begin to denature and coagulate, a gradual, slow rise in temperature is seen until coagulation occurs. At this point, the slow increase either stops abruptly or the temperature drops slightly for a short time before it begins to rise again. This plateauing is evidence that coagulation is an endothermic reaction requiring additional heat input.

If heating continues beyond the point at which coagulation occurs, undiluted egg proteins draw together tightly and become increasingly tough. Diluted protein mixtures curdle as the proteins tighten and separate from the liquid in the systems. To avoid these problems, eggs and egg-thickened products should be served as soon as they are coagulated or else cooled promptly to eliminate residual heat that can cause internal temperature to rise and curdle the food.

***Custards.***    Stirred and baked custards are sweetened milk mixtures thickened with egg and usually flavored with salt and vanilla. Although the ingredients and proportions are the same for both (one egg or two yolks per cup of milk), either a sol or a gel (see Chapter 6) may form when the mixture is heated, depending on the manipulation during heating. Continuous

**Figure 16.7**    The coagulation temperature of egg proteins in a baked custard is raised because of the added sugar and milk that dilute the protein. (Courtesy of the American Egg Board.)

**Stirred custard**
Sweetened milk and egg mixture that is heated to form a sol; agitation during heating prevents formation of sufficient intermolecular linkages to form a gel.

**Baked custard**
Sweetened milk and egg mixture that is baked without agitation until the egg proteins coagulate and form a gel.

stirring results in a sol, called a **stirred custard** or a soft custard. When the mixture is baked in the oven without any agitation, the product gels and is termed a **baked custard**.

In both instances, the proteins of the whole eggs provide most of the thickening, with the milk protein contributing only an insignificant increase in viscosity. However, milk provides the mineral salts needed for coagulation of the egg protein. Without salts, precipitation of the proteins to form a gel does not occur, as is evident when a baked custard is made with distilled water in place of milk.

Several factors influence the coagulation temperature of custards. Ordinarily custards are made with whole eggs, but yolks or whites sometimes are substituted for the whole eggs. Not surprisingly, custards made with egg whites coagulate at a lower temperature than those made with whole eggs, whereas use of yolks only results in a still higher temperature of coagulation.

The amounts of milk and sugar added to custard recipes also have an effect on coagulation temperature due, at least in part, to the diluting effect they have on protein content. The more milk and the more sugar added, the higher the coagulation temperature.

The rate of heating is yet another factor influencing coagulation temperature. As noted previously, a faster rate of heating results in a higher temperature of coagulation. However, less time is required even though the custard must be heated to a higher temperature when the rate of heating is fast than is needed for coagulation when the heating rate is slow and the coagulation temperature is somewhat lower. This time factor is of particular concern in heating custards.

A moderate to high heat makes removal of either baked or stirred custards from the heat at the right point extremely difficult, and overheating (with its accompanying curdling) likely is the result. In other words, residual heat is of far more concern than the actual temperature required for coagulation when rate of heating is being considered. Stirred custards ordinarily are prepared over a very low heat or in a pan surrounded by simmering water, and baked custards are set in a pan of boiling water (Figures 16.8) and heated in an oven at 177°C (350°F).

***Cooked Salad Dressings and Sauces.***    Occasionally, recipes for salad dressings include the use of egg yolk or whole egg as a thickening agent, either as the sole thickener or in concert with starch. The advantage of this type of salad dressing is the low fat content and

**Figure 16.8**    Custard cups are surrounded by very hot water during baking to help insulate the egg proteins from intense heat, thus allowing more time for heat to penetrate into the custard before the outer portions coagulate and begin to toughen. (Courtesy of the American Egg Board.)

consequently low caloric contribution. When both starch and egg proteins are included in the same recipe, the egg is withheld until the starch mixture has been thickened and then is added and heated to coagulate the protein. This procedure is based on the fact that optimum thickening from the gelatinization of starch requires a higher temperature than is appropriate for egg yolk proteins.

The addition of egg to a hot mixture requires care to avoid creating lumps of coagulated protein. By quickly stirring a spoonful of the hot sol into the egg and repeating this process about three times, the egg protein will be diluted enough to raise the coagulation temperature a bit, making it possible to then stir the diluted mixture into the hot sol without forming lumps. This egg-containing sol now needs to be heated enough to coagulate the egg proteins, but not so much that fine lumps of overcoagulated egg develop.

Salad dressings and some egg-thickened sauces contain fruit juice or other acidic liquid that reduces the pH of the system, usually bringing it closer to the isoelectric points of the egg proteins. Denaturation and coagulation occur at a somewhat lower temperature at the isoelectric point than at a pH somewhat remote from it. This increased ease is due to the decreased repulsion between molecules that is present when the electrical charges on the surfaces are at a minimum.

*Cream Pies and Puddings.*   Cream pie fillings and cream puddings are similar to salad dressings and sauces that are thickened by use of both a starch and egg protein. However, they differ in that cream pie fillings need to form a soft gel, rather than a sol, when they are chilled. Recipes for cream pies incorporate sufficient starch and egg yolk to achieve the desired gel when starch gelatinization and egg yolk protein coagulation are accomplished.

Unfortunately, inadequate heating of the filling after the yolk has been added can cause a seemingly thick filling to become quite fluid after chilling. The thinning is the result of the action of $\alpha$-amylase from the yolk, which digests some of the starch that has been included as a thickening agent. This decreases the size of the starch molecules and reduces the amount of amylose available to provide the network needed for gel structure.

Denaturation of $\alpha$-amylase must occur in the preparation of cream pie fillings to achieve a gelled filling. The factors that influence denaturation and coagulation of egg proteins also are applicable to $\alpha$-amylase specifically. Of particular concern is that the sugar content of cream pie fillings is quite high, which raises the coagulation temperature of the proteins. The process of stirring spoonsful of the hot starch-thickened filling into the egg yolks before returning the mixture to the rest of the filling cools the total filling a bit. Adequate heating is necessary after adding the egg yolk to ensure that the $\alpha$-amylase has been inactivated by coagulation. However, this must be achieved without overheating and curdling the yolk proteins.

Recipes often call for heating the filling over very low heat and stirring frequently for 5 minutes or until the mixture loses its gloss. The reason for this direction is that this technique causes the temperature to rise to slightly more than 85°C (185°F), which is hot enough to denature the $\alpha$-amylase.

Custard-type pie fillings are another example of the use of eggs as thickening agents. Unlike the cream pie fillings, which usually use only the yolks (the whites are reserved for the meringue), custard-type pies include the whole egg as the sole source of thickening. Problems with $\alpha$-amylase are not found in custard-type pies because starch is not included as a thickening agent.

The critical aspect of preparing custard-type pies is to ensure sufficient heating to coagulate the egg proteins without overheating them to the point at which curdling and syneresis are evident. Basically, preparation principles for these types of pies are the same as for baked custards; the only difference is the insulating effect of the bottom crust and the additional ingredients used to achieve different flavors and textures. Examples include pumpkin and pecan pies.

*Hard-Cooked Eggs.*   Hard-cooked eggs immortalize the quality of an egg; the peeled and sliced egg reveals the size of the air space, the location of the yolk (which indirectly reveals the amount of thin and thick white in the egg), and often even the freshness. A large air space indicates that considerable carbon dioxide and water have been lost through the pores since the egg was laid. A well-centered yolk indicates a large amount of thick white, whereas a yolk that has drifted up to the top of the egg while being heated, leaving only a curtain of white at the edge, is proof that there was little thick white and a considerable proportion of thin white. A darkish green ring of ferrous sulfide around the yolk is more likely to form in an older egg than in one that is fresher. This is due to the somewhat higher pH that is found in eggs as they lose carbon dioxide.

The firmness of the coagulated white in a hard-cooked egg is influenced by the final temperature reached in the egg and also by the temperature of the water in which it was cooked (Figures 16.9). Boiling eggs for 20 minutes causes them to be less tender than those that are either steamed or simmered for the same length of time. The method preferred by the American Egg Board is to cover the eggs with water at room temperature, heat the water to boiling, and then let the eggs sit in the water without additional heating for 15–17 minutes before cooling under cold, running water. For eggs larger than medium, the time in the hot water needs to be extended about 3 minutes. Eggs heated by placing them in boiling water and then simmering for 18 minutes are less likely to crack and are easier to peel than those prepared from a cold water start.

Ferrous sulfide is an undesirable compound that can form on the surface of the yolk. Prolonged heating, whether from too extended a cooking period or too slow a rate of cooling, promotes development of ferrous sulfide. This combination of iron from the yolk with sulfur occurs with increasing ease as the pH of the egg rises. Increased alkalinity is evidence of an egg that is staling. By using only fresh eggs, cooking them just until the yolk is set completely, and then cooling rapidly under cold, running water, ferrous sulfide formation can be avoided completely.

Companies that are marketing hard-cooked eggs for commercial uses are confronted by these same problems, but in greatly increased quantities. The fact that hard-cooked eggs with whites that are at a pH of at least 8.8 are easier to peel than less basic eggs is appealing, and techniques are being tested to attempt to manipulate egg storage to achieve this pH level. Hard-cooked eggs can be held in a brine or organic acid solution to help preserve them, but the solution used will modify flavor somewhat if the eggs are held more than a day.

**Figure 16.9**   Hard-cooked eggs are heated enough to coagulate the yolk and white proteins completely and the egg slices easily, but the white is not tough.

***Poached Eggs.***   Poached eggs should have the white congealed tenderly surrounding the yolk, which is slightly coagulated and still not set, even at the edges. Gently slipping a high-quality egg into salted water heated to just under boiling and then simmering until the desired endpoint is achieved produces a pleasing poached egg. To assure destruction of *Salmonella enteritidis* that might be present in the yolk, the Food and Drug Administration recommends poaching eggs for 5 minutes in boiling water.

Whites of lower-quality eggs spread and give a jagged, rather feathery appearance when they are poached; the yolk tends to separate or pull away from the white. The yolk also is likely to break when the poached egg is spooned from the water. Addition of a small amount of acid (vinegar or lemon juice) to the water when low-quality eggs are poached can reduce spreading of whites. Adding salt to the water is helpful in poaching a low-quality egg, because it provides sodium and chloride ions to cancel the electrical charges on protein molecules and thus facilitates the coagulation process slightly.

The addition of acid when poaching low-quality eggs speeds the coagulation and minimizes spreading of the whites because the acidified water is closer to the isoelectric point of the albumen proteins, thus favoring hastened coagulation and diminished spreading. The reduced surface ionization of the albumen proteins near the isoelectric point enables the molecules to aggregate fairly quickly, thus reducing the time when the albumen can spread freely in the water.

Eggs can be confined in an egg poacher to avoid the problem of spreading in the poaching water. The small cups in a poacher eliminate this problem because the egg never has the opportunity to enter the water. Another physical approach to limiting spreading is to swirl the pan containing the simmering water until the water develops a whirlpool-like movement. Then the egg is slipped into the water in harmony with the movement of the water. The water helps hold the egg together when this motion is accomplished correctly, but it can be detrimental to limiting the spreading if the egg is moving in opposition to the water.

Even when the spreading problem of whites has been limited as much as possible in an egg of low quality, the flavor of such an egg will still be a bit sulfury and strong. The vitelline membrane also is likely to be too weak to withstand the stress of being spooned out of the poaching water or the poaching cup. A broken yolk is considered a clear sign of lack of quality in a poached egg.

***Fried Eggs.***   Egg quality and controlled heating are the keys to achieving a pleasing fried egg. The excessive amount of thin white and weak vitelline membrane found in eggs of less than AA quality allow a fried egg to cover a large surface area, with the white tending to drain from around the yolk and the yolk flattening and spreading. The risk of breaking the yolk while turning or removing it from the frying pan is great when the vitelline membrane is weak.

The Food and Drug Administration has recommended that sunnyside-up eggs should be fried at 121°C (250°F) for 7 minutes uncovered or for 4 minutes covered. Over-easy fried eggs should be cooked for 3 minutes at 121°C (250°F) before they are turned and fried for 2 more minutes. These methods are designed to assure that the yolks are heated sufficiently to kill any *Salmonella enteritidis*.

***Scrambled Eggs.***   Prior to heating scrambled eggs, milk usually is added to dilute the egg mixture. Sufficient beating is done to mix the egg albumen, yolk, and milk completely, but without creating a foam. Unless adequate mixing is done at this time, streaks of yellow or white may be seen in the finished product. Liquid (about 20 milliliters per egg) is added to dilute the egg proteins and promote tenderness in the finished product.

To assure that scrambled eggs do not contain viable *Salmonella enteritidis,* they need to be heated at 121°C (250°F) for 1 minute. If working in large quantities, no more than three quarts of eggs should be scrambled at a time. Eggs on a steam table need to be held at 60°C (140°F) or hotter.

Overcooking scrambled eggs prepared with added liquid will cause syneresis as the protein in the eggs draws tighter together and toughens from the heat. Without added liquid, the egg mass can become quite tough with overheating.

If scrambled eggs are held on a steam table for an extended period, they begin to turn an unsightly greenish-gray as ferrous sulfide forms. This is the same compound that forms in hard-cooked eggs with extended heating.

Microwave heating is useful when previously cooked scrambled egg products are reheated. The flavor is preferred over scrambled eggs that are reheated conventionally. French omelets and scrambled eggs that have been prepared and then chilled prior to being reheated and served are finding considerable use in settings such as airline meals and institutions because of their ability to be reheated satisfactorily in a microwave oven.

## Foams

*Yolk Foams.* Occasionally, yolk foams are utilized in certain baked products, notably sponge cakes and puffy omelets. Egg yolks can be beaten into somewhat heavy foams with considerable effort. However, yolk foams are quite compact compared with white foams. As the yolks are beaten, a concomitant increase in volume and lightening of foam color can be observed. After several minutes of beating at top speed with an electric mixer, a moderately thick, light yellow yolk foam forms. This yolk foam is ready for use and usually is folded into an egg-white foam.

Yolk foams are sufficiently stable to permit them to be folded with other ingredients and then baked, thus contributing to the textural characteristics and volume of the baked product. They do not become stiff enough to form peaks, but they do pile when beaten enough. Whole egg also can be beaten to form a soft foam.

*White Foams.* Foams (see Chapter 6) need to be stable to have a role in food preparation. Stability is enhanced if the surface tension is low, if vapor pressure is low, and if a substance solidifies on the surface of the bubbles. Egg white meets all of these requirements, making it a particularly useful ingredient when a foam is desired. With agitation, the egg albumen can be spread over a large surface area, and air can be incorporated into the bubbles created by beating the proteins. Some of the proteins are denatured by the beating action, and then they aggregate to enhance stability of the developing foam.

The two factors of utmost importance in egg white foams are stability and volume. Several factors influence one or both of these characteristics and are of significance regardless of the product into which the foam is ultimately incorporated. Rate of formation is of interest, although the use of mechanical means to develop the foams makes speed somewhat less important than stability and volume.

*Stability.* Stability of a foam can be judged by measuring drainage of liquid from a given quantity of foam over a specified period of time. The conditions for such a measurement need to be determined for a given experiment so that results from variations can be compared, but the specifics may differ from one research project to another, making it impossible to compare results from different projects.

A convenient technique is to place a weighed amount of foam in a funnel of known capacity and bore which drains into a graduated cylinder. The funnel and foam need to be covered with plastic wrap to prevent evaporation. The drainage is read after the specified length of time. The more drainage, the less stable is the foam.

The extent of beating is an important factor in stability of an egg white foam. As beating progresses, the foam becomes increasingly stable up to a critical point, after which continued beating decreases stability. Maximum stability is reached when the whites just bend over, but before maximum volume has been reached. If beating continues beyond the point of maximum stability, the surface begins to look slightly dry, and the foam exhibits some brittleness. Foam formation is delayed when the whites are well below room temperature.

The addition of other ingredients also influences stability. Sometimes salt is added to an egg white foam for flavor, but this addition reduces stability slightly. Occasionally recipes include some added liquid in making an egg white foam. This dilutes the proteins in the foam and decreases stability. If yolk happens to contaminate the white at all, as can happen during the separation of yolks and whites, stability of the foam formed from the whites is reduced.

Not all ingredients reduce stability. In fact, the addition of sugar has a positive effect on foam stability. A possible explanation is that the addition of sugar delays foam formation significantly, which means that considerably more beating is necessary to reach the proper stage of foam development. This increased beating results in a foam with a finer texture and more surface area; this foam is stabilized with protein that has been partially denatured by beating.

Acidic ingredients, commonly either cream of tartar or lemon juice, are useful stabilizing agents when making egg white foams, particularly when added early in the formation of the foam (Figures 16.10). Although stability is promoted by reducing the pH of the egg white foam, formation of the foam is delayed by this addition. Again, the delay in reaching the desired endpoint in whipping the foam results in increased total agitation and a finer, more stable foam. Cream of tartar is particularly effective as the acid ingredient when the pH of the white foam approaches 6.0, whereas citric acid and cream of tartar are about comparable in their effect at pH 8.0.

*Volume.*   The temperature of the whites influences volume, because surface tension is greater in whites just removed from the refrigerator than in whites warmed to room temperature. The quality of the whites plays a significant role in determining the volume of the foam. When there is a comparatively large amount of thin white, the foam forms quickly and reaches a large volume, but that volume is reduced when beating is extended. In

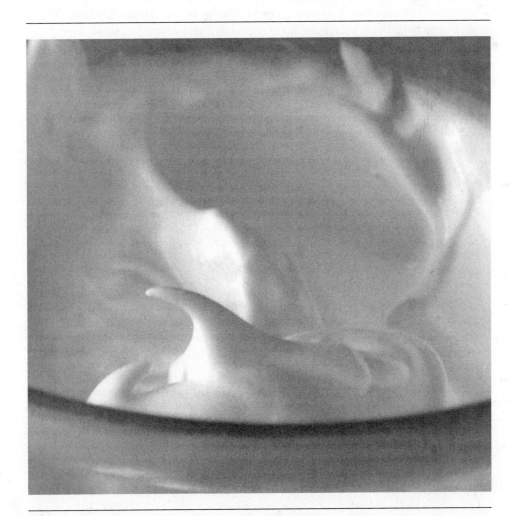

**Figure 16.10**   This egg white foam has a fine texture, stability, and volume because it has been stabilized by adding cream of tartar and has been beaten to the point where the peaks just bend over.

contrast, thick white requires more beating to reach the desired endpoint, but the foam achieved holds its volume well.

The type of beater is yet another factor influencing volume. A wire whisk can create a foam with larger volume than is achieved using a rotary hand beater if the whisk is in the hands of an experienced chef, but lack of experience can have a detrimental effect on volume.

Electric mixers can achieve foams with high volumes, but they may produce foams that are overbeaten and consequently of low volume, unless sugar or acid is added to delay foam formation. When these ingredients are added, an extended beating time is required to reach the desired endpoint, at which the peaks just bend over; this makes use of an electric mixer desirable.

The extent of beating has a definite influence on volume, as well as stability. Volume increases gradually as beating progresses to the foamy stage, the point at which the whites are still transparent and fluid, yet definitely are whipped into large bubbles throughout. With further beating beyond the foamy stage, the foam continues to expand, gradually becomes more opaque, and exhibits the ability to be pulled up into peaks. At first these peaks sag quickly; then rounded, soft peaks form that progress gradually to stiff peaks as beating continues. At this point, maximum volume is achieved. Additional beating beyond stiff peaks results in a dry, rather brittle foam of smaller volume (Figures 16.11). Foams beaten to the dry stage are not useful in food preparation.

Added ingredients definitely influence volume. Acid ingredients reduce volume a little (but the increased stability helps offset this negative effect). Sugar has a similar impact on volume. However, sugar has a positive effect by promoting elasticity and stability. These foams are easy to fold with other ingredients. Little volume is lost during careful folding if the egg white foams have been beaten sufficiently to attain peaks that just bend over.

Contamination of the whites with even a trace of yolk has a negative impact on the volume of an egg white foam. This is due to the presence of fat in the yolk in the compounds lipovitellenin and lipovitellin, both of which interfere with the foaming of two

**Figure 16.11**    When egg white foams are beaten excessively, the foam is too brittle to be folded with other ingredients, which results in pockets of egg white and poor volume.

key foaming proteins in the white, ovomucin and lysozyme. Salt is yet another ingredient with a slightly detrimental effect on volume.

Water is the one ingredient that has a positive effect on volume of egg white foams while they are being beaten. However, the effect is lost almost immediately because of lack of stability. Unfortunately, the dilution of the protein resulting from the addition of water decreases foam stability, which limits the usefulness of water in increasing volume.

***Meringues.*** **Meringues** are egg white foams containing sugar. The quantity of sugar used in their formulations determines whether the meringue is classified as a soft or a hard meringue. **Hard meringues** contain the most sugar (50 g or 4 tablespoons) per white. The extremely large quantity of sugar in hard meringues increases the amount of beating required, but beating the whites to the soft peak stage before adding any sugar, minimizes the time. Then the sugar should be added gradually while operating the electric mixer at full speed. Beating should be continued until stiff peaks can be pulled up (Figures 16.12); then dollops are swirled into a nest-like shape on parchment paper for baking at a low temperature until dry, but not browned. Overbeating is virtually impossible because of the large amount of sugar. These crisp shells are used as the base for a scoop of ice cream or custard and fruit.

Hard meringues are totally different from soft meringues in their characteristics. Instead of being soft and pleasingly browned, hard meringues are considered ideal when they are crisp and dry, with an extremely pale surface. They should cut easily because of their crispness and tenderness. If they are not dried sufficiently, they will be tough and slightly sticky to cut.

**Soft meringues**, the topping often used on cream pies, usually contain 25 g (2 tablespoons) of sugar per egg white, which is a volume approximately equivalent to that of the white (Figures 16.13). Soft meringues are delicate to make because of their propensity for beading and leaking (drainage of liquid from the meringue to the surface of the pie filling on standing). An increase to 32 g $\left(2\frac{1}{2}\text{ tablespoons}\right)$ per egg white results in a soft meringue with excellent appearance and superior cutting quality (but with extra calories).

**Leaking**, sometimes seen in soft meringues, can be caused by failure to coagulate all the proteins in the foam. Placing the meringue on the filling while the filling is still hot [60–77°C (140–171°F)] and then baking immediately usually eliminates this problem. Heat

**Meringue**
Egg white foam containing sugar.

**Hard meringue**
Egg white foam containing about 50 g (4 tablespoons) of sugar per egg white; baked to a dry, brittle cookie or dessert shell.

**Soft meringue**
Egg white foam containing about 25 g (2 tablespoons) of sugar per egg white; topping on cream pies.

**Leaking**
Draining liquid from a soft meringue to the surface of the filling of a cream pie.

**Figure 16.12** Because of their high sugar content, hard meringues require extensive beating to form a stable foam that can be shaped and baked at a low temperature to denature the protein.

**Figure 16.13**    A cloud of poached soft meringue floats in a stirred custard garnished with sliced fresh strawberries.

from the filling helps bring the temperature of the foam at the bottom of the meringue to a high enough point that the egg white proteins can coagulate. This is necessary because the foam itself conducts heat poorly during baking. Baking at 177°C (350°F) for 15 minutes is sufficient to kill any *Salmonella enteritidis* present.

**Beading** is the formation of droplets of a golden-brown color when the finished pie has had an opportunity to stand for a while. It results from the over-coagulation of some of the protein at the surface of the baked meringue. The problem is most likely found when the meringue has been placed on a hot pie filling because the additional heat contained may bake the meringue more quickly than anticipated, leading to overheating.

One effective technique to use when the meringue has been placed on a hot filling is to bake it in a hot oven [218°C (425°F)] until a pleasing golden-brown color develops, which takes a little more than 4 minutes. This quick heating limits the amount of residual heat accumulated in the interior and reduces the likelihood of over-baking and beading. However, leakage may be greater when baked at that temperature than if the meringue is baked at 163°C (325°F), the temperature ordinarily suggested for baking meringues. Meringues baked at 218°C (425°F) are more tender and less sticky than those baked at 163°C (325°F). Regardless of the oven temperature used, any regions of a meringue that are over-baked not only are prone to beading, but also are likely to become sticky and difficult to cut. Despite the baking conditions, any soft meringue will be sticky and hard to manage within 24 hours after baking. They need to be consumed the day they are made.

Commercial bakers are confronted with the difficulty of making meringue pies that are still pleasing when the customer is ready to cut and eat the pie. Meringues for commercial use may have carrageenan or some other type of gum or stabilizer added to improve the storage characteristics of the meringue portion of the pie.

Puffy or fluffy omelets and soufflés are other egg dishes in which egg foams are used. The egg white foam in both omelets and soufflés can acquire optimum stability if cream of tartar, lemon juice, or other compatible acidic juice is added to reduce the pH of the egg

**Beading**
Small drops of amber-colored droplets on the surface of an over-baked soft meringue.

whites. The yolk used in making the sauce for soufflés is important to help emulsify the comparatively large amount of fat in the thick sauce that is folded into the egg white foam.

*Foam Cakes.*   Egg foams figure prominently in three types of foam cakes: angel food, sponge, and chiffon. The formation of high-quality egg white foams is critical to the final product in all three cakes. In addition, egg yolk foam is an important component of sponge cakes. Chapter 17 discusses these cakes in detail.

## EGG SUBSTITUTES

Concern over the cholesterol content of egg yolks has prompted the development of egg substitutes. The basic goal is to simulate whole egg without any cholesterol. In other words, the product is based on the use of egg white and a synthesized yolk mixture. Because the yolk is replaced, the color of the final product depends on the inclusion of a coloring agent in the formulation. Usually, a carotenoid in the form of one of the carotenes is added for this purpose.

Egg substitutes are available in either the frozen form or a fresh, fluid product requiring refrigeration. They provide an alternative to abstention from eggs for people who must limit their cholesterol intake, but are not identical to whole eggs in their potential uses. Because egg substitutes cannot be separated into whites and yolks, they cannot be used in products such as sponge cakes and soufflés. Also, their aroma and flavor are not as good as when fresh whole eggs are used. Volume is greater in cakes made with eggs than in those made with egg substitutes. Fortunately, many whole egg products can be made with egg substitutes to obtain results that are satisfactory even though they are not quite equal to the same products made with fresh eggs.

## SUMMARY

An egg is a complex structure consisting of a multilayered yolk encased in the vitelline membrane and a chalaziferous layer, all of which is surrounded by three layers of albumen (thin, thick, and thin layers). An inner membrane and an outer membrane separate the contents of the egg from the porous shell. The white is largely water, with a useful amount of protein. The yolk contains somewhat less water, a significant amount of lipids, and about half as much protein as the white. Among the lipids in the yolk is lecithin, an effective emulsifying agent. Cholesterol also is found in the yolk. Egg substitutes have been developed and marketed to enable consumers to avoid cholesterol while eating egg-containing foods.

Twelve egg white proteins have been studied. Ovalbumin, by far the most abundant of the proteins in the white, and lysozyme are especially important because of their foaming ability, although the others also contribute to the unique foaming qualities of egg white. Conalbumin is the protein that binds with ions of copper and iron in complexes that produce undesirable color changes when present. Ovomucin contributes much of the fibrous character to the thick albumen.

The prominent yolk proteins in the plasma are livetin (found in three fractions) and the low-density lipoproteins; the proteins prominent in the granules of egg yolk are lipovitellin (two forms, both of which are high-density lipoproteins), phosvitin, and low-density lipoprotein.

Egg quality is determined in the shell by candling and out of the shell by measurements such as yolk index and albumen index or Haugh units. Considerable deterioration of egg quality is likely to occur between the grading and marketing of eggs. The pH, particularly of the white, rises significantly with loss of carbon dioxide through the shell. Some moisture loss during storage contributes to the growth of the air cell at the large end of the egg. The white becomes thinner, and the vitelline membrane weakens.

Because of the easy deterioration of eggs, they need to be stored carefully under refrigeration or else processed for longer storage. Processing may include pasteurizing,

**Table 16.7** Functional Properties of Eggs and Some Applications

| Coloring Agent | Emulsification | Thickening Agent | Foam |
|---|---|---|---|
| Vanilla cream pudding | Mayonnaise | Cream pie fillings | Meringue |
|  | Cream puffs | Custards | Soufflés |
| Lemon meringue pie filling | Hollandaise sauce | Quiche | Angel food cake |
|  |  | Cream puddings | Sponge cake |
|  |  |  | Chiffon cake |

drying, or freezing. The glucose in whites must be digested prior to drying to avoid serious problems with discoloration. Either sugar or salt must be added to yolks if they are to be frozen, because of the gummy character that develops during frozen storage.

Eggs are an excellent medium for the growth of microorganisms. To avoid bacterial contamination, shells should be uncracked and should be cleaned and sanitized prior to commercial refrigerated storage. *Salmonella enteritidis* sometimes may be present in the yolk of an egg if the hen producing the egg is infected. Adequate heat treatment is needed for all forms of egg preparation to assure safety from this bacterium.

Eggs are important because of their functional properties (Table 16.7).

## STUDY QUESTIONS

1. Sketch the cross section of an egg, naming the parts and describing the changes that take place as egg quality declines.

2. What is the significance of the presence of lecithin in egg yolk? Describe its role in at least five different food products.

3. What occurs when egg white comes into contact with (a) copper ions and (b) iron ions? How can this happen?

4. Identify six important albumen proteins and describe unique features of each.

5. What changes are made when an egg substitute is formulated (in comparison with the natural egg)? Why are these changes made?

6. Identify five factors that influence the coagulation temperature of the proteins in a custard and explain the action of each factor.

7. Compare the use of eggs in custards with the use of eggs in cream puddings and pie fillings. How are the eggs combined with other ingredients in the two types of products?

8. What factors influence the stability of an egg white foam? Explain the effect of each factor.

9. What factors influence the volume achieved in an egg white foam? Explain the effect of each factor.

10. True or false. For good nutrition, it is important to buy eggs that are marketed as high in omega-3 fatty acids.

11. True or false. Grade AA eggs are guaranteed to be better in quality than grade B when purchased.

12. True or false. As eggs age, the amount of thick white increases.

13. True or false. *Salmonella enteritidis* contamination usually occurs during storage of eggs.

14. True or false. Excessive heating has very little effect on egg-thickened sauces.

15. True or false. The addition of baking soda promotes stability of egg white foams.

## BIBLIOGRAPHY

Bell, D. D. and Weaver, W. D., Jr. 2001. *Commercial Chicken Meat and Egg Production*. 5th ed. Springer. New York.

Clemens, R. and Pressman, P. 2005. Avian flu, Chicken little, Owl wise. *Food Technol.* 59 (11): 20.

Damodaran, S. 1997. *Food Proteins and Their Applications*. CRC Press. Boca Raton, FL.

Janky, D. M. 1986. Variation in pigmentation and interior quality of commercially available table eggs. *Poultry Sci.* 65: 607.

Kim, K. and Setser, C. A. 1982. Foaming properties of fresh and commercially dried eggs in presence of stabilizers and surfactants. *Poultry Sci.* 61: 2194.

Leutzinger, R. L., et al. 1977. Sensory attributes of commercial egg substitutes. *J. Food Sci.* 42: 1124.

Miraglio, A. M. 2005. Return of the egg. *Food Product Design 15* (1): 93.

Powrie, W. D. and Nakai, S. 1985. Characteristics of edible fluids of animal origin: Eggs. In Fennema, O. R., ed. *Food Chemistry.* 2nd ed. Marcel Dekker. New York, p. 829.

Richardson, T. and Kester, J. J. 1984. Chemical modifications that affect nutritional and functional properties of proteins. *J. Chem. Educ. 61* (4): 325.

Sauter, E. A. and Montoure, J. E. 1972. Relation of lysozyme content of egg white to volume and stability of foam. *J. Food Sci. 37*: 918.

Schmidt, R. H. 1981. *Gelation and Coagulation.* ACS Symposium Series 147; American Chemical Society. Washington, DC.

Shapiro, L. S. 2001. *Introduction to Animal Science.* Prentice Hall. Upper Saddle River, NJ.

Sheldon, B. W. 1986. Influence of three organic acids on quality characteristics of hard-cooked eggs. *Poultry Sci. 65*: 294.

Sheldon, B. W. and Kimsey, H. R., Jr. 1985. Effects of cooking methods on chemical, physical, and sensory properties of hard-cooked eggs. *Poultry Sci. 64*: 84.

Sikorski, Z. E. 2001. *Chemical and Functional Properties of Food Proteins.* CRC Press. Boca Raton, FL.

Spencer, J. V. and Tryhnew, L. J. 1973. Effect of storage on peeling quality and flavor of hard-cooked shell eggs. *Poultry Sci. 52*: 654.

Stadelman, W. J., et al. 1995. *Egg Science and Technology.* 4th ed. CRC Press. Boca Raton, FL.

Whitaker, J. R. and Tannenbaum, S. R., eds. 1977. *Food Proteins.* AVI Publishing: Westport, CT.

Woodward, S. A. and Cotterill, O. J. 1987. Texture and microstructure of cooked whole egg yolks and heat-formed gels of stirred egg yolk. *J. Food Sci. 52*: 63.

Woodward, S. A. and Cotterill, O. J. 1986. Texture and microstructure of heat-formed egg white gels. *J. Food Sci. 51* (2): 333.

Yang, S. S. and Cotterill, O. J. 1989. Physical and functional properties of 10% salted egg yolk in mayonnaise. *J. Food Sci. 54*: 210.

## INTO THE WEB

*www.incredibleegg.org/egg-facts/eggcyclopedia*—Extensive site on eggs, including a dictionary of terms and videos about egg production.

*www.aeb.org*—American Egg Board site for multiple topics about eggs.

*http://chemistry.about.com/od/howthingsworkfaqs/f/copper bowl.htm*—Role of copper bowl in making egg white foams.

*http://www.ams.usda.gov/AMSv1.0/getfile?dDocName=ST ELDEV3004690*—Information on federal egg grading requirements.

*www.fsis.usda.gov/*—Search by typing in eggs.

*http://www.cdc.gov/ncidod/dbmd/diseaseinfo/salment_g. htm*—Background information on *Salmonella enteritidis* infection in eggs.

*http://www.cfsan.fda.gov/~dms/fs-toc.html#eggs*—Center for Food Safety and Applied Nutrition and FDA information on egg safety.

*http://www.safeeggs.com/eggs/how-eggs-are-pasteurized. html*—Information on pasteurizing eggs in the shell.

*http://www.safeeggs.com/National-Pasteurized-Eggs.html*—Information on National Pasteurized Eggs.

*http://www.fsis.usda.gov/factsheets/Egg_Products_and_ Food_Safety/index.asp*—Information on processed eggs.

*http://www.incredibleegg.org/egg-facts/eggcyclopedia/e/egg-products*—Description of various egg products.

*http://www.chm.bris.ac.uk/webprojects2002/rakotomalala/ maillard.htm*—Chemistry of the Maillard reaction.

*http://www.unu.edu/Unupress/food/8F032e/8F032E03. htm*—Information about Chinese methods for preserving eggs.

*http://www.eggbeaters.com/healthcare/index.jsp*—Information on egg substitutes.

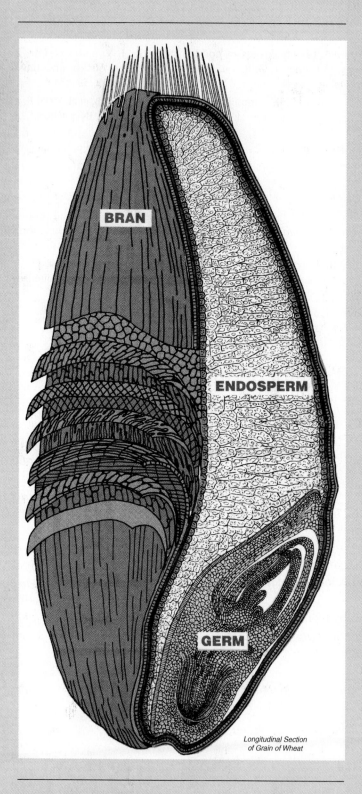

BRAN

ENDOSPERM

GERM

*Longitudinal Section
of Grain of Wheat*

Diagram of a wheat kernel. (Courtesy of Wheat Flour Institute.)

# CHAPTER 17

# Dimensions of Baking

## Chapter Outline

## OBJECTIVES

After studying this chapter, you will be able to:

1. Describe the production of flours from grain.
2. Discuss gluten, its role and appropriate development in baked products.
3. Contrast use of various flours in baked products.
4. Explain the role of key ingredients in baked products.
5. Identify appropriate ingredient choices for specific baking applications.

## INTRODUCTION

Baked products range in complexity from the simple ingredients of a plain pastry to the numerous components of a shortened cake. Despite the many differences that exist, both in ingredients and in techniques for combining and baking them, baked products are based on the use of certain basic ingredients and on the roles that these ingredients can perform, both in mixing and in baking. This chapter presents these key dimensions of baking to provide the scientific foundation for the study of diverse baked products.

## WHEAT FLOUR

### Milling

**Flour**
The fine particles of wheat (or other grain) produced by milling.

**Milling**
Grinding and refining of cereal grains.

**Tempering**
In the context of flour milling, this is the steam treatment preliminary to grinding and is the means of facilitating removal of outer bran layers.

**Grinding**
Abrasive action by corrugated rollers or stones to crack the grain so bran and germ can be split from the endosperm.

**Bran**
Outer layers (fibrous and very high in cellulose) encasing the interior endosperm and germ of cereal grains.

**Endosperm**
Large inner portion of cereal grains composed largely of starch and some protein.

**Germ (embryo)**
Small portion of cereal grain containing fat and a small amount of protein as well as thiamine, riboflavin, and other B vitamins.

**Whole wheat flour**
Wheat flour containing most of the bran and shorts.

**Refined flour**
White flour resulting from removal of at least 25 percent of the bran and shorts during milling.

**Extraction**
Removal of bran and shorts during milling of wheat.

Although the word **flour** ordinarily means wheat flour specifically, technically *flour* is the product from milling any type of grain. Usually the name of the grain is included (e.g., rice flour) if the flour is made from a grain other than wheat to eliminate ambiguity. In this book, *flour* means wheat flour unless another grain is indicated.

Flour production involves grinding and refining cereal grains by a process called *milling*. **Milling** consists of many steps, but the basic objective is to produce a comparatively fine powder consisting of some portion or almost all of the grain, depending on the specific product desired. Milling begins with the whole grains of wheat (or other cereal), which are first subjected to a brief **tempering** treatment with steam to ease removal of the outer bran layers. At this point, more than one variety of wheat (or other grain) may be blended together and any unsound grains removed (Chart 17.1).

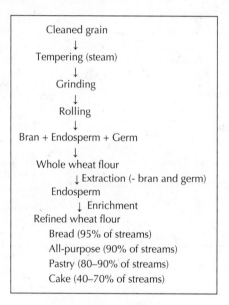

Cleaned grain
↓
Tempering (steam)
↓
Grinding
↓
Rolling
↓
Bran + Endosperm + Germ
↓
Whole wheat flour
↓ Extraction (- bran and germ)
Endosperm
↓ Enrichment
Refined wheat flour
Bread (95% of streams)
All-purpose (90% of streams)
Pastry (80–90% of streams)
Cake (40–70% of streams)

**Chart 17.1    Overview of milling**

**Grinding** is the process that begins to split the grains in preparation for passage through corrugated rollers that break the grain into coarse particles. This action results in splitting of the grains into their components, as well as a breaking up of those parts. The outer layers, which are somewhat brittle because of their high fiber content, are ground into moderately small, thin, and jagged pieces of **bran**. The major portion of the grain, the **endosperm**, is broken into coarse particles, as well. At this point, the small **germ (embryo)** may still be adhering to the endosperm.

Following grinding, much of the bran and shorts (layer just inside the outer bran layer) may be left with the endosperm to be made into **whole wheat flour** (Figure 17.1). The greater the portion of bran retained, the darker the whole wheat flour will be.

To make white, **refined flour** from wheat, the bran and shorts are removed, a process called **extraction**. Extraction is accomplished by sifting or by an air classification system that blows the bran up from the heavier endosperm and germ portions, allowing the bran to be directed out of the system for subsequent processing and packaging as bran or shorts.

The remaining endosperm and germ are passed through reducing rollers to press the germ into flakes and to continue breaking the endosperm into fine particles. This enables

**Nutrition Facts**
Serving Size ¼ cup (30g)
Servings Per Container about 75

| Amount Per Serving | |
| --- | --- |
| **Calories** | 100 |
| Calories from Fat | 5 |

| | % Daily Value* |
| --- | --- |
| **Total Fat** 0.5g | 1% |
| Saturated Fat 0g | 0% |
| Trans Fat 0g | |
| Polyunsaturated Fat 0g | |
| Monounsaturated Fat 0g | |
| **Cholesterol** 0mg | 0% |
| **Sodium** 0mg | 0% |
| **Potassium** 110mg | 3% |
| **Total Carbohydrate** 21g | 7% |
| Dietary Fiber 3g | 12% |
| Other Carbohydrate 18g | |
| **Protein** 4g | |

| | | |
| --- | --- | --- |
| Iron 2% | • | Thiamin 10% |
| Riboflavin 2% | • | Niacin 6% |

Not a significant source of sugars, vitamin A, vitamin C and calcium.

* Percent Daily Values are based on a 2,000 calorie diet. Your daily values may be higher or lower depending on your calorie needs:

| | | Calories | 2,000 | 2,500 |
| --- | --- | --- | --- | --- |
| Total Fat | Less than | | 65g | 80g |
| Sat Fat | Less than | | 20g | 25g |
| Cholesterol | Less than | | 300mg | 300mg |
| Sodium | Less than | | 2,400mg | 2,400mg |
| Potassium | | | 3,500mg | 3,500mg |
| Total Carbohydrate | | | 300g | 375g |
| Dietary Fiber | | | 25g | 30g |

INGREDIENTS: WHOLE WHEAT FLOUR.

DISTRIBUTED BY **General Mills Sales, Inc.**
GENERAL OFFICES, MINNEAPOLIS, MN 55440 USA

© 2008 General Mills

**Figure 17.1**   Whole wheat flour is made using some of the bran and germ, as well as the endosperm of wheat.

the germ and endosperm to be separated by sifting, and the germ is removed, which enhances the shelf life of the flour because the fat that would eventually become rancid is not present.

## Types of Milled Wheat Flour

It is possible to make several different refined flours by milling wheat and removing the bran and shorts. The character of these flours is altered by the amount of the total flour used. The flour made using only a small portion of the total stream of flour is extra-short or fancy-patent flour. **Cake flour** often is made from this type of flour (Table 17.1) or from **short-patent flour**, which is a flour comparatively high in starch and low in protein. **Medium-patent flour** contains 90 percent of the flour streams, resulting in a protein level that is greater than cake flour and a starch content that is a bit less. Bread flour uses between 95 and 100 percent of the flour streams, producing a flour with still higher protein content.

The most common wheat flour on the retail market is **all-purpose flour**, which contains all or mostly hard wheat (Figure 17.2). The protein level in all-purpose flour is

**Cake flour**
Soft wheat, short-patent wheat flour (about 7.5 percent protein).

**Short-patent flour**
Wheat flour comparatively high in starch and low in protein, the result of using the very fine particles of flour from the center of the endosperm (e.g., cake flour).

**Medium-patent flour**
Wheat flour using about 90 percent of the flour streams, resulting in a somewhat higher protein content and relatively less starch (e.g., all-purpose flour).

**All-purpose flour**
Multiuse flour made from hard wheat or a mixture of hard and soft wheat; contains about 10.5 percent protein.

**Table 17.1**  Profile of Milled Wheat Flours

| Type of Flour | Patent | Mill Stream (percentage) | Protein (percentage) | Type of Wheat |
|---|---|---|---|---|
| Cake | Fancy and/or short | 40–70 | 7.5 | Soft |
| Pastry | Short and medium | 80–90 | 7.9 | Soft |
| All-purpose | Medium | 90 | 10.5 | Hard or hard and soft blend |
| Bread | Long | 95 | 11.8 | Hard |

**Bread flour**
Hard wheat, long-patent flour with a protein level of about 11.8 percent.

**Long-patent flour**
Wheat flour made from 95 to 100 percent of the flour streams, yielding flour of rather high protein content (e.g., bread flour).

approximately 10.5 percent, which accounts for its comparatively strong textural properties. Home baking of cookies, pastries, and quick breads usually is done using all-purpose flour. Cakes, however, preferably are made with cake flour, the flour made from soft wheat. Usually the protein content of cake flour is only about 7.5 percent.

Commercial bakers have more options in protein levels in flours than do home bakers. **Bread flour**, one of these options, is the flour obtained by using almost all of the streams and is designated as a **long-patent flour**. Including such a large proportion of the streams incorporates the edges of the endosperm, resulting in greater gluten (protein) content. This type of

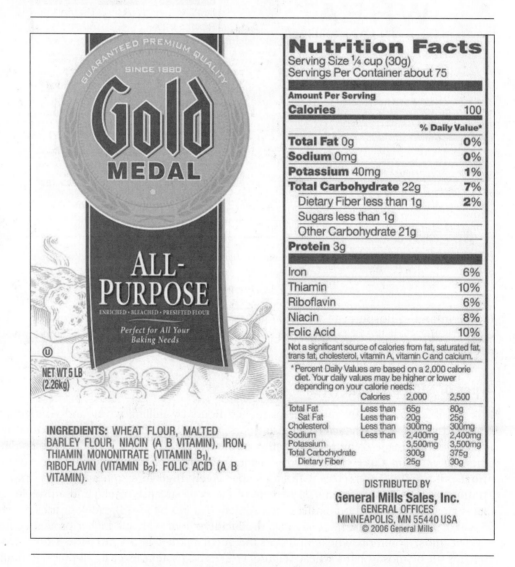

**Figure 17.2**  All-purpose flour is about 10.5 percent protein and can be used in a number of different baking applications including breads, some cakes, cookies, and pastries.

flour, made entirely from hard wheat, has a protein level of about 11.8 percent. The strength of the protein in bread flour is excellent for the physical properties desired in yeast breads.

Another option available to commercial bakers is **pastry flour**. Its protein level of about 9.7 percent is intermediate between all-purpose and cake flours. Production of pastry flour involves use of 80–90 percent of the streams (short- and medium-patent) of milled soft wheat. As the name implies, this flour is tailored specifically to the requirements for making a tender pastry.

**Self-rising flour** is a unique flour in the retail market that may be considered as a "semi-mix" or shortcut to baking because it contains both leavening and salt (Figure 17.3). The acid salts (monocalcium phosphate and sometimes sodium acid pyrophosphate and sodium aluminum phosphate) and baking soda are the leavening ingredients added in amounts sufficient to generate carbon dioxide equivalent to the use of $1\frac{1}{2}$ teaspoons of baking powder per cup of flour. The salt (NaCl) level is equal to $\frac{1}{2}$ teaspoon per cup of flour.

Most commonly, self-rising flour is used in making quick breads, particularly sour milk biscuits. Its inherent leavening action makes it an unsuitable choice for pastry and other unleavened products or for yeast breads in which yeast provides the leavening needed. If self-rising flour is substituted in a recipe for regular all-purpose flour, the baking powder and salt in the recipe should be altered to compensate for the presence of these ingredients in the self-rising flour. Conversely, all-purpose flour can be substituted for self-rising flour if baking powder and salt are added ($1\frac{1}{2}$ teaspoons of baking powder and $\frac{1}{2}$ teaspoon of salt per cup of flour).

**Pastry flour**
Soft wheat, short- and medium-patent flour with a protein level of about 9.7 percent.

**Self-rising flour**
Flour (usually soft wheat) to which baking powder and salt have been added during production.

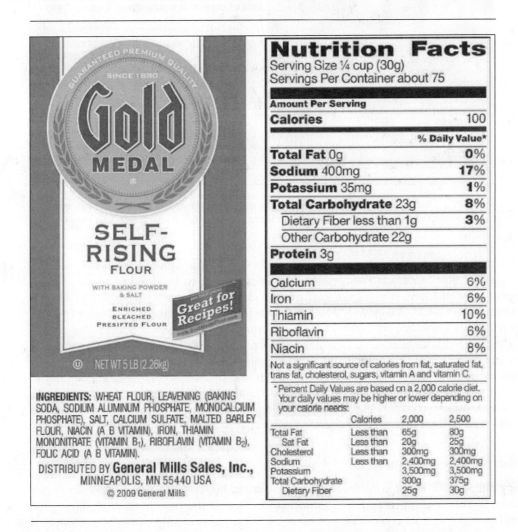

**Figure 17.3**  Self-rising flour contains leavening agents; if it is substituted for regular flour in a recipe, adjustments in salt and leavening need to be made.

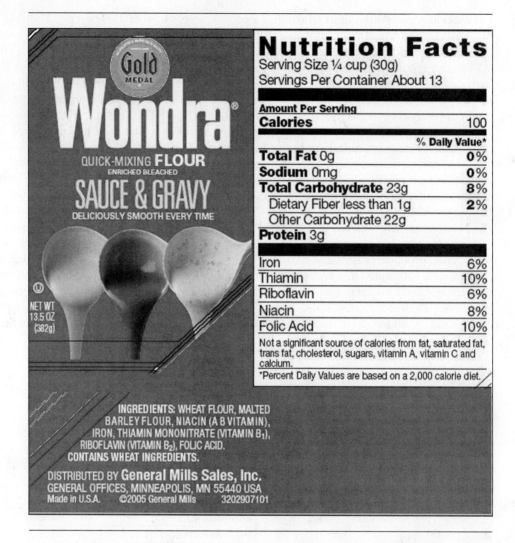

**Figure 17.4**   Instantized flour is made by adding enough moisture so that tiny clumps roll together during production, which eliminates the need for sifting.

**Instantized flour**
Flour made of sand-like granules produced by adding moisture to agglomerate the flour, then drying to produce the granules.

**Instantized flour** is produced by passing wheat flour through moist air, causing it to stick together (agglomerate). This agglomerated flour can then be dried to produce small pellets of flour that neither pack nor contain fine dust of flour particles (Figure 17.4). The resulting product blends easily with water and does not require sifting, thus overcoming two inconveniences associated with regular flour. Unfortunately, gluten does not develop readily when instantized flour is used, which limits the usefulness of this type of flour in making batter and dough products, particularly breads.

**Gluten flour**
Specialty wheat flour made by adding vital wheat gluten to increase the protein level to about 41 percent.

Some specialty stores sell **gluten flour**, which also is used occasionally in commercial breads in which a high protein content is desired. **Vital wheat gluten** (concentrated protein from wheat) is added to flour to bring the protein level to about 41 percent. Bread made with this type of flour is quite chewy and tough, yet very palatable.

**Vital wheat gluten**
Dried crude gluten.

## Modifying Wheat Flour

Freshly milled flour is not considered ideal for baking, but two procedures commonly are followed to enhance appearance and performance. One of the objections to freshly milled flour (Figure 17.5) is its somewhat yellow color, resulting primarily from the presence of xanthophylls (carotenoid pigments). If flour were allowed to stand exposed to air for several weeks, the xanthophylls would be bleached, and the flour would develop the desired

**Figure 17.5** Unbleached flour is slightly darker than bleached flours and produces baked products with a slightly more compact texture.

white color. This is not a practical procedure when dealing with large quantities of flour. However, addition of a small amount of **benzoyl peroxide** to flour (a practice approved by the U.S. Food and Drug Administration) bleaches the xanthophylls rapidly and at little cost. Furthermore, benzoyl peroxide is an additive that seems to have no effect on the baking quality of the flour to which it is added.

The baking performance of wheat flours improves during **aging (maturing)**, apparently due to changes in the protein during storage. Such changes can be accelerated greatly by the addition of chemical compounds classified as maturing agents, which include chlorine (as gaseous chlorine, chlorine dioxide gas, or a mixture of nitrosyl chloride and chlorine), acetone peroxide, and azodicarbonamide.

As flour is aged or matured, some of the sulfhydryl groups (—SH) in the protein are altered. Apparently, cross-linking occurs between some sulfhydryl groups to form disulfide (—S—S—) bonds. This change alters the ability of the flour proteins to stretch and makes doughs containing aged flours exhibit reduced extensibility, which improves the adhesive quality needed in baked products. Bread doughs made with flour with too many sulfhydryl bonds and too few disulfide bonds are sticky and inelastic.

Cake flour especially depends on the addition of gaseous chlorine to mature the flour and lower the pH. Particularly, rich cakes are quite susceptible to falling during baking if the

**Benzoyl peroxide**
Food additive approved by the FDA for bleaching the xanthophylls in refined flours.

**Aging (maturing)**
Chemical process used to alter the physical behavior of flour proteins by modifying some of the sulfhydryl groups to disulfide linkages; maturing agents include gaseous chlorine, chlorine dioxide gas, a mixture of nitrosyl chloride and chlorine, acetone peroxide, and azodicarbonamide.

---

### FOOD FOR THOUGHT: Innovative Wheat Flour

USDA researchers working in North Dakota to develop unique strains of wheat from durum wheat have produced one described as a waxy durum wheat. Durum wheat, which is an important crop in North Dakota, is the favored type of wheat for making pastas. Its starch composition is approximately 76 percent amylopectin and 24 percent amylose. The starch in this new waxy durum wheat is essentially 100 percent amylopectin. Tests seeking possible uses for this flour include exploring its merits as an ingredient in breads.

This research revealed an interesting property of this flour. The relatively small amount of shortening (about 2 tablespoons) used in making a loaf of bread could be omitted if 20 percent of the dough weight was waxy durum wheat flour. This ability to replace shortening is of particular interest because of the recent concern about *trans* fatty acids in the diet. Shortenings often are used in formulas for bread and they contain *trans* fatty acids.

Although there is considerable interest in waxy durum wheat flour, it is not yet available commercially. Scientists in Fargo at North Dakota State University and at the USDA-ARS Red River Valley Agricultural Research Center have been conducting this research project for five years. More time is needed to develop this crop to the point of which farmers can obtain the seed and begin to grow waxy durum wheat commercially.

---

cake flour has not been treated with gaseous chlorine because of the weak protein in the untreated flour. Lipids also undergo beneficial changes during maturing of (chlorinated) flours.

Some nutrients are lost during processing and refining. Because of these nutrient losses and the fact that refined wheat-containing products are eaten in significant quantities, federal regulations require the enrichment of wheat products and also other cereal grains with thiamin, riboflavin, niacin, folic acid, and iron. The addition of these B vitamins and iron is important nutritionally but does not have negative effects on products made from enriched flours.

## Composition of Wheat Flours

Although the presence of proteins in various wheat flours was mentioned earlier, proteins are but part of the total composition of these products. Proteins are emphasized because of their key role in providing structure in baked products, yet they are present in far smaller amounts than starch (Table 17.2). Fat content is somewhat variable, but even at its maximum (in whole wheat flour), it contributes only 2 percent of the total components. The moisture level is relatively constant from one type of flour to another; ordinarily about 12 percent of the weight is due to moisture, and it may rise to a maximum of 15 percent.

Wheat kernels contain several carbohydrates. Cellulose is found primarily in the bran and is eliminated to a large extent during milling of refined flours. A small amount of cellulose is present in the endosperm, although at such low levels that its presence is not evident.

**Table 17.2**    Composition of Various Wheat Flours (average values, in percentage)

| Type of Flour | Carbohydrate | Protein | Water | Fat | Ash |
|---|---|---|---|---|---|
| Whole wheat (hard wheat) | 71.0 | 13.3 | 12 | 2.0 | 1.7 |
| Straight, hard wheat | 74.5 | 11.8 | 12 | 1.2 | 0.46 |
| Straight, soft wheat | 76.9 | 9.7 | 12 | 1.0 | 0.42 |
| All-purpose | 76.1 | 10.5 | 12 | 1.0 | 0.43 |
| Cake | 79.4 | 7.5 | 12 | 0.8 | 0.31 |

Watt, B. K. and Merrill, A. L. 1963. *Composition of Foods, Agricultural Handbook No. 8*. U.S. Department of Agriculture. Washington, DC.

**Pentosans**, polymers of arabinose and xylose (pentoses), are hemicelluloses occurring in cell walls in the endosperm. Both water-soluble and water-insoluble pentosans account for about 2–3 percent of the weight of refined flour. Although somewhat less than half of the pentosans are classified as water soluble, these pentosans retard the development of a cohesive dough during mixing. Dextrins, polymers of glucose just a bit shorter than amylose and amylopectin (see Chapter 7), occur in trace amounts (less than 0.2 percent of total flour weight). Maltose and lesser amounts of glucose, fructose, and sucrose contribute somewhat less than 2 percent of total flour weight.

By far the most abundant carbohydrate is starch, which is found in granules embedded in a protein matrix in the endosperm. On a dry weight basis, starch constitutes up to 80 percent of the total weight of flour. Amylopectin content averages 75 percent and amylose, 25 percent. The high concentration of starch in flour has significant implications in the structure of baked products because of the imbibition of water by the granules in an attempt to gelatinize the starch during baking. This uptake of water results in much less free water, the presence of more bound water, and increased rigidity of structure as a consequence of the reduced free water and the swelling of the granules during gelatinization.

Even in whole wheat flour, the lipid content is low; much of the total lipid is found in the germ and, therefore, is removed when refined flours are produced. In the endosperm, about 1 percent or slightly more of the total weight represents the lipid component. Lipids occur in a variety of forms, the most important of which is the free polar lipid fraction. Included among the individual compounds in the free polar lipid fraction are lecithin (phosphatidylcholine), glycolipids (notably monogalactosyl glyceride), and cephalins.

The proteins in flour include soluble proteins (subdivided into albumins and globulins), gluten proteins (glutenins and gliadins), and enzymes. Gluten, because of its singular importance, is discussed in the next section. Albumins contribute to the structure of baked products even though they constitute around only 10 percent of the total protein in flour. The other water-soluble proteins are the globulins, which are present in similar amounts. Their role in developing the structure of baked products appears to be quite minor, certainly less than the role of albumins.

The enzymes in wheat probably are soluble proteins classified as either albumins or globulins. However, enzymes represent only a small portion of these proteins in flour. Despite their small quantity, enzymes are of interest because of their potential for altering the characteristics of flour as they catalyze chemical changes. Maltose units are split from starch molecules as $\beta$-amylase attacks 1,4-$\alpha$-glucosidic linkages. Action halts at branch points in amylopectin when $\beta$-amylase encounters a 1,6-$\alpha$ linkage, leaving smaller amylopectin-type compounds called **limit dextrins**.

Dextrins result when $\alpha$-amylase cleaves amylose molecules randomly at various 1,4-$\alpha$-glucosidic linkages to create linear fragments of the former amylose chain. The fact that $\beta$-amylase is most active at about 50°C (122°F) limits the action that can occur during baking, whereas $\alpha$-amylase has more time to act before being denatured, because its active temperature range is 60–65°C (140–149°F). In ordinary circumstances, changes resulting from action of wheat amylases are undetectable. In the baking industry, fungal $\alpha$-amylase may be added to improve dough handling properties; its action occurs well below the warm temperatures needed for wheat amylase activity.

Two types of enzymes in wheat contribute toward the problems of rancidity in flours containing the germ and its fat, notably in whole wheat flour. Free fatty acids form during storage as a consequence of the action of lipases on glycerides. Lipoxidase is the most active of the oxidases in wheat flours, causing oxidative rancidity as it catalyzes formation of peroxides in unsaturated fatty acids.

## Gluten

*Gliadin and Glutenin.*    **Gluten**, the protein complex formed when wheat flour is manipulated with water (Figure 17.6), has two fractions—gliadin and glutenin—in

**Pentosans**
Polymers of the pentoses xylose and arabinose.

**Limit dextrins**
Smaller polysaccharides formed during hydrolysis of starch to glucose.

**Gluten**
Complex of gliadin and glutenin that develops in wheat flour mixtures when water is added and the batter or dough is manipulated by stirring, beating, or kneading.

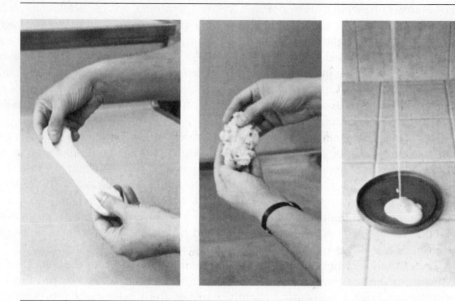

**Figure 17.6**   When manipulated with water, the proteins in wheat flour form an elastic and cohesive complex called gluten (left). Glutenin (center), part of the gluten complex, contributes elasticity. Gliadin (right), the other part of the complex, adds cohesiveness. (Courtesy of Dimler, R. J. 1963. *Bakers Digest* 37(1):52.)

**Gliadin**
Protein fraction in wheat gluten that is soluble in alcohol, compact and elliptical in shape, and sticky and fluid.

approximately equal amounts (Chart 17.2). **Gliadin** is the fraction soluble in 70 percent alcohol and is characterized as rather sticky and fluid. The proteins in the gliadin complex are probably elliptical single polypeptide chains, resulting in quite compact molecules that are held in this shape by internal (intramolecular) disulfide bonds.

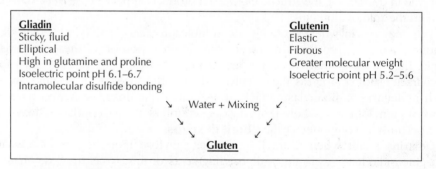

**Chart 17.2**   Overview of gluten

**Glutenin**
Alcohol-insoluble protein fraction in wheat gluten that is characterized by its fibrous, elongated shape and elastic quality.

**Glutamine**
Amino acid prominent in gliadin protein molecules and important in intermolecular hydrogen bonding.

H₂N—C—CH₂—CH₂—CH—COOH
  ‖                     |
  O                   NH₂

**Proline**
Amino acid prominent in gliadin with a cyclic ring structure that restricts protein shape.

        CH₂
  H₂C ⟋    ⟍
      |     HC—COOH
  H₂C ⟍    ⟋
        NH

**Glutenin** is the fraction consisting of the alcohol-insoluble proteins in gluten. Additional differentiation of the proteins in glutenin is accomplished by the addition of dilute acetic acid, which acts as a solvent for one subfraction and fails to dissolve the other subfraction. The combination of these two subfractions results in glutenin, the gluten fraction that is elastic. Glutenin has a fibrous nature, providing a sharp contrast to the elliptical character of gliadin.

The chemical compositions of the gliadin and glutenin fractions are different, thus helping to explain their distinctly different behavior. Gliadin consists of various molecules of differing molecular weight. The majority of the gliadin proteins are single chains of amino acids with a molecular weight around 36,500, although some polypeptide chains may be so small that they have a molecular weight of only about 11,400, and others so complex that they approach a molecular weight slightly more than 78,000. Of particular importance internally in the gliadin proteins is the presence of disulfide bonds.

**Glutamine**, an amino acid particularly prominent in the makeup of gliadin proteins, is thought to be of importance in effecting hydrogen bonding between molecules in developing gluten. **Proline** is another fairly prominent amino acid; its unique structure inhibits the shape that proteins may take wherever a proline residue occurs in the backbone chain, thus contributing to the elliptical shape that predominates in gliadin proteins.

The isoelectric point of the gliadin fraction is reported by different investigators to range from pH 6.1 to 6.7. Solubility of gliadin is least at pH 6.5 and increases with either increasing acidity or increasing alkalinity; solubility is greater at high pH levels than it is at low levels.

Compared to gliadin, the molecular weights of the various proteins constituting glutenin are much greater, ranging from about 100,000 to as high as 15 million. The very high molecular weights of some of the molecules in glutenin appear to be the result of disulfide bonding between polypeptide subunits. This disulfide linkage is one of the differences between glutenin and gliadin, for in gliadin the disulfide linkages apparently are confined to intramolecular bonding. It should be noted that intramolecular disulfide bonding also occurs in glutenin. The isoelectric point of glutenin (between pH 5.2 and 5.6) is lower than that of gliadin. Even as the pH of a batter or dough shifts away from the isoelectric point, glutenin remains somewhat difficult to dissolve and is far less soluble than gliadin.

***The Gluten Complex.***   Only when flour (with its components of gliadin and glutenin) and water are manipulated together does the gluten complex begin to form. Formation of this complex is the key to successfully producing baked products; variations in the extent of development of this complex are responsible for much of the difference in textural characteristics observed in baked products containing exactly the same ingredients.

For the gluten complex to develop, the individual proteins must be hydrated by the addition of water or an aqueous liquid. Approximately twice as much water (by weight) as gliadin and glutenin is needed to hydrate these proteins completely; hard wheat flours bind more water than soft wheat can. On initiation of mixing, the hydrated proteins begin to be altered in their physical relationships to each other. In the dry flour, the proteins occur in aggregates with starch, in endosperm cells, and in wedges of the protein matrix in which starch is embedded in the endosperm. Mixing the hydrated protein disrupts these associations and breaks many inter-molecular secondary bonds and forms new bonds, resulting in the development of a cohesive gluten matrix that provides the foundation of the structure of baked products (Figure 17.7).

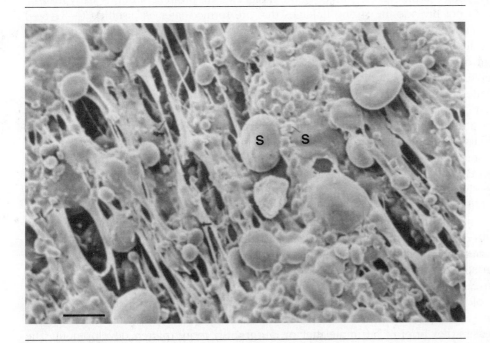

**Figure 17.7**   Scanning electron micrograph showing incubated optimally mixed flour–water dough after stretching. Thin strands are gluten fibrils formed on stretching; s = starch granules. Bar is 20 μm. (Reprinted from Scanning Electron Microscopy III: 583.1981. Courtesy of L. E. Peters and C. M. Pearson.)

The variety of functional groups—for example, the disulfide in cystine and the amide in glutamine—provides ample opportunity for the formation of new intra- and intermolecular bonds as the gluten complex develops during mixing. Flours with lower levels of gliadin require more mixing but result in a more stable gluten matrix than is true for flours with more gliadin. Flours with a comparatively high level of gliadin characteristically produce a weaker structure than those produced when less gliadin is contained in the gluten complex.

Various theoretical structures for the gluten complex have been developed over the years, but agreement has yet to be reached regarding the actual structure. One clear fact is that the lipids that naturally occur in flour are necessary for normal gluten development. One theory is that glycolipids (bound carbohydrate–lipids attracted to water) are bound to glutenin by hydrophobic bonds and to gliadin by hydrogen bonds as the hydrated gliadin and glutenin molecules are manipulated to form the gluten complex.

Evidence of the importance of lipids is clear when bread made with defatted flour is compared with a comparable loaf made with regular flour that contains the lipids normally present; the loaf made with defatted flour is significantly smaller in volume. Cakes made with defatted flour also demonstrate the importance of lipids in gluten development.

The appearance of the hydrated mixture of gliadin and glutenin changes with manipulation. At first, the surface is quite rough and some sharp points protrude. Gradually, strands or stringy film-like areas appear as manipulation continues. Eventually, the surface becomes quite smooth, and the individual rope-like strands are no longer visible. This transformation occurs as the protein-containing, hydrated flour mixture is manipulated, causing the protein to develop into thin sheets that break into fibrils that tend to stick to each other and stretch as the glutenin strands touch each other during kneading or other manipulation.

The process of gluten development is perhaps easier to understand by visualizing gliadin as having some flow properties, while also being a bit sticky. In contrast, glutenin is a protein that has regions that are somewhat like string interspersed with coiled or kinky regions that are responsible for its elastic nature.

The number of functional groups on the surface and available for interaction on a gliadin molecule is far more limited than is true for the very long strands of glutenin molecules. Sulfhydryl groups (—SH) on the surface of the molecules are prominent during early mixing, but they do not provide the strong linkages that are formed later as disulfide (—S—S—) bonds form and begin to lock the network of glutenin and gliadin molecules into the stretchy network called gluten. Because of the presence of gliadin molecules in the network, some movement of the network is still possible, making the dough mixtures containing gluten manipulable, yet comparatively elastic.

The rate at which gluten development takes place depends on several different factors. The gluten complex formed in hard wheat flours develops more slowly and can be manipulated for a longer period without breakage of protein strands than can the gluten in soft wheat flours. This phenomenon can be seen by comparing farinograms recorded when doughs containing hard wheat are manipulated in a **farinograph** operated under the same conditions used to develop a dough made with soft wheat. Gluten from soft wheat develops more rapidly and starts to break down more quickly than does gluten from hard wheat (Figures 17.8 and 17.9).

The temperature of the dough mixture influences the rate of gluten development. As the temperature increases, the rate of hydration of the proteins also accelerates, a change accompanied by an increased rate of gluten development.

Sugar added to a batter or dough retards gluten development. This effect is caused by the intense competition between sugar and the flour proteins for water in the mixture. Because of its hygroscopic nature, sugar effectively attracts water, resulting in reduced hydration of flour proteins and, consequently, slower development of gluten.

Increasing the level of fat in a batter or dough also delays gluten development. The inhibiting effect of fat is due to the ability of fat to coat the surface of gluten, thus making hydration of the flour proteins quite difficult. There may be an additional effect resulting from the ability of the fat to serve as a lubricant, thereby reducing the tendency of the strands to stick to each other during mixing.

**Farinograph**
Objective testing equipment that measures the resistance of stirring rods moving through a batter or dough and records the results graphically.

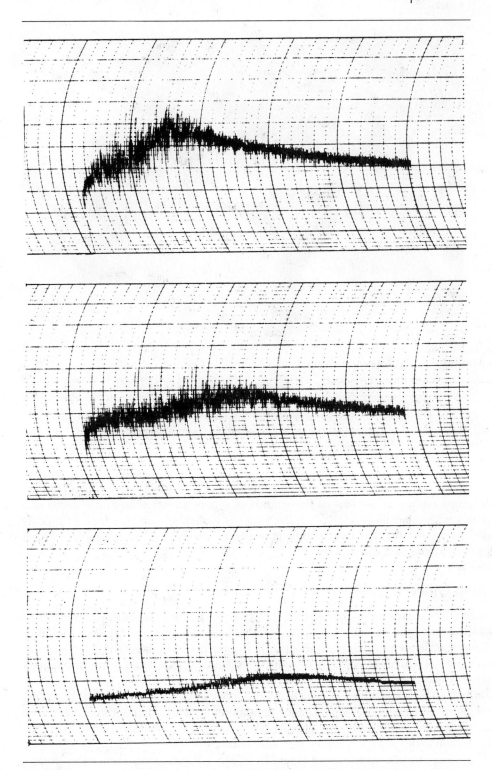

**Figure 17.8** Gluten development, as recorded in mixograms. Gluten is developing in the ascending pattern on the left and breaking down as the pattern descends toward the right. Mixograms are for (a) baker's hard wheat flour, (b) all-purpose flour, and (c) cake flour. (Courtesy of H. Charley.)

The viscosity of a batter—that is, the amount of water in relation to flour—is yet another factor influencing gluten development. When equal amounts of flour and liquid are included in a batter, an excess of fluid is available, resulting in a fluid mixture, such as popover batter. The protein molecules are diluted so much by the liquid that it is quite difficult for strands to cling to each other even when they happen to touch during mixing.

**Figure 17.9**   Cake flour (soft wheat) gluten ball (left) reveals the smaller quantity and weaker nature of its gluten, as contrasted with the gluten ball made with all-purpose flour (right).

In contrast to the 1:1 ratio of flour to liquid in popovers, muffins have a ratio of approximately twice as much flour as liquid (a 2:1 ratio). This is a very sticky ratio, and the gluten develops rapidly in muffins as the strands of gluten are stretched on coming into sticky contact with other molecules when the batter is mixed. Gluten develops so quickly when muffins are mixed that overmixing may occur readily, and undesirable textural changes can be noted (see Chapter 18).

## Flours from Other Sources

Dietary Reference Intakes recommended for dietary fiber by the Food and Nutrition Board of the National Academy of Science–Institute of Medicine are about double the amount of dietary fiber typically consumed by Americans. Recognition of this apparent deficiency has motivated many consumers to select breads and other baked goods made using whole-grain flours, either wheat or other fiber-rich flours.

According to the Whole Grains Council, grain flours that contain the essential parts of a processed grain and include the same balance of nutrients contained in the original grain are whole-grain products even though the grains have been cracked, crushed, rolled, extruded, lightly pearled, and/or cooked. Some of the other sources of flour for making baked products high in fiber include whole-grain flours from rye, triticale, rice, corn, oats, potato, quinoa, and soy. Wheat berries, cracked wheat, and grano (special durum wheat bread flour including all but a small amount of the outer bran) are whole-grain ingredients for making bread. Ultragrain® (see Chapter 1) yields a high-fiber, white wheat flour (the result of genetics and a special milling technique) that can be identified on the ingredient label as whole-wheat flour. Sustagrain® is a waxy barley with a low starch and high fiber content; its flour can be used in formulating high-fiber breads.

***Rye Flour.***   Flours can be milled from any of the cereal grains, but their proteins lack the ability to develop the necessary plasticity and elasticity that are needed for producing the texture desired in baked products and that are uniquely available in wheat gluten. Even rye, the cereal grain that most closely approaches the characteristics of wheat gluten, cannot match the performance of wheat flour in making bread products. The proteins in rye flour include glutenin and gliadin, but they cannot be manipulated into a gluten complex with sufficient elasticity to produce light-textured baked products. Bread made with rye flour is dense. Breads made with a combination of rye and wheat flours have a better texture than those made only with rye.

***Triticale Flour.***   **Triticale** is a unique cereal grain developed by crossing wheat with rye and bears some of the characteristics of each of the parent grains. It has nutritional merit because its protein content ranges to more than 16 percent, and the lysine content is higher than that of wheat. The yield per acre also is high. Although triticale flour can be used in baking, its texture is not as acceptable as that provided by wheat flour. There is potential for satisfactory use of this type of flour either by combination with wheat flour or by use of dough conditioners. Actual characteristics of triticale are somewhat variable, because the strains of this hybrid grain differ in the relative contribution of wheat and rye in the crossbreeding. The greater the wheat contribution, the better are the baking characteristics.

**Triticale**
Hybrid grain produced by crossing wheat and rye.

***Rice Flour.***   Rice flour is another kind of cereal flour, and it sometimes is substituted for wheat flour when making baked products for people with Celiac disease (Figure 17.10). Its use in baked products (as is true for flour from corn, barley, and oats) is frustrating because the name suggests that it should be possible to make a satisfactory baked product substituting rice or any of these other flours for wheat flour. Unfortunately, all of these grain flours lack the elastic and cohesive qualities needed to provide a framework for baked

**Figure 17.10**   Rice grain can be milled into rice flour, but this flour is challenging to use in baking because it lacks gluten.

**Table 17.3**   Profile of Selected Alternate Flours

| Flour | Protein in Flour (percentage) | Acceptable Replacement Level (percentage) | Characteristics Noted in Baked Products |
|---|---|---|---|
| Rye | 12.1 | 20–40 | Heavy texture, smaller volume |
| Triticale | 10.7–16.3 | 40–50[+] | Limit mixing and fermentation time for optimal result |
| Oat | 17 | 30 | Heavy in bread |
| Rice | 6.5–7 | 5–30 | Satisfactory at 5%; somewhat acceptable at 30% |
| Corn (maize) | 7–8 | n/a[a] | Used as masa[b] to make tortillas |

[a]Not available.
[b]Corn cooked and steeped in lime, then washed and ground.

products. In addition, the somewhat gritty texture of rice flour is quite apparent, seemingly defying hydration. However, some of the wheat flour can be substituted fairly satisfactorily by one of the alternative grain flours (Table 17.3).

Commercial baking operations are able to add carboxymethyl cellulose (CMC) to batters and doughs to help complement the bit of structure that is available from the starch in these flours. Baked products made from these other grain flours are particularly important to people who have an allergy to gluten. Satisfactory alternatives for baked products made with wheat flour are being explored, but the ultimate objective of a totally satisfactory replacement has not yet been achieved.

*Corn Flour.*   Corn flour, like rice flour, is limited in its usefulness by its inability to form a useful protein network to provide the structure required in baked products. Somewhat larger granulation also may be used in the form of cornmeal. Either corn flour or cornmeal can supplement wheat flour in making baked products, such as corn bread, but the texture will be quite crumbly because of the lack of gluten in the cornmeal. Flavor and color are benefits gained from using cornmeal or corn flour, but the texture is compromised a bit.

*Potato Flour.*   Potatoes are a tuber, not a grain, but they can processed into flour, flakes, or granules which can add some interest and variety in selected baked products, particularly breads. The component that makes some contribution to the texture of baked products is starch. The flavor of baked products is influenced subtly by the use of potato flour or related potato products. No gluten is provided, which makes it important to include sufficient wheat flour as an ingredient in baked products that include potato flour or granules.

*Soy Flour.*   Soy flour differs from the other flours mentioned in this section because it is produced from a legume, not a grain or tuber. However, its common availability and nutritional interest make it an appropriate flour to include here. Nutritionally, soy flour has some importance because its high lysine content provides a ready complement to wheat flour, which is limited in its nutritional benefit by its comparatively low content of lysine, an essential amino acid. Soy flour aids in forming emulsions in batters.

Unfortunately, soy flour lacks the cohesive and elastic qualities of gluten, but it does provide both protein and starch to strengthen the structure when it is combined with wheat flour in baked products. Supplementation with soy flour can add to the nutritive value of baked products, although its distinctive flavor becomes objectionable at high levels. Soy flour, when added to doughnut dough, reduces oil absorption during frying, which is a useful health benefit in addition to the nutrients provided by the soy flour.

The starch added when soy flour is included as an ingredient makes it necessary to increase the amount of liquid in recipes. A satisfactory result can be obtained if the added liquid equals the weight of the soy flour. The proofing time for yeast breads needs to be lengthened when soy flour supplementation is a part of the recipe.

## ROLES OF INGREDIENTS

### Wheat and Other Flours

Wheat flour is the basic structural component of most batter and dough products (Table 17.4). It is able to perform this textural function because of its gluten content, which allows expansion of cells and provides rigidity of structure after baking. Tenderness and texture of baked products are defined in large measure by the gluten provided by wheat flour.

Mixing should be done to develop the gluten appropriately for the product being prepared. The type of flour is one of the factors of key importance in making high-quality baked products. Doughs made with whole-wheat flour need to be mixed less than those made with refined wheat flour because the jagged edges of the bran cuts into the developing gluten fibers and shortens them, thus negatively impacting bread texture. Moisture, fat, and sugar levels also influence the amount of mixing required to develop gluten optimally for the product.

Starch is another important compound in not only wheat, but other flours as well (Figure 17.11). Water is bound by the starch as it gelatinizes during baking, thus contributing significantly to the remarkable change from a somewhat fluid batter or dough to the firm, rather rigid structure of the baked product. Clearly, starch in flour plays an important role in baked products, albeit a somewhat less dramatic one than that of gluten.

Flour performs other minor roles. It provides some of the sugar needed by yeast as food during fermentation in making yeast breads. Also, crust browning during baking is dependent on the combination of proteins with sugars, and flour provides the protein needed.

### Liquid

Water or other liquid in batters and doughs serves as a solvent, dissolving sugar, baking powder, salt, or other dry ingredients and hydrating yeast when it is present. As mentioned

**Table 17.4**   Functions of Ingredients in Baked Products

| Ingredient | Function(s) |
| --- | --- |
| *Flour* | Gluten for major structure |
| | Starch for some structure |
| *Liquid* | Gluten development |
| | Dissolve sugar |
| | Dissolve baking powder to activate it |
| | Starch gelatinization |
| | Steam for some leavening |
| *Eggs* | Some liquid |
| | Protein structure |
| yolk | Emulsifying batters and/or doughs |
| | Color |
| white | Texture (fine and expanded cells; foam) |
| *Fats* | Flavor |
| | Tenderness |
| | Influence texture |
| | Color |
| *Sugar* | Flavor |
| | Color—Maillard reaction |
| | Tenderness |
| | Volume |
| | Food for yeast in yeast-containing products |
| *Salt* | Flavor |
| | Limit yeast growth |
| *Leavening agents* | Volume |

**Figure 17.11**   Oatmeal bread and most other baked products rely on gluten in wheat flour for their structure, but other grain flours may be added for variety. (Courtesy of Fleischmann.)

earlier in this chapter, liquid must be present for gluten to develop, because the proteins must be hydrated before they can be manipulated to adhere to each other and to form the elastic, plastic network needed for structure. Until water or other aqueous liquid is present, no gluten develops.

During baking, available liquid is bound in starch granules as they gelatinize. Although batters and doughs usually do not contain enough liquid to permit complete gelatinization of starch, at least partial gelatinization is possible because of the liquid in the recipe. Furthermore, liquids serve as a source of steam during baking, and steam is important as either an auxiliary or primary source of leavening in baked products. The significance of this role is discussed under "Leavening" later in this chapter.

When milk is the liquid used, it also contributes to crust browning because of its protein and sugar content and softening of crumb texture because of its fat content (whole and low-fat milk). Fruit juices reduce the pH of batters and doughs when they are the liquid, which alters the flavor and modifies the solubility of the gluten proteins slightly.

## Eggs

As discussed in Chapter 16, eggs contribute to baked products in several ways. In a batter or dough, they contribute liquid during mixing and in baking. Although the quantity of liquid eggs provide usually is rather small, this role cannot be ignored. Of course, the proteins in eggs are of particular importance when they are used in batters and doughs. Coagulation of the proteins during baking contributes to the structure of the finished product and reduces tenderness.

Eggs, particularly the yolks, serve as an emulsifier during mixing of batters and doughs. Formation of an emulsion promotes more uniform dispersion of liquids and fats and favors a fine texture. Yolks also contribute color to light-colored baked products.

Egg foams, particularly egg white foams, contribute significantly to the texture of cakes and a few quick breads. The abundance of air trapped within the bubbles of egg protein creates many small cells that expand from the heat during baking and then are permanently set in their extended position when the egg (and flour) proteins are denatured and coagulated late in the baking period.

## Fats

Depending on the type used, fats play various roles in baked products. Butter and margarine contribute a pleasing and rather distinctive flavor, whereas other fats add to the richness of flavor. Lard has a mild, unique flavor that may be appreciated in certain products, such as pastry. Yellow fats (butter, margarine, or colored shortenings) add a creamy color that subtly implies richness.

Fats also influence texture. For example, oil in a pastry promotes a mealy texture, whereas particles of fat made by cutting a solid fat into the flour yield a flaky, layered texture. Shortenings with added mono- and diglycerides favor development of a fine texture in cakes because of their emulsifying action. Regardless of the type used, fats are effective in promoting the tenderness of baked products. Their hydrophobic action inhibits gluten development; the extent of this inhibition is determined by the type, temperature, and amount of fat used, as well as by the method of incorporation.

## Sugar

The obvious role of sugar in baked products is as a sweetener. However, this is but one of its functions. Browning of the crust is due in part to caramelization, as well as to the sugar-amine (Maillard) reaction. Sugar serves as a tenderizing agent first by retarding gluten development during mixing and then by elevating the coagulation temperature of the structural proteins so that there is more time for cell walls to stretch and volume to increase before coagulation occurs to define the final volume of baked products. Up to a certain critical point, sugar increases volume of baked products, but cakes fall if the coagulation temperature is so high that the gluten strands break and release the trapped gases before the protein structure is set.

Sometimes sugar crystals form as cakes and cookies stale and lose moisture. This tendency can be controlled in commercial cakes by including corn syrup or a polyol as an ingredient.

## Salt

Salt is a flavoring agent at the levels used in most baked products. The sole exception occurs when considering salt in yeast breads. In yeast doughs, salt limits the growth rate of yeast, which serves as a check against excessive carbon dioxide production resulting from the action of yeast and sugar during fermentation and the early period of baking.

## Leavening Agents

Leavening agents contribute significantly to the textural properties of baked products by expanding the batter or dough, sometimes during mixing and always during baking. The physical behavior of gases generated by leavening agents was explained in Chapter 6. The next section discusses the various sources of leavening in batters and doughs.

### Commercial Baking Challenges

Commercial bakers are confronted with palatability factors that are influenced by the time elapsing between baking and reaching the consumer's table. Xanthan gum, the gum produced when *Xanthomonas campestris* ferments corn sugar, is commonly added to cake formulas to improve texture. It serves as a binder in the dough and promotes a fine grain in the baked cake.

Awareness of the increasing numbers of Americans who are overweight and obese has heightened interest in developing commercial baked products with reduced calories. Reduced and low-calorie sweeteners are used in cake and cookie formulations as a means of reducing calories. The other focus is the fat selected in these products. Various combinations of ingredients are used to attempt to create baked temptations for the dieters and health-conscious consumers seeking foods with increased levels of fiber.

Several ingredients are incorporated into commercial baked products in attempts to create acceptable products that are compatible with the health objectives of consumers. Among these are

- gums
- resistant tapioca and resistant high-amylose cornstarches
- various other fiber sources such as Citri-Fi™ (from citrus pulp)
- milled and whole flaxseed
- inulin (one form is Frutafit®, marketed as a prebiotic obtained from chicory root, and another is BakeFlora™)
- fructooligosaccharides
- whey protein concentrates
- bamboo fiber

Additional discussion of the use of various commercial ingredients is included in Chapters 8, 10, and 12.

## LEAVENING

### Air

Air is always a leavening agent in baked products because some air always is trapped within the mixture while the ingredients are blended. This air is essential to the formation of the many cells needed to produce light and pleasing baked products with the expected cell size. However, it is insufficient to achieve the volume usually desired. Sometimes egg white and/or egg yolk foams are prepared and gently folded into a batter to increase the amount of air that is available for leavening the mixture.

Considerable variation in volume is found when air is the principal leavening agent, even when an egg white foam is included. To achieve good volume in an angel cake, a cake that relies heavily on air for leavening, a stable egg white foam of high volume must be folded very gently and efficiently with the other ingredients and then baked promptly in a preheated oven.

There are two critical operations in which individual technique can make a significant difference in the volume of the batter prior to baking. First, the air must be incorporated into the foam to produce a large volume. This requires knowledge of the appropriate times at which to add stabilizing agents (acid and sugar) and of the appropriate point at which to stop beating. Overbeating or underbeating has a deleterious effect on foam volume. An over-beaten foam is rather rigid and brittle and hence difficult to fold in gently with the other ingredients, which leads to increased folding and concomitant loss of air.

Second, careful folding is imperative, for there still is ample opportunity for loss of air during folding. The operator who is rough and very vigorous will break many of the cells and release air from the mixture. A slow worker will lose air unnecessarily because of some evaporation from the surface of the foam and the collapse of some cells while the batter is folded slowly. Inefficient and excessive folding can result in considerable loss of air and a decreased volume. These losses are compounded if the cake must stand before being baked. In short, air is an essential, but only partially reliable, leavening agent.

## Steam

Steam is a far more effective leavening agent than air, because volume is increased 1600-fold as water is converted to steam. Compare this with air, which expands volume by 1/273 for each degree Celsius that the temperature increases. Popovers usually about triple in volume during baking, and most of this leavening is caused by steam (Figure 17.12). Cream puffs are also leavened mostly with steam. Both of these products have equal amounts of liquid and flour by volume, a ratio that is favorable to considerable steam leavening. Even when a baked product contains as little water as is found in pastry, steam can cause appreciable leavening.

**Figure 17.12**   Popovers illustrate the remarkable leavening action of steam. The very fluid batter and hot oven, combined with a small amount of air whipped into the batter, create a large interior cavity and about a threefold increase in volume during baking.

**Table 17.5**    Relative Leavening Action of Air and Steam in Fat-Containing Cakes (in percentage)

| Fat in Cake | Relative Leavening | |
| --- | --- | --- |
| | Air | Steam |
| Butter | 19.8 | 80.2 |
| Oil | 11.4 | 88.6 |
| Hydrogenated lard | 25.0 | 75.0 |

Adapted from Hood, M. P. and Lowe, B. 1948. Air, water vapor, and carbon dioxide as leavening gases in cakes made with different types of fat. *Cereal Chem. 25*: 244.

Air is important as a leavening agent in foam cakes and especially in angel cakes, which are predominantly egg white foams that contribute a large volume of air to the batter. Although this trapped air in the foam expands during baking, steam actually provides between two and three times as much leavening action as air.

In cakes containing fat, the type of fat influences the relative amount of leavening contributed by air and steam. Air is trapped more effectively in batters containing solid fats than in those containing oil, and hydrogenated lard retains more air for leavening than does butter (Table 17.5). Regardless of the type of fat used, far greater leavening is provided by steam than by air. The fluidity of the oil or firmness of a fat is a factor in the retention of air in the batter during mixing and baking. The difference ranged from almost eight times as much action from steam as from air in the oil-containing cakes to about four times as much with butter and three times as much with hydrogenated lard as the fat.

## Biological Agents

*Yeast.*    Carbon dioxide, a gas that provides very effective leavening in baked products, comes from either a biological or a chemical source. The usual biological source is fermentation of sugars by yeast, one-celled plants capable of actively producing carbon dioxide in a dough at room temperature and briefly at oven temperatures during the early phase of baking. The strain normally used in making yeast breads is *Saccharomyces cerevisiae*. Because this yeast must be alive for the fermentation to proceed so that carbon dioxide is produced, temperature must be controlled carefully to achieve satisfactory production (Figure 17.13).

Even the conditions under which the yeast is added to the dough influence the final product, and the form of yeast used determines the optimal conditions. Three forms of *S. cerevisiae* are available: compressed yeast, active dry yeast, and quick-rise active dry yeast. **Compressed yeast** cakes contain cornstarch and about 72 percent moisture; the refrigerated shelf life is about 5 weeks. Compressed yeast needs to be dispersed in a small amount of lukewarm water [32–38°C (89–100°F)] before it is added to a dough. This active form of yeast is convenient to use, but its short shelf life and the need for refrigeration during storage limit its market.

**Active dry yeast** is a granular form with a very low moisture (8 percent) content, which makes it possible to store this form of yeast in tightly sealed containers at room temperature for at least 6 months or in a freezer for as long as two years. Rehydration of this form of *S. cerevisiae* requires a higher temperature than is used for dispersing the compressed cake; the preferred temperature range for rehydration is 40–46°C (104–115°F). This high a temperature is needed to keep glutathione from leaving the yeast cells and subsequently causing breakage of disulfide bonds in the dough, which would reduce the elastic nature of the dough and increase stickiness.

For convenience, active dry yeast sometimes is combined with the flour without rehydration of the yeast. The liquid must be at even higher temperature [49–54°C

*Saccharomyces cerevisiae*
Strain of yeast commonly used as the source of carbon dioxide in yeast-leavened products.

**Compressed yeast**
*Saccharomyces cerevisiae* in a cornstarch-containing cake with a moisture level of 72 percent; requires refrigerated storage; dispersion is best at 32–38°C (89–100°F).

**Active dry yeast**
Granular form of dried *Saccharomyces cerevisiae* (8 percent moisture); storage is at room temperature and rehydration at 40–46°C (104–115°F).

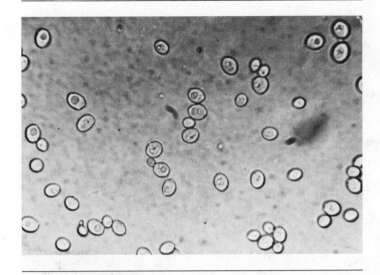

**Figure 17.13** *Saccharomyces cerevisiae*, as seen under a microscope.

(120–129°F)] for this technique because the other ingredients in the dough mixture will cool the liquid to a safe temperature range for the yeast. **Quick-rise active dry yeast** is very similar to active dry yeast, but the strain of *S. cerevisiae* used in this product produces carbon dioxide particularly rapidly in doughs and is capable of saving approximately an hour in fermentation time.

**Quick-rise active dry yeast**
Special strain of *Saccharomyces cerevisiae* available as the active dry yeast and capable of producing carbon dioxide so rapidly in the dough that fermentation time is cut in half.

Even after yeast has been combined with the other ingredients in a dough, care still must be taken to ensure that the yeast remains viable and produces carbon dioxide. Temperature control and time are the factors necessary for adequate leavening to occur when yeast is used. After it is mixed, the dough is left to rest to allow time for the *S. cerevisiae* to metabolize glucose and release carbon dioxide (Figure 17.14). The overall reaction is shown here.

$$\text{glucose} \xrightarrow{\textit{S.cerevisiae}} 2C_2H_5OH + 2CO_2\uparrow$$

ethyl alcohol     carbon dioxide

Temperature of the dough during the fermentation period has a tremendous influence on the rate of gas production in the dough. Fermentation ideally is accomplished between 25 and 27.7°C (77 and 82°F), but the production of carbon dioxide actually can occur at refrigerator temperatures or as high as 35°C (95°F) or even slightly higher. The rate of fermentation increases rapidly with the temperature, tripling as the temperature rises from 20 to 30°C (68 to 86°F) and doubling between 25 and 35°C (77 and 95°F). Despite the rapid fermentation occurring at 35°C (95°F), such a warm temperature is not recommended because of the development of a sour, yeasty flavor as other microorganisms also flourish at the elevated temperature. A fermentation temperature of 43°C (110°F) kills yeast in an hour, thus eliminating production of carbon dioxide.

Glucose is the form of sugar preferred by *S. cerevisiae,* and a small amount of glucose is present in the flour used in the dough. Most yeast bread recipes include sucrose as the significant source of sugar for metabolism by the yeast. The inversion of sucrose to glucose and fructose, which is necessary before the yeast can utilize this source of sugar, is catalyzed by the sucrase present in yeast.

**Figure 17.14** *Saccharomyces cerevisiae* generates carbon dioxide in a batter at room temperature.

Maltose becomes available by enzymatic action on starch, and this sugar can be metabolized by yeast if no other source of sugar is available. The principal enzyme catalyzing production of maltose from starch to provide food for yeast during fermentation is β-amylase. Some action may also occur as the result of α-amylase. If the weight of the sugar exceeds 10 percent of the weight of the flour, some osmosis will occur, and the yeast will metabolize sugar slowly.

Salt has a negative effect on metabolism of sugar by yeast because of its influence on osmotic pressure in yeast. On the other hand, salt is beneficial in a yeast dough because its presence enhances the activity of β-amylase and α-amylase in breaking starch to sugars to provide food for the yeast. It also aids in reducing the breakdown of some of the flour proteins by proteases. Too little salt allows so much protease activity that the gluten weakens to the point that many of the cells break as pressure from carbon dioxide is generated, and the resulting bread has an uneven, porous texture. Salt usually should not exceed 2 percent of the weight of flour.

An appropriate balance between sugar and salt levels is needed for good gas production in sweet yeast doughs. Gas production is retarded in sweet doughs by the high level of sugar, an effect that can be offset to some extent by reducing the level of salt.

The pH of a yeast dough influences the rate of gas production during the fermentation period. *S. cerevisiae* activity is best between pH 4 and 6. Fortunately, fermentation produces some organic acids as well as carbon dioxide, which lowers the pH of the dough from its initial pH of about 6 to between 5 and 5.5, a range favorable for good gas production. The activity of *S. cerevisiae* is reduced if the pH of the dough drops below 4.

**Saccharomyces exigus and Saccharomyces inusitatus**
Yeasts used to produce carbon dioxide in acidic bread doughs

**Lactobacillus sanfrancisco**
Bacterium used to produce lactic acid in some bread doughs.

***Bacteria and Yeast.*** Two other strains of yeast, ***Saccharomyces exigus*** and ***Saccharomyces inusitatus***, produce carbon dioxide in acidic (sour) bread doughs, such as sourdough bread. These two yeasts are active at or below pH 4.5 and function well in conjunction with the use of ***Lactobacillus sanfrancisco***. *L. sanfrancisco* produces lactic acid, which reduces the pH of the dough to the range where *S. exigus* and *S. inusitatus* can generate carbon dioxide effectively, but where *S. cerevisiae* is not the preferred yeast.

Sourdough starters contain a mixture of these microorganisms. The resulting bread using a sourdough starter has a pleasingly sour flavor (the result of the acid produced by the *L. sanfrancisco)* and good volume because of the carbon dioxide produced by the *S. exigus* and *S. inusitatus.* Other starters also can be effective in generating carbon dioxide by the action of assorted microorganisms. Care must be taken to avoid off flavors and odors resulting from the incubation of undesirable microorganisms in uncontrolled cultures.

## Chemical Agents

A chemical reaction between an acidic and an alkaline ingredient in a baked product can generate carbon dioxide for leavening. The effectiveness of various ingredients or compounds in gas production depends not only on the total amount of gas generated, but also on the speed of the reaction. When chemical compounds react, no more gas is produced after either of the original reactants disappears. A rapid reaction means that all of the gas is generated quickly (probably during mixing), and some (or much) of the gas escapes from the batter or dough before baking sets the structure.

*Acid Ingredients.*   Various acid ingredients may be included in a batter or dough recipe to add flavor variety as well as to serve as a component of a leavening reaction. Soured dairy products, including cultured buttermilk, yogurt, sour milk, and sour cream, are able to react with an alkaline ingredient to release carbon dioxide as soon as the alkaline ingredient is dissolved in the milk or cream. Usually about a cup of the sour milk or cream and half a teaspoon of baking soda produce enough leavening for a batter or dough containing two cups of flour.

Other acidic ingredients include honey, molasses, fruit juices, and cream of tartar. The acidity of these ingredients is somewhat variable, which makes it difficult to be certain that the amount of carbon dioxide produced will be optimal. All of these acidic ingredients react with baking soda quickly, which necessitates prompt mixing and baking once the acid and alkali have reacted.

*Alkaline Ingredient.*   Acidic ingredients alone cannot generate carbon dioxide. They must react with an alkaline ingredient, and the common alkaline ingredient used in food preparation is **bicarbonate of soda** (also called sodium bicarbonate or baking soda). This soluble powder must be dissolved before it is capable of reacting with acids. Solution and reaction occur quickly when baking soda comes in contact with the liquid ingredients in a recipe.

**Bicarbonate of soda**
Alkaline food ingredient ($NaHCO_3$) used to react with acids to form carbon dioxide.

Unless the batter or dough contains an acid, baking soda will produce only a little gas for leavening. If there is insufficient acid to react with all of the bicarbonate of soda, the mixture will be alkaline, causing a soapy flavor due to formation of sodium carbonate ($Na_2CO_3$) and a yellowish hue in light-colored baked products. The volume also may be poor.

## Baking Powders

**Baking powders** contain at least one acid salt in addition to the alkaline baking soda (Figure 17.15). In short, baking powder is a complete mixture that provides both the acid and the alkali needed for the reaction to produce carbon dioxide for leavening in a batter or dough.

**Baking powder**
Mixture of acid and alkaline salts and a standardizing agent to produce at least 12 percent of the carbon dioxide available for leavening.

Commercial baking powders prepared for the retail market were actually one of the first convenience items or mixes. The baking powders on the market today are mixtures of acid salts and baking soda carefully formulated to produce between 12 and 14 percent carbon dioxide by weight; the minimum legal level is 12 percent. The leavening results obtained with using baking powder are more reliable and reproducible than those obtained by use of separate acid and alkali in a recipe, because of the precision possible in manufacturing the powder and the delay in the release of carbon dioxide that is possible in formulating baking powders. When baking powder is used to supplement the leavening provided by air and steam in fat-containing cakes, the relative amount of leavening

**Figure 17.15**  Baking soda and sour cream react to release carbon dioxide as soon as the baking soda dissolves during mixing; double-acting baking powder has two acid salts, one reacting at room temperature and the other in the hot oven during baking.

provided by the carbon dioxide from the baking powder is far greater than that provided by either air or steam alone and even more than the leavening provided by the combined effect of air and steam (Table 17.6).

The essential ingredients in a baking powder are bicarbonate of soda, an acid salt, and cornstarch. The acid and alkaline salts are required for the chemical reaction that produces carbon dioxide. Cornstarch does not produce gas but does perform two valuable functions in the powder. It helps extend the shelf life of the powder by absorbing moisture that may enter the can, thus preventing the moisture from dissolving the acid and soda and blocking reaction of the active ingredients. Another role of cornstarch is that of a standardizing agent. Cornstarch is added at the level needed to dilute the active ingredients so that the correct amount of carbon dioxide is generated by a measured amount of baking powder. This function is important because various amounts of the different acid salts are required with a given amount of baking soda. If a large amount of acid is required, a small amount of cornstarch is needed to standardize the baking powder, whereas a small amount of an acid salt in a baking powder formulation is balanced using an increased amount of cornstarch as filler.

**Table 17.6**  Relative Leavening Action of Air, Steam, and Carbon Dioxide in Fat-Containing Cakes (in percentages)

| Fat in Cake | Relative Leavening | | |
| | Air | Steam | Carbon Dioxide |
| --- | --- | --- | --- |
| Butter | 8.0 | 32.3 | 59.7 |
| Oil | 5.6 | 44.1 | 50.3 |
| Hydrogenated lard | 6.7 | 20.7 | 73.1 |

The acid salts available for commercial applications are much greater in number than are those found in baking powders for retail sale. Most baking powders now in grocery stores are double-acting baking powders, meaning that carbon dioxide is produced at two different times when the powder is used in a batter or dough product. First, some carbon dioxide is produced from one of the acid salts during mixing to help develop a light-textured product. The acid salt responsible for this aspect of leavening must react at room temperature with dissolved bicarbonate of soda. Monocalcium phosphate monohydrate is the acid salt that acts at room temperature during mixing. It reacts with bicarbonate of soda according to the following reaction:

$$3CaH_4(PO_4)_2 + 8NaHCO_3 \longrightarrow Ca_3(PO_4)_2 + 4Na_2HPO_4 + 8CO_2 + 8H_2O$$

monocalcium phosphate   bicarbonate of soda   tricalcium phosphate   disodium phosphate   carbon dioxide   water

*Sodium aluminum sulfate* (SAS) is the second acid salt in double-acting baking powders. A two-step reaction is required for this salt to produce carbon dioxide in the oven during baking. The first reaction produces sulfuric acid which then reacts with the bicarbonate of soda:

$$Na_2Al_2(SO_4)_4 + 6H_2O \longrightarrow 2Al(OH)_3 + Na_2SO_4 + 3H_2SO_4$$

SAS   water   aluminum hydroxide   sodium sulfate   sulfuric acid

$$3H_2SO_4 + 6NaHCO_3 \xrightarrow{H_2O} 3Na_2SO_4 + 6H_2CO_3$$

sulfuric acid   soda

$$6CO_2 + 6H_2O$$

**SAS-phosphate baking powder**
Leavening with sodium aluminum sulfate and monocalcium phosphate as the acid salts; double-acting baking powder.

The usual ratio of monocalcium phosphate to sodium aluminum sulfate is one part monocalcium phosphate monohydrate to two parts sodium aluminum sulfate.

Cream of tartar (potassium acid tartrate) is an acid salt that can be used effectively in formulating baking powder for home use to make a tartrate baking powder. Tartrate reacts rather completely in about 2 minutes, making it difficult for slow workers to use effectively. The major advantage of a tartrate baking powder is the lack of an aftertaste from the sodium potassium tartrate residue remaining after baking. The reaction of cream of tartar with sodium bicarbonate is as follows:

$$NaHCO_3 + KHC_4H_4O_6 \xrightarrow{H_2O} KNaC_4H_4O_6 + CO_2 + H_2O$$

sodium bicarbonate   potassium acid tartrate   sodium potassium tartrate   carbon dioxide   water

Although tartrate baking powder may not be available locally, it is available on the Internet. Another possibility is to thoroughly mix together 2 parts (by volume) cream of tartar, 1 part baking soda, and 1 part of cornstarch. A teaspoon of this mixture provides leavening equivalent to 1 teaspoon of double-acting baking powder.

The amount of leavening provided by baking powder varies with the recipe, for recipes utilizing egg white foams do not need to rely so heavily on carbon dioxide. Between 1 and 2 teaspoons of baking powder per cup of flour provides an appropriate leavening action in most breads, whereas slightly less can be used in shortened cakes because of the leavening they gain from the air incorporated in creaming the fat and sugar. Too much baking powder in baked products results in a coarse and slightly harsh texture.

The type of baking powder used influences the pH of the resulting batter or dough. Tartrate baking powders produce the most acidic (pH 6.4) biscuit dough, with phosphate producing an intermediate value (pH 6.7), and double-acting (SAS-phosphate) being the least acidic (pH 7). The more acidic the mixture, the lighter in color and the more tender is the final product.

A double-acting baking powder leaves residue salts, including aluminum hydroxide. The aluminum hydroxide promotes increased elasticity and viscosity in batters and doughs and contributes to the cracking in the sides of biscuits.

An ideal baking powder would release some carbon dioxide during mixing at room temperature but would provide most of its carbon dioxide at oven temperatures, particularly in the early phase of baking. It also should not have an aftertaste. Ready availability and low cost are other criteria.

No ideal baking powder exists. Tartrate powders give off too much carbon dioxide at room temperatures to make them suitable for slow workers or times when the mixture may need to stand before baking. SAS-phosphate, because of its double action from two acid salts, provides leavening in both mixing and baking. Its real disadvantage is the rather metallic aftertaste. Although this aftertaste is not ordinarily a problem in most recipes, it can be unpleasant when increased levels of baking powder are used.

## Baking Ammonia

**Baking ammonia**
Ammonium salt of carbonic acid [$(NH_4)_2CO_3$] that dissolves to release ammonia, carbon dioxide, and water for leavening springerle and other rather dry cookie doughs, crackers, and biscotti.

**Baking ammonia** [$(NH_4)_2CO_3$], the ammonium salt of carbonic acid, is a unique chemical leavening agent that has certain baking applications. Its use originated in the Nordic countries where antlers of male deer were grated to provide leavening, which explains its alternative name "hartshorn." This salt produces ammonia, carbon dioxide, and water when it is moist and is being heated. Unlike baking powder, this salt contains the components needed to generate gas when it is an ingredient in moist dough that is baked. Little gas is produced until the salt is heated. Ammonia has a flavor that is detected easily, which limits the quantity of baking ammonia that can be used. However, in rather dry cookie doughs, such as springerle and lebkuchen, or in crackers and biscotti doughs requiring limited leavening, baking ammonia is a suitable leavening agent. The limited reaction at room temperature is of value in doughs that involve extended bench time before baking.

## SUMMARY

Cereal grains (composed of bran, endosperm, and germ) can be milled into flour, which may be the whole grain or refined product depending on whether the bran and germ are removed. Among the types of flour available from the milling of wheat are whole wheat, all-purpose, bread, pastry, self-rising, gluten, cake, and instantized flours. Benzoyl peroxide is used to accelerate the bleaching of the xanthophylls to produce a white flour. Maturing agents (chlorine, acetone peroxides, and azodicarbonamide) are added to increase the disulfide bonding and enhance the adhesive, elastic properties of the flour gluten.

Starch is the most abundant component of flours, but it is the protein properties that are of special interest in baking. Fat content is low, yet it does contribute to the gluten complex that develops in wheat flour when batters and doughs are mixed. Gliadin, the alcohol-soluble protein fraction in gluten, is elliptical in shape, sticky, and fluid. Glutenin, a fibrous and elastic protein complex, constitutes the other fraction of gluten.

The gluten complex formed when wheat flour is mixed with water is unique and is not duplicated in any other cereal flour, although rye has a limited ability to form this type of structure. When the flour proteins are hydrated, intermolecular bonds break, and new bonds form as the rather sticky glutenin fibers touch each other during mixing. Gliadin also becomes enmeshed and bonded within this network, although it does not provide the strength of structure that is contributed by the glutenin molecules.

Hard wheat flours have a stronger gluten and can be mixed longer than soft wheat flours without breaking strands of the gluten. Sugar and fat delay gluten development; the effect of water depends on the amount of water in relation to the flour. Equal parts of liquid and flour delay gluten development, whereas twice as much flour as liquid (2:1) facilitates rapid development of gluten.

Other flours are available for special applications, and they may be used in baked products with a fair degree of success when combined with wheat flour at a sufficient level to provide the gluten structure. These flours include rye, triticale (a cross between wheat and rye), soy, rice, corn, and potato.

In baked products, flour provides the basic structure. Liquid hydrates the gluten, dissolves sugar and other dry ingredients, gelatinizes starch, and provides steam for leavening. Eggs provide liquid, protein for added structure, emulsification, flavor, color, and a lighter texture when added as a foam. Fats have a tenderizing effect, enhance flavor, influence texture, and may contribute to color. Sugar sweetens, increases volume, tenderizes, and promotes browning. Salt adds flavor, and it slows carbon dioxide production in yeast doughs. Leavening agents provide expansion to give large volumes and tender cells in baked products.

Leavening occurs in any baked product as the result of the presence of air and steam, although the importance of these two ever-present leavening agents varies with the particular product being prepared. Air expands its volume by 1/273 for each degree Celsius rise in temperature during the early phase of baking, but steam expands to 1,600 times the volume occupied by the water from which it forms. Steam is a far more effective leavening agent than air, although air also is essential.

Carbon dioxide, an extremely effective gaseous leavening agent, can be introduced into batters and doughs through the metabolism of *Saccharomyces cerevisiae*, the baker's yeast commonly used in yeast breads. Glucose, sucrose (inverted to glucose and fructose), and maltose are converted slowly to ethyl alcohol and carbon dioxide by yeast during the fermentation period. Salt regulates the speed of production of carbon dioxide. *S. cerevisiae* produces carbon dioxide effectively at the usual pH of doughs (between pH 5 and 6). If bacteria such as *Lactobacillus sanfrancisco* are added to a dough, lactic acid production drops the pH of the dough below the range where *S. cerevisiae* is a good yeast to use; *S. exigus* and *S. inusitatus* are excellent yeasts to add for carbon dioxide production in sourdough breads.

Chemical reactions can produce carbon dioxide as a result of the inclusion of either separate acids and alkali or baking powder. Acids and alkali (baking soda) react as soon as the soda is dissolved, limiting their effectiveness as leaveners unless mixing and baking occur quickly. Baking powders contain at least one acid salt, soda, and a standardizing agent to take up moisture, prevent reaction in the can, and provide at least 12 percent carbon dioxide from a specified measure of the powder.

SAS-phosphate, a double-acting baking powder, is the baking powder available in grocery stores today; it has the ability to provide carbon dioxide during both mixing and baking. It has a rather harsh, metallic aftertaste when used in excess. Tartrate baking powders can be made using cream of tartar as the acid salt to produce a leavening agent without distinctive aftertaste, but one that requires fast mixing and baking to avoid excessive loss of carbon dioxide.

## STUDY QUESTIONS

1. What are the merits and disadvantages of whole wheat flour in baked products?

2. Describe the milling process and explain how the streams used in making various flours influence the characteristics of the resulting flours.

3. Identify the three overall categories of proteins in flour. What proteins are included in each category? What is the contribution of each category to baked products?

4. Describe both fractions of protein constituting gluten and explain how each contributes to the behavior of gluten in batters and doughs.

5. How may flours from cereals other than wheat be used in baked products? Why might there be an advantage in including triticale flour, cornmeal, rye flour, soy flour, or potato flour?

6. Briefly explain the roles of each of the following in baked products: flour, liquid, eggs, fats, sugar, salt, and leavening agents.

7. Compare the effectiveness of air, steam, and carbon dioxide as leaveners in baked products.

8. Although most yeast breads are made with *Saccharomyces cerevisiae*, a few are made with

*Saccharomyces exigus* and *Saccharomyces inusitatus.* Why?

9. Why is *Lactobacillus sanfrancisco* not used alone as a source of leavening?

10. What are the characteristics of an ideal baking powder? Evaluate SAS-phosphate baking powder against the criteria.

11. True or false. Gluten is present in most flours except rice flour.

12. True or false. Unbleached flour produces lighter, higher quality baked products.

13. True or false. Starch contributes to the structure of baked products.

14. True or false. *Saccharomyces cerevisiae* is the type of yeast used as a leavening agent because it produces carbon dioxide from glucose.

15. True or false. Baking powder and baking soda can be substituted in equal amounts for each other in cakes.

## BIBLIOGRAPHY

Berry, D. 2004. Breads on the rise. *Food Product Design 14* (7): 106.

Bhattacharya, M., et al. 2002. Staling of bread as affected by waxy wheat flour blends. *Cereal Chem. 79* (2): 178.

Cauvain, S. P. 2003. *Bread Making: Improving Quality.* CRC Press. Boca Raton, FL.

Decker, K. J. 2005a. High-profile flatbreads. *Food Product Design 15* (1): 97.

Decker, K. J. 2005b. Looking at the whole-grain picture. *Food Product Design 15* (9): 49.

Deis, R. C. 2005. How sweet it is—using polyols and high-potency sweeteners. *Food Product Design 15* (7): 57.

Delcour, J. A. and Hoseney, R. C. 2010. *Principles of Cereal Science and Technology.* 3rd ed. American Association of Cereal Chemists: St. Paul, MN.

Figona, P. 2007. *How Baking Works.* 2nd ed. John Wiley & Sons. New York.

Foster, R. J. 2005. pHood phenomena. *Food Product Design 14* (11): 61.

Fretzdorff, B. and Brummer, J. M. 1992. Reduction of phytic acid during breadmaking of whole-meal breads. (Abstract) *Cereal Foods 37* (2): 228.

Grant, L. A., et al. 2001. Starch characteristics of waxy and non-waxy tetraploid (*Tritium turgidium L.* var. durum) wheats. *Cereal Chem. 78* (5): 590–595.

Grant, L. A., et al. 2004. Spaghetti cooking quality of waxy and non-waxy durum wheats and blends. *J. Sci. Food Agric. 84:* 190.

Grenus, K. 2005. Maintaining texture. *Food Product Design 15* (5): 47.

Hazen, C. 2006a. New fiber options for baked goods. *Food Product Design 15 (10):* 80.

Hazen, C. 2006b. Adding soy ingredients for health. *Food Product Design 16* (2): 63.

Lafiandra, D., et al. 2010. Approaches for modification of starch composition in durum wheat. *Cereal Chem. 87* (1): 28.

Monari, A. M., et al. 2005. Molecular characterization of new waxy mutants identified in bread and durum wheat. *Theoretical and Applied Genetics 110* (8): 1481.

Ohr, L. M. 2006. Go with the grain. *Food Technol. 60* (9): 63.

Pyler, E. J. and Gorton, L. A. 2009. *Baking Science and Technology.* 4th ed. Sosland Publishing. Merriam, KS.

Pszczola, D. E. 2004. Fats: In *trans*-ition. *Food Technol. 58* (4): 52.

Pszczola, D. E. 2006. Reaping a new crop of ingredients. *Food Technol. 60* (7): 51.

Romano, A., et al. 2007. Description of leavening of bread dough with mathematical modeling. *J. Food Engineering 83* (2): 142.

Schultz, M. 2005. Xanthan and foods: Bonded for life. *Food Product Design 15* (7): 17.

Shewry, P. R. and Tatham, A. S. 2000. *Wheat Gluten.* Royal Society of Chemistry. London.

Sikorski, Z. E. 2006. *Chemical and Functional Properties of Food Components.* 3rd ed. CRC Press. Boca Raton, FL.

Spano, M. 2009. Celiac disease feeds gluten-free need. *Food Product Design 19* (10): 28.

Tenbergen, K. and Eghardt. H. B. 2004. Baking ammonia: The other white leavening agent. *Food Product Design 14* (6): 110.

Turner, J. 2005. Everybody loves carbs. *Food Product Design 15* (May Supplement): 12.

Vignaux, N., et al. 2005. Quality of spaghetti made from full and partial waxy durum. *Cereal Chem. 82* (1): 93.

Zandonadi, R. P., et al. 2009. Psyllium as a substitute for gluten in bread. *J. Amer. Dietet. Assoc. 109* (10): 1781.

## INTO THE WEB

*http://www.cwb.ca/public/en/*—Canadian Wheat Board; flour-milling and bread-making information.

*http://www.pastryitems.com/baking_information.htm*—Information on flour.

*http://www.kswheat.com/videogallery.php?id=24*—Kansas Wheat Commission site with information and videos on wheat.

*http://wbc.agr.mt.gov/Consumers/Nutrition/wheat_flours.html*—Information on Montana wheat, barley, and flour.

*http://www.ultragrain.com/what_is_ultragrain.jsp*—Information on Ultragrain®.

*http://www.wheatfoods.org/UrbanWheatfield-25/Index.htm*—Extensive site on wheat from farm to table.

*http://www.fabflour.co.uk/home.asp*—Flour Advisory Bureau, United Kingdom; European perspective about flour.

*http://www.wholegrainscouncil.org/*—Consumer information by the Whole Grains Council.

*http://www.cargill.com/food/emea/en/products/proteins/gluvital-vital-wheat-gluten/index.jsp*—Information on vital wheat gluten.

*http://www.kingarthurflour.com/*—Source for a variety of specialty flours.

*http://www.fda.gov/Food/FoodIngredientsPackaging/FoodAdditives/FoodAdditiveListings/ucm091048.htm*—Listing of approved additives, including flour bleaching agents.

*http://www.wheatfoods.org/Is-bleached-white-flour-harmful.8.10.htm*—Discusses bleaching of flour.

*http://www.nutraingredients.com/Research/Waxy-wheat-produces-low-calorie-bread*—Article on bread and durum wheat quality and utilization.

*http://www.ars.usda.gov/research/projects/projects.htm?ACCN_NO=417937*—Report on research on enhancing hard red spring and waxy durum.

*http://cat.inist.fr/?aModele=afficheN&cpsidt=16579065*—Article describing limitations in using waxy durum wheat in pasta.

*http://www.aaccnet.org/cerealchemistry/articles/2002/0201-02R.pdf*—Article on bread staling.

*http://library.wur.nl/wda/dissertations/dis3414.pdf*—Thesis on effect of pentosans on gluten formation and properties.

*http://www.aaccnet.org/*—American Association of Cereal Chemists journal articles.

*http://www.breadworld.com/Video.aspx?id=M9KX4KFBj5w*—Video on kneading.

*http://www.hort.purdue.edu/newcrop/AFCM/rye.html*—Information about rye.

*http://www.hort.purdue.edu/newcrop/afcm/triticale.html*—Information about triticale.

*http://food.oregonstate.edu/ref/ref_leavening.html*—References on leavening.

*http://www.redstaryeast.com/science_of_yeast/information_for_students.php*—Information on yeast.

*http://www.breadworld.com/Video.aspx?id=7ilQLDcK8U0*—Video on proofing yeast doughs.

*http://www.breadworld.com/Video.aspx?id=ksziuzwLuGY*—Video on using active dry yeast.

*http://www.breadworld.com/Video.aspx?id=jCcf4rY-Zs4*—Video on using rapid-rise yeast in doughs.

*http://www.ncbi.nlm.nih.gov/pmc/articles/PMC205458/*—Discussion of metabolism of *Lactobacillus sanfrancisco* in sourdough bread.

*http://www.ncbi.nlm.nih.gov/pmc/articles/PMC168117/*—Discussion of proteolytic enzymes in sourdough bread.

*www.sourdo.com*—Sourdough bread information.

*http://whatscookingamerica.net/History/BakingPowderHistory.htm*—Brief history of baking powder.

*http://www.enotes.com/how-products-encyclopedia/baking-powder*—Commercial production of baking powders.

*http://www.foodproductdesign.com/articles/2004/09/baking-ammonia-the-other-white-leavening-agent.aspx*—Article about baking ammonia.

Pastry for pies requires careful handling to minimize gluten development and create tender, flaky crusts.
(Courtesy of Pat Chavez.)

# CHAPTER 18

# Baking Applications

## Chapter Outline

## OBJECTIVES

After studying this chapter, you will be able to:

1. Explain the preparation and evaluation of various types of quick breads.
2. Describe the preparation and evaluation of yeast breads.
3. Discuss the preparation and evaluation of foam and conventional cakes.
4. Define the roles of ingredients and the effects of technique on pastry characteristics.
5. Compare the characteristics and components of cookies of various types.

## INTRODUCTION

Despite countless variations, baked goods that are popular throughout this country and in many other parts of the world can be divided into breads (quick and yeast), pastry (plain and puff), cakes (foam and shortened), and cookies (drop, rolled, and bar). This chapter explores the unique aspects of these categories from the context of the ingredients, their functions, the techniques used to make the batter or dough, and changes related to baking.

## QUICK BREADS

### Types

**Quick breads** are breads that can be prepared quickly and do not rely on yeast or other microorganisms for leavening (Figure 18.1). Types commonly included in this category are muffins, biscuits, popovers, cream puffs, waffles, pancakes, and cake doughnuts; many variations of each kind are familiar. This variety of products is the result of differences in ingredients and their ratios, methods of mixing, and ways of baking.

### Ingredients

Variations in the ingredients and their proportions are responsible for many of the unique qualities of specific quick breads. Only three ingredients (flour, liquid, and salt) are common to all quick breads, and the ratio of flour to liquid is quite different from one type of quick bread to another (Table 18.1).

In most instances, all-purpose flour is used in quick breads, but some rich coffee cakes use cake flour. A variety of grain flours may be used to supplement wheat flour, but they will modify texture and flavor, as well as color. These flours are most likely to be found in muffins, but pancakes and waffles sometimes include them as well.

The liquid used most commonly in quick breads is milk, although fruit juices, water, or even moisture-laden vegetables such as zucchini may be the liquid used in a recipe. Coffee cakes, muffins, and loaves are the types of quick breads most likely to incorporate these variations in liquid.

The type of fat used and the way it is incorporated vary considerably from one type of quick bread to another. Butter or margarine frequently is the choice because of the pleasing color and flavor contributions. When the muffin method of mixing is used, this hard fat is melted. As a convenience in making muffins, pancakes, loaves of quick breads, or other

**Figure 18.1**    Scones are a quick bread that is particularly popular in the United Kingdom.

**Table 18.1**   Relative Proportions of Ingredients in Basic Quick Breads on the Basis of Volume

| Quick Bread | Flour (c) | Liquid (c) | Fat (tbsp) | Sugar (tbsp) | Eggs | Baking Powder (tsp) | Salt (tsp) |
|---|---|---|---|---|---|---|---|
| Popovers | 1 | 1 | 0[a] | 0 | 2 | 0 | 0.5 |
| Cream puffs | 1 | 1[b] | 8 | 0 | 4 | 0 | 0.25 |
| Waffles | 1 | 0.9 | 4.25 | 0.4 | 0.8 | 1.6 | 0.2 |
| Pancakes | 1 | 0.8 | 1.66 | 0.5 | 0.8 | 2.4 | 0.5 |
| Muffins | 1 | 0.5 | 2 | 2 | 0.5 | 1.5 | 0.5 |
| Biscuits | 1 | 0.38 | 2.66 | 0 | 0 | 1.5 | 0.25 |
| Cake doughnuts | 1 | 0.14 | 0.09 | 0.2 | 0.9 | 0.9 | 0.14 |

[a]Oil is used to grease cups if needed but is not in the batter.
[b]Liquid is boiling water; all other liquids are milk.

types of quick breads for which liquid fat is needed, salad oil may be the choice rather than melted butter or margarine. Hydrogenated shortening usually is chosen for making biscuits because it can be cut into pieces. It also is the choice in making some quick bread loaves in which the fat is creamed with sugar.

## Mixing Methods

Most recipes for quick breads call for one of three basic methods: biscuit method, muffin method, or conventional method (Chart 18.1). The method used is dictated by the manner in which the fat is distributed. In the biscuit method, a solid fat is cut into small pieces. The liquid fat is dispersed with the other liquid ingredients in the muffin method. A plastic, hard fat is creamed with the sugar in the conventional method.

***Biscuit Method.***   The first step in the **biscuit method** is sifting the dry ingredients together and then adding shortening or other solid fat, which is cut in with a pastry blender until particles are about the size of rice grains (using a light, lifting motion to cut in the fat). Next, the liquid is added all at once, and the mixture is stirred carefully with a fork to just moisten the dry ingredients. In doughs made by this method, gentle kneading is the final step prior to shaping and baking. The final product has a flaky texture.

**Biscuit method**
Mixing method in which the dry ingredients are combined, the fat is cut into small particles in the flour mixture, all of the liquid is added at once and stirred in just to blend, and the dough is kneaded; a flaky product results.

---

**Muffin Method**
1. Combine dry ingredients in one bowl and liquid ingredients in another
2. Make a well in dry ingredients and pour liquid ingredients in
3. Stir together just until moistened

**Biscuit Method**
1. Sift dry ingredients together
2. Add solid fat and cut in
3. Add liquid and stir to moisten
4. Knead, roll, and shape

**Conventional Method**
1. Sift dry ingredients together
2. Cream sugar and fat together
3. Beat in eggs
4. Add 1/3 of dry ingredients and stir
5. Add 1/2 of liquid and stir
6. Sequentially add and stir second third of dry ingredients, last half of liquid, and last third of dry ingredients

---

**Chart 18.1**   Basic methods of mixing quick breads

Fat is cut in to promote a flaky texture, but much of the tenderizing potential from the fat is not available. All of the fat on the interior of each flour-coated piece is unable to come in contact with and lubricate the flour proteins during mixing. The kneading action in making biscuits is basically a folding action, which also promotes the development of flaky layers. In the oven, the fat gradually melts and spreads out, leaving a space where steam and carbon dioxide (from baking powder) can collect and expand.

*Muffin Method.*    Muffins and some other comparatively simple quick breads are made by an extremely quick method—the **muffin method**. All of the dry ingredients are sifted together into one mixing bowl. The liquid ingredients (which include either oil or melted fat) are blended together thoroughly in a second bowl prior to being added all at once to the dry ingredients. A brief stirring, done efficiently with a wooden spoon, completes the preparation for baking. Baked products prepared by the muffin method have a somewhat coarse, open texture and may be slightly crumbly.

*Conventional Method.*    For a fine texture, tender crumb, and good keeping qualities, the **conventional method** of mixing usually is best. The first step, creaming of the solid fat and sugar, is followed by addition of the beaten egg. The dry ingredients are added, a third at a time, followed by brief mixing and alternated with two additions of milk (half of the liquid at each addition). This method is definitely more labor intensive than the muffin method, but the results are optimal, especially for quick breads containing fairly high proportions of sugar and fat.

## Selected Examples of Quick Breads

*Popovers.*    **Popovers** are a quick bread with a high liquid-to-flour ratio (in fact, equal volumes are used); the result is a very fluid batter. In such a dilute flour mixture, gluten does not develop readily with mixing so there is no reason to be concerned about the amount of beating that is required to make a perfectly smooth batter.

The large amount of liquid allows gelatinization of the starch in the flour during baking, evidenced by the slightly gel-like interior walls in baked popovers. This liquid also is critical to the production of a large quantity of steam to help create the large cavity expected in the interior of popovers. Although some air is incorporated into the batter during beating, the fluid state of the batter releases most of the air, leaving almost all of the leavening task to steam, which is created rapidly during the early part of the baking period in a preheated, very hot 218°C (425°F) oven.

The structure of popovers is due largely to the high content of egg proteins and starch from the flour. Expansion is possible because the steam generated in baking presses against the cell walls while the egg proteins remain uncoagulated and the small amount of gluten that is present is still extensible. If an inadequate quantity of egg is used, popovers will be quite compact and spongy in the center and will lack the expected large cavity. This problem clearly illustrates the importance of egg proteins in the structure of popovers. Egg white proteins, in particular, are important to the popping action in the center of popovers. A practical rule of thumb is that at least two large or three medium eggs are needed per cup of flour in a popover batter (Figure 18.2).

The depth of pans is important to successful popping action. Popover pans are approximately twice as deep as muffin pans. This depth permits generation of steam to create interior pressure before the egg proteins become rigid and before the starch is gelatinized. In the absence of popover pans, custard cups provide a suitable alternative. The oven should be preheated until it reaches 218°C (425°F) to ensure rapid generation of steam and popping of the uncoagulated popovers.

Other factors that may interfere with achievement of the desired large interior cavity are related to ingredients. If too much liquid is present in relation to the amount of flour, the protein structure will be too weak and will release the expanding steam without having gotten hot enough to coagulate the protein and maintain the expanded structure. The presence of too much fat in the greased cups can also weaken the structure of popovers and allow the steam to escape prematurely.

**Muffin method**
Mixing method in which the dry ingredients are sifted together in one bowl, the liquid ingredients (including fat) are mixed in another bowl and then poured into the dry ingredients, and the mixture is stirred briefly and baked; the result is a rather coarse, slightly crumbly product that stales readily.

**Conventional method**
Mixing method in which fat and sugar are creamed, beaten eggs are added, and dry ingredients (a third at a time) and liquid ingredients (half at each addition) are added alternately; fine texture and excellent keeping qualities are advantages of this laborious method.

**Popover**
Quick bread made with equal volumes of flour and liquid, including at least two large eggs per cup of flour, and baked in deep cups in an oven preheated to 218°C (425°F).

**Figure 18.2**  Eggs, particularly the whites, are essential to the formation of a large cavity in popovers during baking.

When ingredient proportions are right and the batter is beaten until smooth and baked in appropriate pans in a preheated oven, the resulting popover should be ideal. It will have a large volume and a large interior cavity. The interior will be slightly moist, and the popover will be crisp.

***Cream Puffs.***  **Cream puffs** are classified as a quick bread despite their normal function as an edible container for either an appetizer or a dessert. Although their appearance is somewhat reminiscent of that of popovers and their liquid-to-flour ratio is the same as that for popovers, there are some very significant differences between these two distinctive quick breads. Cream puffs, unlike popovers with essentially no fat, contain half as much fat as liquid (see Table 18.1). To bind this large quantity of fat, a large number of eggs (four per cup of flour) must be included in the recipe to aid in emulsifying the fat and the boiling water. Much of the liquid is bound in the gelatinizing starch during preparation of the paste; the gelatinized starch paste also helps prevent the oil-in-water emulsion from breaking.

Preparation of cream puffs begins by adding the flour and salt to the mixture of boiling water and melted fat, stirring vigorously while heating the paste until the starch gelatinizes and a ball of paste forms. Care must be taken to avoid having excessive evaporation of water during this period, for inadequate water will cause the emulsion to break and the fat to separate from the paste. Each egg is beaten into the cooling paste to enhance the emulsion (see Chapter 16) that already exists in the paste and to increase the liquid level slightly. If the emulsion breaks, it must be reformed by replacing the moisture that has been lost. By adding water slowly and stirring it in until the paste once again assumes its shiny, smooth surface appearance, a broken emulsion can be reformed with no damage to the quality of the finished cream puff.

Baking takes place after small mounds of dough have been piped onto a nonstick baking sheet and placed in a preheated oven at 232°C (450°F) for 15 minutes, followed by an additional 25 minutes at 163°C (325°F) to allow time for heat to penetrate to the interior and coagulate the protein structure. The large amount of steam generated in the puffs by this intense heat is able to expand the puffs, creating a large interior cavity before the proteins (particularly those in the egg whites) lose their ability to stretch as they are coagulated.

**Cream puff**
Quick bread containing a great quantity of fat and eggs that puffs remarkably as a result of the steam generated during baking at 232°C (450°F) for 15 minutes and at 163°C (325°F) for 25 minutes.

Cream puffs should have a pleasing, golden-brown surface, and the volume should be large in proportion to the amount of dough used to form them; the interior should be only moderately moist, and the cavity should be large. The most likely problem in making cream puffs is failure to puff and create the desired interior cavity. This most commonly is the result of a broken emulsion, a problem evidenced either before or during baking by fat visibly oozing from the dough.

The usual reason for a broken emulsion is too much evaporation from the dough, probably during the gelatinization of starch before the eggs are added. It is imperative that the emulsion be reformed by stirring in enough water to re-establish the emulsion prior to baking. If a cream puff dough is too soft, the steam cannot be trapped in the baking puff long enough for the expanded structure to set; escape of the steam through broken cell walls deflates the cavity, and a compact puff is the result. Too low an oven temperature also can cause problems because of failure to generate enough pressure from steam before the proteins coagulate and lose their elasticity.

**Muffin**
Small, rounded quick bread baked from a batter containing a 2:1 flour-to-liquid ratio, plus egg, fat, sugar, baking powder, and salt.

***Muffins.*** **Muffins** contain a ratio of 2 parts flour to 1 part milk (see Table 18.1); other ingredients are fat, sugar, egg, baking powder, and salt. These ingredients are mixed by the muffin method described earlier in this chapter.

Eggs need to be beaten thoroughly, yet not whipped into a foam. Thorough beating ensures that the eggs blend completely with the other liquid ingredients, thus avoiding areas in the baked muffin where the egg protein is concentrated. Inadequate beating of the egg causes some thick cell walls and a somewhat waxy character where the concentrated egg protein is found. There is no need to beat the eggs to a foam because adequate leavening is available from the steam and carbon dioxide generated from the baking powder.

The extent of mixing is of paramount importance in determining the characteristics of muffins. The very sticky nature of the batter facilitates rapid gluten development when the ingredients are stirred together. The appearance of the batter changes quickly as gluten develops during mixing. It begins with a lumpy batter in which big clumps of dry ingredients are encased by the liquid and developing gluten. If the batter containing these large lumps is baked, the volume of the muffins will be poor because not all of the baking powder will have been moistened enough to react completely. They also will be very crumbly as the result of inadequate gluten development to hold the structure together (Figure 18.3).

With only a bit more stirring with a wooden spoon, the batter will still be noticeably lumpy, but there will not be any large aggregates of dry ingredients. After a total of about 25 strokes, sufficient mixing will achieve this appearance, and the batter is ready to be spooned into greased muffin cups and baked at 218°C (425°F) until the crust is a golden brown (usually about 20 minutes).

Excessive gluten development results if stirring is continued beyond this point. The batter gradually becomes smooth, and development of tenacious gluten strands can be

**Figure 18.3** Effects of the extent of mixing of muffin batter: (left) undermixed; (center) optimum mixing; (right) overmixed.

noted if the batter is allowed to flow from the spoon. The volume of the batter begins to reduce as a result of the loss of some air and the release of a little carbon dioxide from the baking powder at room temperature.

With overmixing, the surface of muffins begins to be increasingly smooth as mixing proceeds until the baked crust looks very much like the crust of yeast rolls. Instead of having a nicely rounded top, overmixed muffins exhibit a peaked or pointed appearance. When they are cut in half vertically at the highest point, tunnels converging on the peak are evident, and the texture between them is fairly fine.

Tunnels develop as steam and carbon dioxide collect in a few areas and begin to exert pressure against the cell walls. The batter heats most quickly near the outer edges and on the bottom, causing the protein to coagulate and lose its ability to stretch in these regions, while the center and upper region remain elastic until somewhat later. This continuing elasticity permits the gluten in the center and above to stretch upward with the pressure from the leavening gases; the strength of the gluten strands is sufficient to channel the expansion upward, resulting in the development of tunnels when the gluten is overdeveloped.

A deep pan is a bit more likely to promote tunnels than is a regular muffin pan. Sometimes muffins baked at excessively high temperatures have an increased tendency to develop tunnels, but this does not always happen. Overmixing is guaranteed to create tunnels. The likelihood of tunnels is reduced by the use of either whole wheat or some other type of flour to supplement all-purpose flour. The increased cellulose and other fiber from whole wheat and the lack of gluten in the other cereal flours interfere with gluten development a bit, making the formation of tunnels unlikely.

**Biscuits.**   In **biscuit** dough, the ratio of flour to liquid varies somewhat with the flour but is usually around 3 parts flour to 1 part liquid (3:1), making this dough much less sticky than muffin batter. This means that gluten does not develop as readily in biscuits as in muffins. Because of this difference, biscuits ordinarily are stirred about 25 strokes with a table fork and then gently kneaded 10–20 times with the fingertips prior to rolling and baking.

Unless the dough is stirred adequately and kneaded sufficiently, the desired flaky texture may not be achieved because of a lack of gluten development. This amount of manipulation, though definitely greater than that recommended for muffins, is necessary because of the comparatively slow rate of gluten development in biscuit dough. However, too much mixing causes toughness and humping of the crust.

Sour milk and soda are the source of carbon dioxide in some biscuit recipes, especially for drop biscuits. These recipes produce a slightly acidic dough, causing the biscuits to be snowy white inside, in contrast to the somewhat yellow color that develops in a slightly alkaline dough. Use of buttermilk fosters the desirable white color. On the other hand, high levels of baking powder promote a somewhat creamy color because of the mildly alkaline residue. Thiamine retention is enhanced by using either buttermilk or sour milk with no excess of soda or a minimum of baking powder to keep the dough close to a neutral pH.

**Biscuit**
Quick bread with a ratio of about 3:1 (flour to liquid) and also containing fat cut into small particles, baking powder, and salt; usually kneaded and rolled, but sometimes dropped.

# YEAST BREADS

## Ingredients

The basic ingredients in yeast breads are simply flour (usually all-purpose or bread flour), liquid, salt, sugar, and yeast, yet together they form a food so basic that it is referred to as "the staff of life." Their proportions, expressed as percentages relative to the weight of flour, are defined in Table 18.2.

Yeast breads can be made without any fat, but the crust and the crumb will be a bit hard and crisp. This characteristic is sought deliberately in formulating French bread and a few related specialty breads, but some fat is added to most other breads to tenderize the product. Up to a maximum of 3 percent, adding fat increases volume.

Shortening improves the keeping quality of bread as it is spread throughout the mixture and coats the gluten and starch. Emulsifying agents may be used at levels up to 20 percent

**Table 18.2**    Amounts of Ingredients Relative to Flour in Yeast Breads

| Ingredient | Percentage of the Weight of Flour (range) |
| --- | --- |
| Flour (all-purpose) | 100 |
| Fat | 2–6 |
| Liquid | 60–65 |
| Sugar | 2–6 |
| Salt | 1.5–2 |
| Yeast | 1–6 |

of the weight of the fat as an aid in dispersing the fat on the gluten strands and starch during mixing. Mono- and diglycerides and lecithin are effective for this role in bread.

The optimum amount of liquid in proportion to flour varies with the individual flour. Flours made from soft wheat require less liquid than do those from hard wheats. Elasticity, extensibility, and strength of bread doughs increase with increasing liquid up to about 60 percent, after which these qualities decline.

Whole milk often is the liquid used in making yeast breads. Although milk is more than 87 percent water, its fat, protein, and lactose content adds pleasing textural qualities to the bread. However, milk used in making bread dough should be scalded to at least 92°C (198°F) for a minute or held at a slightly lower temperature for a longer time.

Failure to heat milk sufficiently results in a sticky dough and a baked product low in volume and rather coarse in texture. Apparently, this brief heating denatures some of the protein, possibly one in the serum protein fraction, that is responsible for the negative effect that unscalded milk has on yeast breads. Bakers who add dry milk solids experience similar problems unless the milk has been heat processed adequately prior to drying. It is imperative that scalded milk be cooled to a temperature that is safe for the yeast before the two are combined.

The content of sugar in yeast breads varies but may be as high as 16 percent in some sweet bread recipes. When the level of sugar is between 3 and 6 percent of the weight of the flour, fermentation is optimum, and the volume is at a maximum. Below or above this percentage, volume decreases. Sucrose, fructose, glucose, and invert sugar ferment at about the same rate, but maltose and lactose do not ferment as well by yeast.

Inclusion of between 1.5 and 2 percent salt in a yeast dough reduces the rate of fermentation appropriately, allowing time for development of flavor in the dough and strengthening of the gluten. Too much salt, because of its osmotic effect, impairs the yeast fermentation process and slows carbon dioxide production. The result is a compact, firm bread. If salt is not added, fermentation proceeds too rapidly. This produces a bread that is coarse.

The form of the yeast dictates the temperature to use to disperse and hydrate the yeast in water prior to its addition to the other ingredients (Table 18.3). Compressed yeast cakes are rather heat sensitive, and the temperature of the hydrating water should be about 30°C (86°F) and must not exceed 37°C (98.6°F), which is body temperature.

When active dry yeast is manufactured, the outer membrane enclosing the contents of the yeast cell alters and becomes rather permeable. This facilitates loss of the components within the cells unless the rehydration temperature is controlled carefully. The water should be between 40 and 46°C (104 and 115°F) to retain key cellular components.

Within this range, permeability of the cell wall is controlled and prevents the loss of **glutathione**, the sulfhydryl-containing compound thought to be responsible for the sticky

**Table 18.3**    Recommended Temperatures for *Saccharomyces cerevisiae*

| Type of Yeast | Temperature | Range |
| --- | --- | --- |
| | Dissolved in Liquid | Directly in Dry Flour |
| Compressed | 26.7–32°C (80–90°F) | — |
| Active dry | 40–46°C (104–115°F) | 48.9–54.5°C (120–130°F) |
| Quick-Rise™ | 40–46°C (104–115°F) | 48.9–54.5°C (120–130°F) |

nature of yeast doughs made with active dry yeast rehydrated below 40°C (105°F). Temperatures above 46°C (115°F) are likely to kill the yeast, which will cause either very slow or no fermentation, depending on the severity of the temperature insult. Quick-rise active dry yeast requires the same temperature controls as active dry yeast.

## Dynamics in Mixing and Kneading

Mixing is done to distribute ingredients as uniformly as possible into the dough and to permit incorporation of an adequate amount of flour in relation to the level of liquid in the mixture. This liquid combines with flour to begin to hydrate damaged starch granules and also to lubricate gliadin and glutenin in preparation for developing the gluten complex needed for optimal texture and volume in the finished product. Glutenin (the elongated and quite fibrous protein in wheat flour) molecules appear to join to each other by forming disulfide (—S—S—) linkages. These polymerized units of glutenin are called **concatenations**. However, too much mixing causes some of the polymers of glutenin to dissociate, which reduces elasticity and results in a less desirable texture in the baked bread.

Mixing brings glutenin concatenations not only into contact with each other but also into contact with gliadin, the sticky, fluid protein. These two components then become intermeshed to form a continuous network, the gluten complex. Starch granules then are embedded in this matrix as kneading progresses. Although it is possible for machine kneading eventually to physically break up part of the matrix, kneading by hand is not vigorous enough to cause this problem (Figure 18.4).

Flour tortillas often are used to make wraps, which are becoming popular as replacements for sandwiches made with bread. Gluten needs to be developed sufficiently during mixing so they can be rolled around other foods to create burritos or other types of wraps. However, the dough for wheat tortillas also needs to have gluten that is sufficiently relaxed so it will form a perfect circle without fracturing when it is rolled by machine. A dough conditioner can be added to create a more linear protein structure and promote the desired handling characteristics in flour tortillas. Conditioners that act as a reducing agent cause displacement of some of the cross-linking sulfur bridges with thiol (—SH) groups, a change that promotes extensibility of the dough during rolling.

Bread doughs are usually made using the **straight-dough method** when prepared in homes and restaurants. This method requires scalding the milk and temperature control during hydration of the yeast and preparation of the dough. All of the ingredients are combined prior to the fermentation and proofing (rising time) when the straight-dough method is used. Considerable kneading is required to develop the gluten adequately for optimum texture when making bread by this method. A strong mixer with a dough hook or a food processor also works well.

**Glutathione**
Peptide (γ-glutamylcysteinylglycine) that occurs in yeast cells and is capable of passing through the cell walls of active dry yeast that has been hydrated below 40°C (105°F), causing stickiness in the yeast dough.

**Concatenation**
A linking together; nonspecific description of the association of glutenin molecules by disulfide linkages.

**Straight-dough method**
Method of mixing yeast breads in which all of the ingredients are added, mixed, and kneaded prior to fermentation and proofing.

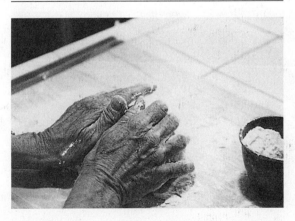

**Figure 18.4**   Yeast dough requires vigorous kneading to develop the gluten.

**Sponge-and-dough method**
Method of mixing yeast breads in which the yeast, liquid, and part of the flour are mixed and fermented to make a sponge before the rest of the ingredients are added; used commercially with strong flours.

Commercial bakeries often use the **sponge-and-dough method** to prepare breads. This mixing method begins with the mixing of liquid, yeast, and part of the flour (as well as possibly part of the sugar). This mixture is allowed to ferment until it becomes quite spongy. Then the remaining ingredients are mixed in.

Proofing time for breads made by the sponge method is somewhat shortened because of the fermentation that occurs when the initial mixture is developing its spongy character as the yeast produces carbon dioxide (Figure 18.5). Strong flours (e.g., bread flour) are necessary for successful use of the sponge method. The flavor of breads made with the sponge method is a bit more yeast-like than is true for those made by the straight-dough method.

## Changes during Baking

Remarkable changes occur in the dough during baking as a result of the impact of a hot oven [204–218°C (400–425°F)]. The initial phase of baking effects a rather dramatic increase in volume, the result of increased production of carbon dioxide because of the acceleration of yeast growth. This rather abrupt expansion, called **oven spring** also is caused by the expansion of the existing carbon dioxide in the dough. The volume increase from this phenomenon is appreciable; the final volume represents an increase of as much as 80 percent beyond the volume of the shaped dough when it was placed in the oven.

**Oven spring**
An increase in volume (usually about 80 percent) of yeast breads during the early part of baking resulting from the expansion of carbon dioxide and the increased production of carbon dioxide stimulated by the oven heat.

If the oven temperature is significantly above 218°C (425°F), the final volume will be reduced because the crust denatures before the gases in the dough have had sufficient time to achieve their maximum leavening. Too cool an oven temperature allows time for too much expansion. The product has the potential for a very porous texture or for a quite compact structure as a result of gluten strands being stretched so far before the protein structure is rigid that the gluten strands break and much of the leavening is lost.

In the early part of baking, oven heat causes the dough to become softer, making it easier for the leavening gases to stretch the gluten. This softening is due in part to increased amylase activity in converting some starch molecules to dextrins. Some shifting of water from gluten to starch occurs when the baking dough temperature rises above 60°C (140°F) and gelatinization is initiated. Gelatinized starch contributes significantly to the structure of the baked bread. Another effect of the heat during baking is killing of the yeast at about 60°C (140°F). This prevents additional generation of carbon dioxide. Amylase also is inactivated, which halts the degradation of starch to dextrins.

Crust browning, an aesthetically important change during baking, is largely the result of the Maillard or carbonyl-amine reaction. Milk in the formula enhances browning because lactose is available for the reaction, whereas water does not provide this added sugar. Browning is hastened once the surface of the baking bread becomes dryer due to

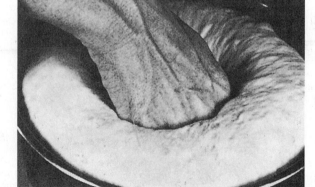

**Figure 18.5**  After yeast dough has doubled in volume, it is punched down to distribute the carbon dioxide more evenly before being shaped.

evaporation; the dry surface increases in temperature much more than it would if still moist. In fact, the crust temperature may approach 150°C (300°F), which accelerates the carbonyl-amine browning.

## Flour Options

Although breads have been made with flours from various cereals for centuries, new impetus was given to the use of flour alternatives in bread making in the 1980s, and the present concern over fitness and fiber has extended the variety still more. Some of the flour options available today are rye, rice, **triticale**, **spelt**, barley, millet, oat, **teff**, and **kamut**®. Fiber from a variety of sources can be added to bread by various alterations in their formulas. Various flours can substitute for part of the flour, but flavor and texture will change. It is possible to add wheat protein isolate or vital wheat gluten at levels up to 15 percent of the substituted flour to enhance the structural properties of these breads.

The quality of bread made entirely with triticale flour is better than that of bread made entirely with rye flour, but it still is not satisfactory. If sodium stearoyl-2-lactylate is added as a dough conditioner and mixing is reduced, the quality of triticale bread improves. A shorter fermentation period also is helpful in enhancing the quality of bread made with triticale flour.

Triticale bran and flax seed meal are products that may be added to bread dough to increase fiber content. Finely ground triticale bran can be added to a level of 15 percent of the wheat flour and still result in a bread that is satisfactory. Bran helps retain moisture longer and thus keeps the crumb softer. Coarse-textured bran has a greater effect on texture than does a fine grind; its roughness can be detected on the tongue, and it also interferes somewhat with gluten development because of its cutting action during mixing. Other options for increasing fiber include wheat bran and microcrystalline celluloses at a level of 7 percent (Figure 18.6).

**Triticale**
Hybrid grain made by crossing wheat and rye.

**Spelt**
Grain traced to ancient Iran; similar to wheat, but with tougher bran.

**Teff**
Cereal with very small grains; origin is Ethiopia where it has long been used to make injera, the traditional bread there.

**Kamut**®
Type of wheat grown in ancient Egypt.

**Figure 18.6**  Slices of whole wheat bread (1) and whole wheat bread supplemented with 7.5 percent of the following fibers: (2) flax hulls; (3) pea hulls; (4) coarse wheat bran; (5) sunflower hulls (coarse); and (6) cellulose. Flax was weak, darkest, reddest, and smallest. Pea hulls produced a light color; cellulose was similar to whole wheat, according to judges. (Courtesy of A. M. Cadden, F. W. Sosulski, and J. P. Olson, University of Alberta. *Journal of Food Science*. 1983. Vol. 48: 1151. Copyright © Institute of Food Technologists.)

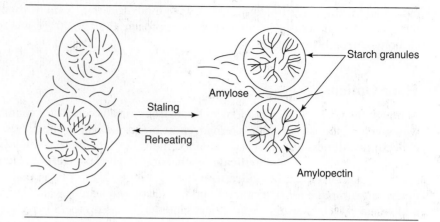

**Figure 18.7**    Diagram of possible changes during staling and reheating of bread.

Soy flour is another product of nutritional interest in bread making. This flour may be either full-fat or defatted. If sodium stearoyl-2-lactylate is added to full-fat soy flour, as much as 24 percent soy flour can be used with acceptable results. In fact, the water-binding capability of the soy flour improves the softness of the crumb during storage compared with bread made using only wheat flour.

One of the limiting factors in substituting with soy flour is the distinctive flavor, often described as beany, that can be detected at levels above 24 percent. **Sucroglycerides** help off-set the impact of soy flour in bread. Even with the use of dough conditioners, mixing time and fermentation time need to be shortened when soy flour is part of the formula. The remarkable ability of soy to absorb liquid also alters the amount of liquid required in relation to the flour.

**Sucroglycerides**
Sucrose esters (sucrose and glycerides or free fatty acids) used as a dough conditioner with soy flour.

## Staling

Bread begins to undergo deteriorative changes as soon as it is removed from the oven. One of the first changes is retrogradation (increasing crystallinity caused by cross-linkage of molecules) of amylose released from starch granules during gelatinization in the oven. The rather crystalline and firm character of the crumb that develops as bread is stored for a day or more may be the result of retrogradation within the starch granules and probably involves changes in the physical configuration of amylopectin (perhaps some folding of the branches and cross-linking of molecules). This retrogradation is reversible on reheating the bread, but recurs when it cools again (Figure 18.7).

Moisture levels also are involved in the staling process. There is a reversal in the location of water; some of the water that migrated to starch from gluten during baking returns to the gluten proteins. Water also migrates from the center of the loaf toward the crust, causing the crust to gradually increase in leatheriness, while remaining rather soft. This migration occurs even when evaporation from the crust is prevented.

Storage temperature has an effect on staling. Refrigerator storage accelerates firming of the crumb compared with storage at a warm room temperature; the firmness after a day at 8°C (46°F) is about the same as that after six days at 30°C (86°F). The only advantage of refrigerator storage is inhibition of mold growth in breads made without preservatives. Freezer storage is effective in inhibiting the firming of the crumb.

**Foam cake**
Cake featuring a large quantity of foam (usually egg white), which results in a light, airy batter and a baked cake with a somewhat coarse texture with moderately large cells; angel food, sponge, and chiffon cakes.

**Angel food cake**
Foam cake containing an egg white foam, sugar, and cake flour.

## FOAM CAKES

Three subcategories of **foam cakes**, each with unique characteristics, are familiar in the United States. The simplest of these is angel food cake. The others are sponge and chiffon cakes. The differences among these cakes result from ingredient variations, as Table 18.4 shows. **Angel food cake** basically is simply an egg white foam to which sugar and cake

**Table 18.4**   Comparison of Ingredients in Foam and Shortened Cakes

| Ingredient | Foam Cakes | | | Shortened cakes | |
|---|---|---|---|---|---|
| | Angel Food | Sponge | Chiffon | Layer | Pound |
| Cake flour | 1 c | 1 c | 1 c | 1 c | 1 c |
| Liquid | None | 5 tbsp water | $5\frac{1}{3}$ tbsp water | 8 tbsp milk | 5 tbsp milk |
| Eggs | 12 whites as foam | 4 yolks; 4 whites as 2 foams | 2 yolks; 4 whites as foam | 1 whole beaten | 2 whole, beaten |
| Fat | None | None | $3\frac{1}{3}$ tbsp oil | 4 tbsp shortening | $8\frac{1}{2}$ tbsp butter |
| Sugar | 12 tbsp | 8 tbsp | 11 tbsp | 8 tbsp | 9 tbsp |
| Baking powder | None | None | $1\frac{1}{4}$ tsp | 1 tsp | None |

flour are added. **Sponge cake** contains two foams—an egg yolk foam and an egg white foam. Some liquid, sugar, and flour are other key ingredients. **Chiffon cakes** are the most complex of the foam cakes, for they contain oil and baking powder, in addition to the ingredients in sponge cake.

Angel food cakes are white, whereas sponge cakes are yellow because of the egg yolk foam. Both of these foam cakes are leavened by air and steam only; thus, the volume of the egg foams is of great consequence in determining the size and uniformity of the cells and the final cake volume. All of the foam cakes are quite delicate structurally when removed from the oven; therefore, they should be inverted (suspended) during cooling to stretch the cells until the cell walls become cool and more rigid. Angel food and sponge cakes are fairly tender; sponge cakes are a little less tender than angel food cakes because of the slight toughening provided by the yolks. Chiffon cakes are more tender than the other foam cakes because of the oil, which is very effective in tenderizing them.

## Angel Food Cake

The basic ingredients in an angel food cake are simply cake flour, sugar, egg white, and cream of tartar, and yet considerable variation in quality can result under varying conditions. When all-purpose flour is substituted for cake flour, the volume is smaller, the texture is more compact, and the cake may be less tender than one made with cake flour. This latter difficulty can be overcome by slightly increasing the sugar (Figure 18.8).

Cream of tartar is important because of its action in stabilizing the egg white foam and in effecting a small reduction in the pH of the cake batter (between pH 5.2 and 6). As discussed in Chapter 16, adding cream of tartar at the foamy stage of beating egg whites delays foam formation and stabilizes the resulting foam. The smaller air cells that develop in a foam made with cream of tartar result in an angel food cake with cells of moderate size rather than the coarse texture that results when cream of tartar is omitted.

The acidity of cream of tartar bleaches the flavonoid (anthoxanthin) pigments in flour, thus enhancing the whiteness desired in angel food cake and avoiding the yellowish tint that is evident when the pH of the batter is higher. Cream of tartar also exerts a tenderizing effect on angel food cake, probably because of the improved stability of the foam and volume of the cake.

Sugar also helps to stabilize the foam and promote a finer texture in angel food cakes. The levels of the ingredients in an angel food cake influence the final cake. The larger the amount of flour relative to egg white, the less tender and drier is the resulting angel food cake. The recommended weight of flour is between 0.2 and 0.4 g for each gram of egg white.

Sugar usually is recommended at a level of 1 g for 1 g of egg white, although the level can be raised to a maximum of 1.25 g if the cake is prepared at an altitude no higher than 1,000 feet and the maximum amount of flour is used to offset the tenderizing influence of the extra sugar. Too much sugar results in a somewhat crisp crust, with crystals of sugar giving a crystalline, rather shiny appearance to the crust.

**Sponge cake**
Foam cake containing an egg yolk foam, an egg white foam, sugar, and cake flour.

**Chiffon cake**
Foam cake that includes oil and egg yolk as liquid ingredients, an egg white foam, baking powder, sugar, and cake flour.

**Figure 18.8**   Angel cakes ranging from 0 to 100 percent high-fructose corn syrup (HFCS) in combination with sucrose ranging from 100 to 0 percent. At 25 percent HFCS the cake was still satisfactory, but increasing levels caused browner crust, yellow crumb, firmer texture, and decreased sweetness. The beating time for the foams was decreased with higher levels of HFCS. (Courtesy of P. E. Coleman and C. A. Z. Harbers, Kansas State University. *Journal of Food Science.* 1983. Vol. 48: 452. Copyright © Institute of Food Technologists.)

Volume is an important attribute of angel food cakes. The quality and temperature of the egg whites are key factors determining the foam volume and, consequently, the final cake volume. Fresh eggs (see Chapter 16) produce angel food cakes of larger volume than do older eggs. The reason for this difference has not been identified with certainty, although the likely factors are the rise in pH as an egg ages and the reduced amount of thick white and increased amount of thin white. Whites warmed to room temperature [21°C (70°F)] yield angel food cake of maximum volume.

Oven temperature also influences the volume of angel food cakes made with fresh egg whites. Volume is greatest when the cakes are baked at 218°C (425°F). At even higher temperatures, the volume is reduced, as is palatability. Between 177°C (350°F) and 218°C (425°F) volume increases consistently as oven temperature increases. However, the opposite effect is noted in angel food cakes made using a commercial mix; an oven temperature of 177–191°C (350–375°F) yields better volume and palatability scores than are obtained at 204–218°C (400–425°F). Preheated ovens produce angel food cakes of larger volumes than are reached by baking from a cold start.

The extent of beating of the egg white foam is of utmost importance in determining the volume of the foam and, ultimately, the cake itself. Underbeating of the whites causes a cake with a smaller volume than would be possible if the whites had been beaten more. This is due to the reduced amount of air introduced into the foam and the somewhat lower stability of the rather fluid foam, the result of limited denaturation of the egg white proteins during the brief period of beating.

If the amount of beating is optimal so that the peaks just bend over, the foam will have good stability, yet it will retain the extensibility of the egg white proteins needed to stretch to a maximum during baking. Beating the egg whites to this point gives maximum volume in the baked cake, but it does not give the greatest volume possible during beating. However, it is the volume in the final product that is of greater importance.

The various proteins that comprise egg white have somewhat different foaming properties and stability when beaten. Globulins foam the best and also give the largest volume when used to make angel food cake. Ovalbumin also foams very well and results in an angel food cake that is somewhat smaller than one made with the globulin foam, but larger than one made with egg white (natural mixture of all white proteins).

Angel food cakes made with foams of lysozyme, ovomucoid, ovomucin, or conalbumin are smaller than ones made with the egg white mixture of proteins. However, a combination of lysozyme and ovomucin produces a larger angel food cake than is obtained using either of these proteins separately.

If egg whites are beaten until the peaks stand up straight, the volume of the foam will be greater than if they are beaten until the peaks bend over, but the stiff peak stage has reduced ability to stretch during baking because of the large amount of denaturation that has occurred in the egg white proteins by this stage of beating. Egg whites beaten to the stiff peak stage and beyond not only have reduced extensibility, but they also are difficult to fold with other ingredients. Their rather brittle nature at this stage causes many of the cells to break during the prolonged mixing needed to eliminate pieces of the egg white, and considerable air is lost from the foam, air that would have contributed to leavening during baking.

The level of sugar relative to flour influences volume, both because of the ability of the sugar to stabilize the egg white foam and because of the elevation of coagulation temperature of the egg white and gluten proteins as the level of sugar increases. However, there must be a balance between these two ingredients, for too much sugar in relation to flour will elevate the coagulation temperature of the proteins to the point at which some of the gluten strands break, and the cake begins to sag. The ratio of sugar to flour ordinarily should not exceed 3:1; a ratio of 2.5:1 is adequate but will give a smaller volume than the higher level of sugar.

Use of all-purpose rather than cake flour reduces the volume and tenderness of angel food cakes. The recommended amount of cake flour per gram of egg white is 0.2–0.4 g, with the smaller amount promoting moistness and tenderness. The weight of sugar should not exceed the weight of the whites unless the flour is increased to the upper end of this range.

## Sponge Cake

Angel food and sponge cakes have many similarities, but the use of an egg yolk foam in addition to the white foam in sponge cakes adds unique aspects to sponge cakes. Sponge cakes contain both a very viscous egg yolk foam and a white foam. Proper preparation of the egg yolk foam is a vital step in producing a high-quality sponge cake, and this foam requires considerably more beating than is required for the whites.

Optimal results are obtained when the eggs are at room temperature or above. The specific gravity of the foam prepared from eggs that are at refrigerator temperature [1.7°C (35°F)] will never be as low as can be achieved if the eggs are warmer when beating is initiated. The desired stage of beating an egg yolk foam is reached about twice as fast when the eggs are at 27°C (80°F) as at 4.4°C (40°F). Because a considerable amount of beating with an electric mixer is required to reach the appropriate foam stage for the yolks, there is a definite advantage in removing the eggs from the refrigerator well in advance of using them for beating the yolk foam.

The important steps to achieving quality in a sponge cake are the beating of the egg yolk foam and the subsequent beating of this foam after the addition of sugar and liquid until the mixture forms a very light foam. By extensive beating at this point, the yolk foam becomes sufficiently viscous to remain suspended in the white foam when they are folded together. The yolk foam also entraps a valuable quantity of air to aid in achieving a good volume and thin cell walls, a necessity for a tender sponge cake.

After preparation of the egg yolk foam, the cake flour is folded into the yolks. Then the white foam is beaten until the peaks just bend over, at which point the yolk mixture is folded gently into the whites and transferred to a tube pan for baking, just as is done for angel food cakes. The weak structure of sponge and other foam cakes makes it necessary to cool them in an inverted position so that the rather spongy cell walls are extended to their maximum. This enables the structure to become firm, with the cell walls stretched as thin as possible.

## Chiffon Cake

In some ways, chiffon cakes are hybrids between foam cakes and shortened cakes, for they do contain baking powder and fat (but in different forms). These ingredients promote tenderness and volume, yet the texture of the baked chiffon cake is neither as fine in cell size as in shortened cakes nor as large as in angel food and sponge cakes.

The factors influencing the quality of angel food cakes also are applicable to the quality of chiffon cakes. Much of the quality, particularly volume and tenderness, is influenced greatly by the quality of the egg white foam prepared in making a chiffon cake.

One of the important differences between chiffon cakes and the other two types of foam cake is the extent to which the egg whites are beaten. In both angel food and sponge cakes, the tips of the whites should just bend over, but the peaks in a chiffon cake foam should stand up straight, yet not be dry or brittle. Extended beating is needed because the yolk mixture that is to be folded into it is extremely fluid and requires a great deal of folding before it can be dispersed uniformly and held within the white foam. Unless the white is beaten sufficiently and the yolk mixture is suspended throughout the whites, the yolks will drain to the bottom of the pan and form a rubbery layer.

## SHORTENED CAKES

### Ingredients

Shortened cakes are complex mixtures, and many variations are possible. Bleached cake flour is the flour of choice to produce a tender cake with a fine crumb. The smaller amount of protein and the more tender nature of this protein in comparison with all-purpose flour are compatible with the qualities desired in a shortened cake. Sugar, as discussed in Chapter 17, performs several critical roles in shortened cakes. Of particular interest is its role in influencing the volume of shortened cakes (Figure 18.9). Because of its role in delaying

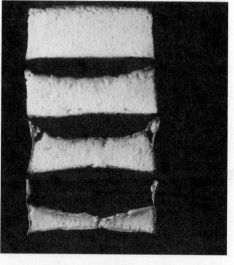

**Figure 18.9**    Cross section of cakes containing between 0 and 240 percent sugar (flour basis): (left, top to bottom) 9 percent, 40 percent, 80 percent; (right, top to bottom) 120 percent, 160 percent, 200 percent, 240 percent. (Courtesy of M. Mizukoshi. 1985. *Cereal Chem.* 62(4): 242. American Association of Cereal Chemists.)

gluten formation during mixing, the amount of mixing must increase when the sugar level increases. By increasing the mixing, the strength of the structure is enhanced, which helps prevent the structure from falling.

Shortened cakes have a delicate texture that is influenced significantly by the ratios of flour, fat, sugar, and liquid. The following guidelines work well for preparing shortened cakes in large quantities:

1. The weight of the fat should not be over one-half the weight of the sugar.
2. The weight of the fat should not exceed the weight of the eggs.
3. The weight of the sugar should be between 15 and 40 percent greater than the weight of the flour. (If used, the weight of chocolate or cocoa is added to the weight of the flour.)
4. The weight of the liquid (milk plus eggs, not weight of dried milk or eggs) should at least equal the weight of the sugar.

The color of chocolate cake depends on the color of the cocoa or chocolate used as well as on certain other variables. The color of cocoa and chocolate is influenced by the variety of the cacao beans from which they come, the extent of the roasting of the beans, the addition of alkali, and oxidation. As would be anticipated, the deeper the roast, the darker is the color of the resulting chocolate or cocoa. On the basis of processing, natural-processed cocoas and chocolates range between pH 5.1 and 6.2, but Dutch-processed products range between pH 6 and 7.8.

Oxidation of cacao polyphenol to form a **phlobaphene** is another factor determining the color of cocoa and chocolate. The phlobaphene is responsible for the reddish color seen in cocoa and chocolate to varying degrees, depending on the extent of oxidation. The presence of oxygen also influences the shelf life of cocoa and chocolate because of the potential for oxidative rancidity of the fat in these products.

**Phlobaphene** Derivative of a polyphenol in cacao that is formed in the presence of oxygen and is responsible for the reddish color sometimes noted in cocoa and chocolate.

The pH of chocolate or cocoa-containing cakes differs as a result of the cocoa or chocolate as well as the presence of leavening ingredients. For a desirable flavor, the pH of the batter should be no higher than 7.9. Chocolate-containing cakes range in color from a definite brown at a pH between 6 and 7 to mahogany between pH 7 and 7.5 and increasing redness above pH 7.5.

Sometimes honey is the sweetening agent in shortened cakes. Actually, the use of honey influences more than simply the sweetness of the cake due to its liquid content and acidity. Honey is a liquid, so it does not provide crystals to aid in formation of a foam during creaming of the fat. This causes honey-containing cakes to be a bit coarser than those made with only granulated sugar.

On the basis of sweetness, a cup of honey provides essentially the same sweetening as a cup of sugar. To compensate for the liquid that honey contributes, one-fourth cup of liquid should be deleted from the recipe for each cup of honey used. The acidity is quite variable in different honeys; between $\frac{1}{12}$ and $\frac{1}{2}$ teaspoon of soda may be needed to neutralize a cup of honey, but the batter can remain more acidic in most cases.

Honey promotes rapid browning because of the abundance of reducing sugars it adds to the batter. Reducing sugars are very susceptible to carbonyl-amine browning reactions. The high fructose content of honey is noteworthy in its effect on browning.

## Methods of Mixing

***Conventional Method.***   The large number of ingredients in cakes makes possible a variety of mixing methods. However, the method used most frequently because of its excellent effect on both texture and keeping quality is the conventional method. The dry ingredients, with the exception of the sugar, are sifted together and set aside until needed. The fat and sugar are then creamed together in a mixing bowl, either by hand or with an electric mixer, to produce a light, fluffy foam. This foam is possible because the physical manipulation causes the sharp sugar crystals to dig into the soft fat and create small pockets

## FOOD FOR THOUGHT: Chocolate in the Spotlight

For centuries, chocolate has been a source of pleasure to people in countries around the world. It also has been an important cash crop in tropical climates where *Theobroma cacao* trees are grown on family plots as well as on larger farms (Figure 18.10). In the last decade of the twentieth century, a fungus that causes witches' broom attacked cacao trees in Bahia, the major area in Brazil where a significant quantity of the world's chocolate supply grows. Crop yields dropped to about a third of previous production. Such a dramatic threat to one of the world's favorite foods triggered international cooperation to attempt to develop strains that would have improved resistance to the witches' broom fungus (*Crinipellis perniciosa*) and other pests.

International experts from around the world have been working together on the agricultural challenges involved in consistently growing a reliable crop of cacao to

**Figure 18.10**   Cacáo pods grow on a *Cacao theobroma* tree in southern India.

meet the world's demand for chocolate (see Photo C.64 in the color insert). The USDA's Subtropical Horticulture Research Station (SHRS) has been working cooperatively with institutions from other nations, including the Tropical Agricultural Research and Education Center (CATIE) in Turrialba, Costa Rica; Instituto Nacional de Investigaciones Agricolas y Pecuarias (INIAP) Estacion Experimental Pichilingue (EET Pichilingue) in Quevedo, Ecuador; International Institute for Tropical Agriculture (IITA) in Nigeria; Cocoa Research Institute of Ghana (CRIG); and the Coconut and Cacao Institute (CCI) in Papua New Guinea. Coordination of efforts is under the leadership of the International Group for Genetic Improvement of Cocoa (INGENIC).

Genome mapping and selective breeding of cacao, experimental farming techniques, biocontrol as a pest control measure, and reforestation programs planting cacao trees are facets of the international consortium's work. Reforestation in tropical areas is an important ecological objective, and planting of disease-resistant cacao trees is important not only from this perspective, but also because of the income cacao can bring to many families with small farms.

The world is counting on the success of these efforts, not only for the pleasure chocolate brings to candy lovers, but also for the potential health benefits that are being attributed to flavanols in chocolate. Because flavanol content may be reduced when cocoa is manufactured, food researchers have been developing ways of minimizing this loss. Extensive research efforts are underway to attempt to understand the various ways in which cocoa flavanols promote health. Cardiovascular benefits have been reported, and these are attributed in part to promoting blood flow by helping relax blood vessels and reducing the tendency to form blood clots. Antioxidant capability is another potential role in promoting health.

Chocolate lovers who are eager to find reasons to justify eating their favorite food can rejoice in the positive health benefits that are being identified. However, the cocoa/flavanol message is tempered by the chocolate's high caloric content. Moderation clearly is essential if people are to benefit from eating chocolate.

of air, which then serve as the cells where steam and carbon dioxide can collect and expand during baking.

A maximum volume of this creamed mixture results from having the fat at an optimum temperature and beating at a medium speed on an electric mixer. The optimum temperature for creaming butter is 25°C (77°F). Creaming should stop before the appearance of the fat foam changes toward a frothy or runny consistency.

Eggs are added to the creamed fat-sugar foam and creaming is continued to establish a water-in-fat emulsion superimposed on the sugar-fat foam. Ideally, the air bubbles are uniform, small, and surrounded by fat. This type of dispersion results in a shortened cake with the desired fine and uniform texture, plus a velvety crumb. Insufficient creaming at this point leads to a coarse and uneven texture with thick cell walls.

Usually, the sifted mixture of flour and other dry ingredients is divided into thirds; the first third is added to the egg-fat-sugar mixture and mixed at a low setting on an electric mixer or beaten by hand with a wooden spoon. Then, the first half of the liquid is added, and mixing follows to blend the ingredients. The second third of the dry ingredients is mixed in, followed by the last half of the liquid ingredients. Mixing follows this final addition of liquid ingredients before the last third of dry ingredients is added and beaten in. The amount of beating that is needed during these alternating additions varies, depending on the ratio of ingredients in the formula. Rich batters that are high in fat and sugar require more mixing than do leaner formulas. In any event, cakes must be mixed more than muffins so that sufficient gluten is developed to provide the necessary structure.

**Modified conventional method**
Method of making cakes similar to the conventional method except that the whites are added as a foam at the end of mixing.

**Conventional sponge method**
Method of making cakes in which part of the sugar and all of the egg are withheld to make a sugar-stabilized meringue that is folded into the cake batter as the final step in preparation.

**Muffin method (cakes)**
Method in which the liquid ingredients (including oil or melted fat) are combined, the dry ingredients are sifted together, and then the two mixtures are stirred together until blended sufficiently to develop the necessary gluten; results in a cake with coarse texture and limited keeping qualities.

*Modifed Conventional Method.*   In the **modified conventional method**, creaming is done in the same fashion as for the conventional method. At this point, the yolks are added, but the whites are saved for addition as a foam after the dry ingredients and the liquids have been added and mixing has been completed. The whites are beaten to a foam in which the peaks just bend over, and this foam is folded gently and efficiently into the batter just prior to baking. The modified conventional method produces a cake very similar to that made by the conventional method, although the egg white foam has the potential to produce a cake of slightly larger volume.

*Conventional Sponge Method.*   Sometimes the whole egg or the egg white is not added until the rest of the ingredients (except part of the sugar) have been mixed to form the batter. At this point, a meringue of either whole egg or egg white is made with the sugar that was withheld. This meringue is then folded into the batter, and the cake is baked promptly.

The **conventional sponge method** (also called the conventional meringue method) avoids dissolving some of the sugar crystals that would otherwise be dissolved when the egg is added immediately after creaming. If sugar crystals dissolve in the fat-sugar foam, the volume of the resulting cake is reduced because of loss of air from the foam. For this reason, the conventional sponge method is desirable for obtaining maximum volume in shortened cakes. The amount of sugar used in making the meringue should not exceed the equivalent of two tablespoons of sugar per egg white.

*Muffin Method.*   The **muffin method** for making shortened cakes is the same as for making muffins. In other words, the liquid ingredients are mixed together in one bowl, and the dry ingredients are mixed in a second bowl. Then the liquid ingredients are poured into the dry ingredients and mixed to the desired end point. Appreciably, more mixing is required than is done for muffins to compensate for the delayed development of gluten resulting from the comparatively high ratios of fat and sugar to flour. The fat must be either oil or melted fat if this method is used.

---

### FOOD FOR THOUGHT: A "Prime" Ingredient

Commercial bakers usually select the shortening used in their cakes on the basis of the type of crystals contained in the fat. Shortening containing beta prime ($\beta'$) crystals (see Chapter 11), when creamed with sugar, traps air in very small pockets. This rather fluffy creamed foam of fat and sugar with air creates the fine cell structure desired in shortened cakes.

Because of cost, cottonseed or canola oil is often the base oil used for creating a shortening for commercial cake production. However, the structure of shortened cakes requires a creamed mixture of a solid fat and sugar. Therefore, this oil needs to be transformed into a solid that can be creamed. The addition of a fully hydrogenated shortening to the oil that is at a level of between 10 and 15 percent of the total mixture achieves the necessary firmness. Sometimes coloring and flavoring also may be added to complete the engineered shortening.

The desired physical properties for a shortening tailored for making shortened cakes are created by the presence of $\beta'$ crystals. Fully hydrogenated fat from palm or cottonseed oil is used to provide $\beta'$ crystals in the shortening because $\beta'$ is the dominant form of crystal formed from these oil sources. Tallow also promotes $\beta'$ crystal formation, but it is used less frequently because some consumers avoid animal fats for religious or health reasons. A side benefit of using these fully hydrogenated products in combination with canola or cottonseed oils is that fully hydrogenated fats do not contribute *trans* fatty acids.

The muffin method produces a batter that lacks the air that is incorporated when fat and sugar are creamed together. As a result, the volume of cakes made by the muffin method is not optimal. In addition, cakes made by the muffin method are rather coarse in texture and may not be as tender. This more open texture causes them to stale more rapidly than cakes in which the fat and sugar have been creamed.

***Pastry-Blend Method.*** Cakes can be made by creaming the flour and fat together, a procedure called the **pastry-blend method**. As gluten does not develop during this creaming period, creaming can be done until the fat has been dispersed extensively and considerable air has been incorporated into the foam. The baking powder and salt are dispersed in the sugar before being added with the liquid and stirred into the foam mixture. This pastry-blend method produces an excellent shortened cake even when ingredient ratios may vary rather widely.

***Single-Stage Method.*** The **single-stage method** (sometimes called the quick-mix method) is a rapid method of mixing in which all of the ingredients, with the possible exception of part of the liquid and the egg, are placed in a bowl and mixed together vigorously to disperse all of the ingredients and develop the gluten. Then, the egg and any remaining liquid are added and the total batter is beaten. Fats must be comparatively soft, which usually means at least at room temperature. Shortenings containing mono- and diglycerides are important with this mixing method because these emulsifiers help trap air in the mixture and promote a finer texture. Even then, the single-stage method is likely to produce a cake with a somewhat coarse texture and limited keeping qualities.

**Pastry-blend method**
Method in which the flour and fat are creamed (first step); sugar, baking powder, salt, and half the liquid are added (second step); and the last half of the liquid and the egg are combined (third step).

**Single-stage method**
Mixing method in which all of the ingredients except the egg and half the liquid are added and beaten before the egg and the last of the liquid are beaten in.

## Baking

Cakes undergo remarkable changes during baking. Oven temperatures cause expansion of the air and carbon dioxide already in the batter and generate steam from the liquid. The heat also generates additional carbon dioxide production to further increase the volume. Pressure from these leavening agents expands the cells and holds them in their stretched, thin state while the proteins denature and coagulate and the starch gelatinizes. In addition, some moisture evaporates. This truly is an extremely dynamic state, one that is very fragile at the critical point just before proteins in the cell walls coagulate. Cakes will fall unless pressure is maintained in the cells until the structure of cell walls is set by permanent changes in proteins (coagulation) and carbohydrates (gelatinization).

The depth of batter in baking pans influences final outcome. Shallow pans enable the heat to penetrate to the center of the batter more quickly, resulting in a cake of optimal volume. The surface is flatter and lighter in color than is produced when the batter is baked in a deeper pan. A cake baked in a deep pan also has a greater tendency to crack in the center than does one baked in a shallow pan. Regardless of the depth of the pan, a cake should almost fill the pan when baking ends. This usually means filling the pans about half full with the batter. Tunnels are more likely to develop in cakes baked in 8-inch rather than in 9-inch cake pans (Figure 18.11).

Oven temperature influences the quality of shortened cakes. At too low a temperature, the volume is poor because some of the cells collapse while others become quite large with rather thick walls. As the temperature increases, the volume improves and so does the texture. However, if the oven temperature is too high, the crust sets while the interior is still fluid. The pressure generated within the fluid mass presses against the crust, causing it either to begin to hump or even to peak, depending on the temperature.

A higher temperature can be used for a cake that is high in fat and sugar because these ingredients elevate the temperature at which the proteins in the crust coagulate. This allows a little more time for the crust to remain somewhat flexible and accommodate the pressures produced within the batter. An oven temperature of at least 185°C (365°F) is suitable for baking most shortened cakes, but some are better when baked at 190°C

**Figure 18.11** Too small a baking pan (top) results in batter running over the edge and cracking on top. Proper pan size (center) permits maximum volume and uniform browning. Too large a pan (bottom) causes poor browning because hot air cannot circulate easily over the top. (Courtesy of General Mills.)

(375°F). Regardless of baking temperature, preheating of the oven is recommended (Figures 18.12 and 18.13).

Even the material of the baking pans influences the quality of shortened cakes. Heat penetration is best if the baking pan is dark and/or dull so that the heat is absorbed efficiently to promote rapid heating of the batter. However, rapid heating causes the sides of the cake to set quickly before the interior has had time to generate much of its potential gas.

**Figure 18.12** Too low a baking temperature causes a coarse-textured, heavy cake with poor volume and a pitted surface. (Courtesy of General Mills.)

**Figure 18.13**   Cake baked at too high a temperature is cracked and humped because the structure sets at the edges while gas continues to exert pressure and expand the interior. (Courtesy of General Mills.)

Humping and cracking because of this lack of uniform heating within the batter can be a problem, but total volume is increased.

Shiny pans absorb heat more slowly because they reflect the heat so that there is time for the heat to spread into the interior of the cake before the sides of the cake set. This leads to a cake that has a gently rounded or flat surface rather than one that is humped. Browning is more delicate and uniform as a consequence of the slower rate of baking. However, volume is reduced a little, and the cells are a bit coarser and their walls a little thicker.

## Causes of Variations

Cakes are very sensitive to variations in formulation and baking conditions (Table 18.5). The results of these variations range from pleasing to disastrous. For example, alterations in pH as a consequence of variations in the acidity of the liquid can alter flavor, the color of both the crumb and the crust, cell size, and volume. Substitution of buttermilk for sweet (regular) milk reduces the pH, which tends to promote a light color and a fine grain. However, the taste may be detectably sour, and the volume may be slightly smaller. Sour milk gives similar results.

If soda is added to neutralize the buttermilk or sour milk, the potential exists for an excess of soda. The higher pH promotes excessive browning of the crust, a yellowing of the interior crumb, a coarser texture (both larger cells and thicker walls), and possibly a slightly soapy flavor; these problems far outweigh the possibility of a small increase in volume. For optimum browning and optimum volume, the pH of the batter should be a bit acidic (about 6.3).

Alterations in the amount of sugar in a cake recipe can cause wide variations in the characteristics of cakes. As the content of sugar increases, the volume of the cake increases (because of the longer time required to reach the elevated coagulation temperature of gluten) up to the point at which the volume is so great and the gluten so weak that the gluten strands snap and the cake falls in the center.

The exact amount of sugar that will cause a cake to fall is influenced by the amount of egg and shortening. Egg adds protein to help strengthen the cake when a large amount of sugar is included in the formula. Added fat increases the aeration of the batter to help increase viscosity, but it may weaken the structure, especially if mixing is inadequate. Liquid also usually is increased with increased sugar because of the need for additional liquid to dissolve the sugar and to compensate for the liquid that is adsorbed by the added sugar. To offset the effect on volume of the increased sugar, the level of baking powder may need to be reduced.

**Table 18.5**    Some Causes of Variations in Shortened Cakes

| Variation | Possible Cause |
|---|---|
| Yellowing | Alkaline batter (excess soda) |
| Fallen center | Excess sugar |
| | Excess fat |
| | Inadequate mixing |
| | Excess baking powder |
| | Too low a temperature during baking |
| | Opening of oven door too early in baking |
| Tough, dry crumb | Too much flour |
| | Too much egg |
| | Too little fat |
| | Too little sugar |
| | Too much mixing |
| | Over-baking |
| Dark crust | Alkaline reaction (excess soda) |
| | Fructose (promotes rapid browning) |
| | Honey (fructose in it promotes rapid browning) |
| | Improper placement in oven (too near top or bottom) |
| | Over-baking |
| | Too much sugar |
| Gummy, crystalline appearance | Too much sugar |
| Coarse texture | Too much baking powder |
| | Too much sugar |
| | Too low a temperature during baking |
| | Inadequate mixing |
| Poor volume | Too little baking powder |
| | Too low a temperature during baking |
| | Improper level of sugar |
| | Improper level of fat |
| Humped | Too much flour |
| | Too much mixing |
| | Too deep a pan |
| | Too hot an oven |
| | Too little sugar |
| | Too little fat |
| | Too little liquid |

## Altitude Adjustments

At high altitudes, the reduction in atmospheric pressure requires adjustments of cake formulas to avoid the collapse of the structure. Strengthening of gluten by a modest increase in mixing may be helpful in producing shortened cakes with an acceptable texture at elevations above 2,500 feet. This effort provides increased resistance to the pressure generated in the cells to offset the decreased resistance provided by the lower atmospheric pressure at low mountain elevations. However, at 3,000 feet and above, more aggressive steps need to be taken, and the formulations need to be adjusted (Table 18.6).

The problem in adjusting cake formulas for higher elevations lies in providing an appropriate balance between the internal pressure generated in the cells and the opposing, rather low atmospheric pressure. Reducing the amount of baking powder in the formula will reduce the internal pressure. A reduction of one-eighth of the baking powder is recommended at 3,000 feet and one-fourth at 5,000 feet. At 7,200 feet, baking powder needs to be reduced by one-third.

Sugar also needs to be decreased, at the rate of approximately one tablespoon per cup of sugar at 3,000 feet, two tablespoons at 5,000 feet, and three tablespoons per cup of sugar at 7,000 feet. Another alteration that helps strengthen the cell walls is reduction of the fat by one and two tablespoons per cup of fat in the original recipe.

**Table 18.6**   Suggested Alterations in Cake Formulas Prepared at Altitudes of 3,000 or More Feet[a]

| Ingredient | Altitude | | |
| | 3,000 feet | 5,000 feet | 7,000 feet |
| --- | --- | --- | --- |
| Baking powder | $-\frac{1}{8}$ tsp/tsp | $-\frac{1}{4}$ tsp/tsp | $-\frac{1}{3}$ tsp/tsp |
| Sugar | $-1$ tbsp/cup | $-2-3$ tbsp/cup | $-3$ tbsp/cup |
| Fat | $-1$ tbsp/cup | $-1\frac{1}{2}$ tbsp/cup | $-2$ tbsp/cup |
| Liquid | $+1$ tbsp/cup | $+3$ tbsp/cup | $+5$ tbsp/cup |

[a]Increased mixing is needed to strengthen gluten.

At high elevations, cakes lose more liquid than normal during baking because water boils at a lower temperature. To compensate for this increased evaporation, liquid can be increased by a tablespoon per cup of liquid at 3,000 feet, and another tablespoon of liquid is needed for each 1,000-foot gain in elevation. To illustrate, at 6,000 feet an extra four tablespoons (one-fourth cup) of liquid need to be added for each cup of liquid in the original recipe.

## PASTRY

### Ingredients

Whether pastry is in the form of piecrust or puff pastry, it is a simple baked product that contains few ingredients yet requires some skill for successful results. Piecrust is made with fat, flour, salt, and water—the same ingredients used for preparing puff pastry, the multilayered pastry in products such as napoleons. The difference between these two forms of pastry is the fat (see Chapter 12). The fat differs in kind, amount, and method of incorporation into the dough.

Pastry for pies usually is made with shortening, but sometimes oil, lard, or butter is selected. Although the ratio of flour to pure fat may vary from one pie dough recipe to another, a volume ratio commonly used is 3:1 (three times as much flour as fat). Skilled workers can make a satisfactory pastry with the leaner ratio of 4:1.

The fat usually selected for making puff pastry is butter, and it is used in a flour-to-fat ratio of 2:1 (on a volume basis)—that is, twice as much flour as fat—which is a very rich dough.

Fats are measured by volume when pastry is prepared in the home; even when the same volume is measured carefully, there is some difference in the amount of fat used in the recipe because of the physical differences among fats. For example, lard and shortening contain only fat, but a cup of lard contains more fat than a cup of shortening because of the aeration of the shortening during its manufacture. This difference is of greater academic interest than practical significance, because either fat can be measured satisfactorily by volume to obtain excellent results. However, a cup of butter or margarine cannot be used interchangeably with a cup of lard or shortening in making pastry because butter contains a little more than 80 percent fat and about 16 percent water, differences that require adjustments in the recipe if butter is to be used successfully. Oil flows readily when added to the pastry mixture, which makes it possible to use a 4:1 ratio (the leanest of the flour-to-fat ratios).

The fat to be used in making a pastry influences the character of the baked product. A yellow-hued fat such as a yellow-tinted shortening, butter, or margarine gives a somewhat golden color to the baked crust, which subtly suggests richness. Any flavor unique to the fat is imparted to the pastries. For instance, the distinctive flavor of lard can be detected in pastry even when the lard is very fresh. Similarly, butter or butter-flavored shortening or margarine adds flavor to a pastry.

**Figure 18.14**    The correct amount of water is essential to make a tender pastry.

The texture and tenderness of pastries are influenced by the physical properties of the fat selected. Shortening and lard favor the production of a flaky and tender crust, whereas oil produces one that is tender even when the flour-to-oil ratio is only 4:1. However, oil ordinarily causes a **mealy**, rather than a **flaky** texture. Unless the ingredient proportions are altered, butter and margarine produce rather tough pastries because they contain extra water and too little fat.

Another key factor influencing tenderness is the amount of water in relation to the flour and fat (Figure 18.14). Although flour varies a bit depending on its source, a tender pastry can usually be made using a ratio of 6 parts flour:2 parts fat:1 part water. Excess water promotes gluten development and toughness, while too little causes the dough to crumble and be difficult to handle.

## Causes of Variations

Piecrusts vary in tenderness as a result of several possible modifications. As noted earlier, oil has a greater ability to coat flour and prevent water from hydrating the protein to enhance gluten development than does any other form of fat. Therefore, the use of oil definitely enhances tenderness.

Among the solid fats, those containing only fat have a greater shortening or tenderizing effect than those containing some water (butter and margarine). Soft fats spread more readily than hard fats and are more effective tenderizing agents. Therefore, the warmer the temperature of a solid fat, the more effective it will be as a tenderizing agent in pastry. The smaller the particles of fat are cut, the greater is the tenderizing effect of that fat on the pastry.

The time elapsed between addition of water to the dough and baking of the crust influences the tenderness of pastry. When dough is mixed and then allowed to stand before being rolled, the gluten has an opportunity to hydrate to a greater extent, and the pastry is less tender than if the dough is rolled immediately and baked.

The level of fat has an important effect on tenderness, with high levels creating a more tender pastry than do low levels of fat. Conversely, higher water content produces a tougher pastry because of the increased development of gluten. Pastry flour promotes more tenderness in pastry than all-purpose flour does; an increased level of either type of flour causes the pastry to be less tender.

Flakiness in pastry is the result of cutting solid fat into pieces the size of uncooked rice grains, which are coated with flour. During baking, the fat melts, leaving spaces between gluten strands. Flakiness is enhanced by cutting fat into pieces large enough to leave spaces that will create the desired thin, short layers. Smaller fat particles mean less flakiness because the resulting breaks in the gluten are very small.

Tenderness and flakiness are opposing characteristics, yet both are considered desirable. The finer the particles of fat are cut, the more the flour is coated with the fat and the more tender the crust because of the greater restriction on gluten development. All-purpose flour yields a pastry that is more flaky than one made with pastry flour. This difference is due to the stronger character of the gluten in all-purpose flour. Apparently, the tenacious nature of the hard-wheat gluten is effective in retaining steam fairly efficiently in pastry. This trapped steam creates sufficient pressure during baking to separate the layers and develop the desired flaky texture.

It should be noted that oil ordinarily produces a mealy texture rather than a flaky one. However, with a considerable amount of mixing of the ingredients, enough air can be incorporated and sufficient gluten can be developed to produce some flakiness. This technique reduces tenderness, for the greater the mixing after the liquid has been added, the less tender the product.

## Puff Pastry

Puff pastry is prepared to only a limited extent in American menus, but it is a very interesting pastry because of the unique way in which it is prepared. Only a portion of the fat is used in the first step, the preparation of a stiff dough. This dough is chilled thoroughly before being rolled out to a thickness of about a quarter of an inch. This chilling period increases hydration of the gluten, which helps explain the distinctly tough nature of puff pastry despite the very high fat content.

Half of this rectangle is spread with the rest of the fat. The dough is then folded to cover the butter and then this smaller rectangle is folded into thirds before being chilled again. When chilled, the dough is once again rolled to the size of the original rectangle and folded in the same fashion as described before, but no more butter is added.

This process is repeated at least three times. With each folding and rolling, more layers are imprinted in the dough. Finally, the dough is baked, at which time the intense oven heat generates the necessary steam from the water in the butter and in the dough to force apart the layers. This flakiness causes a significant increase in the volume of the baked puff pastry, thus explaining its name.

## COOKIES

Cookie types include bar, drop, rolled, pressed, and refrigerated cookies. This diversity results from using various proportions and types of ingredients to create cookies with the specific characteristics desired. Bar cookies have a comparatively high liquid content, which results in a batter that can be poured into a baking pan. Drop cookies have less liquid, which creates a slightly sticky dough that can be dropped as individual cookies onto a cookie sheet and retain their general shape during baking. Rolled, pressed, and refrigerated cookie doughs are rich enough to stick together when the flour, liquid, and other ingredients are mixed, yet the dough is so stiff that it can be manipulated without sticking to the pastry board, hands, press, or cutters.

Most cookies contain a rather rich mixture featuring fat, sugar, and flour. The functions of these three basic ingredients are essentially the same as in other baked products, but most cookies contain a high ratio of sugar (up to 25 percent) and fat and a small amount of liquid. These differences create cookies with a relatively dense, but tender texture.

The type of fat has a significant effect on the texture, shape, and color of cookies. Butter contributes a wonderful flavor, but its low melting point causes cookies to spread considerably during the early baking period before the protein structure has set. Thin, wide cookies are the result.

In contrast, cookies made by creaming shortening and sugar together will be aerated somewhat because the air that is trapped in the air-fat foam during creaming will expand the cookie dough during baking. The higher melting point of shortening also means that the dough will have a limited tendency to spread during baking because it will remain fairly solid in the early period of baking. The negative side of using shortening is that it does not make as pleasing a flavor contribution as does butter (unless butter flavoring has been added). One way of capitalizing on the benefits of both butter and shortening is to use a mixture of the two types of fat, with shortening predominating in quantity.

If cookies are made with melted shortening or with oil, the texture tends to be compromised because it is not possible to create the numerous tiny pockets of air that are formed by creaming with a solid shortening. The texture of the baked cookies will be rather compact. On the plus side, calories can be reduced somewhat because less oil is needed for tenderness compared with the amount of solid shortening required.

Granulated sugar is particularly useful in creating the desired texture in baked cookies because the relatively coarse crystals help dig many tiny pockets in the shortening when these ingredients are creamed. Brown sugar often is used, at least as part of the sugar, because it adds flavor and a slight acidity while also participating in the creaming process. Honey adds sweetness and possibly distinctive flavors, but it accelerates browning and increases spreading during baking (resulting at least partially from its liquid content). The high fructose content of honey explains why honey in cookies serves as a **humectant**.

**Humectant**
Substance that helps retain moisture.

The principal structural component in cookies is flour. Consumers ordinarily use all-purpose flour made from hard wheat, but commercial bakers sometimes opt to use soft wheat flours, depending on the specific textural characteristics desired. Flour that has been treated with bleaching and maturing agents is useful in cookies to promote starch gelatinization. Chlorination of flour helps reduce the spread of cookies.

The limited amount of liquid in cookies (often only the liquid from eggs) means that little starch gelatinization occurs during baking, but this change does contribute to the desirable texture of cookies. Increasing water in a cookie mixture generates more steam during baking and a lighter, somewhat more porous texture. Eggs also contribute by helping to emulsify the dough. The presence of an emulsion in the early part of baking helps prevent the liquid in the dough from dissolving the sugar crystals and collapsing the dough structure.

Some leavening always occurs during baking because of air and steam. Usually, at least a small amount of chemical leavening is added. If acidic ingredients are a part of the formula, a small amount of baking soda can be added to neutralize the reaction and generate a little leavening. Baking powder often is the leavening agent selected for making cookies.

## SUMMARY

Quick breads include a wide variety of types and ingredients, but flour, liquid, and salt are common to all of them. The type of flour, liquid, and fat can be altered to add variety; addition of fat, egg, sugar, and ingredients for flavor or texture heightens the array of quick breads. These breads are made by the biscuit method, the muffin method, or the conventional method.

Yeast breads rely on the use of wheat flour with a strong gluten because of the stretching required during the fermentation and baking of the dough. Fat is an optional ingredient but often is included to promote tenderness and softness. The liquid may be water or milk. If milk is used, it should be scalded to at least 92°C (198°F) for 1 minute and cooled prior to contacting the yeast. Sugar content varies considerably, although some usually is included to facilitate yeast fermentation. Doughs with more than 6 percent sugar are chosen for some sweet breads, but fermentation time is extended by the high levels. Salt slows fermentation, helping to produce a bread with desirable cell size. The level of yeast in the formula determines the length of time required for fermentation and proofing.

Baking at 204–218°C (400–425°F) results in excellent volume and a pleasingly browned crust unless the sugar content is high, thus requiring a lower temperature. Oven spring occurs as the result of increased carbon dioxide production in the early part of baking and the expansion of existing gases at the hot oven temperature. Gluten stretches, the water shifts toward the starch to gelatinize some of the starch, and amylases convert some starch to dextrins. Then yeast is killed, enzymes are inactivated, and the protein structure sets as the starch becomes more rigid as a result of gelatinization during baking. The carbonyl-amine reaction on the crust as the crust dries in the oven accounts for much of the browning.

Substitution of a portion of the wheat flour with other grain flours such as triticale or rye and addition of some bran are ways of adding variety to wheat breads. Addition of sodium stearoyl-2-lactylate to triticale or soy flour improves the quality of bread baked from these flours. Sucroglycerides are other additives effective in enhancing the usefulness of soy flour in breads.

Staling of bread is accompanied by retrogradation of free amylose immediately after baking and by subsequent retrogradation of amylose and amylopectin in the gelatinized granules. Moisture migrates from starch to gluten and also from the interior to the surface of the bread. At refrigerator temperatures, the crumb becomes excessively firm within a day, but staling is much slower at warm room temperatures and is delayed considerably by freezing.

Foam cakes, so-called because of their dependence on egg foam to achieve good volume, include angel food, sponge, and chiffon cakes. Angel food cakes are made with egg white foam. Sponge cakes contain both yolk foam and white foam. Chiffon cakes use liquid yolks and egg white foam; they also include baking powder as a leavening agent and oil for tenderizing the cake.

Shortened cakes contain a comparatively high amount of fat. Egg white foam may be included in a shortened cake, but it is not the primary source of leavening. Layer cakes are leavened with baking powder as the source of carbon dioxide. Pound cakes traditionally are rather compact, because they are leavened only by the air creamed into them and by some steam generated during baking.

The ratios of ingredients in shortened cakes are critical to the quality of the baked cake. In proper proportions, shortened cakes can be made with a fine, velvety texture and a good volume. Too much sugar, fat, liquid, egg, or baking powder can have serious consequences. The methods of mixing also influence the characteristics of shortened cakes. Cakes of high quality can be produced using the conventional method, conventional sponge method, modified conventional method, and pastry-blend method. The muffin method and single-stage method create shortened cakes with a rather coarse texture and a tendency to stale readily.

The conditions of baking also influence cake quality, with shallow pans and a baking temperature of 185°C (365°F) generally giving good results. Varying the level of the different ingredients and using natural-processed and Dutch-processed chocolate or cocoa affect the final product. Formulas for shortened cakes baked at altitudes of 3,000 feet or higher need to be adjusted by reducing baking powder, sugar, and fat and increasing the amount of liquid. More mixing also is needed.

Tenderness and flakiness are contradictory goals in making pie crust. Tenderness is enhanced by using oil, soft fats, or fat cut into small pieces. Minimal mixing also increases tenderness. Flakiness is promoted by leaving the fat in coarse particles, which block water from some of the gluten and thus create areas where pressure from the steam evolved during baking forces apart layers or flaky areas of the dough.

Puff pastry is a flaky pastry with slightly tough layers. To make puff pastry, a large portion of butter is spread on the rolled dough and the dough is folded and chilled, a process that is repeated several times. This procedure creates many thin layers of dough, which then are separated readily by steam during baking.

Cookies are high in fat and sugar and low in liquid, with the levels of ingredients varying from bar cookies (comparatively high moisture content to create a rich batter) to the low moisture in rolled, pressed, and refrigerated doughs that can be handled easily without sticking. Drop cookies contain an intermediate level of liquid. The type of fat used in cookies is particularly important because of the impact on texture and spreading.

## STUDY QUESTIONS

1. Describe the changes that occur in muffin batter and the baked muffins as a result of stirring 10 strokes, 25 strokes, and 200 strokes and explain why these changes happen.

2. Explain the reasons for the steps used in making biscuits by the biscuit method.

3. Describe the changes that occur during the baking of bread and during staling.

4. Carefully describe the desirable characteristics of angel food cakes, sponge cakes, and chiffon cakes. What causes the differences among these three foam cakes?

5. Identify the factors that influence volume of an angel food cake and discuss the effect of each factor.

6. What changes need to be made when honey is substituted for part of the granulated sugar in a shortened cake? Why is each change necessary?

7. Outline the steps in preparing shortened cakes by the conventional, pastry-blend, and single-stage methods. What are the advantages and disadvantages of each method?

8. Cite the factors that can cause a shortened cake to fall and explain why each factor has this result.

9. What roles are performed by each of the ingredients in pastry?

10. What factors influence tenderness of pastry? Explain the effect of each.

11. What factors influence flakiness of pastry? Explain the effect of each.

12. True or false. Yeast breads require more time to prepare than quick breads because carbon monoxide needs to be produced in the dough.

13. True or false. Biscuits are a type of yeast bread.

14. True or false. The method that is considered best to use to make a shortened cake with optimal qualities is the muffin method.

15. True or false. When making pastry, the ratio of flour to fat is much more important than the amount of water.

## BIBLIOGRAPHY

Berry, D. 2004. Breads on the rise. *Food Product Design* 14 (7): 106.

Cauvain, S. P. and Young, L. S. 2007. *Technology of Bread Making*. 2nd ed. Springer. New York. NY.

Decker, K. J. 2002. The gourmet cookie experience. *Food Product Design* 11 (10): 34.

Decker, K. J. 2005. High-profile flatbreads. *Food Product Design* 15 (1): 97.

Decker, K. J. 2005. Looking at the whole-grain picture. *Food Product Design* 15 (9): 49.

Deis, R. C. 2005. How sweet it is—using polyols and high-potency sweeteners. *Food Product Design* 15 (7): 57.

Foster, R. J. 2005. pHood phenomena. *Food Product Design* 14 (11): 61.

Grant, L. A., et al. 2001. Starch characteristics of waxy and nonwaxy tetraploid (*Tritium turgidium L*. var. durum) wheats. *Cereal Chem*. 78 (5): 590.

Grenus, K. 2005. Maintaining texture. *Food Product Design* 15 (5): 47.

Hazen, C. 2006. New fiber options for baked goods. *Food Product Design* 15 (10): 80.

Hazen, C. 2006. Adding soy ingredients for health. *Food Product Design* 16 (2): 63.

Hui, Y. H. 2006. *Handbook of Food Science and Technology*. Vol. 4. CRC Press. Boca Raton, FL.

Hui, Y. H., et al. 2006. *Bakery Products: Science and Technology*. Blackwell Publishing. Ames, IA.

Kobs, L. 2001. "C" is for cookie. *Food Product Design* 11 (9): 31.

Kuntz, L. A. 2002. Designer fats for bakery. *Food Product Design* 12 (8): 55.

Manley, D. J. R. 2000. *Technology of Biscuits, Crackers, and Cookies*. 3rd ed. CRC Press. Boca Raton, FL.

Martin, M. L., et al. 1991. A mechanism of bread firming. I. Role of starch swelling. *Cereal Chem*. 68: 498.

Pszczola, D. E. 2004. Fats: In *trans*-ition. *Food Technol*. 58 (4): 52.

Pszczola, D. E. 2006. Reaping a new crop of ingredients. *Food Technol*. 60 (7): 51.

Schultz, M. 2005. Xanthan and foods: Bonded for life. *Food Product Design* 15 (7): 17.

Seguchi, M., et al. 1999. Breadmaking properties of triticale flour with wheat flour and relationship to amylase activity. *J. Food Sci*. 64 (4): 522.

Sivisankar, B. 2002. *Food Processing and Preservation*. Prentice Hall of India. New Delhi, India.

Tenbergen, K. and Eghardt. H. B. 2004. Baking ammonia: The other white leavening agent. *Food Product Design* 14 (6): 110.

Vollmar, A. and Meuser, F. 1992. Influence of starter cultures consisting of lactic acid bacteria and yeasts on the performance of a continuous sourdough fermenter. *Cereal Chem*. 69 (1): 20.

# INTO THE WEB

http://www.foodproductdesign.com/topics/bakery-and-cereal. aspx—Articles on various baked products.

http://www.foodtimeline.org/foodfaq2.html—Brief look at the history of muffins.

http://www.breadworld.com/—Overview of yeast bread making.

http://www.aaccnet.org/cerealchemistry/articles/1998/1006-06R.pdf—Article on soft and durum wheat flour in bread making.

http://www.dakotayeast.com/product_compressed.html—Information on compressed yeast.

http://www.muehlenchemie.de/downloads-future-of-flour/FoF_Kap_14.pdf—Article with illustrations of gluten development.

http://www.aaccnet.org/approvedmethods/summaries/10-10-03.aspx—AACC-approved conditions using the straight-dough method for testing breads.

http://www.bakingandbakingscience.com/princ. htm—Overview of bread making.

http://www.dict.uh.cu/Bib_Dig_Food/ift/jfs/jfs64/jfsv64n4 p582-586ms2838.pdf—Article on baking qualities of triticale and wheat flours.

http://www.exploratorium.edu/cooking/bread/recipe-injera. html—Discussion on making injera.

http://www.splenda.com/cooking-baking—Guidance in using Splenda as an ingredient in baking.

http://www.ghirardelli.com/chocopedia/making.aspx—Back ground information about cacao and processing.

http://www.worldcocoafoundation.org/who-we-are/partnership-meetings/pdfs/HBernaert_Fermentation. pdf—Fermentation of polyphenols in chocolate production.

http://www.rain-tree.com/chocolate.htm—Information on chocolate and its history.

http://flandersbio.be/files/5_Patrick_Hautphenne_voor_web.pdf—PowerPoint showing some aspects of chocolate production.

http://food.oregonstate.edu/learn/cake.html—Summary of making fat-containing cakes.

http://www.msu.edu/~lentnerd/NewFiles/Cake.html—Describes various methods of making shortened cakes.

http://www.swcoloradohome.com/articles/food/020114_b. asp—Recommendations for adjustments in baking at high altitudes.

http://www.foodtimeline.org/foodpies.html—Background information on different pastries.

http://www.youtube.com/watch?v=Yg-zXn_YpLI—Video on making Danish puff pastry.

# 6

## Food Supply Perspectives

Microbiologist Terry Arthur (left) and USMARC director Mohammed Koohmaraie examine petri dishes for *Salmonella* growth. (Courtesy of Agricultural Research Service.)

# CHAPTER 19

# Food Safety Concerns and Controls

## Chapter Outline

## OBJECTIVES

After studying this chapter, you will be able to:

1. Identify the microbiological agents, contaminants, and natural toxicants of importance in food safety.

2. Describe common food-borne illnesses, their potential food sources, and symptoms.

3. Discuss handling procedures needed to keep foods safe for human consumption.

4. Identify the person or agency responsible for oversight of food safety, including international, national, commercial, and consumer.

## DEFINING THE PROBLEM

Thoughts of food conjure up different images for all of us, but probably most are pleasurable. However, the truth of the matter is that food presents a curious paradox. We all must eat food to live, but sometimes food actually causes illness and even death. The U.S. Public Health Service cited in its *Food Code 2009* that annually approximately 76 million people in the nation suffer from a food-borne illness (325,000 require hospitalization, and 5,000 die) at a cost estimated to be between 10 and 83 billion dollars. The negative aspect of food simply cannot be ignored, despite its very distasteful nature. The good news is that food safety can be assured if the integrity of the food supply is protected and if food is handled correctly from the farm to the dinner table. Education regarding food safety and constant vigilance throughout the entire sequence by all those involved in production, marketing, importing, preparation, and service of food are needed to bring safe food to diners.

---

### Waste Management Challenges

Livestock and poultry produce manure throughout their lives, and this waste must be removed. Small farms can use it as fertilizer on fields, but the problem multiplies rapidly as the numbers of animals and fowl increase in large commercial operations. Various harmful microorganisms can thrive on manure, so careful disposal is essential. One of the main problems that occur when manure is stored in piles or holding areas is the runoff when water drains through and subsequently enters streams and rivers, carrying these microorganisms all along the route. Food grown in near-by fields and fish in the streams can become vectors to spread them. A tragic example of an outbreak of food-borne illness due to improper disposal of animal waste occurred in 2006 when baby spinach contaminated with *Escherichia coli 0157:H7* was bagged and marketed for use in fresh salads.

The Environmental Protection Agency (EPA) is the federal agency charged with overseeing safety from pollution, and enforcement has resulted in some lawsuits against major producers. In some cases, airplane reconnaissance has been used to spot large piles of manure that were in areas obscured from view. This approach was used in a case that involved a large hog-producing company with 275 hog farms in North Carolina. The potential for food contamination from animal waste is significant and needs to be controlled, but producers also need to be able to make a profit so that meat will continue to be available at a price consumers can afford. Technology has an important role to play in the interface between the environment and the producer.

---

Considerable safeguards are in place because of homeland security measures (see Chapter 1) and laws and inspections. Nevertheless, outbreaks of food-borne illnesses occur each year and are reported to the Centers for Disease Control and Prevention (CDC) in Atlanta, Georgia. Doubtless, far more individual episodes of food-borne illness occur, but they are not diagnosed by physicians or reported. These illnesses may be the result of errors anywhere in the sequence from the farm to the table.

The problems may occur in the commercial arena, but the handling of food in homes can often be the source. Food-borne illnesses stemming from errors in the commercial handling of food grab news headlines—for example, the outbreak of hepatitis A resulting from use of frozen strawberries in school lunches in Michigan in 1997 (apparently resulting from illegal use of imported fruit that had been harvested in fields lacking toilets for the pickers).

Food-borne illnesses that result from poor sanitary practices at home ordinarily are not publicized, and little has been done in the past to reduce these risks. However, attention toward improving food-handling practices in homes is increasing as a result of consumer surveys that showed many unsafe food practices in the majority of households interviewed.

## MICROBIOLOGICAL HAZARDS

Various kinds of microorganisms can cause food-borne illnesses. These include many different kinds of bacteria, viruses, molds, yeasts, protozoa, and algae (Table 19.1). Particularly, bacteria are some of the more common causes of food-borne illnesses. This chapter highlights bacterial infections from food and also other types of microorganisms causing illnesses in humans. Five microorganisms have been identified by the Centers for Disease Control as having high infectivity via food contaminated by infected food handlers: Norovirus, *Salmonelli typhi*, enterohemorrhagic or *Shiga* toxin-producing *E. coli*, *Shigella* spp., and hepatitis A virus.

**Table 19.1**   Classification of Some Disease-Causing Microorganisms

| Bacteria | Virus | Mold | Microalgae | Parasites | Protozoa |
|---|---|---|---|---|---|
| Clostridium botulinum | HAV | Aspergillus flavus | Gonyaulax catanella | Trichinella spiralis | Giardia lamblia |
| Staphylococcus aureus | | | | Ascaris lumbricoides | Entamoeba histolytica |
| Clostridium perfringens | | | | | |
| Yersinia enterocolitica | | | | | Cestodes |
| Salmonella enteritidis | | | | | |
| Salmonella typhi | | | | | |
| Escherichia coli | | | | | |
| Campylobacter jejuni | | | | | Flukes |
| Listeria monocytogenes | | | | | |
| Shigella boydii | | | | | |
| Vibrio cholerae | | | | | |
| Morganella morganii | | | | | |

## Bacteria

Bacteria are extremely tiny, single-celled organisms measuring an average of between 1 and 3 μm and weighing only about $1 \times 10^{-12}$ g. Their ability to reproduce by binary fission to achieve large populations in a short time (generation times are sometimes as brief as 20 minutes) can result in dangerous levels of bacteria in some foods. Bacteria are abundant in nature and are simple **prokaryotes** that survive under extremely harsh conditions by forming **spores** (a structure that is very difficult to kill, even with the use of high heat, abrasion, dehydration, chemicals, or freezing). These qualities (e.g., rapid reproduction and spore formation) help explain why it is so important to eliminate or at least minimize bacterial contamination initially, to maintain high levels of sanitation, and to control temperature outside the favorable range for reproduction when working with food.

The conditions influencing the viability of **Enterobacteriaceae** and their reproductive rates include the requirements for oxygen and temperature. Bacteria that require oxygen are classified as **aerobic**; those that fail to grow if oxygen is present are **anaerobic**. Bacteria thriving below 15°C (59°F) are said to be **cryophilic**, while those preferring temperatures between 15 and 45°C (59 and 113°F) are **mesophilic**, and those living at 45–95°C (113–203°F) are **thermophilic**. These criteria dictate the handling recommendations to minimize bacterial risks in certain foods.

Bacterial infections of different types are all too familiar to consumers and public health workers because of their frequency and health consequences that sometimes even include death. Among the toxin-producing bacteria capable of causing food-borne illnesses are various strains of *Salmonella, Shigella, Campylobacter, Listeria, Clostridium, Vibrio, Yersinia,* and *E. coli.*

Bacterial toxins are classified as *exotoxins* or *endotoxins.* **Exotoxins** are formed as waste products when bacteria thrive and multiply. They attack specific parts of the body and trigger the production of antibodies in the host. This makes it possible to produce immunity against such exotoxin-producing bacteria as **Clostridium botulinum** (Figure 19.1).

**Endotoxins** are actually part of the cell wall of some bacteria, but they are released into the host by enzymes following ingestion of the bacteria. The actions of endotoxins are quite diverse and impact various physiological functions. *Salmonella typhi* produce endotoxins, as do several other bacteria discussed later in this chapter.

*Clostridium botulinum.*   The exotoxin produced by *C. botulinum* causes **botulism**, which is highly lethal unless antitoxin is administered promptly to the infected person (Table 19.2). The scenario for botulism to occur begins with the fact that *C. botulinum* is quite widespread in soils and casts spores into the soil and water. The development of toxin

**Prokaryotes**
Cellular organisms without a distinct nucleus.

**Spore**
Specialized structure of bacteria capable of retaining viability under extremely adverse conditions.

**Enterobacteriaceae**
Bacteria that can go through the stomach and be viable in the intestines, reproducing there to cause illness.

**Aerobic**
Requiring oxygen for survival and growth.

**Anaerobic**
Requiring an oxygen-free environment for survival and growth.

**Cryophilic**
Microorganisms with optimal reproduction and survival below 15°C (59°F).

**Mesophilic**
Microorganisms with optimal reproduction and survival between 15 and 45°C (59 and 113°F).

**Thermophilic**
Microorganisms with optimal reproduction and survival between 45 and 95°C (113 and 203°F).

**Exotoxin**
Poison formed as waste when certain bacteria thrive and multiply.

**Clostridium botulinum**
Anaerobic, spore-forming bacteria that can produce a highly poisonous toxin capable of killing people.

**Endotoxin**
Poison in cell walls of certain bacteria.

**Botulism**
Potentially fatal food poisoning resulting from ingesting even a minuscule amount of toxin produced by *C. botulinum.*

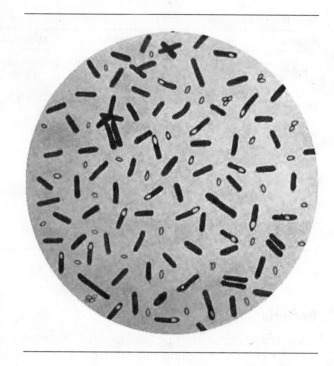

**Figure 19.1** *Clostridium botulinum* is an anaerobic bacterium that produces an exotoxin capable of causing death if ingested. (Courtesy of Agricultural Research Service.)

occurs if the spores are in an anaerobic and nonacidic environment for an extended period of time, as it might be the case in canned vegetables and meat that have not been heat processed adequately (Figure 19.1).

Pressure canning is required to process canned vegetables and meats to ensure the killing of any spores of *C. botulinum* that might be present. The bacteria are killed at lower temperatures than the spores, which are so heat resistant that they must be heated under pressure [at least 15 pounds of pressure, a temperature of 115°C (239°F)] for an extended period that is determined by the size of the containers.

**Table 19.2**    Summary of Some Bacterial Food-Borne Illnesses

| Agent | Incubation Time | Symptoms | Duration | Host Foods |
|---|---|---|---|---|
| *Campylobacter jejuni* | 2–5 days | Diarrhea, cramps, vomiting, fever | 2–10 days | Raw/undercooked poultry, raw milk, contaminated water |
| *Clostridium botulinum* | 12–72 hours | Blurred vision, diarrhea, vomiting, muscle weakness | Days or months; can be fatal | Improperly canned low-acid foods |
| *Clostridium perfringens* | 8–16 hours | Abdominal cramps, nausea, watery diarrhea, no fever | 1–2 days | Meat, poultry, gravy, precooked foods |
| *Escherichia coli O157:H7* | 1–8 days | Severe diarrhea, vomiting, no fever, abdominal pain | 5–10 days | Rare beef, raw milk, raw produce, contaminated water |
| *Listeria monocytogenes* | 9–48 hours | Fever, nausea, or diarrhea, muscle pain | Varies | Fresh soft cheeses, raw milk, hot dogs |
| *Salmonella* spp. | 1–3 days | Diarrhea, fever, cramps, vomiting | 4–7 days | Contaminated eggs, poultry, fecally contaminated water |
| *Shigella* spp. | 1–2 days | Diarrhea, fever, abdominal cramps | 4–7 days | Fecally contaminated food or water |
| *Staphylococcus aureus* (preformed enterotoxin) | 1–6 hours | Sudden onset of severe nausea, vomiting; fever and diarrhea | 1–2 days | Inadequate refrigeration of meats, egg and potato salads, cream pastries |
| *Yersinia enterocolytica* | 1–2 days | Fever, diarrhea, vomiting | 1–3 weeks | Raw milk, undercooked pork, contaminated water |

Adapted from American Medical Association (www.ama-assn.org/ama/article/3707-3891.html), September 5, 2003.

The botulism cases that occur usually are the result of improperly processed home-canned vegetables or other low-acid products. Commercial canning operations are monitored and almost always produce foods free of viable *C. botulinum* spores. However, low-acid foods canned at home may not have been processed adequately to inactivate any spores. To eliminate the potential risk, home-canned vegetables and meats should be boiled actively for at least 15 minutes before they are even tasted.

***Staphylococcus aureus.***   Staphylococcal infections are the result of contact with toxins produced by **Staphylococcus aureus**. Among its 10 different toxins is an exotoxin that is responsible for more cases of food-borne illness than are caused by any other bacteria. The symptoms include vomiting, diarrhea, nausea, weakness, headache, and fuzzy vision; some of these may be apparent within 2 hours of ingestion of the toxin and disappear within 12 hours, a short, albeit seemingly endless time for patients in the grip of truly violent symptoms.

**Staphylococcus aureus**
Bacteria capable of producing an enterotoxin as it grows, which can lead to food-borne illness.

The key to preventing infection of this type is to ensure that nobody with boils or other possible Staph infections comes in contact with food being prepared and/or served to others. Active bacteria that are introduced into the food can grow and produce the enterotoxin, which is the actual agent causing the illness. The other precaution is to be sure that prepared foods (dishes containing vegetables, meat, fish, poultry, milk, and fruits) are either served promptly or kept chilled below 4°C (41°F). This prevents active growth of the bacteria, thus blocking production of the illness-triggering enterotoxin.

Heat treatment of food that has been infected with *S. aureus* and contains enterotoxin will kill the bacteria, but not the enterotoxin unfortunately. Heating at a temperature of 100°C for 10 minutes, refrigerating for weeks, or drying will not destroy the enterotoxin that may have been formed in the food when the bacteria were viable. The only protection against this type of infection is to avoid the original contamination.

***Clostridium perfringens.***   **Clostridium perfringens** is an anaerobic, spore-forming bacteria. The toxin formed by *C. perfringens* during growth and spore formation is stored in the spore's coat. The toxin may be released into the infected food or into the intestines of the person eating the food.

**Clostridium perfringens**
Anaerobic, spore-forming bacteria that can produce a toxin capable of causing a mild food-borne illness.

Whether the toxin is already in the food or is released into the diner's intestines is not important. In either case, the unlucky person will experience discomforts such as mild diarrhea, headache, nausea, and stomach pain, all of which will usually be gone in between 12 and 24 hours. These problems are uncomfortable, but far less distressing than those caused by *S. aureus*. Nevertheless, avoidance of the toxin clearly is preferred. Meat-containing dishes should be served immediately or else refrigerated promptly to avoid the growth of *C. perfringens*. Such foods are attractive to this microorganism because they provide the desired anaerobic circumstance; room temperatures promote rapid growth of *C. perfringens* if any are present.

***Salmonella enteritidis.***   **Salmonella enteritidis** was recognized during the last quarter of the twentieth century as a cause of salmonellosis, an illness that can be transmitted through ingestion of raw or under-cooked eggs even if sanitary handling conditions prevailed. The problem has been traced to intact eggs containing *S. enteritidis* in the yolk (Figure 19.2). Infection of the egg occurs during its formation and before the shell is deposited. Occurrence of these contaminated eggs is rare (estimated to be about 1 in 10,000 eggs in the supermarket), but steps are taken in egg-laying operations to attempt to reduce this figure still more by maintaining extremely sanitary conditions and combating infections in the flocks.

**Salmonella enteritidis**
Type of *Salmonella* sometimes found in the yolks of unbroken eggs; capable of causing salmonellosis (a food-borne illness).

Consumers need to be sure either to use pasteurized egg products or to cook eggs to a temperature of at least 71°C (160°F) before eating them or incorporating them into products that will not be heated. Refrigeration of eggs at all times during storage and avoidance of room temperatures except when actually preparing products containing eggs are effective measures to retard bacterial growth. Hot dishes made with eggs need to be served promptly or held at a temperature of 57°C (135°F) or hotter to retard growth of any viable *S. enteritidis*.

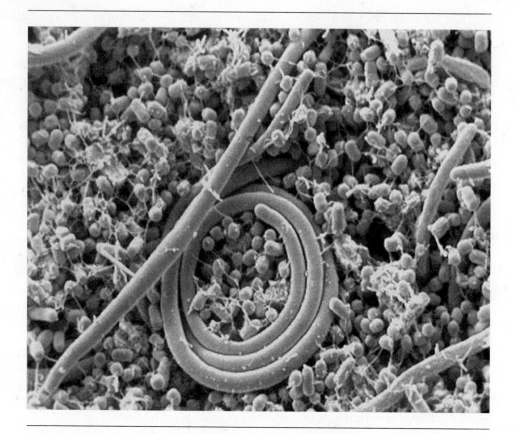

**Figure 19.2** *Salmonella enteritidis* can be inside a fresh egg if the hen is infected. (Courtesy of Agricultural Research Service.)

The symptoms of *S. enteritidis* infection include fever, chills, diarrhea, vomiting, abdominal pains, and headache. These usually are apparent between 12 and 36 hours. Deaths have occurred each year from this type of *Salmonella* infection, most commonly among the frail elderly in nursing homes so pasteurized egg products are required there. This type of *Salmonella* infection is responsible for more than half of the reported bacterial outbreaks in the United States.

*Salmonella typhi* Type of *Salmonella* causing typhoid fever.

**Salmonellosis** General name for illness caused by *Salmonella*, regardless of the specific species.

*Salmonella* Genus name for several species of gram-negative bacteria that can cause gastrointestinal illnesses, including typhoid fever.

***Salmonella typhi.*** A particularly virulent type of *Salmonella* is ***Salmonella typhi***, which causes typhoid fever. Like other *Salmonella* infections, onset occurs as early as 10 hours after ingestion, but the symptoms are more severe than other types of **salmonellosis**. Prompt medical attention with antibiotic therapy can overcome the infection in about two to three weeks, but death can result without treatment. Fortunately, *S. typhi* infections do not occur commonly in the United States, and public health officials track any occurrence to contain potential outbreaks.

Other species of ***Salmonella*** also are capable of causing food-borne infections, although they are not nearly as virulent as *S. typhi*. Nevertheless, any *Salmonella* infection can be extremely unpleasant. *Salmonella* are ubiquitous in the environment, which makes sanitation measures to avoid fecal contamination of water and food essential to reduce the possibility of consuming contaminated products.

Thorough washing of hands with hot water and soap must be a strong habit for all food handlers. This measure also is needed for all cutting boards and other surfaces coming in contact with food. The other essential is temperature control, with time at temperatures between 4°C (41°F) and 57°C (135°F) kept to an absolute minimum.

## FOOD FOR THOUGHT: Preamble to Safety

Preventing *Salmonella* in poultry is the first line of defense against salmonellosis in humans. Researchers in the U.S. Department of Agriculture knew that chickens at least three weeks old had some resistance to Salmonella so they identified the types of bacteria in these healthy older chickens. Then, they created a blend of 29 kinds of live bacteria that they had found.

The question then was one of how to administer this bacterial potion to baby chicks to impart the protection of these live benign bacteria. Their goal was to get the live bacteria into the intestines of the chicks where they could compete against *Salmonella* and overcome that bacterial hazard and establish a healthy gut. Various ideas were tried, but the mode of delivery that was chosen was a mist sprayed over the baby chicks. Chicks get the bacteria into their gastrointestinal tract because they habitually preen their down and feathers using their beaks; grooming ensures that the bacteria get into their mouths. Since the mist falls over all the birds, they all receive the benign bacteria that will prevent colonization by *Salmonella* in the chicks regardless of how much food they eat.

This benign live bacterial spray was named Preempt™ to indicate its preemptive action in preventing colonization of *Salmonella* in chickens. It has been available for use since 1998 and has helped bring safer poultry to the marketplace. Although this is an effective means of reducing the hazard of *Salmonella* in the poultry people eat, it does not mean that safety can be ignored. Adequate refrigeration of poultry is essential. During preparation, steps to keep surfaces clean and prevent cross-contamination from cutting boards, knives, and other surfaces still need to be taken. Poultry needs to be heated to a temperature of 74°C (165°F).

*Escherichia coli.*   One of the more highly publicized bacteria causing food-borne illness is *Escherichia coli*, commonly referred to as *E. coli* (Figure 19.3). This **coliform** occurs in a great many different **serotypes** (closely related organisms having a common set of antigens or toxins capable of stimulating production of antibodies).

The particular serotype that popped into the news in the early 1990s was *E. coli 0157:H7*. This specific *E. coli* was the causative agent in numerous outbreaks of food-borne illness; the sources were traced to a range of foods, including undercooked ground beef, unpasteurized milk, low-acid apple cider, mayonnaise, unchlorinated water, poultry, and uncooked fruits and vegetables.

The presence of *E. coli* in food or water can be traced to fecal contamination at some point around the farm or on through the sequence that moves food and water through the marketing and preparation steps that precede consumption. Fecal waste can contaminate food any time workers with unclean hands and/or fingernails handle any kind of food or utensil that comes in contact with food. Unsanitary slaughter and food-processing plants also afford unfortunate opportunities for contamination of food with *E. coli* (as well as other pathogenic microorganisms).

Ideally, *E. coli* will not be introduced into food and water, but sometimes people and other vectors such as flies and cockroaches do cause contamination. Actually, the wise thing to do when handling food is to assume that *E. coli* or other bacteria are present. Fresh produce should be washed thoroughly in chlorinated water. Storage of protein-rich foods (meats, poultry, fish, milk and milk products, and eggs) should be at temperatures below 4°C (41°F) or above 57°C (135°F).

Illnesses caused by ingestion of viable *E. coli* and the presence of its toxins include watery diarrhea. In some cases, blood and some flecks of mucus are expelled in the stool. Vomiting, nausea, and a low fever also may be present. *E. coli* also can cause meningitis.

**Escherichia coli**
Type of coliform bacterium with many different serotypes, some of which produce toxins that cause food-borne illnesses.

**Coliform**
Colonic bacterium.

**Serotypes**
Closely related organisms, such as *E. coli*, having a common set of antigens.

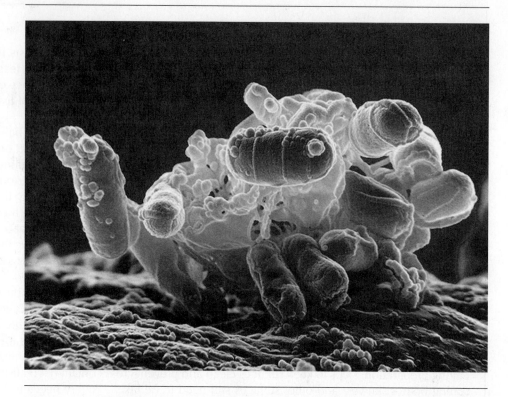

**Figure 19.3**    A cluster of *E. coli* bacteria. Individual bacteria in this photo are oblong and dark. (Courtesy of Agricultural Research Service.)

**Campylobacteriosis**
Food-borne illness caused by toxin from *Campylobacter jejuni* (or other toxin-producing strains).

**Campylobacter jejuni**
Common strain of *Campylobacter* that can cause campylobacteriosis.

***Campylobacter jejuni***.    Although several strains of *Campylobacter* are known (Figure 19.4), the one responsible for many of the cases of **campylobacteriosis** is ***Campylobacter jejuni***. The most likely sources of this type of bacterial contamination are raw milk, undercooked or raw meats or poultry, and contaminated water. Development of symptoms can occur between 2 and 10 days following ingestion; symptoms start with a fever, muscle pains, and headache, with stomach discomfort, nausea, and diarrhea following.

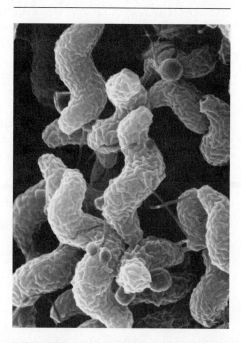

**Figure 19.4**    *Campylobacter* bacteria are the number-one cause of food-related gastrointestinal illness in the United States. To learn more about this pathogen, ARS scientists are sequencing multiple *Campylobacter* genomes. This scanning electron microscope image shows the characteristic spiral, or corkscrew, shape of *C. jejuni* cells and related structures. (Courtesy of Agricultural Research Service.)

---

## FOOD FOR THOUGHT: Activated Lactoferrin

Lactoferrin, a protein that occurs naturally in milk, is isolated from whey proteins and marketed in the form of activated lactoferrin. This product, which has been given GRAS status by the FDA, is an effective agent against a wide range of bacteria that can cause food-borne illnesses in humans. The hazards of food-borne infections caused by *E. coli 0157:H7*, *Campylobacter*, and *Salmonella* and many other pathogens can be reduced by the use of activated lactoferrin during food production.

This protein is applied to meats (particularly beef) as a spray on the surfaces. Lactoferrin fights bacteria thereby detaching microorganisms that may already have attached to the meat, blocks attachment of free bacteria, and inhibits growth of colonies. (This surface action on meat or other foods on which activated lactoferrin has been sprayed is similar to the protective action in an infant's small intestine when milk passes through it.) After the activated lactoferrin has been in contact with the meat long enough to release the bacteria, the surface is rinsed with water to remove the pathogen–lactoferrin complex that forms when activated lactoferrin comes in contact with bacteria.

Activated lactoferrin is particularly valued for its effectiveness in reducing the risk of infections caused by *E. coli 0157:H7* in meats, especially ground meats. This protein treatment is certainly not intended to eliminate the need for very careful sanitation measures when working with meat at all times until it is eaten. Cleanliness, sanitary handling practices, and careful temperature controls are still essential. However, activated lactoferrin provides another important tool in protecting consumers from food-borne illnesses.

---

Even though campylobacteriosis is usually less dangerous than some other infections caused by contaminated food, it still is quite uncomfortable and should be avoided. Adequate heating of meats [71°C (160°F)] and poultry [82°C (180°F)] and use of only pasteurized milk and chlorinated water are useful measures in avoiding *C. jejuni*.

***Listeria monocytogenes***.   Food-borne illnesses associated with the presence of viable ***Listeria monocytogenes*** were not diagnosed until the 1980s, although the microorganism had been identified in farm animals early in the twentieth century. The illness is **listeriosis**, a condition that can develop within a few hours following consumption or may require as long as six weeks before truly serious symptoms are evident. Death occurs in some cases, but fatigue, fever, nausea, vomiting, and diarrhea are the early symptoms and may be the only problems.

The obvious way of avoiding the hazards of *L. monocytogenes* is to prevent contamination of foods, particularly of meats, meat products, and milk, including soft cheeses. Unfortunately, cold merely slows multiplication; even commercial freezing at −18°C (0°F) does not does not kill them, but they no longer multiply.

In the temperature range between 4 and 45°C (41 and 120°F), *L. monocytogenes* will flourish. Storage at room temperature should be avoided. Refrigeration helps slow bacterial growth, and holding temperatures above 60°C (135°F) actually destroy *Listeria*. They also are destroyed by heating meats to 71°C (160°F) and poultry to 82°C (180°F) and by pasteurizing milk.

***Shigella***.   One type of dysentery, **bacillary dysentery**, is caused by ingestion of *Shigella;* the most common cause is ***Shigella boydii***. The condition develops in one to five days, when fairly sudden abdominal pains and extreme diarrhea and an accompanying high fever occur. The extreme loss of fluids eliminates electrolytes as well, which can quickly lead to shock and even death unless fluids and electrolytes are aggressively and quickly replaced. Infection from *Shigella* usually is transmitted from person to person via the route of fecal contamination of food and/or water.

---

**Listeria monocytogenes**
Type of bacteria sometimes found in meats, meat products, unpasteurized milk, and other foods; toxin causes listeriosis.

**Listeriosis**
Food-borne illness caused by *Listeria*.

**Bacillary dysentery**
Food-borne illness characterized by severe diarrhea and electrolyte loss, accompanied by intense abdominal cramps and a high fever and capable of ending in death.

**Shigella boydii**
Type of *Shigella* most likely to cause the food-borne illness designated as bacillary dysentery.

*Vibrio cholerae*
Bacteria carried by fecal contamination to food and causing cholera when ingested.

**Cholera**
Food-borne illness especially problematic in tropical areas around the world; caused by *V. cholerae.*

**Scombrotoxin poisoning**
Allergic-type response to ingestion of high levels of histamine, saurine, and other metabolites produced by the action of *Morganella morganii* on tuna and related fish.

*Morganella morganii*
Bacteria producing scombrotoxin from histidine on the surface of tuna and related fish.

**Yersiniosis**
Food-borne illness focusing on the intestines caused by consuming infected raw milk or undercooked pork.

*Yersinia enterocolitica*
Bacteria sometimes found in raw and undercooked pork and raw milk, which cause yersiniosis.

**Virus**
Chemical macromolecule capable of being engulfed by a body cell and eventually altering or killing the host cell.

**Hepatitis A**
Food-borne illness characterized by inflammation of the liver; caused by hepatitis A virus.

**HAV**
Hepatitis A virus, the cause of a food-borne illness that attacks liver cells.

*Vibrio cholerae*. Water that is heavily contaminated with human fecal matter unfortunately is a favored site for *Vibrio cholerae* bacteria to flourish and ultimately infect humans via either food and/or water. India has been plagued by **cholera** caused by this microorganism. Cholera also is a health problem throughout Southeast Asia, the northern area of Africa, and South America.

Severe diarrhea, as well as nausea and vomiting abruptly announce the presence of cholera. If antibiotics, glucose, electrolytes, and adequate water to replace the lost fluids are administered promptly, the prognosis for recovery is excellent; without timely and appropriate treatment, the death rate can be as high as 75 percent. The key to overcoming the threat of cholera in very hot countries where it poses health risks is to improve sanitation. Safe water, sanitary food handling, and sound personal sanitary habits by food handlers are essential to avoid fecal contamination and the related outbreaks resulting when *V. cholerae* contaminates the food supply.

*Morganella morganii*.   **Scombrotoxin poisoning** is caused by ingestion of fish having an elevated level of histamine, as much as 50–100 milligrams per 100 g of tuna, mahi-mahi, or bluefish. Actually, histamine is produced from histidine on the surface of these fish by *Morganella morganii*, a type of bacteria. Allergic responses to the histamine, saurine, and other metabolites may include a rash and gastrointestinal upset. The problems occur about an hour after eating and usually are gone in about half a day. Control of this possible problem is accomplished simply by keeping these types of fish refrigerated to minimize the formation of histamine; cooking has no effect.

*Yersinia enterocolitica*.   **Yersiniosis** is another food-borne illness of bacterial origin. The most common species causing this disease is *Yersinia enterocolitica*, which sometimes is found in pork and raw milk. Usually symptoms develop between four and seven days and include diarrhea, fever, and abdominal pain. This disease, which can last three weeks and even longer, can be avoided by adequately cooking pork products; by using milk and milk products that have been pasteurized; and by avoiding cross-contamination of cutting boards, counters, and cooking utensils in the kitchen.

## Viruses

**Viruses** are macromolecules that humans can ingest via water and/or food. The harm viruses cause in the body occurs when a viral macromolecule is engulfed by a cell in the host and then eventually interacts with DNA in the cell to form viral genes capable of altering or even killing the host cell. Hepatitis A and poliomyelitis are viral infections that can sometimes be traced to contaminated water or food.

*Hepatitis A*.   **Hepatitis A** is caused by **HAV**, the hepatitis A virus. This type of hepatitis virus is sometimes present after fecal contamination of food and/or water (Table 19.3). Tracing the source of the HAV infection may be quite difficult because the incubation period

**Table 19.3**   Summary of Some Viral Food-Borne Illnesses

| Viral Agent | Incubation Time | Symptoms | Duration | Host Foods |
|---|---|---|---|---|
| Norwalk-like | 1–2 days | Nausea, vomiting, watery diarrhea (large amount) | 1–5 days | Foods uncooked after touching by infected food handler, salads, sandwiches, ice, fruit, inadequately cooked shellfish |
| Rotavirus | 1–3 days | Vomiting, low fever, watery diarrhea | 4–8 days | Fecally contaminated foods |
| Hepatitis A | 15–50 days | Dark urine, flu-like symptoms, diarrhea, headache, abdominal pain | 2–12 weeks | Shellfish from contaminated waters, raw produce, cooked foods or foods not reheated after handling by infected food handler |

Adapted from American Medical Association (www.ama-assn.org/ama/pub/article3707-3897.html).

lasts from two to four weeks from ingestion before symptoms are noted. Among the symptoms are an enlarged liver and jaundice, because the liver is the focus of the virus's action. Abdominal pain, loss of appetite, and nausea also are evident. Recovery in 8–12 weeks is usual, although a few people may die if the illness is not treated.

Inadequately cooked shellfish (mussels, clams, oysters, and shrimp) are the most likely food sources of HAV contamination. Prevention and control of possible hepatitis A outbreaks require careful control of sanitation and an uncontaminated water supply.

*Norwalk-Like Viruses (NLV)*.   Headlines in travel sections in 2003 publicized the outbreak of a food-borne illness caused by **Norwalk-like viruses (NLV)**. *Caliciviridae* is the family causing many of the episodes ultimately attributed to a Norwalk-like virus. NLV apparently are responsible for many outbreaks of food-borne illnesses, although underreporting of cases and difficulty identifying the causative agents make it impossible to determine how many are infected each year.

A Norwalk-like virus infection results in an acute onset of nausea, vomiting, diarrhea, and stomach cramps within 12–48 hours. These symptoms last between 12 and 60 hours in most cases. This infection usually originates from eating food with some fecal contamination, but contaminated water also can cause the illness.

Salad bars may be a source of contaminated food because these raw foods are consumed without any cooking after being prepared in the kitchen and placed in the salad bar. Food handlers obviously need to be extremely careful to wash their hands thoroughly and frequently.

**Norwalk-like virus (NLV)**
Virus causing sudden onset of acute gastrointestinal problems in 12–48 hours after eating food contaminated with feces.

## Molds

Molds are composed of many cells, in contrast to single-celled bacteria. The usual structure of molds begins with roots that are tiny threads deriving the nutrients required for their survival from the food host. A stalk with spores attached rises from the root, and it is these spores that vividly declare the presence of the mold. When disturbed, the spores may be wafted delicately in the air to another site, but the invasive, barely discernible root remains behind to continue to contaminate the food. Here, the hazards of molds are the concern. However, it is appropriate to note that some molds are used to create interesting cheeses (see Chapter 14).

The transported spores begin a new growth cycle of mold on any suitable new host they may happen to land on. Molds grow particularly well in warm, shady areas with high humidity, either indoors or outdoors. Grains, peanuts, silage, and similar plant crops provide ideal hosts for molds when stored under warm, moist conditions. Some types of molds form **mycotoxins** when crops such as these are stored inappropriately for a period of time.

Patulin, ochratoxin, aflatoxins and fumonisins are mycotoxins that may be formed by certain molds contaminating various foods (Table 19.4). Prevention of mycotoxin contamination of foods is important, particularly in hot, humid locales in developing countries where crop storage conditions may promote mold growth and resultant mycotoxin formation.

**Mycotoxin**
Poison produced by some molds that can be lethal when consumed.

**Aspergillus flavus**
Mold that thrives when peanuts, grains, and other field crops are stored under warm and moist conditions.

**Aflatoxin**
Mycotoxin produced by *Aspergillus flavus*.

*Aspergillus flavus*.   **Aspergillus flavus** is a mold that produces **aflatoxin**, a poisonous mycotoxin (Figure 19.5). The condition resulting from ingesting aflatoxin is called **aflatoxicosis**. This poisoning is evidenced first by nausea and vomiting but then progresses to the central nervous system to cause convulsions, coma, and finally death.

**Aflatoxicosis**
Condition caused by ingesting aflatoxin; begins with nausea and vomiting but ends with convulsions, coma, and death.

**Table 19.4**   Mycotoxins in Some Foods

| Mycotoxin | Mold Source | Food Sources |
|---|---|---|
| Patulin | *Aspergillus, Penicillium* | Cereals; low-acid fruit juices; damaged grapes, apples, peaches |
| Ochratoxin | *Aspergillus* spp., *Aspergillus* | Infected grains and meat and cheese from infected animals |
| Aflatoxins | *Aspergillus flavus* | Peanuts, corn, copra, and so on stored in hot, humid, unclean places |
| Fumonisins | *Fusarium verticilloides,* | Maize |
| | *Fusarium proliferatum* | Maize |
| Ergot | *Claviceps purpurea* | Infected grains |

**Figure 19.5** ARS chemist Kenneth Ehrlich isolates different types of *Aspergillus* spores.

**Ergotism**
Potentially fatal food-borne illness caused by eating grain that is contaminated with ergot, the toxin produced by the fungus *Claviceps purpurea*.

**Paralytic shellfish poisoning**
Often fatal, paralytic condition caused by ingesting shellfish containing high levels of toxin in their flesh, the result of red-tide feeding.

***Aflatoxin***.  Prevention of aflatoxin in the food supply is based on avoiding formation of visible mold in stored grains, peanuts, and other field crops. If mold is detected, the moldy product must be removed from the food supply to prevent deaths. Vigilance in searching for any mold is essential to protect consumers from aflatoxin in any of these crops.

***Ergot***.  **Ergotism** is an illness caused by a toxin produced by a fungus (*Claviceps purpurea*) that can infect wheat, rye, triticale, and other cereal grains. Spores produced by the fungus are blown by winds and may invade the embryo of developing grain kernels where they flourish and develop visible purple or black sclerotia (ergot bodies). The toxin contains some alkaloids that can be harmful to humans if ergot-infected grain is eaten.

## Microalgae

Some microalgae are capable of causing **paralytic shellfish poisoning**, a condition characterized by tingling sensations and then numbness and finally paralysis of the tongue, lips, and mouth. Subsequent symptoms are vomiting and diarrhea, with difficulty in breathing, cardiac arrest, and loss of consciousness preceding death if the level of ingested toxin is high.

The route to an episode of paralytic shellfish poisoning begins when such dinoflagellates or microalgae as ***Gonyaulax catanella*** reproduce in such unusually large numbers that a so-called **red tide** can be seen in the sea during the day and a phosphorescent glow is discernible at night. These dinoflagellates produce a toxin descriptively named paralytic shellfish poison. This is ingested when shellfish eat the microalgae, and the toxin is concentrated in their tissues.

When red tide occurs, shellfish gorge on the huge population of *G. catanella* and other dinoflagellates. The toxin causes no symptoms in the shellfish, but the high concentrations in their flesh are capable of causing paralytic shellfish poisoning in humans eating the shellfish. For this reason, it is imperative that shellfish not be harvested or consumed when red tide occurs.

## Parasites

**Parasites** differ from other agents causing food-borne illnesses because these organisms have to be ingested by humans or other suitable hosts for them to proceed through their life cycles. The sources of parasites that may invade humans are contaminated food and water. Usually, the symptoms associated with ingestion of viable parasites are most evident at times of growth or reproduction of the parasites.

*Nematodes*.   Of the four types of parasites (nematodes, protozoa, cestodes or tapeworms, and trematodes or flukes) that have been identified in patients in the United States, the most common are **nematodes** (also referred to as roundworms) and protozoa. Adequate cooking of food that is infected with roundworms eliminates the risk for humans even if they happen to ingest infected food.

***Trichinella spiralis*** is a roundworm occasionally found in pork from pigs that have been fed uncooked garbage (Figure 19.6). This is the reason for recommending that all

**Gonyaulax catanella**
Type of microalgae capable of producing paralytic shellfish poison.

**Red tide**
Visible evidence of unusually large populations of dinoflagellates (microalgae) capable of causing paralytic shellfish poisoning.

**Parasite**
An organism living in or on a host organism.

**Nematodes**
Roundworms that are parasitic.

***Trichinella spiralis***
Parasite sometimes found in pork or wild game.

**Figure 19.6**   An innovative, on-farm program to certify pigs as free of trichina parasites could become a model for excluding other meat-borne diseases from foods. (Courtesy of Agricultural Research Service.)

**Ciguatera poisoning**
Food-borne illness caused by ciguatoxin produced in fish that have eaten the algae *Gambierdiscus toxicus*.

**Scombroid (histamine) poisoning**
Food-borne illness caused by eating spoiled fish with high levels of histamine.

***Gambierdiscus toxicus***
Algae that sometimes are eaten by red snapper and other finfish and then produce ciguatoxin in the fish.

---

## FOOD FOR THOUGHT: Something Fishy!

The nutritional merits of fish are prompting people to consume fish more frequently than they did a few years ago. However, fish can present health hazards as well as benefits. Among the various food-borne illnesses that fish can cause are ciguatera poisoning, paralytic shellfish poisoning, and scombroid (histamine) poisoning.

Finfish (e.g., red snapper and grouper) and shellfish can cause ciguatera poisoning if the fish have been eating *Gambierdiscus toxicus*. This type of algae produces ciguatoxin, which causes upsets in the gastrointestinal tract and neurological symptoms such as numbed lips and tongue, blurred vision, and paralysis. The incidence of ciguatera poisoning is not high, but it is an example of a natural cause of a food-borne illness.

Paralytic shellfish poisoning may occur when shellfish are living in seawater that has a high population of *G. catanella*. The population of this type of algae may multiply rapidly when the water is about 8°C (46°F) and contains fairly rich levels of nutrients from runoff water. This situation is evidenced as red tide, in which the reddish tint and slightly luminescent quality of the water serve as a warning to avoid eating shellfish from the area.

Scombroid (histamine) poisoning is a food-borne illness caused by eating fish (particularly tuna, bonito, and mackerel) that contain high levels of histamine. This amino acid is formed as proteases (autolytic and bacterial) break down muscle proteins in the fish while they are stored before being cooked. Among the symptoms of scombroid poisoning are gastrointestinal problems, tingling and burning or itching skin, and palpitations.

The key to avoiding scombroid poisoning is to prevent histamine formation. The best means of avoiding food spoilage is to clean fish promptly after they are caught and to store them frozen or for a relatively short time at < 1°C (< 34°F). Unfortunately, histamine is so stable in heat that it will still cause poisoning if it has been formed in the fish prior to cooking. Therefore, spoiled fish should be discarded without being tasted.

---

**Trichinellosis**
Illness caused by eating meat containing viable *T. spiralis*; sometimes called trichinosis.

**Ascaris lumbricoides**
Type of parasitic roundworm that may be present in some seafood and may cause illness leading to possible pneumonia if food is not heated enough to kill it.

**Protozoa**
Parasitic single-celled complex microorganism or simple animal; some cause illnesses in humans when fecal contamination of water or food occurs and live organisms are ingested.

***Giardia lamblia***
Type of protozoa that, if ingested when they are alive, causes giardiasis in humans.

**Giardiasis**
Parasitic illness caused by *G. lamblia*, a type of protozoa, and characterized by severe diarrhea.

pork be cooked to an internal temperature of 77°C (170°F). This also is the temperature recommended for cooking wild game because of the possible presence of *T. spiralis*. The illness resulting from ingestion of viable *T. spiralis* is **trichinellosis**, a condition that damages the intestinal lining, reducing absorption and triggering muscle pain when larvae invade them.

***Ascaris lumbricoides*** is another roundworm or nematode. The dietary source for humans is contaminated seafood. The practice of eating raw or undercooked seafood is not recommended because of the possibility of ingesting viable roundworms, such as *A. lumbricoides*, along with delectable morsels. Detection of roundworms in fish is difficult, which means that prudence requires killing any that might be viable in seafood. If fish is cooked until it flakes, this parasite will be killed. Ingesting live *A. lumbricoides* not only leads to damage to the intestinal lining but also may make its host susceptible to developing pneumonia because of possible damage in the lung and its capillaries.

*Protozoa.* Only a few **protozoa** (single-celled complex microorganism or simple animal) are likely to cause food-borne illnesses in people. Two protozoa that may be ingested from contaminated food or water are *Giardia lamblia* and *Entamoeba histolytica*, both of which can cause serious diarrhea.

The ingestion of viable ***Giardia lamblia*** is a particular problem in regions where poor sewage management and general lack of personal sanitation result in fecal contamination of food and water (Table 19.5). The numbers of viable protozoa present determine the severity of **giardiasis**, the condition caused by ingesting *Giardia*. Symptoms of giardiasis range from vague stomach ache, some tenderness in the abdomen, and mild

**Table 19.5**  Summary of Some Parasitic Food-Borne Illnesses

| Parasitic Agent | Incubation Time | Symptoms | Duration | Host Foods |
|---|---|---|---|---|
| *Giardia lamblia* | 1–4 weeks | Acute/chronic diarrhea | Weeks | Drinking water, other foods |
| *Trichinella spiralis* | 1 day–8 weeks | Nausea, vomiting, diarrhea, abdominal discomfort, fever, muscle pains | Months | Raw or undercooked infected pork or wild game |
| *Cyclospora cayatenesis* | 1–11 days | Fatigue, protracted diarrhea | Weeks/months | Imported berries, contaminated water, lettuce |

diarrhea to a severe diarrhea. *G. lamblia* can be eliminated by repeated doses of quinacrine hydrochloride.

**Entamoeba histolytica** causes **amebic dysentery**, which means a malfunctioning caused by an amoeba. The presence of *E. histolytica* is a particular problem in the hot regions near the equator, especially in areas where fecal contamination of water and food due to faulty sewage treatment and general lack of good sanitation are the norm. The long delay between ingestion of *E. histolytica* and appearance of symptoms (between 10 days and up to 4 months) complicates the identification of the actual source of contamination.

Some people have mild symptoms of flatulence and diarrhea or constipation, while others have severe abdominal pains, fever, chills, and diarrhea containing some mucus and blood. Very severe cases result in ulceration of the intestine, which often leads to peritonitis, hepatitis, and even involvement of the brain, any of which may cause death unless treated with an effective medication, such as chloroquine.

*Cestodes (Tapeworms).*  Tapeworms or **cestodes** are yet another type of parasite, some of which use humans as their hosts. *Taenia solium* is the name of the tapeworm sometimes found in pigs that can be transmitted into humans via infected, undercooked pork. In beef and a variety of wild game (e.g., deer, buffalo, and antelope), *Taenia saginata* and *Taeniarhynchus saginatus* are examples of tapeworms that might be present.

If cestodes are present in meat that is eaten without sufficient heating, their eggs will develop within the host and thrive to produce tapeworms as long as 3 meters (~10 feet). This remarkable growth is possible because the head of the worm attaches to the intestinal lining and obtains the nutrients it needs from the host.

The usual type of tapeworm from fish is *Dibothriocephalus latus*, which is found in fish from the Great Lakes, Scandinavia, and Russia. This tapeworm can reach lengths of as much as 12 meters (~39 feet). Prevention of this problem hinges on sanitary handling of uncooked fish from areas of possible infestation and then thorough cooking of the fish before even tasting it to assure that no larvae are still viable.

*Flukes.*  Trematodes, or **flukes**, are a class of flatworm capable of invading certain tissues of human hosts. The life cycle of various flukes may involve invasion of snails and then a period in fish before ultimately arriving in a human host. Generally, the cycle involves an opportunity for fecal contamination in water, such as might occur in a rice paddy or small stream passing through an area exposed to human or animal waste. Even plants grown in such water (e.g., water chestnuts) may harbor flukes that then can be ingested by humans.

*Opisthorchis sinensis*, also called the Chinese liver fluke, can cause serious health problems and even death when many of them invade the liver as they mature following ingestion of contaminated raw or inadequately cooked fish. *Fasciola hepatica* is another example of a liver fluke that can be found in many parts of the world. *Fasciolopsis buski* is yet another fluke that can invade humans; in this case, the effects are in the small intestine. Control of snails and avoidance of uncooked water plants in addition to sanitary management of human and animal waste are important measures in avoiding this intestinal fluke.

**Entamoeba histolytica**
Type of protozoa that can cause amebic dysentery in humans.

**Amebic dysentery**
Protozoan infection characterized by diarrhea of varying severity (sometimes containing mucus and blood) and sometimes more severe symptoms (also called amebiasis).

**Cestodes**
Parasitic tapeworms, some of which use humans as hosts; possible food sources are various meats and fish that have been undercooked or eaten raw.

**Flukes**
Flatworms (parasites) that can invade the liver, small intestine, or lungs.

Southeast Asia is faced not only with the possibility of Chinese liver flukes, but also with a lung fluke called *Paragonimus westermanni*. The problem arises from the common practice in this region of eating raw crabs, crayfish, and shellfish. Uncooked juice from infected crabs and crayfish also must be avoided. In contrast to many of the other flukes that may be only an inch or two long, *P. westermanni* may grow as long as 4 inches. The lung invasion by this fluke leads to coughing, chest and throat discomfort, and flecks of blood in the sputum. The coughing forces the flukes into other tissues. The situation becomes critical when the brain and central nervous system become involved.

## CONTAMINANTS AND NATURAL TOXICANTS IN FOODS

### Contaminants

Harmful contaminants (e.g., lead, cadmium, and mercury) may enter the food supply from environmental contamination and from handling, processing, and cooking foods. Metals may present long-term problems because of their impact on polluted environments, both on land and in freshwater streams and lakes, and even the sea. Lead may occur now in plants, animals, and fish (freshwater and saltwater) and other aqueous species as a result of careless handling of lead-containing industrial materials and products. This heavy metal accumulates in humans over many years if they consume foods containing it and may result in lead poisoning if toxic levels are reached.

Mercury levels in fish have been reported rather widely. The levels increase as this metal moves through the food chain in the ocean. When fish eat other fish contaminated with mercury, the mercury in the fish that was eaten is retained, thus increasing the mercury content of the victorious fish. Swordfish have been cited as often being high in mercury because they are at the upper end of the food chain and eat other fish that have been concentrating mercury in their bodies.

Cadmium is another metal that is a possible environmental contaminant. It may be found in chocolate, meats, leafy and other vegetables, rice, wheat, and peanuts.

This identification of contaminants is certainly not complete, because many unique circumstances occur during the production and manufacturing of food and food products. However, this does serve as a useful reminder of the importance of careful attention to all aspects affecting the safety of the food supply.

### Natural Toxicants

**Linamarin**
Cyanogenic glycoside in cassava that is toxic to humans; can be removed by careful processing.

**Amygdalin**
Cyanogenic glycoside in apricot and peach pits that can be a source of cyanide poisoning if consumed.

**Solanine**
Poisonous glycoalkaloid that forms in potatoes that have areas of the skin starting to turn green or are sprouting.

**Glucosinolates**
Sulfur-glucose ethers in *Brassica* vegetables that may promote goiter but help prevent colon and rectal cancers.

Some foods naturally contain compounds that are toxic to humans. Cassava needs to be processed very carefully to remove **linamarin**, the cyanogenic glycoside in cassava root and leaves. Usually, cassava root is ground under running water and then drained and squeezed to remove excess water. The final step is drying. Linamarin content must be reduced before gari (the form of cassava commonly eaten in Africa) and tapioca are eaten because this glycoside can be converted to cyanide by the action of an enzyme, and it is the cyanide that is dangerous to humans.

**Amygdalin** is the cyanogenic glycoside that is found in the kernel that is inside the pits of apricots and peaches. The flesh of these fruits does not contain amygdalin so these fruits can be eaten without risk of cyanide poisoning from the amygdalin in the kernels contained in the pits.

Several vegetables are sources of glycoalkaloids that are natural toxicants. Green areas and sprouts in potatoes contain **solanine**, a glycoalkaloid that can cause vomiting, labored breathing, weakness, and even paralysis. *Brassica* vegetables (e.g., Brussels sprouts, cauliflower, broccoli, cabbage, turnips, and kale) contain **glucosinolates**, compounds that inhibit thyroid function and promote goiter but may play a helpful role in prevention of colon and rectal cancers. Table 19.6 presents information about some natural toxicants in foods.

**Table 19.6**  Some Natural Toxicants in Food

| Compound | Food Source | Characteristics |
|---|---|---|
| Unidentified | Fava beans | Hemolysis, vomiting, dizziness, prostration |
| Aflatoxin | Moldy peanuts with *Aspergillus flavus* contamination | Liver damage, chronic consumption may lead to cancer |
| Ergot | Moldy rye | Severe muscle contraction, serious to fatal involvement of nervous system |
| Antitrypsin | Legumes | Blocks protein digestion, inactivated by heat |
| Goitrogen | Cabbage | Blocks thyroxine synthesis |
| Phytin | Cereals | Restricts utilization of calcium and iron |
| Solanine | Sunburned potato | Vomiting and diarrhea |
| Oxalic acid | Spinach, rhubarb | Binds calcium |
| Caffeine | Coffee, tea | Possible teratogen and carcinogen |
| Benzopyrene | Charbroiled meat | Carcinogen |
| Nitrates and nitrites | Some vegetables | Potential carcinogen |
| Gossypol | Cottonseed | Toxic to animals |
| Cyanogens (amygdalin) | Almonds, peach and apricot pits | Headache, heart palpitations, weakness; can be fatal |
| Pressor amines (histamine, tyramine) | Camembert cheese, bananas | Elevated blood pressure resulting from constriction of blood vessels |

## CONTROLLING FOOD SAFETY

In *Food Code 2009*, the five key areas where food safety may be impacted are:

- improper holding temperatures,
- inadequate cooking, such as undercooking raw shell eggs,
- contaminated equipment,
- food from unsafe sources, and
- poor personal hygiene.

These problems need to be addressed at all levels to help reduce the incidence of food-borne illness nationally and around the world. The following discussion identifies responsibility and preventive measures that can be taken all along the food chain.

### Personal Responsibilities

Since food has the potential to become unsafe to eat all along the way from farm to table, avoidance of contaminants, including microorganisms, is a complex matter. Governmental agencies, the food industry (both manufacturers and food service), and consumers need to work together because food safety is achieved only when consumers obtain food that is safe and then continue to maintain adequate standards until it is eaten. Any errors in handling along the route can result in food-borne illnesses, some of which can be fatal.

Throughout production, marketing, and in the home, food must be treated with respect to assure that it is safe to eat. This dictates a need to assure a sanitary environment at all times, including clean floors, counters, sinks, utensils, and dishes. Tasting spoons must not be used more than one time. Even more important is the cleanliness of the food handler: Hands and fingernails must be washed thoroughly with soap and hot water before handling food and any time when one has used the bathroom, sneezed, coughed, or touched either face or hair. Hair should be covered, and clothes need to be clean. Obviously, food handlers need to be in good health and not carriers of illnesses.

Temperature control also is vital to the quality and safety of many foods. Foods from animal sources are particularly susceptible to spoilage because they provide an ideal environment for many microorganisms to thrive. Particular attention needs to be given to

> ## SUSHI AND SASHIMI— Gambling with Safety?
>
> Is that tempting bite of sushi or sashimi going to fill you with regret later? These popular appetizers have climbed to the list of favorite foods, yet they also carry an unspoken element of risk. They contain uncooked ingredients, and *Salmonella, Staphylococcus aureus,* and *Vibrio parahaemolyticus,* and sometimes other microorganisms, find raw fish as tempting as diners do. The slogan for anybody involved in handling ingredients and serving sashimi and sushi needs to be "Cool, clean, and quick." The fish have to be very fresh and clean; they must be kept cold and served as promptly as possible to minimize the risk of food-borne illness. This means that all aspects of the process from the time they are caught until they are consumed must be committed to maintaining a very clean environment at 4°C (40°F) or cooler and with quick preparation and service. Acidifying the rice to pH 4.8 or lower is a further deterrent to growth of micro-organisms. Sushi bars that maintain these standards are able to serve safe sushi and sashimi. Beware of venues that do not.

assuring that meat, fish, poultry, and dairy products are kept at temperatures below 4°C (41°F) or above 57°C (135°F). Microorganisms grow much more slowly when they are either below or above this danger zone than when they are at temperatures between 4 and 57°C (41 and 135°F).

## Federal Regulators

Authority to regulate the food supply and to provide correct information on food handling to consumers is spread across three departments at the federal level: U.S. Department of Health and Human Services, U.S. Department of Agriculture, and U.S. Department of Commerce. In some cases, as in the development of MyPyramid, departments and/or agencies have worked together relatively effectively. In many other cases, they have worked either in isolation or even against each other. Despite internal problems, the nation's food supply clearly is generally safe. It is the occasional episode that soars into the headlines. Nevertheless, there is need for continued vigilance to avoid possible problems through proactive measures. The Department of Homeland Security interacts with these agencies where food safety issues are related to defense and emergencies.

The Food and Drug Administration in the Department of Health and Human Services has particularly heavy responsibilities for a safe food supply. Labeling requirements for the majority of foods, regulation of food additives, and inspection of many food plants are under the jurisdiction of the FDA. Programs focusing on consumer education about food safety also are found in its domain. The Centers for Disease Control is another powerful agency in the Department of Health and Human Services.

The National Marine Fisheries Service is housed in the Department of Commerce. Inspection and safety of seafood and freshwater fish are the responsibilities of this agency.

Inspection of meat and poultry are the domain of the Food Safety and Inspection Service (FSIS) of the Department of Agriculture. Other agencies in the same department include the Agricultural Marketing Service and the Agricultural Research Service.

Although microorganisms and their control are a primary interest of these agencies, their attention extends to other areas that impact the food supply. For example, pesticide and chemical residues and environmental contaminants need to be monitored and eliminated or significantly reduced if problems exist. New products in the marketplace often mean new uses of approved additives or petitions to use new substances in formulating and packaging foods. Indeed, the area of packaging is extremely dynamic and represents the potential for significant modifications in the food supply. Developments in biotechnology and also in food processing are other areas requiring surveillance.

This plethora of agencies may seem somewhat overwhelming if specific information is needed. Table 19.7 is a guide to contacting some of the agencies.

**Table 19.7** Some Sources of Food Safety Information

| Agency | Phone |
|---|---|
| Food Safety and Inspection Service (www.fsis.usda.gov/) | (800) 256–7072 |
| Meat and Poultry Hot Line (FSIS) www.fsis.usda.gov/ . . . meat_&_poultry_hotline | (888) 674–6854 |
| Center for Food Safety and Applied Nutrition (www.fda.gov/aboutfda/centersoffices/cfsan/default.htm) | (888) 723–3366 |
| U.S. Fish and Wildlife Service | (800) 344–9453 |
| Meat and Poultry Hot Line (USDA) | (800) 535–4555 |
| Food Information and Seafood Hotline (FDA) | (800) 332–4010 |

## Federal and Industrial Cooperative Efforts

Much of the current focus on safety in the food supply chain is on a program often referred to simply as **HACCP**. This acronym translates to the **Hazard Analysis and Critical Control Point System**. Although publicity has been given to HACCP only recently, the system actually was developed in the 1960s through the joint efforts of The Pillsbury Co., the National Aeronautics and Space Administration, and the U.S. Army Natick Research and Development Laboratories.

The following seven principles of HACCP have evolved over time:

1. Conduct hazard analysis and risk assessment.
2. Determine critical control points (CCP).
3. Establish critical limits.
4. Establish monitoring procedures.
5. Establish corrective actions.
6. Establish verification procedures.
7. Establish record-keeping and documentation procedures.

Development of a HACCP plan for a food company or other food handlers involves careful study and documentation. Requirements mandated by the *Food Code* must include flow diagrams, product formulations, training plans, and a corrective action plan.

This system is implemented in many food and food-service companies throughout the nation. Each company needs to do a detailed assessment and then identify the points where control is critical. Workable plans for maintaining high-quality food products from start to finish are essential to making HACCP work. The entire workforce in a food company needs to be aware of its HACCP system and to be conscientious about conformance with the plan as it impacts the responsibilities of individual employees. Continual vigilance is needed to assure that HACCP is working effectively within each company employing this system.

HACCP also is being used by federal agencies involved in monitoring the food supply. The Food Safety and Inspection Service and the Food and Drug Administration are basing inspection plans on the HACCP principles. Although development and implementation of regulations are underway, use of HACCP by federal agencies is still in an evolutionary phase. However, HACCP does seem to be a foundation on which future inspection will be based.

## International Food Safety Control Efforts

In today's global economy, food safety is an international issue, which has led to the establishment of **ISO 9000** quality standards. These standards are under the **International Organization of Standardization (ISO)**, which is based in Geneva, Switzerland, and is a federation of the standards boards of 91 countries. The European Community (EC) has given

**Hazard Analysis and Critical Control Point System (HACCP)** System for analyzing, monitoring, and controlling food safety during production.

**ISO 9000** Overall document of standards for food quality, established by the ISO.

**International Organization of Standardization (ISO)** International federation of standards boards from 91 participating countries.

considerable impetus toward developing and using standards for food that is to pass between and into other countries. The United States is gradually moving into involvement with ISO 9000 quality standards, a move that doubtless will grow stronger in future years.

Actually, ISO 9000 quality standards are not a single document, but actually five standards: ISO 9000, ISO 9001, ISO 9002, ISO 9003, and ISO 9004. ISO 9000 provides an overview of the rest of the documents in the series to direct users to the pertinent one(s) for specific users. Companies that do product development and manufacturing of products use ISO 9001. Companies doing only manufacturing use ISO 9002; companies supplying commodities use ISO 9003. ISO 9004 deals with quality management and quality system elements.

Registration is done by a registering organization in each of the participating countries, so registration can mean differences in quality in some countries versus others. The standards that are easiest for a company to meet are in ISO 9003, but this registration is only valid for commodity suppliers. The most comprehensive and the most stringent standards are those in ISO 9001.

## SOME COMMERCIAL APPROACHES TO EXTEND SHELF LIFE

### Use of Antimicrobial Agents

Spoilage of foods has concerned people for centuries, and such familiar additives as sugar, salt, vinegar, and smoke from various woods were used long ago and still are used to extend shelf life and reduce food losses due to spoilage. Some spices have antibacterial properties that help slow growth of potentially harmful microorganisms. Black cumin, garlic, cloves, basil, black thyme, cinnamon, mustard, oregano, and rosemary are particularly protective spices and herbs, and many others also have some effect.

The palette of additives (particularly acids) has been expanded greatly by food scientists. Commercial food products often contain not only vinegar but also many other organic acids and their derivatives. Particularly prominent are benzoic and sorbic acids and their related compounds. Citric acid also is used in numerous products. Carbon dioxide, ethylene, and propylene oxide are examples of gases that can be used to extend shelf life. These are but a few of the additives that may be used in various foods to help maintain food quality during the marketing process. Chapter 21 identifies many others.

Another approach to food safety is the use of antimicrobial rinses for poultry during processing. This is one means of reducing the microbial contamination prior to marketing.

### Energy

Pulsed light treatment of food includes intense flashes of broad-spectrum "white" light that are capable of killing microorganisms. The pulsed light, which is about 20,000 times the intensity of sunlight, uses long wavelengths that do not cause ionization of the small molecules that may be in a food. Tests have shown pulsed light treatment to be an effective means of killing dangerous microorganisms such as *E. coli 0157:H7* and *L. monocytogenes*. Bread treated with pulsed light was still free of mold after 11 days of storage at room temperature. Pulsed light treatment of shrimp makes it possible for shrimp to remain edible after seven days of refrigerated storage. Request for FDA approval for pulsed light treatment was initiated in 1994.

Irradiation is another means of extending the shelf life of foods (see Chapter 6) by killing microorganisms. The FDA has approved exposure of some foodstuffs to gamma rays at specific dosages. Wheat and spices can be irradiated to eliminate insect problems. Potatoes and fresh fruits will retain their fresh and desirable qualities in the market for significantly longer times than they would without irradiation. *Salmonella, Campylobacter, Listeria,* and other bacteria can be destroyed in poultry and meats. However, it is important to recognize that irradiation only kills microorganisms that were present at the time of treatment. Contamination that occurs subsequently can mean that poultry or meats may be

**Table 19.8**   Foods Approved for Irradiation in the United States

| Food | Dosage (kGy) | Approved Use |
|---|---|---|
| Spices, dry vegetable seasoning | 30 | Decontaminate, control insects and microorganisms |
| Dry or dehydrated enzyme preparations | 10 | Control insects and microorganisms |
| Beef, lamb, pork | 4.5 fresh, 7 frozen | Control spoilage and disease-causing microorganisms |
| Poultry | 3 | Control disease-causing microorganisms |
| All foods | 1 | Control insects |
| Fresh foods | 1 | Delay maturation |
| Wheat, wheat flour | 0.2–0.5 | Control insect infestation |
| White potatoes | 0.05–0.15 | Sprout inhibition |

Adapted from Henkel, J. 1998. Irradiation: Safe measure for safer food. *FDA Consumer 32* (3): 16.

carrying viable microorganisms at the time of purchase and preparation. Refrigerated storage still is required.

Foods that have been irradiated must be identified with the international logo that was first used in the Netherlands (Figure 19.7). The FDA has required that irradiated foods be labeled since 1966. Internationally, the Committee on the Wholesomeness of Irradiated Food has been studying food irradiation issues for many years and has deemed that irradiation of any food up to a level of 10 kiloGrays produces no toxicological hazard. This level is far greater than the 3 kiloGrays the FDA has approved for irradiating poultry. Almost 40 nations now utilize irradiation as a means of making the food supply safer and/or of good quality. This technology has found limited use in the United States to date, largely because of some campaigning on the part of some consumer activists.

**Figure 19.7**   Logo (called radura) used internationally and required in the United States to indicate that food has been irradiated.

The increasing focus on the potential hazards associated with ground meats and poultry, as well as some other foods identified as sources in outbreaks of food-borne illnesses, has been altering public perceptions regarding safeguarding the food we eat. This shift in the late 1990s is helping modify consumer attitudes toward the irradiation of food. Table 19.8 presents a list of the foods approved for irradiation by 1998. Others doubtless will follow fairly quickly. However, approval does not mean that such foods will swiftly appear in the markets as irradiated products. The demonstrated safety of irradiated foods (see Chapter 6) and the potential savings because of extended shelf life are two compelling reasons to promote the entry of more irradiated foods into the marketplace.

## SUMMARY

Food-borne illnesses are a fact of life—and sometimes death. The causes are varied and include bacteria and their toxins, viruses, molds, algae, and parasites. Among the bacteria responsible for food-borne illnesses are *Clostridium botulinum, Staphylococcus aureus, Clostridium perfringens, Salmonella enteritidis, Salmonella typhi, Escherichia coli, Campylobacter jejuni, Listeria monocytogenes, Shigella boydii, Vibrio cholerae,* and *Morganella morganii.* Hepatitis A virus (HAV) is carried in food and can cause hepatitis A. The aflatoxin formed by *Aspergillus flavus* and toxins from other molds can cause serious illness and even death when ingested. The microalgae such as *Gonyaulax catanella* causing red tide produce toxins that are concentrated in shellfish, causing paralytic shellfish poisoning if humans eat the shellfish. Parasites include nematodes such as *Trichinella spiralis,* protozoa (e.g., *Giardia lamblia* and *Entamoeba histolytica*), cestodes or tapeworms (e.g., *Taenia saginata*), and flukes or flatworms.

Control of food-borne illnesses focuses on temperature controls to avoid the danger zone between 4 and 57°C (41 and 135°F) and careful sanitation at all times (from growing the food and continuing throughout marketing and preparation) to be sure that neither water nor food is contaminated with fecal material or other possible hazards. Awareness and control of contaminants and natural toxicants in foods also are factors in food safety.

The Food and Drug Administration in the Department of Health and Human Services, the Department of Agriculture, and the National Marine Fisheries Service in the Department of Commerce all have significant roles in assuring that the nation's food supply is safe when it reaches consumers. These agencies, as well as various food companies and organizations, provide food safety information to consumers. Government agencies also work directly with the food industry to help assure a safe food supply. Hazard Analysis and Critical Control Points (HACCP) is the system being implemented by the government and by many food companies to help assure food safety and quality.

The International Organization of Standardization has developed ISO 9000–ISO 9004 quality standards to help assure a high quality and safe food supply internationally.

Various commercial approaches to help extend the shelf life of fresh food have been developed. Antimicrobial agents in food rinses or as additives are one technique. Gases in packaging also afford longer shelf life for some products. Pulsed light treatment kills some microorganisms to lengthen the normal shelf life of selected products. Irradiation is an even more powerful approach to the use of energy to extend shelf life. Gamma rays can be used to kill microorganisms and insect infestations in various food products.

## STUDY QUESTIONS

1. Have you ever had a food-borne illness? Carefully describe the symptoms you had. If possible, try to identify the food that caused the problem and explain what circumstances caused the microorganisms to be at dangerous levels. What microorganism do you think caused your illness?

2. What are some important control points in your laboratory work to help avoid food-borne illnesses stemming from your food products?

3. Identify a question that you have regarding safe handling of a specific food. Decide what agency might be able to answer your question. Call or e-mail the appropriate agency and obtain the answer. State your question, the agency you contacted, and the way in which your query was handled, and state the answer that was given.

4. What steps would you take to find information to answer a question on food safety that a consumer might ask you?

5. Why is ISO 9000 important to you?

6. True or false. It is the toxin of *Clostridium botulinum* that is toxic to humans.

7. True or false. *Shigella* spp. is the microorganism sometimes found in eggs.

8. True or false. One of the common causes of food-borne illness is *Escherichia coli*.

9. True or false. *Listeria monocytogenes* in pasteurized milk is a health hazard in the United States.

10. True or false. The likely cause of illness from eating shellfish that were caught during a warning period is *Campylobacter jejuni*.

11. True or false. Pork needs to be heated enough to kill a bacterium that pigs may acquire from eating garbage.

12. True or false. HACCP is a professional trade organization for bacteriologists.

13. True or false. *Food Code 2009* contains guidelines for HACCP plans.

14. True or false. Consumers need to monitor their own food-handling practices.

15. True or false. Food manufacturing companies are not required to meet HACCP regulations.

## BIBLIOGRAPHY

Bernard, D. T. and Scott, V. N. 1999. *Listeria monocytogenes* in meats: New strategies are needed. *Food Technol. 53* (3): 124.

Cassens, R. G. 1995. Use of sodium nitrite in cured meats today. *Food Technol. 49* (7): 72.

Dunn, J., Ott, T., and Clark, W. 1995. Pulse-light treatment of food and packaging. *Food Technol. 49* (9): 95.

Durant, D. 1999. *Listeria*: Public health strategies. *Food Safety Educator 4* (1): 1.

Duxbury, D. 2004. Keeping tabs on *Listeria*. *Food Technol. 58* (7): 74.

Floyd, B. 2005. Increasing Listeria awareness. *Food Product Design 15* (2): 93.

Giese, J. 2004. Testing for BSE. *Food Technol. 58* (3): 58.

Han, J. H. 2000. Antimicrobial food packaging. *Food Technol. 54* (3): 56.

Hema, R., et al. 2009. Antimicrobial activity of some South-Indian spices and herbals against pathogens. *Global J. of Pharmacology 3* (1): 38.

Hueston, W. and Bryant. C. M. 2005. Understanding BSE and related diseases. *Food Technol. 59* (7): 46.

Joe, M. M., et al. 2009. Antimicrobial activity of some common spices against certain human pathogens. *J. Medicinal Plants Research 3* (11): 1134.

Keck, A. S. and Finley, J. W. 2004. Cruciferous vegetables: Cancer protective mechanisms of glucosinolate hydrolysis products and selenium. *Integrative Cancer Therapies 3* (1): 5.

Klapthor, J. N. 2004. Wall-to-wall mad cow coverage. *Food Technol. 58* (2): 91.

Kolettis, H. 2004. It's a mad, mad world. *Food Product Design 13* (11): 21.

Krashen, S. 2009. Are apricot kernels toxic? *The Internet Journal of Health 9* (2).

Kurtzwell, P. 1998. Safer eggs: Laying the groundwork. *FDA Consumer 3* (5): 10.

Langen, S., 2004. Conference focuses on ensuring safety of food supply. *Food Technol.* 58 (2): 34.

Lewis, C. 1998. Critical controls for juice safety. *FDA Consumer 32* (5): 16.

Loaharanu, P. 1994. Cost/benefit aspects of food irradiation. *Food Technol. 48* (1): 104.

Looney, J. W., et al. 2001. The matrix of food safety regulations. *Food Technol. 55* (4): 60.

McEntire, J. C. 2004. IFT issues update on foodborne pathogens. *Food Technol. 58* (7): 20

Murphy, P. A., et al. 2006. Food mycotoxins: An update. *J. Food Science 71:* R51.

Newsome, R. 2006. Understanding Mycotoxins. *Food Technol. 60* (6): 50.

Olson, D. G. 2004. Food irradiation future still bright. *Food Technol. 58* (7): 112.

Pommerville, J. D. 2004. *Alcamo's Fundamentals of Microbiology.* Jones & Bartlett. Sudbury, MA.

Pszczola, D. E. 2002. Antimicrobials: Setting up additional hurdles to ensure food safety. *Food Technol. 56* (6): 99–107.

Ray, B. 2001. *Fundamental Food Microbiology.* 2nd ed. CRC Press. Boca Raton, FL.

Schauwecker, A. 2005. To preserve and protect. *Food Product Design 15* (3): 67.

Schildhouse, J. 2002. U.S. Food supply reported safer. *Food Product Design 12* (3): 21.

Smith, M. A. 1998. Minimizing microbial hazards for fresh produce. *Food Technol. 52* (12): 140.

Swientek, B. 2006. Global challenges to food safety. *Food Technol. 60* (5): 123.

Tainter, D. R. and Grenis, A. T. 1993. *Spices and Seasonings.* 2nd ed. John Wiley, and Sons. New York.

U.S. Department of Health and Human Services. 2009. *Food Code 2009.* Public Health Service, Food and Drug Administration. Washington, D.C.

Williams, V. S. 2003. Safety first. *Food Product Design 13* (November Supplement): 51.

## INTO THE WEB

*http://www.foodpoisonjournal.com/tags/spinach/*—Report on the outbreak of food-borne illness in 2006 caused by baby spinach contaminated with *E. coli 0157:H7*

*http://www.fws.gov/laws/lawsdigest/FWATRPO.HTML*—Overview of federal Clean Water Act and amendments.

*http://nationalhogfarmer.com/mag/farming_waterkeeper_lawsuits_target/*—Review of a lawsuit regarding pollution from large hog farms.

*http://www.bt.cdc.gov/agent/botulism/*—CDC information on botulism.

*http://www.cdc.gov/ncidod/dbmd/diseaseinfo/salment_g.htm*—CDC site for information on *Salmonella enteritidis.*

*http://www.cdc.gov/ncidod/dbmd/diseaseinfo/TyphoidFever_g.htm*—CDC site on typhoid fever.

*http://www.cdc.gov/salmonella/*—CDC information on *Salmonella.*

*http://www.fda.gov/AnimalVeterinary/Products/Approved AnimalDrugProducts/FOIADrugSummaries/ucm117 130.htm*—Information about Preempt™.

*http://www.cdc.gov/ecoli/*—CDC site on *E. coli.*

*http://www.fda.gov/food/foodsafety/foodborneillness/food borneillnessfoodbornepathogensnaturaltoxins/badbug book/ucm071284.htm*—FDA information on *E. coli 0157:H7.*

*http://www.ext.colostate.edu/safefood/newsltr/v10n3s01. html*—Discussion on sushi and safety issues.

*http://www.fda.gov/Food/FoodSafety/FoodborneIllness/ FoodborneIllnessFoodbornePathogensNaturalToxins/ BadBugBook/ucm070064.htm*—Bad Bug Book information on listeriosis.

*http://www.textbookofbacteriology.net/Shigella.html*—Information on *Shigella.*

*http://www.fda.gov/Food/FoodSafety/FoodborneIllness/ FoodborneIllnessFoodbornePathogensNaturalToxins/ BadBugBook/ucm070563.htm*—Bad Bug Book information on *Shigella*.

*http://www.cdc.gov/ncidod/diseases/submenus/sub_cholera.htm*—CDC information on cholera and *V. cholera*.

*http://www.fda.gov/Food/FoodSafety/FoodborneIllness/ FoodborneIllnessFoodbornePathogensNaturalToxins/ BadBugBook/ucm070823.htm*—FDA's Bad Bug Book site on scombrotoxin poisoning.

*http://www.cdc.gov/ncidod/DBMD/diseaseinfo/yersinia_g. htm*—CDC site about *Yersinia*.

*http://www.cdc.gov/hepatitis/*—CDC site about hepatitis.

*http://www.cdc.gov/mmwr/preview/mmwrhtml/rr5009a1. htm*—CDC report on Norwalk-like viruses (NLV).

*http://www.fsis.usda.gov/FactSheets/Molds_On_Food/*—USDA site discussing molds and food.

*http://www.whoi.edu/HABs/*—Information on red tide, causes, locations, and hazards.

*http://www.cdc.gov/ncidod/dpd/*—CDC site on parasites.

*http://www.dpd.cdc.gov/DPDx/HTML/Para_Health.htm*—CDC site on parasites.

*http://www.trichinella.org/index.htm*—Information on *Trichinella spiralis*.

*http://www.dpd.cdc.gov/dpdx/HTML/Trichinellosis.htm*—CDC site about trichinellosis.

*http://www.cdc.gov/nceh/ciguatera/*—CDC site on ciguatera poisoning.

*http://www.nlm.nih.gov/medlineplus/ency/article/000628 .htm*—NIH site on *Ascaris lumbricoides*.

*http://www.cdc.gov/ncidod/dpd/parasites/Giardiasis/factsht_ giardia.htm*—CDC information on *Giardia lamblia*.

*http://www.nlm.nih.gov/medlineplus/ency/article/000298. htm*—NIH information on amebic dysentery.

*http://pathmicro.med.sc.edu/parasitology/cestodes.htm*—Information on cestodes.

*http://pathmicro.med.sc.edu/parasitology/trematodes.htm*—Information on flukes.

*http://news.newamericamedia.org/news/view_article.html? article_id=d0376ccd43a9ab3de200bce79e959714*—Report of lead in some spices (e.g., turmeric and chili) from Nepal, Bangladesh, and India.

*http://www.google.com/hostednews/ap/article/ALeqM5b4 BegxZT-6P4D6nC-uoBqbZijaxQD9EOBCTG0*—News report on 2010 legal settlement for mercury poisoning in Japan in 1966.

*http://www.scielo.br/scielo.php?script=sci_arttext&pid= S0104-79301996000100002*—Article about linamarin.

*http://www.ispub.com/journal/the_internet_journal_of_ health/volume_9_number_2_13/article/are-apricot-kernels-toxic.html*—Paper on toxicity of apricot kernels.

*http://www.news-medical.net/news/20100129/Nectar-of-almond-tree-produces-amygdalin.aspx*—Report on amygdalin in nectar from apricot blossoms.

*http://www.inchem.org/documents/jecfa/jecmono/v30je19. htm*—Review of numerous papers on effects of solanine toxicity in many species.

*http://www.chiroonline.net/_fileCabinet/cruciferous_ vegetables.pdf*—Paper on cruciferous vegetables and protection against cancer.

*www.cfs.gov.hk/english/programme/programme_haccp/ . . ./sushi.ppt*—PowerPoint of sushi safety program in Hong Kong.

*www.capitalhealth.ca/ . . ./guidelinesforthepreparationo fsushiproducts.pdf*—Guidelines for making safe sushi.

*http://www.sushifaq.com/sushi-health-risks.htm*—Extensive site on sushi and sashimi.

*http://www.marlerclark.com/press_releases/view/dangers-of-eating-sushi-emphasized-after-e-coli-outbreak-at-reno-restaurant*—News release on cases on illness from eating sushi.

*http://www.dhs.gov/xoig/assets/mgmtrpts/OIG_07-33_Feb07. pdf*—Role of Department of Homeland Security in protecting the food supply.

*http://www.fda.gov/Food/FoodSafety/RetailFoodProtection/ FoodCode/FoodCode2009/*—U.S. Public Health Service 2009 Food Code guidelines governing food services and stores.

*http://www.fao.org/docrep/005/Y1390E/y1390e0a.htm*—Overview of developing a HACCP plan.

*http://www.iso.org/iso/home.html*—Site of International Organization of Standardization.

*http://www.iso.org/iso/iso_catalogue/management_ standards/iso_9000_iso_14000.htm*—Site for information on ISO 9000 standards.

*http://www.mccormickscienceinstitute.com/content.cfm? ID=10503*—Discussion of possible role of spices in food packaging to reduce health risks.

*http://www.news.cornell.edu/releases/July01/veggie_spice. hrs.html*—Article about historical use of spices in various recipes and cultures.

*http://www.foodtechsource.com/emag/008/trend.htm*—Interview of various energy treatments for making food safe.

*http://www.extension.iastate.edu/foodsafety/irradiation/*—Overview of food irradiation.

*http://www.physics.isu.edu/radinf/food.htm*—Overview of food irradiation processes and answers to consumer questions.

*http://ec.europa.eu/food/fs/sc/scf/out193_en.pdf*—2003 report of the Committee on the Wholesomeness of Irradiated Food.

Norway's fishing industry has long relied on large racks for drying its catches.

# CHAPTER 20

# Food Preservation

## Chapter Outline

## OBJECTIVES

After studying this chapter, you will be able to:

1. Discuss why each of the methods used to preserve foods is effective in keeping food safe for later consumption.

2. Explain the methods used in canning and why specific foods are suited to a particular method.

3. Describe the techniques used in drying foods.

4. Define the essential ingredients and their roles in preserving foods with sugar.

5. Outline the preservation methods used in the food industry and the advantages and disadvantages of each.

The dynamic nature of food during storage and the need for a food supply throughout the year require careful management so that the food available for consumption is safe to eat. Emphasis in food preservation is on safety—the ultimate concern. Certainly quality is important, but this facet of preserved foods is secondary to safety. Unsafe foods generate health risks ranging from simple gastrointestinal upset to even death (see Chapter 19).

## METHODS OF PRESERVATION (HOME AND COMMERCIAL)

Several techniques are available for either retarding growth of microorganisms or actually killing bacteria, yeast, and molds that may be present in food to be preserved. Canning sterilizes food by heating it to a high enough temperature to kill all of the microorganisms and to seal that environment against subsequent invasion by microorganisms during storage. Pasteurization is a more moderate heat treatment that kills harmful microorganisms but does not permit long-term storage because viable microorganisms remain and cause spoilage.

Freezing kills some microorganisms and slows the growth of others; nevertheless, frozen foods have finite storage periods because of enzymatic and microorganism actions. This method of preservation is particularly popular today in the home and commercially.

Drying dehydrates microorganisms, as well as the food containing them. At moisture levels below 13 percent, food can be stored at room temperature for extended periods.

Salting is a preservation method that sometimes is used for fish and meats. The heavy concentration of salt draws moisture out of the microorganisms by osmotic pressure, killing the microorganisms and making the food safe when stored at room temperature.

Addition of sugar to fruits (jams and jellies) is a related technique for preservation, for sugar similarly creates unfavorable osmotic pressure to kill microorganisms.

Pickling, which is associated with low pH because acid is used, also provides a hostile environment for microorganisms; this preservation technique usually is combined with heat treatment (canning) to ensure absolute safety.

Irradiation, high pressure, and pulsed electrical fields are additional ways commercial food companies can preserve food. These methods require expensive equipment, but are being used increasingly to help meet the challenges of feeding the world's growing population safely.

---

### FOOD FOR THOUGHT: Kimchi in Today's World

For centuries, Koreans have been fermenting and salting seasonal vegetables (notably cabbage, with other vegetables added for variety in color and flavor) in large storage jars to make kimchi (Figure 20.1). Originally, this technique was developed to make it possible to have vegetables throughout the year until the next crops were available. The salt draws some of the water from the vegetable cells by osmosis and also inhibits growth of some microorganisms. *Leuconostoc mesenteroides* and *Lactobacillus plantarum* are two of the principal bacteria, but several others are also involved in the fermentation and production of carbon dioxide. Sufficient acidity develops as the mixture stands and microorganisms ferment the vegetable mixture.

Koreans eat this traditional dish as often as three times a day 365 days each year. Families used to make and store enough kimchi to feed themselves for many months. Excellent kimchi no longer needs to be made at home. It can be purchased canned or in refrigerated packages, albeit with a shorter shelf life than canned kimchi. The heat required in canning makes it possible to store canned kimchi safely without the microbiological hazards that could develop during extended storage. Nevertheless, it is possible to alter acidity and salt levels safely when making variations of kimchi that will be marketed in refrigerated packages and stored chilled until the date indicated.

Kimchi even is being traveled in outer space when the first Korean astronaut flew on a space mission in 2008. Astronauts find food aromas and flavors generally are not stimulating in space because of lack of air movement. Kimchi, which is noted for its strong and distinctive aroma and flavor, may prove to be a favorite for Russians and American astronauts, too.

Requirements for foods in space are more challenging, but the Korean Atomic Energy Research Institute and CJ Corporation, a Korean company, worked together to develop kimchi for astronauts. Irradiation killed all microorganisms in the kimchi and avoided any possible growth in space during the trip. Then, all gas was evacuated from the package, leaving only kimchi vegetables and their juice. A special package developed by CJ kept juice from squirting out when opened in space.

**Figure 20.1**   Since early times, Koreans have been preserving vegetables by pickling and then storing them outdoors in huge pots.

## FREEZING

Most foods can be held satisfactorily in short-term storage either at room temperature or refrigerated. Foods that are to be stored for more than a few days before they are eaten often require special treatment to control the microorganisms with which they may be contaminated. The treatments commonly available commercially and in the home are freezing, canning, and drying. Two other methods—freeze-drying and irradiation—are used to a limited extent commercially today; irradiation shows some likelihood of increasing in importance in the future (see Chapter 19).

Freezing is a popular method of preserving many foods because of its relative convenience and the quality of foods after freezing. Freezing and frozen storage are effective in permitting safe food storage with only limited quality loss for up to six months. Even at 12 months, there are somewhat more deteriorative changes in quality than in safety. Freezing retards growth of microorganisms, killing some and slowing reproduction of others; the extent of control depends on the temperature maintained during frozen storage.

The textural characteristics of frozen foods may be altered to varying degrees, depending on the food. In some cases, such as strawberries, the changes are quite detrimental; in other products, such as bread, little difference is noted.

Foods to be frozen may be categorized into four classes:

- Perishable raw foods (meats, poultry, fish, fruits and fruit juices, a few vegetables, and fluid eggs)
- Perishable heated or cooked foods (blanched vegetables, TV dinners, pies, cooked deboned meats, and some baked items)
- Semi-perishable foods (breads, partially baked breads, unheated dough products, cheese, butter)
- Nonperishable foods (nuts)

Preparation for freezing differs with the class.

### Produce

Perishable raw foods require limited preparation for freezing (Table 20.1). For fruits, the main action is washing, after which they can be frozen satisfactorily without heat treatment. Washing reduces the microbiological count prior to freezing. Grape juice needs to be

**Table 20.1**   Preparation for Freezing Some Foods

| Food | Preparation | Blanching | Other Treatment |
|------|-------------|-----------|-----------------|
| Asparagus | Wash, trim | 2–4 minutes | |
| Broccoli | Wash, trim | 3 minutes | |
| Cauliflower | Wash, trim in 1″ pieces | 3 minutes | |
| Green beans | Wash, trim | 3 minutes | |
| Peas | Shell, wash | 1/2 minute | |
| Sweet corn | Husk, cut from cob after blanching | 4 minutes | |
| Apples | Wash, peel, core, slice | No | Sugar, vitamin C |
| Cherries | Wash, stem, pit | No | Sugar, vitamin C |
| Peaches | Wash, peel, slice | No | Sugar, vitamin C |
| Strawberries | Wash, remove stem | No | Sugar |
| Whipping cream | Whip before freezing | | Drop in dollops |

pasteurized at 60°C (140°F) to kill yeasts and molds that enter the juice when it is extracted from the whole grapes, which have these microorganisms on their skins. Rigid sanitation measures are necessary in the commercial production of frozen orange juice to avoid formation of diacetyl and organic acids by a variety of microorganisms (*Lactobacillus, Leuconostoc, Achromobacter, Enterobacter,* and *Xanthomonas*) that can flourish in the equipment in which the juice is processed.

Vegetables are blanched before being frozen to halt the detrimental action of lipoxidase, peroxidases, and catalase. The heat of blanching also is effective in killing many of the microorganisms that are present even after washing. Blanching also sets chlorophyll so that the bright green color remains in the frozen vegetable. Greens that are blanched must be dispersed vigorously to ensure that heat penetrates the greens; clumped greens would impede heat distribution.

Unfortunately, bacterial recontamination occurs very quickly after blanching when vegetables are frozen commercially. Rapid cooling in large quantities of cold water in an air-conditioned facility helps keep the temperature low enough to retard bacterial growth while the vegetables are packaged and frozen. The net result of commercial efforts in processing frozen vegetables has been a great reduction in bacterial counts, and frozen vegetables are not likely causes of food-borne illnesses. Vegetables frozen in the home are processed in such small quantities and frozen so promptly that they do not present a hazard.

Freezing kills most bacteria, and some of those that survive die after frozen storage. However, some freeze-resistant bacteria remain viable even though their reproduction is retarded greatly. As many as 60 percent of the bacteria in a food are killed by freezing; still more die during storage. Even so, there is good reason to maintain excellent standards of sanitation when preparing any food for freezing.

Although most vegetables are blanched in boiling, salted water to retain chlorophyll and halt enzymatic action, fruits usually are not blanched in preparation for freezing because of the change in flavor and soft texture generated by the blanching process. Enzymatic browning does present problems in fruits such as peaches and apples during frozen storage. To halt this action, ascorbic acid (sometimes mixed with citric acid) or a coating of sugar or sugar syrup may be used to prevent oxidation and browning.

Dry sugar causes fruits to become limp because it creates unfavorable osmotic pressure that draws juice from the fruit, collapsing many of the cells. Sugar syrup, because of its somewhat lower concentration of sugar, is less detrimental to texture. The texture of frozen strawberries can be improved if the berries are frozen whole and then coated with a 60 percent sugar solution.

Ice crystals that form during freezing rupture the comparatively delicate structure of fruits, and the juice tends to drain out, leaving a rather flabby texture. This textural problem is reduced if frozen fruits are served while they still have a few ice crystals in them.

**FOOD FOR THOUGHT: Current Crisis**

In the summer of 2003, electrical power suddenly stopped flowing throughout a huge region in the Northeast because of a power grid failure. Severe summer storms and heavy ice and snowstorms also cause power outages, although not usually for such a large area. Unfortunately, refrigerators and freezers without power present potential problems to anybody affected by an electrical outage, regardless of the cause.

Frozen foods need to be stored at a temperature of −18°C (0°F) or colder, and protein-containing moist foods such as meats, poultry, fish, dairy products, and eggs must be kept at temperatures below 4°C (40°F) for safety. Without power, freezers and refrigerators are unable to maintain their expected temperatures. If the doors are not opened, refrigerators can be expected to maintain food safely for 4 hours, and freezers usually can keep food safely for one to two days. If it is available, dry ice can be placed in refrigerators and freezers to help keep the temperatures low enough. Another possibility in cold winter conditions is to freeze water in milk cartons or other closed containers outside and then place these in the refrigerator or freezer. However, food should not be stored outside because of risk of contamination and the possibility of being too warm at times.

When the power comes back on, immediately check the temperature inside the refrigerator and freezer. If the refrigerator is above 4°C (40°F), the following foods need to be discarded: meats, poultry, seafood, soft cheeses, milk, yogurt, sour cream, eggs and egg-containing foods, sauces, creamy dressings, fresh or cooked pastas, pies and pastries with any filling except fruit, cooked vegetables, and open cans of food.

If frozen foods still have ice crystals in them, it is safe to refreeze them. However, if frozen foods have thawed completely and then been held at 4°C (40°F) or warmer for more than 2 hours, the foods to be discarded are the same as those listed for discard from refrigerators.

## Protein Foods

Meats, poultry, and fish frequently are frozen, but they do not require treatment prior to packaging and freezing (Figure 20.2). The effect of freezing on tenderness of meats has been studied extensively, and the results are contradictory, in some cases indicating increased tenderness, and in other instances decreased tenderness after freezing.

The type of fat appreciably influences flavor changes in meats; ground beef flavor decreases in six weeks; comparable loss of quality can be detected in pork chops after only one to four weeks of frozen storage. The heme pigments in red meats promote the onset of rancidity, particularly in ground meats. The spices used in making sausage and some other highly seasoned ground meat products delay the development of rancidity during frozen storage, whereas sodium chloride promotes rancidity.

Fats that are higher in unsaturated fatty acids undergo oxidative rancidity more rapidly during frozen storage than do fats with more saturated fatty acids. This is demonstrated by the comparatively rapid development of rancidity in fish compared with poultry.

Color and drip loss caused by textural changes also are noted when frozen meats and poultry are thawed. The darkening that sometimes is seen in frozen poultry near the bone is caused by damage to the marrow as a result of freezing and consequent release of some of the hemoglobin. The drip loss from frozen meats during thawing and cooking can result in weight loss and reduced juiciness. This undesirable effect of freezing can be minimized by very rapid freezing so that there is less growth of large ice crystals and hence reduced damage to the cells in the tissues. The amount of drip loss increases gradually as frozen storage is extended, because of the dynamic changes in the size of ice crystals—as small crystals thaw, larger crystals grow and puncture the cell walls.

**Figure 20.2** Food to be frozen needs to be packaged in aluminum foil or plastic that is sealed tightly after expelling as much air as possible so that no areas are exposed and subject to freezer burn.

## Other Foods

Many other products also can be frozen either in the home or commercially. Sugar or salt must be added to whole eggs and egg yolks to avoid the gummy, lumpy character evident when these products are frozen without additives. Whites freeze satisfactorily without additives.

Some baked products are of higher quality if they are frozen after baking than if they are frozen prior to baking (Table 20.2). Breads and quick breads in particular are of better quality when frozen after baking, although yeast dough can be frozen, then shaped and baked after thawing. However, pies are better if they are baked after (placed into a preheated oven directly from the freezer) rather than before frozen storage because the bottom crust is less soggy. Starch-thickened fillings in iced cakes retrograde during frozen storage, but the baked cakes themselves freeze well. Equally good results can be obtained if the batter of shortened cakes is frozen and thawed completely before baking.

Certain foods present unique problems when frozen. Salad dressings and other emulsions tend to break, giving a rather curdled appearance and a watery consistency to foods containing them. This phenomenon makes it unwise to freeze sandwich fillings or other mixtures containing mayonnaise or other emulsified salad dressings unless they are made with an oil that does not crystallize and cause the emulsion to break at freezer temperatures.

**Table 20.2** Freezing of Products

| Food | Bake before Freezing | Bake after Freezing |
|------|----------------------|---------------------|
| Quick breads | Yes | No |
| Yeast breads | Yes | Satisfactory |
| Cookies | Yes | No |
| Shortened cakes | Yes | Thaw completely |
| Icing | On cooled cake | On cooled cake |
| Filling with starch | No | On cooled cake |
| Pies | No | Yes |

Safflower oil has a lower crystallization temperature than the other oils commonly used in salad dressings and does not even crystallize at −7°C (19°F). It can be winterized to a temperature of −12°C (10°F), two degrees below the temperature often used for freezing and frozen storage; however, safflower oil cannot be winterized at −18°C (0°F) because of the limited amount of fluid oil remaining. Peanut oil actually solidifies at a slightly higher temperature than safflower oil, but it can be used satisfactorily if emulsifying agents are added. These ingredients make it possible to create a salad dressing with little tendency to crystallize and break the emulsion during frozen storage. This unique behavior of peanut oil is thought to be due to the gelatinous properties of the oil at freezer temperatures.

Meringues undergo considerable loss of quality during frozen storage. The quality of soft meringues is optimized by a sugar content of at least 46 percent by weight, the amount ordinarily included in their preparation. Frozen baked soufflés have greater volume than soufflés that are baked after frozen storage. Adequate flour in the soufflé (about 8 percent) helps minimize loss of volume during frozen storage.

Starch-thickened mixtures have a strong tendency to curdle during frozen storage as a result of retrogradation of the starch. If the product can be heated and stirred before serving, as can be done with a cream soup, this is not a serious problem. However, this is a definite disadvantage in a chicken pot pie or other item that cannot be stirred.

This problem has been traced to the amylose fraction of starch, which explains why waxy starches (see Chapter 9) are preferable when preparing starch-thickened mixtures for freezing. Of the waxy starches, waxy rice starch is preferred because it can be stored longer than any other waxy starch before retrogradation is noted. Unfortunately, waxy rice starch produces a gel texture similar to the stringy quality characteristic of tapioca starch mixtures. Although pure waxy rice starch has this textural defect, waxy rice flour is effective in delaying retrogradation in frozen storage while also promoting a texture similar to that of wheat flour–thickened items.

The superior storage characteristics of waxy rice starch relative to other waxy starches are attributed to the very extensive amount of branching of the amylopectin constituting this starch and to the very small particle size. Waxy rice flour has a larger particle size than waxy rice starch, and the frozen and thawed sauces made with waxy rice flour will separate when they have been held in frozen storage at −12°C (10°F) for only two months. Surprisingly, the finer particles of the waxy rice starch appear to be responsible for the ability of sauces made with them to be stored for a year or more at −18°C (0°F) without separating on thawing.

Some interesting changes occur in products as they freeze. Freezing does not occur uniformly throughout complex food mixtures. Instead, freezing begins at the edges and proceeds inward toward the center as the mixture chills. Water freezes first, but it actually is **supercooled** to about −8°C (17°F) before crystals start to form. Then, the **latent heat of crystallization** causes the temperature to rise to 0°C (32°F), where it remains until all of the free water is frozen.

At that point, the temperature can be brought well below 0°C (32°F). As water freezes, salts, sugars, proteins, and other components of the food system become more concentrated in the remaining water. The increased concentration of salts and/or sugars lowers the freezing temperature of the remaining solution. The cycle of freezing of water and consequent reduction in freezing temperature of the unfrozen solution continues until all of the water is frozen or until the freezing point drops below the temperature of the environment in which the food is frozen.

This means that items such as frozen orange juice concentrate have a very low freezing temperature and often have a mushy interior, with considerable unfrozen water. The increase in lactose concentration that occurs as ice cream freezes explains why lactose sometimes crystallizes, giving a gritty, granular texture to ice creams lacking stabilizing agents.

In the home, foods are frozen at a slow rate unless only a small amount of food is frozen at one time. The thinner the package of food, the faster the food freezes. Careful attention to the quantity and package dimensions will enhance the quality. Commercially, much larger quantities must be handled.

**Supercooling**
Reduction of the temperature of water below freezing until crystallization begins, after which the temperature rises to 0°C (32°F) because of the latent heat of crystallization.

**Latent heat of crystallization**
Heat released during transition from the liquid state (higher energy) to the frozen state (lower energy).

## Commercial Methods

Creative developments in the frozen food industry have helped speed freezing. **Sharp freezing**, the practice of freezing in still air, used to be the method for freezing foods commercially. This method is quite slow compared with the fast freezing methods currently in use. Today, sharp freezing has been modified. Now **air-blast freezing** forces frigid [−30 to −45°C (−22 to −49°F)] air to circulate at a high velocity in the chamber or tunnel where the food is being frozen. The result is a great acceleration in the rate of freezing and much finer ice crystals in the frozen food.

In addition to **freezing in air**, commercial freezing operations may use indirect-contact freezing techniques or immersion freezing. **Indirect-contact freezing** often is done by placing flat packages of the food on metal shelves that are maintained at the desired freezing temperature by refrigerant circulating through the shelves. In a fast indirect-contact method for liquid, the liquid is forced through a chilled tube and the frozen crystals are constantly removed (scraped off the walls) as they freeze. This intimate contact causes freezing to occur almost immediately.

In **immersion freezing**, either the food itself or packages of food are immersed directly in a refrigerant. Two media may be used—low-freezing-point liquids and **cryogenic liquids**. A 21 percent sodium chloride solution can serve as the immersion liquid, because it remains fluid at −18°C (0°F); however, the food must be packaged to protect it from the extreme saltiness. Other media in which packaged foods can be frozen are glycerol and water, propylene glycol, and, rarely, a 62 percent sugar solution.

The cryogenic liquids are liquid nitrogen, which has a boiling point of −196°C (−321°F), and liquid carbon dioxide, which boils at −79°C (−110°F). Liquid nitrogen frequently is used despite its high cost because it freezes foods at a very rapid rate, resulting in high quality because of small ice crystals. Their fluidity makes it possible for these cryogenic liquids to chill oddly shaped foods quickly. These liquids vaporize while the food freezes and do not remain part of the frozen food. Plate freezers usually can freeze a flat package of food about 1.3 centimeters thick in an hour and reach a temperature of −18°C (0°F) or less, whereas cryogenic liquids can freeze a comparable amount in only a very few minutes.

The final temperature that needs to be reached for successful freezing and then maintained during frozen storage is −18°C (0°F) or below. At this temperature, deteriorative changes do occur, but it represents a satisfactory compromise between quality and cost. Although maintenance of a temperature of −30°C (−22°F) is considerably more costly, the quality remains high for a longer period at this colder temperature than at −18°C (0°F). Temperatures at which frozen foods are stored actually may rise above the recommended level of −18°C (0°F), but the storage life decreases with storage at higher temperatures (Table 20.3).

**Table 20.3**   Approximate Storage Life of Frozen Foods at Three Different Temperatures

| Food | Approximate Storage Life (months) | | |
|---|---|---|---|
| | **−18°C (0°F)** | **−12°C (10°F)** | **−6.7°C (20°F)** |
| Orange juice (heated) | 27 | 10 | 4 |
| Peaches | 12 | <2 | 6 days |
| Strawberries | 12 | 2.4 | 10 days |
| Cauliflower | 12 | 2.4 | 10 days |
| Green beans | 11–12 | 3 | 1 |
| Peas | 11–12 | 3 | 1 |
| Spinach | 6–7 | 3 | 21 days |
| Raw chicken | 27 | $15\frac{1}{2}$ | <8 |
| Fried chicken | <3 | <1 | <18 days |
| Turkey pies | >30 | $9\frac{1}{2}$ | $2\frac{1}{2}$ |
| Beef (raw) | 13–14 | 5 | <2 |
| Pork (raw) | 10 | <4 | $<1\frac{1}{2}$ |
| Lean fish (raw) | 3 | $<2\frac{1}{4}$ | $<1\frac{1}{2}$ |
| Fat fish (raw) | 2 | $1\frac{1}{2}$ | 24 days |

U.S. Department of Agriculture.

## CANNING

Canning is the preservation of food by heat, a method that requires careful control of both time and temperature (Figure 20.3). When food is canned properly, pathogenic microorganisms and their spores are destroyed although some safe microorganisms may still be viable in the sealed container. Canned foods can be stored at room temperature for at least two years and still be safe to eat if they have been processed so that they are **commercially sterile**, the term used to designate that sterilization has killed all microorganisms that are toxic or that form toxins. Unfortunately, some deterioration in flavor, texture, and nutritive content does occur in such long-term storage.

The rate at which bacteria and bacterial spores are destroyed by heat is termed the **logarithmic order of death**, because approximately the same percentage is killed each minute that the food is held at a given temperature. In other words, if the temperature is so hot that 95 percent are killed in the first minute of heating, 95 percent of the remaining viable microorganisms will be killed the next minute, and so on throughout the processing period. This explains why careful timing is so important in canning. A deficiency of even a minute of processing time can make a critical difference in the number of pathogens in a food. It also underlines the importance of beginning with food that has been cleaned carefully so that the original pathogen count is low.

Temperature is another key factor in determining the death rate of microorganisms. The higher the temperature, the more quickly microorganisms die. This is true despite the fact that some microorganisms are much more heat resistant than others, and spores are very resistant. **Thermal death time curves** can be plotted by careful research to determine the rate at which a specific microorganism is destroyed at different temperatures. These thermal death time curves show clearly that the time for heat processing at a high temperature is much shorter than that at a lower temperature.

By processing under pressure to control the processing temperature, the time required to achieve commercial sterility is reasonable—that is, not as many hours are required as would be at atmospheric pressure. This is why pressure canning is the method selected for

**Commercially sterile**
Food that has been heat processed enough to kill all pathogenic microorganisms and spores.

**Logarithmic order of death**
Percentage of bacteria and bacterial spores killed per minute at a constant temperature.

**Thermal death time curve**
Comparison of the rate of death of pathogenic microorganisms over a range of processing temperatures.

**Figure 20.3**    Home canning is a way of preserving fruits and vegetables that are available in abundance seasonally in gardens or farmers' markets.

**Figure 20.4**   Because of their low acidity, vegetables must be canned under pressure following official guidelines for time and pressure at the elevation where the food is being canned to assure safety.

safe canning of vegetables, meats, and other low-acid foods that have the potential for contamination by *Clostridium botulinum* and its heat-resistant spores (Figure 20.4).

Thermal death time curves help determine the processing times and temperatures that are used in the food industry and in home canning. These times are calculated to include a margin of safety because of the possibility of variability in the populations and the types of microorganisms that need to be killed during processing. As *C. botulinum* is quite sensitive to acid, foods with a pH below 4.5 can be processed at a lower temperature or for a shorter period to achieve commercial sterility.

Heating must be adequate at all points in the food, a task complicated by the fact that heat is transferred throughout the food in the can by both conduction and convection and at rates determined by the nature of the food being processed. Convection is more rapid than is conduction. Foods that are rather fluid can be heated comparatively easily because of the convection currents that develop and circulate throughout the material, but solid foods are heated primarily by conduction. The rate of heating by conduction is influenced by the shape and composition of the food. In both types of heating, there will be a **cold point** in the can, and this is the critical point at which the time and temperature required for safe processing should be determined.

**Cold point**
Coldest area of food in a
can being heat processed.

## Methods

Commercial canners choose to do canning using pressure to be sure the food reaches a safe temperature within a reasonable time frame, with the exact procedures being dictated by the product. They also have choices, both in packaging and in equipment for processing. For example, cans may be processed in a retort without movement or they may be agitated to promote conduction within the can during processing. Actually, canning does not even have to be done in cans; **retort pouches** are used for many so-called canned foods because of advantages in processing time and other economies. These are but a few of the innovations in industry.

**Retort pouch**
Flexible packaging, usually
consisting of three layers.

In the home, the choice generally is dictated by the food being processed. The pH of the food dictates whether or not **pressure canning** is necessary. Low-acid foods (above pH 4.5) must be processed in a **pressure canner** to achieve the temperatures needed to prevent survival of spores of *C. botulinum,* which can flourish in low-acid foods if processing is inadequate. Vegetables, meat, fish, and poultry must be canned with a pressure canner to ensure that a sufficiently high temperature is reached and maintained to kill *C. botulinum* spores.

Foods with a pH of 4.5 or lower (fruits and pickles) are classified as high-acid foods and can be canned in a **water bath canner** because they are sufficiently acidic to prevent growth of *C. botulinum* during storage. The approximate pH values of many foods that are canned are shown in Figure 20.5.

Tomatoes vary in their acidity depending on the variety, which makes the suitability of water bath canning questionable. Some varieties, such as Ace, Rutgers, and Beefsteak, are not acidic enough to be processed safely in a water bath. Additional problems occur when other ingredients are added to tomatoes to make items such as stewed tomatoes with green pepper and onion. These added vegetables raise the pH of the mixture and frequently necessitate pressure canning. For home water bath canning of tomatoes, the addition of 30 ml (2 tablespoons) bottled lemon juice per quart (or $\frac{1}{2}$ teaspoon citric acid/quart or 3 tablespoons of 5 percent vinegar/quart) is sufficient to ensure a low enough pH for safety.

**Pressure canning**
Processing done under pressure so that the temperature is elevated enough to kill microorganisms in a reasonable length of time.

**Pressure canner**
Device designed to maintain the selected pressure needed to reach the desired temperature for heat processing low-acid canned foods.

**Water bath canner**
Large, deep kettle for canning high-acid foods at sea level without adding pressure to raise the boiling temperature.

| pH | Food |
|---|---|
| 2.5 | |
| | Plums |
| | Gooseberries |
| | Prunes |
| | Dill pickles, rhubarb, apricots |
| 3.0 | Apples, blackberries |
| | Sour cherries, strawberries |
| | Peaches |
| | Sauerkraut, raspberries |
| | Blueberries |
| | Sweet cherries |
| | Pears |
| 4.0 | |
| | Tomatoes |
| | Pimiento (dry roasted) |
| | Okra |
| 5.0 | Pumpkins, carrots |
| | Pimiento (lye peeled) |
| | Turnips, cabbage |
| | Parsnips, beets, string beans, green peppers |
| | Sweet potatoes, baked beans (tomato sauce) |
| | Spinach |
| | Asparagus, cauliflower |
| | Baked beans (plain sauce) |
| | Red kidney beans |
| | Lima beans |
| | Succotash, meats, poultry |
| 6.0 | |
| | Peas |
| | Corn, hominy, salmon |
| | White fish |
| | Shrimp, wet pack |
| 7.0 | Lye hominy |
| 8.0 | |

**Figure 20.5**   Average pH of selected foods.

Food to be canned is cleaned and trimmed to eliminate bones or other inedible portions. Sometimes spinach or other bulky foods may be cooked briefly to shrink the volume and then placed in the canning jars while hot, a technique termed *hot pack*. However, most foods are packed into jars without being heated. In both hot and cold pack, liquid is added until there is approximately $\frac{1}{2}$ to 1 inch of headspace remaining at the top of the jar. A sugar syrup usually is used with fruits because it reduces osmotic pressure and helps retain the desired slightly firm texture.

Then the jars are sealed according to the directions for the specific type of closure, placed in either a water bath canner (fruits) or a pressure canner (low-acid foods), and processed at the correct time and temperature. A seal develops as the jars cool undisturbed and the gas in the headspace contracts. This seal is essential in preventing the entry of additional microorganisms during storage.

**Water Bath Canning.**   **Water bath canning** is done by immersing the jars and their holding rack in water that is heated to and maintained at boiling throughout the processing period. The time required for processing is determined by the size of the jar, the type of food, and the elevation. At higher elevations, the processing time needs to be increased because the water bath boils at a lower temperature than at sea level. Table 20.4 indicates the alteration in time needed to adjust for the effect of altitude in water bath canning.

## Potential Problems

The distinctive aroma of hydrogen sulfide reveals **sulfide spoilage** in canned foods in which viable *Clostridium nigrificans* bacteria are present. If spoilage proceeds long enough, sufficient hydrogen sulfide gas is generated to bulge the top of the can; but even before bulging occurs, the food is inedible because of the hydrogen sulfide.

Failure to process adequately or to achieve a tight seal in canning allows microorganisms to grow within the jars or cans (Figure 20.6). Four types of spoilage are likely to occur if contaminated food is not commercially sterile when canned. **Flat-sour spoilage** can occur if *Bacillus coagulans* is viable in the container. Lactic acid is formed from sugar by this anaerobic bacillus. This results in increased acidity without the production of any gas to signal spoilage by a bulging lid.

**Hydrogen swells** is a condition seen only in home-canned foods processed in metal cans that have become contaminated with *thermophilic anaerobes*. A bulging can likely is caused by this problem, but it is not a common occurrence.

Botulism is by far the most dangerous of the common forms of spoilage occurring in canned foods. This disease is the result of the presence of a paralyzing, often fatal exotoxin produced by viable spores of *C. botulinum* during storage (see Chapter 19). Three types (A, B, and E) are found most commonly, but two others (C and F) also have been observed. Contamination leading to Type A and B botulism is generally from soil, and these two types

**Water bath canning**
Heat processing of food in containers immersed in water at atmospheric pressure.

**Sulfide spoilage**
Canned food contaminated with hydrogen sulfide produced by *Clostridium nigrificans*.

**Clostridium nigrificans**
Bacteria capable of producing hydrogen sulfide in canned foods to cause sulfide spoilage.

**Flat-sour spoilage**
Increased acidity caused by viable *Bacillus coagulans* digesting sugar without producing gas.

**Bacillus coagulans**
Anaerobic bacillus that digests sugar to lactic acid in canned foods, causing flat-sour spoilage.

**Hydrogen swells**
Spoilage of food and bulging of cans caused by anaerobic, thermophilic microorganisms that produce hydrogen during storage.

**Table 20.4**    Increase in Processing Time for Water Bath Canning at Selected Altitudes

| Altitude (feet) | Increase in Time (minutes) | |
| --- | --- | --- |
| | Processed ≤20 minutes at Sea Level | Processed >20 Minutes at Sea Level |
| 1,000 | 1 | 2 |
| 2,000 | 2 | 4 |
| 3,000 | 3 | 6 |
| 4,000 | 4 | 8 |
| 5,000 | 5 | 10 |
| 6,000 | 6 | 12 |
| 7,000 | 7 | 14 |
| 8,000 | 8 | 16 |
| 9,000 | 9 | 18 |

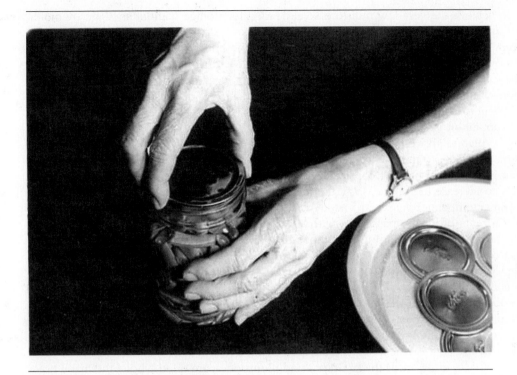

**Figure 20.6** Jars need to be sealed according to directions given for the type of closure being used. The next day the seal should be checked on each jar before storing them.

are the most resistant to heat. Occasionally fish may be contaminated with Type E because of contaminated waters, but its limited resistance to heat makes its destruction complete when proper controls are exercised during commercial canning operations. Home canning of fish requires pressure canning and careful control of both pressure and time to prevent this potential source of food poisoning.

## DRYING

Drying, the original method of preserving food, has been practiced for at least 3,000 years. Even today, some people dry food at home; the food industry uses dehydration rather extensively. Commercial methods include drum drying, spray drying, vacuum shelf drying, vacuum belt drying, atmospheric belt drying, freeze-drying, rotary drying, cabinet drying, and tunnel drying. Obviously, these processes are far more sophisticated than the original, simple sun drying that still persists today. In the home, foods are dried in the sun, in the oven, or in a dehydrator.

Drying is an effective means of food preservation if the moisture level of the dried food is sufficiently low to kill harmful microorganisms. Bacteria generally do not survive at moisture levels below 16 percent (Table 20.5), whereas yeasts require a moisture level of about 20 percent or higher for survival. Molds can survive unless the moisture level is as low as 13 percent and thus are the most likely source of spoilage in dried foods. The foods must be dried sufficiently to drop below 13 percent moisture level and then stored in airtight packaging to prevent resorption of some moisture.

**Table 20.5** Minimum Moisture Levels Required to Maintain Selected Viable Microorganisms (in percentage)

| Type of Microorganism | Minimum Moisture Level for Survival |
| --- | --- |
| Molds | 13 |
| Bacteria | 16 |
| Yeasts | 20 |

To achieve the very low moisture level needed for storage without spoilage by molds, foods must be dried quite thoroughly. This requires mild heating and evaporation. The process is most efficient and effective if the food to be dried is in thin pieces with exposed surfaces; this increases the surface area from which moisture can be removed. As the water in the food changes in state from liquid to water vapor, heat is absorbed—the heat of vaporization—and the surface is cooled. Mild heat, even that from direct sunlight, adds energy to the system to help maintain a temperature warm enough for evaporation to continue.

Evaporation is enhanced when relative humidity is low, air temperature is warm, and air is circulating above the food. Water can be removed from food quite efficiently when these conditions are met. When relative humidity is low and the air temperature is warm, the air is capable of holding considerably more water, and water from the food evaporates readily to increase the saturation of the air.

Warm temperatures also are needed to provide the energy for the heat of vaporization. Currents of air are passed over the food to aid in removing the evaporating water from the food by maintaining a comparatively low humidity directly above the food. These conditions are met effectively by the practice of spreading thin slices of food on a clean towel in direct sunlight in the desert, but machinery designed for commercial drying also is effective.

Food should be as sterile as is practical prior to drying. For vegetables, this means a thorough scrubbing and usually blanching. Blanching is important not only for sanitizing but also for inactivating enzymes that can cause deterioration during storage. Fruits are washed but usually are not blanched. Meats are wiped clean. Usually foods are sliced between 3 and 6 millimeters ($\frac{1}{8}$ and $\frac{1}{4}$ inch) thick. Grapes and plums often are dried whole; the many holes that are poked in their skins allow evaporation to occur and result in raisins and prunes, respectively.

Color changes during drying can present problems in some foods. Apricots and peaches brown rather unattractively unless they have been subjected to sulfuring or blanching prior to drying. **Sulfuring** is accomplished by exposing the cut fruit to smoke from burning sulfur flowers. This treatment produces the rather bright orange usually seen in apricots that have been dried commercially.

**Sulfuring**
Exposure of cut fruits to the smoke fumes created by burning sulfur flowers to retain a bright color during drying.

Blanching minimizes the color change of drying but is not totally effective in retaining the original orange color of these fruits. Blanching does aid in maintaining an acceptable color in dried vegetables. Apples and bananas are examples of fruits that retain their color better during drying if they are dipped in an acidic fruit juice or a solution of ascorbic acid as soon as the cut surfaces are exposed to air. This minimizes browning.

During drying, browning sometimes occurs as a result of the Maillard reaction, particularly during the period when the moisture level drops from 20 to 15 percent. The ease of reaction between aldehydes and amino groups increases during drying because the solids increase in concentration as the moisture decreases. Browning caused by caramelization of sugars also occurs sometimes when fruits reach very low moisture levels after being dried at too high a temperature.

Proteins may become partially denatured during drying because of the high concentrations of salts resulting from the reduced moisture and the heat. Some sugar and salts are lost from damaged cells during drying. Starch and gums may lose some of their ability to absorb moisture during rehydration. The combination of these changes often limits the ease of rehydration of dried foods. Fortunately, simmering in water is quite effective in promoting rehydration and is also a wise technique to avoid growth of microorganisms, which can be a hazard when food is rehydrated for an extended period at room temperature.

Food slices that are dried simply on a clean cloth in a sunny spot must be taken indoors as the temperature drops in late afternoon and then returned to the sun each morning until dried sufficiently. The biggest hazards in sun drying are birds and other pests. The drying food must be protected from such possible sources of contamination.

Oven drying eliminates the environmental hazards posed by sun drying. Good air circulation is needed beneath as well as above the food. A screen or cheesecloth fastened across the oven racks can be used effectively to hold the food being dried. The lowest oven setting, usually 60°C (140°F), is used to minimize browning and achieve drying at a reasonable rate. The oven door can be left ajar to maintain good flow of air over the food.

Dehydrators vary in their design, but they are usually quite simple devices with a source of moderate heat and a fan to promote air circulation and removal of moisture. Trays of screens framed with metal are spaced above each other in the dehydrator to permit the drying of a considerable amount of food at one time.

The exact duration of drying varies with the thickness of the food, the type of food, and the drying conditions. Dehydrators and oven drying provide more rapid dehydration than usually is accomplished by sun drying. In most cases, this equipment can dry foods to the desired moisture level of about 10 percent in a matter of a few hours, whereas sun drying may take several days, depending on weather conditions. As soon as the desired level of drying has been reached, the food needs to be packaged in airtight containers to prevent additional contamination by microorganisms and to block the rehydration possible at moisture levels above 13 percent, which can lead to spoilage by mold growth.

## PRESERVING WITH SUGAR

Jams and jellies represent yet another way of preserving food. The high concentration of sugar establishes unfavorable osmotic pressure that, in combination with the boiling period during preparation, effectively kills microorganisms. Jams and jellies are pectin gels containing fruit pieces (jams) and fruit juice to which sugar and pectin are added at levels enabling a pectin gel to form on cooling. Frequently, acid also is added to enhance gel formation (Figure 20.7).

Each ingredient in jelly plays a crucial role in forming the desired gel. Pectin is a complex carbohydrate polymer of galacturonic acid (an organic acid derivative of galactose). As noted in Chapter 10, pectin is only one of the pectic substances, but it is the key substance in the formation of the gel structure needed in jams and jellies.

In underripe fruit, protopectin is the pectic substance. As fruit ripens, protopectin gradually is methylated by esterification of methanol with the acid radical on the sixth (external to the ring) carbon atom. This methylation converts protopectin to pectinic acid or pectin; these compounds are capable of cross-linking under optimal conditions to form a pectin gel in which the pectinic acid or pectin molecules form the continuous network or solid phase in which the liquid juice is trapped. Hydrogen bonding between these long,

**Figure 20.7**    Jams use sugar to preserve the fruits that are used. The correct concentration of sugar must be reached for jellies and jams to gel.

rather fibrous molecules locks the molecules into a seemingly semi-rigid structure that is capable of trapping water to form a spreadable gel. The pectin in overripe fruit undergoes chemical changes, resulting in shorter chains of pectic acid that do not provide an opportunity for sufficient cross-linkage between molecules and hence gel formation.

Sugar is an essential ingredient in pectin gel formation. Its role appears to be that of binding some of the water to the pectin network so that free water does not flow or move out of the gel. The concentration of sugar needs to be between 60 and 65 percent in the finished gel to achieve the desired firmness without too much free water. This rather high concentration is achieved by using a liberal amount of sugar in the initial recipe and then boiling off the liquid until the proper concentration is achieved.

If the amount of sugar added to the mixture is decreased, longer boiling is required to concentrate the sugar appropriately. This also will concentrate the pectin and cause the resulting jelly to be rather rubbery. In addition, the prolonged cooking darkens the color of the jam or jelly. Addition of too much sugar will dilute the pectin and cause the product to be rather soft or even fluid. In addition to its role in gel formation, sugar at its recommended concentration (60 to 65 percent) is very effective in killing microorganisms. This role of sugar must be remembered in making low-calorie jams and jellies, because the preserving action of sugar will not be available if its concentration is reduced.

For optimal pectin gel formation, the pH of the mixture must be below 3.5 and preferably about 3.3. If the pH is shifted to the optimum pH for a particular type of jelly, the jelly will be firmer. Acid may need to be added to the jelly mixture. Citric acid or lemon juice frequently is the ingredient of choice for modifying pH. Addition of the acid to the jelly prior to boiling has the advantage of hydrolyzing some of the sucrose to produce invert sugar, which reduces the likelihood of crystal formation in the jelly during storage. Unfortunately, the pectin in the mixture is hydrolyzed somewhat at the same time, which reduces its gel-forming capability. However, if the mixture is boiled at a moderate rate, acid can be added prior to boiling without having an appreciable negative impact on the resulting jelly.

The pectin content of fruits is quite variable and is determined both by the type of fruit and by ripeness. When jellies and jams are made at home, the usual practice is to add commercial pectin to ensure that adequate pectin is present to form a gel. Commercial pectins, available in either liquid or powdered form, are prepared by extraction either from the albedo (white portion of the skin) of citrus fruits or from apple cores and skins. The granular pectin products have a longer shelf life than do the liquids.

Low-methoxyl pectin also is available for preparation of low-calorie jams and jellies. Its structure is as follows:

Low-methoxyl pectin

**Figure 20.8**   Jam needs to be boiled until the correct concentration of sugar has been achieved through evaporation.

Interestingly, these pectins are able to gel without addition of sugar to the mixture, which explains why the resulting products using low-methoxyl pectins are low in calories. Bonding, in the form of ionic bonding, is achieved between molecules of this special pectin by an added divalent ion (usually calcium) when the ion cross-links to carboxyl groups on two molecules of pectin.

Preparation of jams and jellies requires washing and preliminary preparation of the fruit, including sorting and cutting. In commercial operations, the level of the pectin in the fruit is determined by measuring the viscosity of the extracted fruit juice. A jelmeter is the device commonly used for this measurement. It is similar to a pipette. If the juice is allowed to drain through the pipette, the relative flow properties can be determined and the amount of sugar needed can be calculated. The more viscous the juice, the greater is the amount of sugar that must be added to achieve the desired gel.

In the home, the length of time the mixture should be boiled is determined with a cooking thermometer or by the sheeting-off test (Figure 20.8). Usually boiling is continued until the temperature is between 104 and 105°C (219 and 221°F); this indicates a sugar concentration of about 65 percent at sea level. At higher elevations, the thermometer should read between 4 and 5°C (7 and 9°F) above the temperature of boiling water at that elevation.

The sheeting-off test is a visual method in which a spoonful of jelly is allowed to flow slowly from the edge of a large spoon. When it is sufficiently viscous to separate into two streams, the sugar concentration is approximately correct.

## ADDITIONAL COMMERCIAL METHODS

### Freeze-Drying

**Freeze-drying** is a commercial technique that is a variation of dehydration. The difference is that the food is frozen, and then the water is sublimated from the frozen food. This procedure results in dried foods that have a volume comparable to the original food and a porous texture because of the loss of water from the food. Storage is comparable to that of other dried foods—that is, airtight packaging and storage at room temperature.

**Freeze-drying**
Preservation of food by first freezing and then dehydrating the product.

**Sublimation**
Transition from the frozen state directly to the gaseous state without liquefaction.

**Sublimation** is accomplished by having the food in a frozen state at a temperature of 0°C (32°F) or lower and then placing it in a vacuum chamber at a pressure of 4.7 millimeters of mercury or less. The low temperature keeps the water in its frozen state as ice, and the low atmospheric pressure enables frozen water molecules to escape quickly as water vapor to dehydrate the food.

Strawberries and other fruits and vegetables are well suited to freeze-drying. This process is effective in retaining much of the original flavor of the food and in maintaining individual pieces that are light, rather than extremely compact. These characteristics make freeze-dried fruits useful ingredients in dry breakfast cereal products. However, the comparatively high cost of energy for freeze-drying has limited the use of this form of food preservation.

## Irradiation

Radiant energy can be used commercially to preserve food (Figure 20.9). Gamma rays radiated from cobalt-60 and beta particles produced by special electronic machines are the sources of energy used for preserving food by irradiation (see Chapter 6). The goal of irradiation is to kill harmful microorganisms and inactivate enzymes without altering the food. Changes in the food itself are minimized if it is irradiated in a vacuum or in an inert gas with ascorbic acid or another scavenger of the free radicals formed by irradiation.

Varying doses of irradiation energy are needed in different foods to achieve the desired end results. A dosage of 7,500 rads halts sprouting of potatoes enough that they can be stored for at least two years. Insects in flour can be killed by 50,000 rads. Berries subjected to 150,000 rads keep for at least three weeks.

Meats sealed in a semi-permeable film and stored at 0°C (32°F) can be held at least 2 months after irradiation with 1,000,000 rads. Because of its great resistance, *C. botulinum* is used to determine the level of irradiation needed to achieve death of bacteria and safety of food. Pork may be irradiated to eliminate this potential hazard. Enzymes usually are even more resistant to change than are microorganisms, so the need to inactivate specific enzymes also can determine the amount of irradiation necessary.

The greater the amount of irradiation, the greater is the cost for preserving food in this way. Cost and public resistance to consumption of irradiated foods because of perceived potential hazards are limiting the use of irradiation in commercial food preservation (see Chapter 19). Although the FDA has approved spices for irradiation, even these rather costly food ingredients rarely are available in markets at the present time in an irradiated form. However, they are used in some processed foods.

At the present time, the FDA has approved irradiation of fresh meat and poultry (whole or cut-up birds, skinless poultry, ground poultry, hamburger and other ground meats, pork chops, roasts, stew meat, liver) white potatoes, wheat and wheat powder, many spices, dry vegetable seasonings, shell eggs, and fresh produce. Any food that has been irradiated must be identified by displaying the **radura**, the symbol representing food in the center and the energy source with rays by the broken line of the circle. The words "Treated with Radiation" or "Treated by Irradiation" must also appear with the logo (see page 477). The levels of irradiation approved by the FDA are presented in Table 20.6.

**Radura**
Logo that must be displayed on irradiated foods.

## High-Pressure Processing

**High-pressure processing (HPP)**
Process using high pressure and water to make food safe and give it a somewhat extended shelf life during refrigerator storage.

**High-pressure processing (HPP)** is a rather recent technique used by a few food processors today. However, it is likely to be used with increasing frequency as research clarifies the mechanisms involved in controlling food spoilage and quality. Special equipment is used in which foods in plastic packages are housed while a small amount of heated water is introduced and pressure is applied (Figure 20.10).

The exact conditions for an individual food product are determined by experiments that are evaluated on both the microbiological safety and the resulting food quality. Appropriate pressure will inactivate vegetative cells, but foods in which bacterial spores may be present require both heat and pressure to inactivate the spores. Some enzymes in foods may be

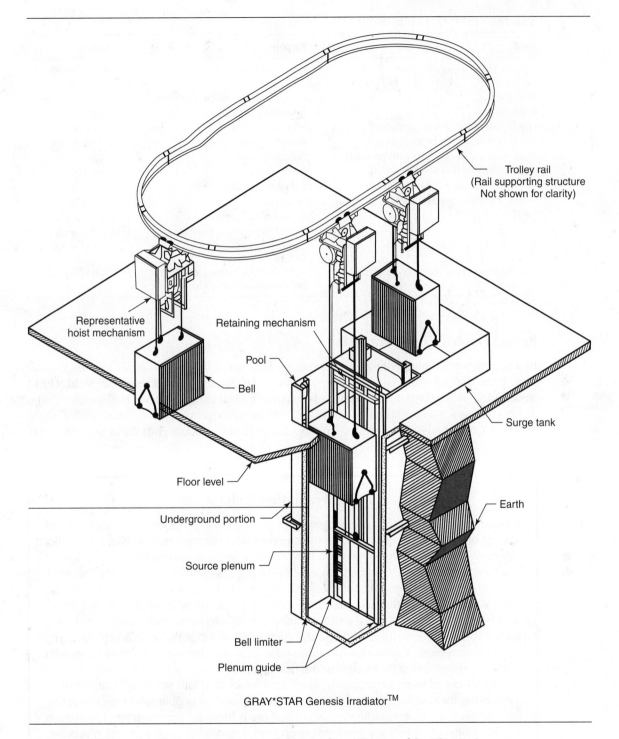

Floor level

Underground portion

Source plenum

Bell limiter

Plenum guide

GRAY*STAR Genesis Irradiator™

**Figure 20.9**    Diagram of the floor plan of a Gray*Star Genesis irradiator. (Courtesy of Gray*Star, Inc.)

altered by high-pressure processing, either by the pressure or by the change in temperature. The objective of high-pressure processing is to extend shelf life, but for a shorter time than is achieved with canning, freezing, or drying. Refrigerated storage is required.

The most successful high-pressure processed product currently available in retail markets is guacamole. Some HPP meats are available for sale in delis. Research efforts are now underway to attempt to use high-pressure processing to produce brie (soft French cheese) that is free of viable *Listeria monocytogenes* although it is made using raw milk.

**Table 20.6**  Levels Approved by the FDA for Irradiating Foods

| Food | Purpose | Dose |
| --- | --- | --- |
| Fresh, nonheated processed pork | Control of *Trichinella spiralis* | 0.3 kGy min. to 1 kGy max. |
| Fresh foods | Growth and maturation inhibition | 1 kGy max. |
| Foods | Arthropod disinfection | 1 kGy max. |
| Dry or dehydrated enzyme preparations | Microbial disinfection | 10 kGy max. |
| Dry or dehydrated spices/seasonings | Microbial disinfection | 30 kGy max. |
| Fresh or frozen, uncooked poultry products | Pathogen control | 3 kGy max. |
| Frozen packaged meats (solely NASA) | Sterilization | 44 kGy min. |
| Refrigerated, uncooked meat products | Pathogen control | 4.5 kGy max. |
| Frozen, uncooked meat products | Pathogen control | 7 kGy max. |
| Fresh shell eggs | Control of *Salmonella* | 3.0 kGy max. |
| Seeds for sprouting | Control of microbial pathogens | 8.0 kGy max. |
| Fresh or frozen molluscan shellfish | Control of *Vibrio* species and other food-borne pathogens | 5.5 kGy max. |
| Fresh iceberg lettuce and fresh spinach | Control of food-borne pathogens and extension of shelf life | 4.0 kGy max. |

## Pulsed Electric Field Processing

**Pulsed electric field (PEF) processing**
Pasteurization of juice by pulsing it 1,000 times per second in an electrical field of about 35,000 V/cm.

Pasteurization is an important step in preparing fruit juices for the consumer market, but the heat treatment may have a detrimental effect on flavor. **Pulsed electric field (PEF) processing** is a recent technological development that can overcome this problem. The process exposes the juice to pulses about 1,000 times per second in an electrical field of about 35,000 V/cm with very little change in temperature of the product. Vegetative cells that may

### The Irradiation Debate

Gradually, irradiated foods are appearing as a choice for consumers, but their advent has not been without considerable opposition from concerned members of the public. Scientists are continuing to research the potential hazards and to identify the lowest levels of irradiation that can be used while assuring destruction of dangerous microorganisms, such as *Escherichia coli 0157:H7*, that might be present in ground meats and other foods. Their findings demonstrate the safety of foods irradiated at the levels approved by FDA. Those on the other side of the debate claim that irradiation is dangerous to consumers and is not necessary for safety reasons because adequate cooking kills dangerous microorganisms. Their focus is on cleaning up the environment where foods are being produced for the market.

Both sides of the issue regarding irradiation have important points to consider; they are seeking the common goal of safe food. The numbers of people who are contracting food-borne illnesses are ample evidence that there is room for improvement. Consumers and the food-service industry need to take on more responsibility for selecting, preparing, and serving safe foods. The FDA is the agency that regulates and oversees irradiation so that these food choices are available for those who wish to buy hamburger or other approved food that has the added safety provided by that treatment. That agency also is charged with monitoring conditions to promote food safety from farm to table.

Recent moves by local groups to block use of irradiated foods in school lunches are intended to protect children from what some people view as the hazards of irradiation, but their arguments ignore the other side of the picture. Hamburger and spinach are two foods that have been the source of several outbreaks of food-borne illness that could have been avoided by serving irradiated foods, and these foods are often served in school lunches.

**Figure 20.10**  Commercial unit for high-pressure processing foods; the operator is looking into the pressure vessel where the food will be placed.

be present in the juice are killed and the juice retains essentially its original flavor. Shelf life is four weeks; refrigerated storage is required, because the juice is pasteurized, not sterilized, by this treatment. The negative aspect is that pulsed electric field processing is expensive and increases the cost of the juice. It also is not suitable for processing many other foods.

## SUMMARY

Spoilage by growth of molds, bacteria, and yeasts limits the time that food can be stored and still be safe and palatable. The environmental conditions optimal for reproduction vary with the type of microorganism. Foods no longer serve as suitable hosts for microorganisms when they are preserved properly. Alterations in moisture level, pH, temperature, and solute concentration are ways of preserving foods.

In addition to the economic losses represented by spoiled food, health hazards may develop while foods are stored unless they are preserved effectively. Salmonellosis, streptococcal infection, staphylococcal poisoning, and botulism are some of the illnesses that can result from contaminated foods.

Freezing protects foods by reducing the rate at which viable microorganisms reproduce by holding the foods below the freezing point of water. Blanching is done prior to freezing vegetables to inactivate enzymes and thus retard deteriorative changes during storage. Browning in sensitive fruits can be avoided by coating them with sugar, a sugar syrup, lemon or other acidic juice, or a solution of ascorbic acid. Specific foods, such as egg yolks, salad dressings, and starch-thickened mixtures, require use of ingredients selected to overcome textural problems that would otherwise occur during frozen storage. Freezing should be done at a very rapid rate to create numerous small ice crystals, thus minimizing the damage to cell walls. Commercial techniques are more effective at accomplishing rapid freezing than are those available in the home.

Canning is somewhat laborious, but this rigorous heat treatment and sealing in airtight containers enable canned foods to be stored at room temperature for a very long time. Time and temperature of processing must be controlled carefully to ensure that spores of *C. botulinum* or other microorganisms that may be present are killed.

Low-acid foods such as vegetables and meats must be processed in pressure canners to reach a high enough temperature. High-acid foods (pH 4.5 or lower) can be processed safely in a water bath canner. Inadequate processing can result in flat-sour spoilage, sulfide spoilage, hydrogen swells, and botulism, as well as other less common problems.

Drying preserves food by reducing the moisture level below 13 percent (usually to 10 percent or slightly less), a level at which microorganisms, even molds, are killed. Cleanliness of food to be dried should be stressed. In the home, foods may be dried in the sun, in the oven, or in a dehydrator. Vegetables should be blanched before drying to inactivate enzymes and reduce the changes in color and flavor that develop during drying. Sulfuring protects the color of apricots and peaches during drying. Browning in apples and bananas can be blocked somewhat by coating the slices with acidic fruit juice or a solution of ascorbic acid before drying.

Sugar at high concentrations (60 percent or more) has a preserving action on foods because of the unfavorable osmotic pressure created by this solute at these levels. Pectin gels often are prepared using fruit juices, sugar, added pectin, and sometimes acid. The pH must be below 3.5 and preferably should be about 3.3 to form a gel of satisfactory strength. Sugar binds much of the water within the gel structure. Pectin forms the solid network by hydrogen bonding between long, fibrous molecules. Acid helps keep enough water away from the pectin molecules for the necessary hydrogen bonds to form. Low-methoxyl pectins cross-link when calcium is present, forming a structure sufficiently strong to eliminate the need for sugar to hold the water in the gel. This makes it possible to prepare low-calorie jams and jellies, but a preservative is necessary because the level of sugar is inadequate to prevent growth of microorganisms.

Freeze-drying differs from other methods of drying in that the food is first frozen to convert water into ice crystals. The ice then is sublimated to water vapor by placing the frozen food in a partial vacuum for dehydration. The result is a porous dried food with good shelf life when stored in an airtight package.

Gamma rays from cobalt-60 and beta particles generated by machines can be used to irradiate food. This method of preservation is quite costly, but it is very effective in preventing deteriorative changes for prolonged periods. At the present time, the FDA has authorized irradiation to a limited extent because of the need for additional research to determine safe conditions for treating many foods by this method.

## STUDY QUESTIONS

1. What conditions generally favor the growth of (a) yeasts, (b) molds, and (c) bacteria?

2. What changes occur during the blanching of vegetables in preparation for freezing?

3. What changes may occur in meats, poultry, and fish during frozen storage?

4. What is likely to occur when mayonnaise is frozen? Explain why this may happen and how it probably can be prevented.

5. Which starch product is best for use in starch-thickened products that are to be frozen? Explain what happens when most starches are used and why this starch is a good choice.

6. What precautions need to be taken when foods are being canned? Why are these precautions necessary?

7. Why is pressure canning used for canning vegetables and meats?

8. Why are fruits canned in a water bath canner, but not in a pressure canner?

9. Describe the drying of (a) onions and (b) apricots at home.

10. What differences would be expected between two apple jellies made with (a) a comparatively low level of sugar and (b) the optimum amount of sugar? Explain why these differences develop.

11. True or false. The easiest way to preserve food at home is drying.

12. True or false. Some tomatoes need to be pressure canned because they are too acidic.

13. True or false. Home canning of vegetables requires careful processing in a pressure canner using correct pressure and time.

14. True or false. Sugar aids in preserving fruits as jams or jellies.

15. True or false. Irradiation is an effective way of helping to assure that hamburger is safe.

## BIBLIOGRAPHY

Brody, A. L. 2002. Food canning in the 21st century. *Food Technol. 56* (3): 75–78.

Brody, A. L. 2005. Extended shelf-life packaging. *Food Technol. 59* (3): 61.

Brody, A. L. 2006. Aseptic and extended shelf-life packaging. *Food Technol. 60* (2): 66.

Choi, H. J., et al. 2002. Weisella kimchii sp. nov., a novel lactic acid bacterium with kimchi. *International J. Systematic Evolutionary Micro. 52*: 507.

Clark, P. 2006a. Evaluating nonthermal processes. *Food Technol. 59* (12): 79.

Clark, P. 2006b. Pulsed electric field processing. *Food Technol. 60* (1): 66.

Clark, J. P. 2006. High pressure processing research continues. *Food Technol. 60* (2): 63.

Clark, J. P. 2007. High pressure effects on foods. *Food Technol. 61* (2): 69.

Clark, J. P. 2010. Considerations on drying. *Food Technol. 64* (3): 70.

Cowell, N. D. 2007. More light on dawn of canning. *Food Technol. 61* (2): 40.

De Heij, W. B. C., et al. 2003. High-pressure sterilization maximizing the benefits of adiabatic heating. *Food Technol. 57* (3): 37.

Donsi, G., et al. 2010. Pasteurization of fruit juices by means of a pulsed high pressure process. *J. Food Sci. 75* (3): E169.

Draughton, F. A. 2004. Use of botanicals as bio-preservatives in foods. *Food Technol. 58* (2): 20.

Fan, X., et al. 2009. *Microbial Safety of Fresh Produce.* Wiley-Blackwell. New York.

Fox, J. A. 2002. Influence on purchase of irradiated foods. *Food Technol. 56* (11): 34.

Meyer, R. S., et al. 2000. High-pressure sterilization of foods. *Food Technol. 54* (11): 67–72.

National Institute of Food and Agriculture. 2009. *Complete Guide to Home Canning.* Agriculture Information Bulletin 539.

Olson, D. G. 2004. Food irradiation future still bright. *Food Technol. 58* (7): 112.

Parnes, R. B., et al. 2003. Food irradiation: An overlooked opportunity for food safety and preservation. *Nutrition Today 38* (5): 174.

Peleg, M. 2006. It's time to revise thermal processing theories. *Food Technol. 60* (7): 92.

Potter, N. N. and Hotchkiss, J. H. 1999. *Food Science.* 5th ed. Aspen Publishers. Gaithersburg, MD.

Schauwecker, A. 2004. New technologies: Under pressure. *Food Product Design 14* (4): 96.

Schauwecker, A. 2005. To preserve and protect. *Food Product Design 15* (3): 67.

Swintek, B. 2006. Global challenges to food safety. *Food Technol. 60* (5): 123.

Teixeira, A., et al. 2006. Keeping botulism out of canned foods. *Food Technol. 60* (2): 84.

Tewari, G. and Juneia, V. 2007. *Advances in Thermal and Non-Thermal Food Preservation.* John Wiley and Sons. New York.

## INTO THE WEB

*http://www.uga.edu/nchfp/*—Web site for National Center for Home Food Preservation.

*http://www.uga.edu/nchfp/multimedia.html#video*— Videos on home food preservation.

*http://ijsb.sgmjournals.org/cgi/reprint/52/2/507.pdf*— Microorganisms in kimchi.

*http://dbs.extension.iastate.edu/answers/projects/answerl ine/questions/FoodPreservation.html*—Information on canning at home.

*http://www.flex-news-food.com/pages/12653/Packag/ retort-pouch-%E2%80%93-fast-growing-packaging- technology-todays-consumer-world.html*— Information on retort pouch development.

*http://www.extension.umn.edu/distribution/nutrition/ 00019.html*—Guide to drying food.

*http://www.cdc.gov/ncidod/DBMD/diseaseinfo/foodirradi ation.htm*—CDC site on food irradiation.

*http://www.fsis.usda.gov/Fact_Sheets/Irradiation_and_ Food_Safety/index.asp*—FDA site on irradiation of foods.

*http://www.fda.gov/Food/FoodIngredientsPackaging/ IrradiatedFoodPackaging/ucm074734.htm*— Levels of irradiation approved by the FDA for various foods.

*http://www.fda.gov/ForConsumers/ConsumerUpdates/ ucm093651.htm*—Irradiation of lettuce.

*http://pubs.cas.psu.edu/freepubs/pdfs/uk108.pdf*— Information on labeling requirements for irradiated foods.

*http://www.food.gov.uk/safereating/rad_in_food/irradfoo dqa/*—British site for information on irradiated foods.

*http://www.fsis.usda.gov/Fact_Sheets/Irradiation_and_ Food_Safety/*—USDA fact sheet on irradiation of foods.

*http://www.iaea.org/programmes/nafa/d5/public/food irradiation.pdf?q=food*—Extensive international report on food irradiation.

*http://ohioline.osu.edu/fse-fact/pdf/0001.pdf*— Information on high-pressure processing.

*http://grad.fst.ohio-state.edu/hpp/*—Site for Ohio high-pressure food processing laboratory.

*http://ohioline.osu.edu/fse-fact/0002.html*—Information on pulsed electric field processing.

Spices are additives that enhance flavors of foods.

# CHAPTER 21

# Food Additives

## Chapter Outline

## OBJECTIVES

After studying this chapter, you will be able to:

1. Discuss the occurrence of accidental additives in foods.
2. Explain the reasons intentional additives are allowed in processed foods.
3. Identify some specific food additives and discuss why they are used.
4. Outline the role of government in overseeing additives in food.

## OVERVIEW

Food additives are a source of considerable controversy in this nation and throughout other industrial nations. They constitute such a debatable topic that even their definition is the subject of argument. The dictionary states that food additives are substances added in relatively small amounts to impart or improve desirable qualities or suppress undesirable ones. Some groups have their own definitions. The Food Protection Committee of the National Academy of Sciences–National Research Council considers an additive to be "a substance or a mixture of substances other than a basic foodstuff, which is present in a food as a result of any aspect of production, processing, storage, or packaging. The term does not include chance contaminants." This definition clearly excludes accidental additives or contaminants even though they may be there because of environmental circumstances. This definition is helpful and clear, but it does not have legal implications.

The definition used by the Food and Drug Administration is the one that has a legal impact in this country. Therefore, the somewhat tortured language was developed to eliminate as many ambiguities and questions as possible. According to the FDA, a food additive is

> any substance, the intended use of which results or may reasonably be expected to result, directly or indirectly, in its becoming a component of or otherwise affecting the characteristics of any food (including any substance intended for use in producing, manufacturing, packing, processing, preparing, treating, transporting or holding a food; and including any source of radiation intended

for any such use), if such substance is not generally recognized, among experts qualified by scientific training and experience to evaluate its safety, as having been adequately shown through scientific procedures (or, in the case of substances used in food prior to January 1, 1958, through either scientific procedures or experience based on common use in food) to be safe under the conditions of its intended use.

This definition is included in the Food Additives Amendment to the Federal Food, Drug, and Cosmetic Act of 1938; the amendment was passed in 1958 and is still in effect. The final parenthetic clause established the concept for creating a list of additives "generally recognized as safe," a list usually referred to simply as the **GRAS list**. In essence, any substance that a manufacturer wants to market to the food industry as an additive has to be proven safe for human consumption by scientific experiments or else has to be on the GRAS list.

The *Food Chemicals Codex* is an international directory of food additives originally published by the Institute of Medicine and presently by the U.S. Pharmacopeia. This comprehensive list is used universally by nations throughout the world. It defines essential criteria and methods of analysis to authenticate and determine the quality of food ingredients so that food in international trade will be held to the same requirements regardless of its source.

The cost of the testing to obtain clearance for a new additive is great, and considerable money has been spent since 1958 to conduct federally funded tests on the additives that were identified on the GRAS list. Food manufacturers wishing to use a new additive must bear the cost of the testing necessary for obtaining FDA approval. Testing is to ensure safety from health risks in the nation's food supply. The motivation for using the additives is to enhance the quality of the foods being consumed.

## RATIONALE

Consumers today are bombarded by reports of hazards and dilemmas regarding what to eat. Accuracy of these stories ranges the gamut from totally accurate to highly emotional and unproven. This situation has created considerable concern about what to eat. One of the issues is food additives; the chemical names sound ominous to those who have not studied the science. Fortunately, laws on labeling require ingredient labeling, which means that all ingredients, including additives, must be listed in descending order by weight (Figure 21.1). As a result, shoppers can read labels and avoid additives they do not wish to eat. For those who wish to minimize additives in their food, the option of preparing their meals using basic ingredients is a suitable choice. Others may prefer the convenience and quality of processed foods. The important point is that consumers have a choice, and the information they need is available when they are deciding.

Food that reaches the typical American table today has not been rushed from the field to the table. Frequently, it has been shipped many miles, sometimes even from other countries. Convenience items range from packaged mixes to fully prepared meals that need only to be reheated are assembled and packaged in factories, and they pass through many hands en route to the consumer. Such foods are subjected to far more stresses than are foods that are served as soon as they are prepared, and yet the quality of the convenience items needs to be competitive with food prepared in the home.

The only way to ensure suitable quality of commercial food products when they actually arrive on the consumer's table is through the use of additives to overcome the problems engendered by the long and complex supply route that convenience foods must travel. The ready acceptance and strong market for these items support food manufacturers in their commitment to using additives to bring the world of food to the consumer.

Although communication between manufacturer and consumer would appear to be direct, there actually is a very prominent, though less visible, voice in the matter of food additives, and that voice is the U.S. Food and Drug Administration. The FDA has the power of law behind its voice. Much of the power of the FDA in relation to additives is contained within the **Food Additives Amendment of 1958**, although the original bill was the **Food, Drug, and Cosmetic Act of 1938**.

---

**GRAS list**
Additives "generally recognized as safe" because of their common and safe use prior to January 1, 1958.

**Food Chemicals Codex**
International directory of food additives.

**Food Additives Amendment of 1958**
Legislation governing food additives that placed the burden of proof of safety on the food manufacturer.

**Food, Drug, and Cosmetic Act of 1938**
Federal legislation establishing the U.S. Food and Drug Administration and defining its responsibilities.

**Ingredients:** Sugar, Wheat Flour Bleached, Wheat Starch, Egg White, Leavening (baking soda, citric acid), **Corn Starch, Calcium Chloride, Modified Soy Protein, Salt, Cellulose Gum, Artificial Flavor, Sodium Lauryl Sulfate** (a whipping aid).

**CONTAINS WHEAT, EGG AND SOY; MAY CONTAIN MILK INGREDIENTS.**

DIST. BY **GENERAL MILLS SALES, INC.,** MINNEAPOLIS, MN 55440 USA

© 2009 General Mills, Inc.

Carbohydrate Choices: 2

# Nutrition Facts
Serving Size ¹⁄₁₂ pkg (38g mix)
Servings Per Container 12

**Amount Per Serving**

**Calories** 140   Calories from Fat 0

|  | % Daily Value* |
|---|---|
| **Total Fat** 0g | **0%** |
| Saturated Fat 0g | **0%** |
| Trans Fat 0g | |
| Polyunsaturated Fat 0g | |
| Monounsaturated Fat 0g | |
| **Cholesterol** 0mg | **0%** |
| **Sodium** 310mg | **13%** |
| **Potassium** 35mg | **1%** |
| **Total Carbohydrate** 32g | **11%** |
| Sugars 23g | |
| **Protein** 2g | |

| Calcium | 4% |
|---|---|

Not a significant source of dietary fiber, vitamin A, vitamin C and iron.

* Percent Daily Values are based on a 2,000 calorie diet. Your daily values may be higher or lower depending on your calorie needs:

| | Calories | 2,000 | 2,500 |
|---|---|---|---|
| Total Fat | Less than | 65g | 80g |
| Sat Fat | Less than | 20g | 25g |
| Cholesterol | Less than | 300mg | 300mg |
| Sodium | Less than | 2,400mg | 2,400mg |
| Potassium | | 3,500mg | 3,500mg |
| Total Carbohydrate | | 300g | 375g |
| Dietary Fiber | | 25g | 30g |

**Figure 21.1**   Food labels must list ingredients and additives in descending order by weight; even print size is mandated for ease of reading while shopping.

The 1958 legislation was of particular importance because it shifted the burden of proof for safety of additives from the federal government to the manufacturers of new additives proposed for use. Proof of safety requires animal feeding tests for an extended period, occasionally as long as seven years. People also are used in a portion of the testing, although human testing is quite limited and is not done in the early phase. The manufacturer usually has these tests conducted by a testing service.

The *Delaney clause* was a highly controversial clause in the 1958 legislation. This rather emotionally charged legislation requires that additives that produce cancer when consumed at any level by animals or by people cannot be added to foods. Cyclamates were banned in 1969 under the Delaney clause, leaving only saccharin as a non-nutritive sweetener. Then, the FDA moved to ban saccharin in 1972. The fact that saccharin had been used by large numbers of people over many years without any evidence of a carcinogenic effect made its banning seem to be an overly zealous and perhaps even totally unnecessary action. The objections to the proposed ban were loud and numerous. The results of this controversy were continuous extensions of permission to use saccharin in foods and preliminary efforts to find a way to introduce logic into the Delaney clause rather than require automatic banning of a substance when there is only limited evidence of possible carcinogenicity, and that evidence is in laboratory animals given totally unrealistic levels of the substance.

The concept of "risk versus benefit" has evolved from the additives controversy. A useful illustration is provided by the controversy over the use of nitrates and nitrites in curing meats. Concern stemmed from evidence that nitrites can combine with amines to form nitrosamines during intensive heating (as in frying bacon to the very crisp stage) and in the intestines when nitrites are eaten. Nitrosamines are of concern because they are quite potent carcinogens. That would seem to be the justification for enforcing the Delaney clause and prohibiting the use of nitrites and nitrates in cured meats. However, this is the risk side of the equation.

On the benefit side is the protection provided by nitrites and nitrates against the risk of botulism from these meats; botulism is a real and very lethal risk without these additives. Consequently, nitrates and nitrites still are permitted in cured meats, but the level used has been reduced somewhat. If an effective replacement can be found, nitrates and nitrites likely will be banned from the food supply. Nevertheless, this controversy illustrates the application of the concept of "risk versus benefit" despite the fact that such action actually violates the Delaney clause.

The Color Additive Amendments of 1960 are superimposed on the other legislation affecting the use of additives in the food industry. This legislation covers the use of all substances used to color food, whether they are additives extracted from other foods or synthetic compounds. This has been an area of considerable interest because some colors that had been used extensively were banned under this legislation when they were found to be carcinogenic as defined under the Delaney clause. The Color Additive Amendments of 1960 formed the basis for banning FD&C Red No. 2 and No. 4, as well as carbon black.

The Miller Pesticide Amendment of 1954 is the legislation regulating the use of agricultural pesticides and their residues in the food supply. The Environmental Protection Agency has the authority to establish the allowable amounts of residues. Enforcement is under the Food and Drug Administration. A recent illustration of the effectiveness of this legislation was the banning of ethylene dibromide, a pesticide that was used extensively in stored grains and citrus fruits to prevent losses from insects during storage.

The *Food Chemicals Codex* is a book of particular interest to the food industry regarding additives. This guide states the standards for purity, clearly indicating the methods of analysis and the allowable levels of trace contaminants. This work was developed by the Food Protection Committee of the National Academy of Sciences–National Research Council and is updated regularly to remain current with the state of the art in the arena of food additives.

Consumers are aided in knowing what they are eating by the legal requirement for ingredient labeling on all food packages. All ingredients contained in a food, including all additives, are identified in descending order by weight. Such labeling is mandatory.

Additives are used for a variety of reasons, including some of the following:

1. To enhance nutritive value
2. To improve flavor
3. To improve color
4. To extend shelf life
5. To improve texture
6. To control pH
7. For leavening
8. For bleaching and maturing
9. To ease manufacturing problems

All of these reasons have economic implications because of the improved sales potential for highly palatable and/or nutritionally enhanced foods and because of modifications that make production and processing easier. Today's health-conscious consumers continue to maintain a high level of interest in the nutritive value of the foods they are eating and may be attracted to products to which specific nutrients have been added. Addition of thiamin, riboflavin, niacin, and iron to breads and cereal products and of iodine to salt has been sanctioned for many years to enrich the intake of nutrients that are inadequate in the diets of a large proportion of the population. More recently, nutrients have been added to a wide array of foods to meet consumer interest and demand for fortified foods.

Pleasure is a vital reason why people eat, and pleasure can be increased if the sensory qualities of a food are enhanced. This compelling rationale explains the motivation for adding ingredients that affect flavor, color, and texture. Additives that have particular appeal to the senses have dominated a considerable amount of research money because of the large economic potential for food products that are able to gain broad consumer acceptance and carve a niche in the highly competitive food industry.

Additives can be crucial to the development of food products with a satisfactory shelf life. Foods with molds or other microbiological contamination not only are offensive to the consumer who has purchased them, but they also may eliminate that consumer as a customer. Food manufacturers cannot afford this loss of market. In addition, harmful food products have the potential for lawsuits, which can be damaging to the company that is the object of the complaint. Additives that interfere with growth of microorganisms can be essential to avoiding such problems. Other ingredients help extend shelf life by aiding in retention of moisture.

Manufacturers may use additives to avoid or at least alleviate production problems resulting from physical properties of a food during its production. For example, the foaming that develops in the preparation of dried beans or in milk is but one illustration of problems faced in processing large quantities of food. Numerous other reasons for the inclusion of additives can be found throughout the food industry.

---

### FOOD FOR THOUGHT: Translating "EAFUS"

The world of food additives and their regulation is full of acronyms that can make visitors to this realm wish for the support of a handy guide or a friendly hand. "EAFUS" provides a fine illustration of the government's "alphabet soup." It really is a Food Additive Database, but the acronym for that would be FAD, and that simply would not be compatible with the serious world of food additive regulation. Therefore, the U.S. database that lists more than 3,000 food additives is called "Everything" Added to Food in the United States, which quickly was dubbed EAFUS.

EAFUS is the responsibility of the Center for Food Safety and Applied Nutrition, better known as CFSAN. This center, which is under the U.S. Food and Drug Administration, meets its responsibilities for EAFUS under its Priority-based Assessment of Food Additives (PAFA) program. [Incidentally, this acronym is not to be confused with PUFA (polyunsaturated fatty acids).] The additives listed in this database belong to one of four categories: direct, secondary direct, color additives, or GRAS.

EAFUS is constantly updated to reflect the current status of the review of toxicological evidence. The listing for each chemical includes a column for the Doc Type, one for the Doc Number, a column for the Mainterm (identifying name), a column for the CAS RN (Chemical Abstract Service Registry Number) or other code, and finally a column headed Regnum (for the section and part numbers under Title 21). The Doc Type is interesting because it identifies the specific status of each additive with regard to approval of toxicology information (e.g., fully updated toxicology information, initial toxicology literature search in progress, and no reported use and no toxicology information available).

EAFUS is an important database for food scientists developing new products that may require various additives in their formulations. The approval status of potential additives provides essential information in making ingredient decisions to create new products that can be manufactured efficiently without causing breakdowns during production.

## ADDITIVES

### Accidental Additives

Additives in foods can be divided into accidental and intentional. The accidental additives that sometimes occur in foods are the result of some unintentional incident, perhaps a hair, oil from a machine, or another contaminant that was not intended to be a part of the product yet is there. Careful monitoring of production from the initial ingredients to the packaged product is essential to elimination of such problems.

### Intentional Additives

Intentional additives represent quite a different type of additive; these are added to the food deliberately and at intended levels to

1. Maintain or improve safety and freshness
2. Improve or maintain nutritional value
3. Improve taste, texture, and appearance

Most additives are added directly to the food; indirect additives may be introduced from packaging. This is why packaging materials also are subject to approval.

Use of additives, whether direct or indirect, is sanctioned by the Food and Drug Administration only if the additive:

1. Performs a useful function
2. Does not deceive the customer by obscuring use of low-quality ingredients or poor manufacturing
3. Does not reduce nutritive value substantially
4. Does not merely accomplish the same result that improved manufacturing techniques could provide
5. Can be measured in the product by a recognized method of analysis

Capricious use of additives in food manufacturing is not allowed by the Food and Drug Administration. Unnecessary use of additives is not favored by the food industry either because of the added cost.

Various food additives were classified in the Federal Register [1974, 39(185), 34175] according to their technical functions in foods. These functions and some specific additives that are used for these purposes are as follows:

1. *Anticaking agents, free-flow agents*—Calcium silicate, dicalcium phosphate, mannitol, silicon dioxide, sodium aluminum silicate
2. *Antimicrobial agents*—Acetic acid, benzoic acid, propylene oxide, sodium nitrate, sodium nitrite, hydrogen peroxide
3. *Antioxidants*—Ascorbic acid, BHA (butylated hydroxyanisole), BHT (butylated hydroxytoluene), propyl gallate
4. *Colors, coloring adjuncts* (*color stabilizers, color fixatives, color retentive agents, etc.*)—Annatto, ultramarine, yellow dye No. 5, titanium dioxide, turmeric, red dye No. 40, FD&C Blue No. 1, FD&C Red No. 3, FD&C Green No. 3, FD&C Yellow No. 5, cochineal, saffron
5. *Curing, pickling agents*—Sodium erythorbate, sodium metaphosphate, sodium nitrate, sodium nitrite
6. *Dough strengtheners*—Ammonium sulfate, locust bean gum, monocalcium phosphate, potassium bromide, glycerol monostearate
7. *Drying agents*—Sodium silicoaluminate, sodium aluminum citrate
8. *Emulsifiers, emulsifier salts*—Lecithin, monoglycerides, potassium pyrophosphate, potassium polymetaphosphate, sorbitan monooleate, polysorbate 60, 65, and 80

9. *Enzymes*—Amylase, rennet, pectinase, peroxidase, papain

10. *Firming agents*—Aluminum sulfate, calcium chloride

11. *Flavor enhancers*—Monosodium glutamate, disodium guanylate, disodium inosinate

12. *Flavoring agents, adjuvants (enhancers)*—Allyl disulfide, black pepper, turmeric, mustard, ethyl vanillin

13. *Flour-treating agents (including bleaching and maturing agents)*—Acetone peroxide, azodicarbonamide, benzoyl peroxide

14. *Formulation aids (carriers, binders, fillers, plasticizers, film formers, tabletting aids, etc.)*—Corn syrup, mannitol, propylene glycol, sodium caseinate, sodium carboxymethyl cellulose

15. *Fumigants*—Phostoxin

16. *Humectants, moisture-retention agents, antidusting agents*—Sodium tripoly phosphate, sorbitol, glycerine

17. *Leavening agents*—Baking powder, sodium carbonate, baking soda (with acid), sodium aluminum sulfate, dicalcium phosphate

18. *Lubricants, release agents*—Propylene glycol, monosodium phosphated mono- and diglycerides

19. *Non-nutritive sweeteners*—Tagatose, sucralose

20. *Nutrient supplements*—Alanine, thiamin, potassium iodide

21. *Nutritive sweeteners*—Fructose, sucrose

22. *Oxidizing and reducing agents*—Acetone peroxide, thiosulfate, α-tocopherol

23. *pH control agents (including buffers, acids, alkalies, and neutralizing agents)*—Acetic acid, adipic acid, aluminum sodium sulfate, baking soda, calcium citrate

24. *Processing aids (clarifying agents, clouding agents, catalysts, flocculents, filter aids, etc.)*—Aluminum phosphate, calcium sulfate, dimethyl polysiloxane

25. *Propellants, aerating agents, gases*—Nitrous oxide

26. *Sequestrants*—Glucono-delta-lactone, potassium gluconate

27. *Solvents, vehicles*—Alcohol, glycerol, isopropanol, propylene glycol

28. *Stabilizers, thickeners (suspending and bodying agents, setting agents, gelling agents, bulking agents, etc.)*—Acacia gum, agar agar, alginatescellulose, karaya gum

29. *Surface-active agents (other than emulsifiers, including solubilizing agents, dispersants, detergents, wetting agents, rehydration enhancers, whipping agents, foaming agents, and defoaming agents)*—Ammonium stearate, silicon dioxide, petroleum waxes, decanoic acid, dimethyl polysiloxane

30. *Surface-finishing agents (including glazes, polishes, waxes, and protective coatings)*—Acacia gum, gum arabic (acacia gum), beeswax

31. *Synergists*—Citric acid, tartaric acid

32. *Texturizers*—Agar agar, gum arabic (acacia gum), locust bean gum, cellulose, starch

In the global world of today, food additives are an aspect of importance to many nations (Figure 21.2). In response to this issue, the functioning agency is the **Joint FAO/WHO Expert Committee on Food Additives (JECFA)**.

Natural additives sometimes are used to enhance palatability at home, but then their use is more likely to elicit praise for the cook rather than controversy (Figures 21.3 and 21.4). The herbs and spices that brighten recipes actually (Table 21.1) are additives, and these also may be added when commercial food products are prepared according to FDA regulations, although they need to be listed on the ingredient label.

Coloring agents provide an illustration of additives that are used to achieve the desired appearance. Some spices, for example paprika and saffron, and such brightly pigmented foods as beets and tomatoes contain natural coloring agents for commercial and home use. Annatto extract, beta-carotene, grape skin extract, cochineal extract or carmine, paprika

**Joint FAO/WHO Expert Committee on Food Additives (JECFA)** International committee charged with overseeing issues on food additives in international trade.

**Figure 21.2**   Spices are imported from Grenada and other places in the Caribbean, as well as from India, Indonesia, and many other countries, including the Middle East.

oleoresin, caramel color, fruit and vegetable juices, and saffron are familiar examples of natural sources of pigments used to enhance specific food products. In addition to natural coloring agents, the food industry has available FD&C Blue Nos. 1 and 2, FD&C Green No. 3, FD&C Red Nos. 3 and 40, FD&C Yellow Nos. 5 and 6, Orange B, and Citrus Red No. 2 to achieve the desired look.

## Manufacturing Applications

Selection of specific food additives for incorporation into commercial food products is based on the characteristics of the food requiring modification and the suitability of a particular additive, in terms of both its effectiveness in the food item and its cost compared with possible alternatives (Table 21.2). Food technologists conduct research to determine the most appropriate additives and the levels to use in formulating new products.

**Figure 21.3**   Cardamom pods form after the bloom fades on the tendril that rests on the ground. The tiny seeds in the pods are used as flavor accents in some Scandinavian cookies.

**Figure 21.4**   Star anise is part of the Chinese five spices and also is often included in garam masalas made for flavoring many Indian dishes.

**Table 21.1**    Additives: Herbs and Spices

| Additive | Some Suggested Uses |
|---|---|
| *Herbs* | |
| Basil | Tomato dishes, salads, meats |
| Bay leaf | Soups, stews, meats |
| Dill weed | Salads, pickles, fish |
| Marjoram | Soups, stews, meats, poultry, fish |
| Oregano | Stews, omelets, chili, tomato dishes |
| Parsley | Soups, sauces, salads |
| Rosemary | Meats, fish, soups, stews |
| Sage | Dressing for poultry, fish, salads |
| Thyme | Soups, stews, tomato dishes |
| *Spices* | |
| Cayenne | Meats, fish, vegetables |
| Cinnamon | Baked products, sweet potatoes, pickles |
| Cloves | Ham, stews, pickled fruit, baked products |
| Ginger | Baked products, stir fries, chutney |
| Mace | Chocolate, cakes, puddings, fish sauce |
| Mustard | Meats, sauces, salads |
| Nutmeg | Baked custard, eggnog, baked products |
| Paprika | Garnish, chicken and meat dishes |
| Pepper | Meats, vegetables, eggs, sauces, salads |
| Saffron | Rice, chicken, baked products |
| Turmeric | Meats, dressings, relishes, salads |
| *Blends* | |
| Chili powder | Mexican dishes, eggs, meats |
| Curry powder | Indian dishes, poultry, fish, rice, vegetables |
| Fines herbs | Vegetables, meats, poultry, fish |
| Poultry seasoning | Stuffing, poultry, fish, rice |

## Safety

The Food Safety and Inspection Service (FSIS) in the U.S. Department of Agriculture and the Food and Drug Administration in the U.S. Department of Health and Human Services share the responsibility for the safety of egg, poultry, and meat products, but the FDA alone is responsible for most other foods. Initial evaluation of the safety of food additives is under the FDA; Labeling and Consumer Protection specialists in FSIS also evaluate the safety, function, and conditions of use for additives proposed for egg, poultry, and meat products. The FDA has the final authority, but FSIS may apply more stringent standards.

The safety of additives in the food supply is a question of personal and social significance, yet one that is proving to be difficult to document. The protocol for testing the toxicity of an additive includes acute tests; prolonged tests; and chronic, long-term or extended tests.

- *Acute tests:* Single massive dose administered to two species, at least one being a nonrodent. This dosage level is lethal to half of the test animals. Survivors are studied for seven days.

- *Prolonged tests:* Three different doses administered daily to two species for a minimum of three months. The control group and the three test groups usually have at least 10 males and 10 females each. Weekly physical examinations are conducted throughout the test period, which is followed by complete autopsies and histologic study of organs.

- *Chronic tests:* Same testing as for prolonged tests, but the time frame is at least a year and often at least 18 months, and the number of animals tested is three or four times greater. Some additives are tested for possible effects on reproduction and may include a three-generation study.

Selection of animals for testing is difficult. Rodents have many more intestinal bacteria than do humans, thus modifying the products available for absorption and creating differences in

**Table 21.2**   Functional Needs and Some Possible Additives to Meet Manufacturing Requirements for Food Products

| Functional Need | Some Possible Additives to Meet Need |
|---|---|
| Enhance nutritive value | Potassium iodide, ferrous gluconate and other iron salts, calcium salts, vitamins, amino acids |
| Improve flavor | Acetanisole, acetophenone, allyl disulfide, amyl propionate, benzoyl acetate, benzoyl isoeugenol, corn syrup, dextrose and other sweeteners, disodium guanylate, and inosinate |
| Improve color | Annatto, canthaxanthin, caramel, carotenes, citrus red No. 2, cochineal, turmeric; FD&C blue No. 1, green No. 2, red Nos. 3 and 40, yellow No. 5; saffron |
| Extend shelf life | Ascorbic acid, bisulfite salts, BHA, BHT, butyl paraben, calcium lactate, sorbic acid, calcium propionate and sorbate, citric acid, EDTA, gum guaiac, heptyl parabens |
| Improve texture | Gums (acacia or arabic, carob bean, carrageenan, guar, ghatti, karaya, tragacanth, xanthan), agar, alginates, aluminum phosphate, aluminum and ammonium sulfates, ammonium and calcium alginates, calcium chloride, calcium salts (stearate, silicate, stearoyl-2-lactylate), cellulose, cholic acid, cornstarch, dextrin, diglycerides, disodium phosphate, lecithin, modified food starch, polysorbates, silicon dioxide |
| Control pH | Acetic acid, adipic acid, baking soda, benzoic acid, calcium salts (carbonate, citrate, gluconate, pyrophosphate), citric acid, dicalcium phosphate, hydrochloric acid, phosphates, phosphoric acid, potassium acid citrate |
| Leaven | Baking powder, baking soda, calcium carbonates, yeasts, glucono delta-lactone |
| Bleach and mature | Azodicarbonamide, benzoyl peroxide, calcium bromate, glycerol monostearate, hydrogen peroxide, potassium bromide |
| Ease manufacturing problems | Aluminum phosphate and stearate, butyl stearate, calcium salts (lactobionate, silicate, stearate, sulfate), decanoic acid, dimagnesium phosphate, dimethyl polysiloxane, dioctyl sodium sulfosuccinate, magnesium carbonate, oleic acid |

the potential risks to the subjects. Additional differences are found in the ways in which various species metabolize substances. Clearly, it is important to determine the safety of substances that are deliberately added to the food supply and to ban the use of those found to be unsafe.

## EDIBLE FLOWERS

Although flowers are not traditionally considered foods, they increasingly are used as garnishes or color and flavor accents. Nasturtiums are one of the flowers commonly used in these ways. Others that are also safe to eat include pansies, violas, chrysanthemum and carnation petals, fuchsias, geraniums, jasmine, lavender, some roses, violets, and squash blossoms (with the stamens removed).

Flowers selected either as a garnish on a plate or as an ingredient must be safe. Flowers from gardens or commercial sources may have been treated with pesticides/herbicides because these likely are not intended for human consumption. Unless flowers are known to be free from these contaminants, they should be used in a centerpiece, not served on plates with food. Some flowers are poisonous to humans. Sweet peas, oleander, many lilies, tuberous begonias, and tulips are some of the flowers that cannot be considered edible because they can cause serious to fatal reactions when eaten.

## HERBS AND SPICES

As people have broadened their dining experiences to include a wide range of foods from other cultures, the use of herbs and spices has increased, both in commercial products and in recipes prepared in many homes. **Herbs** are plants that grow seasonally, often just as annuals, and are valued for the range of flavors they can add to various dishes. **Spices** are gathered from sturdy plants that are not tied to seasonal growth. For instance, cinnamon is the bark of a tree, and it is prized for its ability to blend with fruits and other foods to add to the pleasure of eating. Fresh herbs commonly available in the produce section include basil, chives, dill, oregano, parsley, rosemary, sage, tarragon, and thyme. Even wider selections of dried herbs and spices are available on store shelves.

The strength of the flavors from fresh herbs is more delicate than is found in dried counterparts. A fresh herb can be substituted in a recipe calling for a dried herb by using about eight times more of the fresh herb. Table 21.3 identifies some dishes in which herbs can be used particularly effectively.

**Herbs**
Various parts of seasonal plants used in cooking to add a variety of flavors.

**Spices**
Seasonings harvested from aromatic edible plants and used to add some intensity of flavor.

**Table 21.3** Suggested Uses for Some Herbs and Spices

| Herb/Spice | Suggested Uses |
| --- | --- |
| Basil | Tomato-containing dishes, eggs, meats, vegetables, salads, rice |
| Chile | Chili, Mexican dishes, beans |
| Chives | Stews, yogurt and sour cream dips, vegetables |
| Cinnamon | Fruit desserts, curry, chicken |
| Cloves | Ham, curry |
| Cumin | Curry, meats, soups |
| Dill | Fish dishes, cucumbers, salads, bread, potatoes |
| Ginger | Cookies, curry, stir-fry dishes |
| Mint | Lamb, tea and citrus beverages, fruits |
| Oregano | Tomato-containing dishes, eggs, soups |
| Parsley | Tomato-containing sauces, soups, vegetables |
| Rosemary | Meats, soups, vegetables |
| Sage | Poultry and stuffing, soups |
| Tarragon | Meats, beans, cucumbers, yogurt |
| Thyme | Meats, poultry, eggs, fish |

Some herbs and spices are useful because of their antimicrobial action. In marinades, sauces, and dressings, herbs and spices such as basil, cloves, rosemary, mustard, cinnamon, garlic, and thyme may help to reduce growth of certain microorganisms. They can serve as a helpful protection but cannot replace the need for adequate refrigeration or other techniques, depending on the food. Of course, the flavor contributions of these ingredients are valued and must be considered when selecting herbs and spices in a specific product. Some people have allergic responses to specific herbs and spices, which makes it important for them to be aware of these ingredients.

---

## FOOD FOR THOUGHT: Truffles Treasures

Truffles are very special ingredients that are famous for being rare, expensive, and exotic. They are an edible type of mushroom, a fungus that grows a short distance beneath the earth's surface and has a symbiotic relationship with the roots of certain trees. The hyphae of the truffle are filaments that transfer water and various nutrients from the soil into the roots of the tree to feed it. In return, the tree makes sugar through photosynthesis and passes it from its roots into the truffle. This symbiotic arrangement explains why truffles in the wild are found in forests under the ground among the roots of specific types of trees.

Black truffles (*Tuber melanosporum*) sometimes are found in the ground under oak trees in parts of France and Italy. These are very expensive because of their scarcity and their unique impact on food flavors. Pigs and dogs can detect their aroma as they search the countryside, so truffle hunters often use these animals to lead them to the edible treasure. Truffles can be sliced thinly and eaten raw, or they may be cooked and used to flavor a wide range of recipes. Fortunately, their flavor is sufficiently intense that a small slice can flavor a large quantity of food.

South of the equator in Namibia, the Kalahari truffle sometimes can be found in the spring growing a short distance under Bushman grass in the Kalahari Desert. Growth of the grass and the buried truffles occurs somewhat sporadically and is dependent on rainfall. Good crops usually develop only about every four years. Kalahari truffles resemble round small potatoes and are a light brown.

Unfortunately, Kalahari truffles need to be eaten within a week, which is another reason they are not available in markets in North America. A person fortunate enough to be in Namibia when the truffles are harvested can feast upon them either raw or cooked. The price of this Namibian delicacy is low there. They cannot be bought at any price in Europe.

## DIETARY SUPPLEMENTS

Many extracts and other materials from plants, as well as various nutrients, are produced as dietary supplements and marketed heavily to a health-conscious public. Countless health claims and marketing ploys are used to generate a remarkable business for companies producing a wide range of dietary supplements.

The 1994 **Dietary Supplement Health and Education Act (DSHEA)** limits the FDA's authority over dietary supplements. Manufacturers of these products have considerable freedom to market their products without authorization from the FDA. This is in sharp contrast to the requirements for food ingredients, which must be approved by the FDA. Under DSHEA, the FDA can prove that dietary supplements are unsafe only after the supplements are on the market. This means that unsafe dietary supplements may be on the market for a period of time before sufficient proof of risk can be gathered for the FDA to ban them.

Labeling requirements are defined for dietary supplements and must include

**Dietary Supplement Health and Education Act (DSHEA)**
Act from 1994 that defines dietary supplements and limits the role of the FDA in regulating them.

- Statement of identity (e.g., "ginseng")
- Net quantity of contents (e.g., 60 capsules)
- Structure-function claim and the statement "This statement has not been evaluated by the Food and Drug Administration. This product is not intended to diagnose, treat, cure, or prevent any disease."
- Directions for use (e.g., "Take one capsule daily".)
- Supplement Facts panel (lists serving size, amount, and active ingredient)
- Other ingredients in descending order of predominance and by common name or proprietary blend
- Name and place of manufacturer, packer, or distributor: This is the address to write for more product information.

Dietary supplements are not drugs, and they are not food ingredients so are not under tight regulations the FDA applies to drugs and ingredients. Three of the dietary supplements that are consumed most widely today are gingko biloba, echinacea, and ginseng. Consumers are cautioned to check with their physicians before using dietary supplements and to be particularly wary of inexpensive competitors from uncertain sources.

Ginkgo biloba, an extract from the leaves of the gingko or maidenhair tree, is a dietary supplement that is available in many health food stores. It is suggested to be effective in increasing blood flow to the brain and extremities of the body. It may have side effects, including diarrhea, nausea, vomiting, and restlessness. As is true with other dietary supplements, the quality may vary significantly from one manufacturer to another, and safety is not assured.

Echinacea is produced from snakeroot, also called purple coneflower. Some people take this to help fight the common cold. Although some people are allergic to echinacea, it does help to stimulate the immune system, which is the reason for using it against colds in the initial stage of infection.

Ginseng is a dietary supplement that can be traced back more than 7,000 years to China. Evidence for any benefits from ginseng is contradictory and unconvincing to date. Side effects may include hypertension and gastrointestinal difficulties.

Consumers interested in eating "natural" foods and botanicals for possible health benefits are prompting food manufacturers to consider inclusion of some botanicals to appeal to this segment of the market. Teas and other beverages boasting of such ingredients as yerba mate, ginkgo, and rooibas (from South Africa) are gaining a significant place in the market.

## SUMMARY

Food additives are important constituents of processed and convenience foods in the United States. Regulation of the commercial use of additives is under the U.S. Food and Drug Administration. The Food Additives Amendment to the Federal Food, Drug, and Cosmetic

Act of 1938, passed in 1958, requires that the safety of all additives other than those on the GRAS list be demonstrated by the additive manufacturer and approved by the FDA prior to inclusion in foods being prepared for sale.

Additives are used to improve nutritive value, flavor, color, and texture; to reduce losses during marketing; and to ease manufacturing problems that occur in quantity production. Accidental additives must be avoided by careful control of manufacturing. Intentional additives must serve a useful purpose without being harmful to health. More than 2,000 additives are available for inclusion in foods and serve in improving processed and convenience foods in many ways. The safety of these additives is tested in acute, prolonged, and chronic tests.

## STUDY QUESTIONS

1. What legislation determines the use of additives in foods (include all legislative actions that apply)?

2. What are the pros and cons of the use of additives?

3. What is the Delaney clause and how does it conflict with the concept of "risk versus benefit"?

4. What conditions must an additive meet before it is used in a food?

5. Explain the reason for the inclusion of each food additive listed on the ingredient label of five different food products.

6. Suggest appropriate herbs, spices, and flowers for enhancing colors and flavors in food products.

7. True or false. Food manufacturers in this country have the option of listing only the 10 major ingredients on a food label.

8. True or false. Food packaging materials need to be approved by the CDC.

9. True or false. Additives may be used to enhance color to obscure quality of a food.

10. True or false. Additives are permitted to thicken a sauce.

11. True or false. High-fructose corn syrup is an approved additive.

12. True or false. The GRAS list includes only locally grown produce.

13. True or false. The Delaney clause has been rescinded.

14. True or false. The agency responsible for administering and enforcing the Food Additive Amendment of 1958 and subsequent related legislation is the Department of Homeland Security.

15. True or false. EAFUS stands for Everything added to food in the United States.

## BIBLIOGRAPHY

Ash, M., and Ash, I. 2008. *Handbook of Food Additives*. 3rd ed. Synapse Information Services, Inc. Endicott, NY.

Branen, A. L., et al. 2001. *Food Additives Revised and Expanded*. CRC Press. Boca Raton, FL.

Brennan, J. G. 2005. *Food Processing Handbook*. Wiley. New York.

Brody, A. J. 2009. Food packaging migrants: Hazardous or insignificant? *Food Technol. 63* (10): 75.

Emerton, V. and Choi, E. 2008. *Essential Guide to Food Additives*. 3rd ed. Royal Society of Chemistry. London.

Foster, R. J. 2010. Naturally colorful. *Food Product Design 20* (supplement): 3.

Lioe, H. N., et al. 2010. Soy sauce and umami taste: A link from the past to current situation. *J. Food Sci. 75* (3): R71.

Luck, E. 1997. *Antimicrobial Food Additives: Characteristics, Uses, Effects*. Springer-Verlag. New York.

Matromatteo, M., et al. 2010. Use of lysozyme, nisin, and EDTA combined treatments for maintaining quality of packed ostrich patties. *J. Food Sci. 75* (3): M178.

Molins, R. A. 2001. *Food Irradiation*. Wiley. New York.

Nabors, L. O. 2007. Regulatory status of alternative sweeteners. *Food Technol. 61* (2): 24.

Ohr, L. 2008. Adding spice to life. *Food Technol. 62* (1): 59.

Potter, N. N. and Hotchkiss, J. H. 1999. *Food Science*. 5th ed. Springer-Verlag. New York.

Watson, D. H. 2002. *Food Chemical Safety*. Vol. 2. Additives. CRC Press. Boca Raton, FL.

World Health Organization. 2010. *Evaluation of Certain Food Additives: Sixty-Ninth Report of the Joint FAO/WHO Expert Committee on Food Additives*. (WHO Technical Report Series). World Health Organization. Rome.

## INTO THE WEB

*http://www.fda.gov/opacom/backgrounders/miles.html*—Listing of federal food-related legislation.

*http://www.fda.gov/Food/FoodIngredientsPackaging/ GenerallyRecognizedasSafeGRAS/GRASListings/default. htm*—Summary of GRAS notices.

*www.leffingwell.com/gras19.htm*—Flavor and Extract Manufacturers (FEMA) expert panel; Report on GRAS list for flavor additives (19th listing).

*http://www.usp.org/fcc/*—Information on the *Food Chemicals Codex.*

*http://www.google.com/search?q=Delaney+clause&hl=en &sa=G&tbs=tl:1&tbo=u&ei=BJnES-O5FYP6sgP67-SDDQ &oi=timeline_result&ct=title&resnum=11&ved=0CCs Q5wIwCg*—Time line on actions dealing with the Delaney clause.

*http://www.accessdata.fda.gov/scripts/fcn/fcnNavigation. cfm?rpt=eafusListing*—Everything Added to Food in the United States (EAFUS) list.

*http://www.fda.gov/Food/FoodIngredientsPackaging/ ucm094211.htm#coloradd*—FDA brochure on food additives.

*http://www.foodsubs.com/Chilefre.html*—Descriptions of many types of chiles.

# Glossary

**A band**   Total portion of the sarcomere in which thick and thin myofilaments overlap; includes the H band.

**Acesulfame-K**   Low-calorie sweetener derived from acetoacetic acid.

**α-amylase**   Amylose-digesting enzyme contained in abundance in egg yolk and to a lesser extent in egg white.

**α-lactose**   Less soluble form of lactose, the disaccharide prominent in milk; form of lactose that is largely responsible for the sandy texture of some ice creams.

**Acidophilus milk**   Fermented milk product (usually whole milk) to which *Lactobacillus acidophilus* is added to digest the lactose.

**Acrolein**   A highly irritating and volatile aldehyde formed when glycerol is heated to the point at which two molecules of water split from it.

**Acrylamide**   Carcinogen formed in starchy fried foods and also in baked products.

**Actin**   Myofibrillar protein existing primarily in two forms (F and G).

**Actinidin**   Proteolytic enzyme in kiwi fruit.

**Active dry yeast**   Granular form of dried *Saccharomyces cerevisiae* (8 percent moisture); storage is at room temperature and rehydration at 40–46°C (104–115°F).

**Actomyosin**   Protein complex of actin and myosin that forms to effect contraction of the sarcomere.

**Adenosine triphosphatase (ATPase)**   Enzyme in muscle tissue involved in glycolytic reactions leading to lactic acid formation.

**Aerobic**   Requiring oxygen for survival and growth.

**Affective testing**   Sensory testing to determine acceptability or preference between products.

**Aflatoxicosis**   Condition caused by ingesting aflatoxin; begins with nausea and vomiting but ends with convulsions, coma, and death.

**Aflatoxin**   Carcinogenic mycotoxin produced by *Aspergillus flavus* and *Aspergillus parasiticus*.

**Aftertaste**   The aromatic message of the flavor impression that lingers after food has been swallowed.

**Aging (maturing) of flour**   Chemical process to modify the sulfhydryl groups in flour to disulfide linkages, usually utilizing a chlorine-containing compound.

**Aging (of meat)**   Holding of meat while it passes through rigor mortis and sometimes for a period extending several days or even two weeks, depending on the quality and type of carcass; storage of meat to enhance tenderness.

**Air cell**   Space between the inner and outer shell membranes at the large end of an egg.

**Albumen**   White of an egg; consists of three layers.

**Albumen index**   Grading measurement of albumen to determine quality on the basis of the amount of thick white.

**Aldose**   Hexose with one carbon atom external to the 6-membered ring.

**Alitame**   Sweetener resulting from combining L-aspartic acid, D-alanine, and an amine.

**Alliin**   Odorless precursor in garlic that ultimately is converted to diallyl disulfide.

**Alliinase**   Enzyme in garlic responsible for catalyzing the conversion of alliin to diallyl thiosulfinate, the precursor of diallyl disulfide.

**Allium**   Genus including onions, chives, garlic, shallots and leeks; unique for its sulfur-containing flavor compounds.

**All-purpose flour**   Multi-use flour made from hard wheat or a mixture of hard and soft wheat; contains about 10.5 percent protein.

**Alpha (α) crystals**   Extremely fine and unstable fat crystals.

**Amebic dysentery**   Protozoan infection characterized by diarrhea of varying severity (sometimes containing mucus and blood) and sometimes more severe symptoms.

**Amorphous**   Form of solid lacking an organized, crystalline structure.

**Amorphous candies**   Candies that lack an organized crystalline structure because of their very high concentration of sugar or interfering substances.

**Amphoteric**   Capable of functioning as either an acid or a base, depending on the pH of the medium in which the compound is found.

**Amygdalin**   Cyanogenic glycoside in apricot and peach pits that can cause cyanide poisoning if consumed.

**Amylograph**   Device designed to control the temperature of a starch paste and to measure its viscosity.

**Amylopectin**   Branched fraction of starch consisting primarily of glucose units linked with 1,4-α-glucosidic linkages, but interrupted occasionally with a 1,6-α-linkage resulting in a very large polysaccharide.

**Amylose**   Straight-chain, slightly soluble starch fraction consisting of glucose units joined by 1,4-α-glucosidic linkages.

**Anaerobic**   Requiring an oxygen-free environment for survival and growth.

**Analysis of variance (ANOVA)**   Statistical approach to determining differences between many sets of data.

**Angel food cake**   Foam cake containing an egg white foam, sugar, and cake flour.

**Angstrom (Å)**   Unit of measure; for example, 1 cm = 100,000,000 Å.

**Anthocyanidin**   Anthocyanin-type pigment that lacks a sugar in its structure.

**Anthocyanin**   Flavonoid pigment in which the oxygen in the central ring is positively charged.

**Anthoxanthin**   Flavonoid pigment in which the oxygen in the central ring does not carry an electrical charge.

**Arborio rice**   Medium-grain rice used for making paella and risotto.

***Ascaris lumbricoides***   Type of parasitic roundworm that may be present in some seafood and may cause illness leading to possible pneumonia if food is not heated enough to kill it.

**Aspartame**   Very sweet, low-calorie methylated dipeptide composed of phenylalanine and aspartic acid; used as a high-intensity sweetener.

***Aspergillus flavus***   Mold capable of making aflatoxin in nuts, grains, and legumes stored under moist, warm conditions.

***Aspergillus parasiticus***   Mold capable of making aflatoxin in stored nuts, grains, and legumes.

**Astaxanthin**   Reddish-orange carotenoid pigment in salmon and in cooked crustaceans.

**Astringency**   Puckery feeling in the mouth created by compounds such as tannins.

**Autoxidation**   Oxidation reaction capable of continuing easily with little added energy.

**Avidin**   Albumen protein that binds biotin when in the native state, but not when it is denatured.

**Beta (β) crystals**   Extremely coarse and, therefore, undesirable fat crystals.

**Bacillary dysentery**   Food-borne illness characterized by severe diarrhea and electrolyte loss, accompanied by intense abdominal cramps and a high fever and capable of ending in death.

***Bacillus coagulans***   Anaerobic bacillus that digests sugar to lactic acid in canned foods to cause flat-sour spoilage.

**Baked custard**   Sweetened milk and egg mixture that is baked without agitation until the egg protein coagulates and forms a gel.

**Baking ammonia**   Ammonium salt of carbonic acid [$(NH_4)_2CO_3$] that dissolves to release ammonia, carbon dioxide, and water for leavening springerle and other rather dry cookie doughs, crackers, and biscotti.

**Baking powder**   Mixture of acid and alkaline salts and a standardizing agent to produce at least 12 percent of the carbon dioxide available for leavening.

**Basmati rice**   Long-grain, aromatic rice.

**Benzoyl peroxide**   Food additive approved by the FDA for bleaching the xanthophylls in refined flours.

**Beta prime (β') crystals**   Very fine and reasonably stable fat crystals.

**Beta ray**   Radiant energy that is very slightly longer than gamma rays and that can penetrate food, but not aluminum.

**Betacyanins**   Group of betalains responsible for the reddish-purple color of beets; not an anthocyanin, but behaves colorwise in the same fashion.

**Betalains**   Two groups of pigments (betacyanins and betaxanthins) that contribute the anthocyanin-like color to beets, but differ chemically from the anthocyanins.

**Bicarbonate of soda**   Alkaline food ingredient ($NaHCO_3$) used to react with acids to form carbon dioxide.

**Biotechnology**   Development of new products by making a genetic modification in a living organism.

**Birefringence**   Refraction of light in two slightly different directions.

**Biscuit**   Quick bread with a ratio of about 3:1 (flour to liquid) and also containing fat cut into small particles, baking powder, and salt; usually kneaded and rolled, but sometimes dropped.

**Biscuit method**   Mixing method in which the dry ingredients are combined, the fat is cut into small particles in the flour mixture, all of the liquid is added at once and stirred in just to blend, and the dough is kneaded; a flaky product results.

**Blast freezing**   Freezing in air-blast freezers with frigid (from −30 to −45°C) air circulating at a high velocity; causes rapid freezing and small ice crystals.

**Bleaching**   Refining step in which coloring and flavoring contaminants are removed from fats, often by filtration through active charcoal or other suitable substrate.

**Bloom**   Granular-appearing, discolored areas on the surface of chocolate, the result of melting of less stable crystals and recrystallization as β crystals.

**Bloom (eggs)**   Natural protective coating sealing the shell pores when an egg is laid.

**Bloom gelometer**   Modification of a penetrometer designed especially for measuring the tenderness of gels.

**Boiling point**   Temperature at which vapor pressure of a liquid just exceeds atmospheric pressure.

**Botulism**   Potentially fatal food poisoning results from ingesting even a minuscule amount of toxin produced by *Clostridium botulinum.*

**Bound water**   Water that is bound to other substances and no longer exhibits the flow properties and solvent capability commonly associated with water.

**Bran**   Outer layers (fibrous and very high in cellulose) encasing the interior endosperm and germ of cereal grains.

**Bread flour**   Hard wheat, long-patent flour with a protein level of about 11.8 percent.

**Brix scale**   Hydrometer scale designed to indicate the percentage concentration of sugar in sugar solutions.

**Bromelain**   Proteolytic enzyme in pineapple.

**Bt**   Designation that a seed has been modified by splicing a gene from *Bacillus thuringiensis* to promote resistance to insects.

**Butylated hydroxyanisole (BHA)**   Antioxidant effective in animal fats used in baking.

**Butylated hydroxytoluene (BHT)**   Antioxidant used to retard oxidation in animal fats.

**Cage-free eggs**   Eggs laid by hens raised on the floor of a building rather than in cages.

**Cake flour**   Soft wheat, short-patent flour (about 7.5 percent protein).

**Calcium-activated factor (CAF)**   Proteolytic enzyme activated by calcium; contributes to tenderizing of aging meats.

**Campylobacteriosis**   Food-borne illness caused by toxin from *Campylobacter jejuni* (or other toxin-producing strains).

**Campylobacter jejuni**   Common strain of *Campylobacter* that can cause campylobacteriosis.

**Canola oil**   Oil from rape (*Brassica napus*) seeds, a genetically modified variety of rape.

**Capsaicinoids**   Capsaicin and related compounds responsible for the fiery quality of chili and other peppers.

**Caramelization**   Fragmentation of monosaccharide into a variety of compounds, including organic acids, aldehydes, and ketones, as a result of extremely intense heat.

**Carotenes**   Group of carotenoids containing only hydrogen and carbon in a polymer of isoprene.

**Carotenoids**   Class of pigments contributing red, orange, or yellow color as a result of the resonance provided by the isoprene polymers.

**Casein**   Collective name for milk proteins precipitated at pH 4.6.

**Casein micelle**   Casein aggregate that is comparatively stable and remains colloidally dispersed unless a change such as a shift toward the isoelectric point or the use of rennin destabilizes and precipitates casein.

**Case-ready meats**   Meats that are processed and packaged in retail packaging at a centralized site for distribution ultimately to retail markets.

**Catechins**   Flavonoid pigments that are a subgroup of the flavonols.

**Cathepsins**   Group of proteolytic enzymes that can catalyze hydrolytic reactions leading to the passing of rigor mortis.

**Cellulose**   Complex carbohydrate composed of glucose units joined together by 1,4-β-glucosidic linkages.

**Certified raw milk**   Milk that has a small microorganism population but has not been heat treated, and, therefore, may cause serious illness.

**Cestodes**   Parasitic tapeworms, some of which use humans as hosts; possible food sources are various meats and fish that have been undercooked or eaten raw.

**Chalazae**   Thick, rope-like extensions of the chalaziferous layer that aid in centering the yolk in the egg.

**Chalaziferous layer**   Membranous layer surrounding the vitelline membrane of egg yolk.

**Chiffon cake**   Foam cake that includes oil and egg yolk as liquid ingredients, an egg white foam, baking powder, sugar, and cake flour.

**Cholera**   Food-borne illness especially problematic in tropical areas around the world; caused by *Vibrio cholerae*.

**Chlorophyll a**   Blue-green, more abundant form of chlorophyll; the chlorophyll form in which the R group is a methyl group.

**Chlorophyll b**   Yellowish-green form of chlorophyll in which the R group is an aldehyde group.

**Chlorophyllase**   Plant enzyme that splits off the phytyl group to form chlorophyllide from chlorophyll.

**Chlorophyllide**   Chlorophyll molecule minus the phytyl group; water-soluble derivative of chlorophyll responsible for the light-green tint of water in which green vegetables have been cooked.

**Chlorophyllin**   Abnormal green pigment formed when the methyl and phytyl groups are removed from chlorophyll in an alkaline medium.

**Chloroplast**   Type of plastid containing chlorophyll.

**Cholesterol**   A sterol found in abundance in egg yolk.

**Chromatography**   Separation of discrete chemical compounds from a complex mixture by the use of solvents or gases; separation may be accomplished by use of a GLC or a HPLC or by other somewhat less sophisticated means.

**Chromoplast**   Type of plastid containing carotenoids.

**CIE**   Commission Internationale de L'Eclairage; group that established a system of measuring color based on spectral color, degree of saturation, and brightness.

**Ciguatera poisoning**   Food-borne illness caused by ciguatoxin produced in fish that have eaten the algae *Gambierdicus toxicus*.

**Circumvallate papillae**   Large, obvious protuberances always containing taste buds and distinguished easily because they form a "V" near the back of the tongue.

**Cis configuration**   An arrangement in which the hydrogen is attached to the carbon atoms on either end of the double bond from the same direction, causing a lower melting point than its *trans* counterpart.

**Climacteric**   Period of maximum respiratory rate just prior to the full ripening of many fleshy fruits.

**Climacteric fruit**   Fruit that continues to ripen after it has been picked—for example, bananas and peaches.

**Clostridium botulinum**   Anaerobic spore-forming bacteria that can produce a highly poisonous toxin capable of killing people.

**Clostridium nigrificans**   Bacteria capable of producing hydrogen sulfide in canned foods to cause sulfide spoilage.

**Clostridium perfringens**   Anaerobic, spore-forming bacterium that can produce a toxin capable of causing a mild food-borne illness.

**Coagulation**   Precipitation of protein as molecules aggregate (often as a result of energy input, such as heating or beating).

**Coldpack (club) cheese**   Cheese product made by adding an emulsifier to a mixture of natural cheeses.

**Cold point**   Coldest area of food in a can being heat processed.

**Cold pressing**  Mechanical pressing of olives to express oil without heat, resulting in an oil of excellent purity.

**Cold shortening**  Severe contraction of muscles in carcasses that have been chilled too quickly and severely after slaughter.

**Collagen**  Fibrous protein composed of three strands of tropocollagen.

**Collenchyma tissue**  Aggregates of elongated collenchyma cells providing supportive structure to various plant foods, notably vegetables.

**Colloid**  Material with a particle size between 0.001 and 1 millimicron.

**Colloidal dispersion**  Two-phase system in which the particles in the dispersed phase are between 0.001 and 1 micron in diameter.

**Color-difference meter**  Objective machine, such as the Hunter color-difference meter or Gardner color-difference meter; capable of measuring color difference between samples utilizing the CIE or Munsell color systems.

**Commercially sterile**  Food that has been heat processed enough to kill all pathogenic microorganisms and spores.

**Comminuted meats**  Products made by almost pulverizing meats and adding the desired fat and salts before heating the resulting mixture.

**Complex carbohydrates**  Polysaccharides, such as starch and cellulose.

**Composite ice cream**  Frozen dessert containing at least 8 percent milk fat and 18 percent total milk solids and no more than 0.5 percent edible stabilizer; flavoring particles are not to exceed 5 percent by volume.

**Compressed yeast**  *Saccharomyces cerevisiae* in a cornstarch-containing cake with a moisture level of 72 percent; requires refrigerated storage; dispersion is best at 32 to 38°C (89 to 100°F).

**Compressimeter**  Objective equipment that measures the force required to compress a food sample to a predetermined amount.

**Conalbumin**  Protein in egg albumen capable of complexing with iron ($Fe^{+3}$) and copper ($Cu^{+2}$) ions to form red and yellow colors, respectively.

**Concatenation**  A linking together; nonspecific description of the association of glutenin molecules by disulfide linkages.

**Conduction**  Transfer of energy from one molecule to the adjacent molecule in a continuing and progressive fashion so that heat can pass from its source, through a pan, and ultimately throughout the food being cooked.

**Cones**  Cone-shaped dendrites of photoreceptor neurons that enhance the sharpness of visual images and add the dimension of color to vision.

**Conjugated proteins**  Proteins combined with some other type of compound, such as a carbohydrate or lipid.

**Consistometer**  Device for measuring the spread or flow of semisolid foods in a specified length of time.

**Consumer panel**  Sensory evaluation panel selected from people who happen to be available at a test site and are willing to participate.

**Continuous phase**  Medium surrounding all parts of the dispersed phase so that it is possible to pass throughout the emulsion in the continuous phase without traversing any portion of the dispersed phase.

**Convection**  Transfer of heat by the circulation of currents of hot air or liquid resulting from the change in density when heated.

**Convection oven**  Oven designed with enhanced circulation of heated air to increase heating by convection, reduce baking time, and promote optimal crust browning.

**Conventional method**  Mixing method in which fat and sugar are creamed, beaten eggs are added, and dry ingredients (a third at a time) and liquid ingredients (half at each time) are added alternately; fine texture and excellent storage qualities are advantages of this laborious method.

**Conventional sponge method**  Method of making cakes in which part of the sugar and all of the egg are withheld to make a sugar-stabilized meringue that is folded into the cake batter as the final step in its preparation.

**Cooking losses**  Total losses from meat by evaporation and dripping during cooking.

**Corn syrup**  Sweet syrup of glucose and short polymers produced by the hydrolysis of cornstarch.

**Correlational research**  Research that determines interrelationships between variables.

**Creaming**  Separation of fat from the aqueous portion of milk that takes place when fat globules cluster into larger aggregates and rise to the surface of milk.

**Cross-linked starch**  Starch produced under alkaline conditions, usually in combination with acetic or succinic anhydride; notable as a thickener and stabilizing agent that undergoes minimal retrogradation.

**Cruciferae**  Family of vegetables including Brussels sprouts, cabbage, rutabagas, turnips, cauliflower, kale, and mustard; includes sulfur-containing flavor compounds that differ from those found in *Allium* vegetables.

**Crustaceans**  Shrimp, lobsters, crabs, and other shellfish with a horny covering.

**Cryogenic liquids**  Substances that are liquid (not solids) at extremely cold temperatures.

**Cryophilic**  Microorganisms with optimal reproduction and survival below 15°C (59°F).

**Crystalline candies**  Candies with organized crystalline areas and some liquid (mother liquor).

**Cultured buttermilk**  Low fat or nonfat milk containing *S. lactis* and *L. bulgaricus* that has been incubated to produce some lactic acid.

**Curd**  Milk precipitate that contains casein and forms readily in an acidic medium.

**Cyclamate**  Sweetener widely used in the world but banned in the United States.

**Cyclodextrins (a, b, g CD)**  Cyclic compounds containing 6–8 glucose units derived from starch by bacterial enzymes (cyclodextrin glucosyl transferases).

**Dark-cutting beef**  Very dark, sticky beef from carcasses in which the pH dropped to only about 6.6.

**Degradation**  Opening of the ring structure as the prelude to the breakdown of sugars.

**Denaturation**  Relaxation of the tertiary structure to the secondary structure, accompanied by decreasing solubility of a protein.

**Denatured globin hemichrome**  A gray-brown pigment formed when myoglobin is heated.

**Dendritic**   Branching.

**Deodorizing**   Using steam distillation or other suitable procedure to remove low molecular weight aldehydes, ketones, peroxides, hydrocarbons, and free fatty acids that would be detrimental to the aroma and flavor of fats.

**Dependent variable**   The measured variable of an experiment.

**Dermal system**   Outer protective covering on fruits and vegetables, as well as other parts of plants.

**Descriptive flavor analysis panel (DFAP)**   Thoroughly trained panel that works as a team to describe precisely in words the flavor of a sample.

**Descriptive scale**   Array of words describing a range of intensity of a single characteristic, with each step on the scale representing a subtle degree of intensity.

**Descriptive statistics**   Probability of predicting an occurrence by use of statistical tests such as chi-square, analysis of variance (ANOVA), student's $t$-distribution, or other statistical tools.

**Descriptive testing**   Using descriptive words in sensory evaluation to characterize food samples.

**Designer food**   Manufactured food that has been created to meet consumer demand for a food that may be effective in promoting health and avoiding or minimizing the risk of certain physical problems.

**Dextrans**   Complex carbohydrates in bacteria and yeasts characterized by 1,6-α-glucosidic linkages.

**Dextrinization**   Hydrolytic breakdown of starch effected by intense, dry heat and producing dextrins.

**Dextrins**   Polysaccharides composed entirely of glucose units linked together and distinguishable from starch by a shorter chain length.

**Dextrose**   Synonym for glucose; so named because polarized light bends to the right in a glucose solution.

**Dextrose equivalent (D.E.)**   Measure of the amount of free dextrose (glucose), which parallels glucose formation by hydrolysis of larger carbohydrate molecules; pure dextrose = 100 D.E.

**Diallyl disulfide**   Key flavor aromatic compound from garlic.

**Dietary Supplement Health and Education Act (DSHEA)**   Act from 1994 that defines dietary supplements and limits the role of the FDA in regulating them.

**Dietetic balance**   Single-pan, spring balance suitable for portion control, but not sufficiently accurate for food experimentation.

**Diet margarines**   Spreads made from plant oils that have been partially hydrogenated and then blended with more than twice as much water as is used in stick margarines.

**Difference testing**   Sensory testing designed to determine whether detectable differences exist between products.

**Diglyceride (diacylglyceride)**   Simple fat containing two fatty acids esterified to glycerol.

**Dipole**   Molecule that is electrically asymmetrical—that is, one portion is slightly negative and another part is slightly positive.

**Directed interesterification**   Process of interesterification in which the fat is kept below its melting temperature.

**Disaccharide**   Carbohydrate formed by the union of two monosaccharides with the elimination of a molecule of water.

**Discontinuous (dispersed) phase**   Phase distributed in a discontinuous fashion, making it necessary to pass through at least some of the continuous phase to reach another part of the dispersed phase.

**Docosahexanoic acid (DHA)**   Omega-3 fatty acid containing 22 carbon atoms and 6 double bonds.

**Drip losses**   Combination of juices and fat that drip from meat during cooking.

**Duo–trio test**   Difference test in which two samples are judged against a control to determine which of the two samples is different from the control.

**Dutch-processed cocoa and chocolate**   Cocoa and chocolate produced from cacao with an alkaline treatment to produce a pH of 6–7.8 and a more soluble, darker-colored product than the natural product.

**Dye**   Water-soluble chemical coloring agent certified for use in coloring foods.

**Eggs lower in saturated fat and cholesterol**   Eggs with at least a 25 percent reduction in saturated fat and cholesterol, the result of feeding hens a diet rich in canola oil.

**Eicosapentanoic acid (EPA)**   Omega-3 fatty acid with 20 carbon atoms and 5 double bonds.

**Elaidic acid**   *Trans* isomer (t9-18:1) of oleic acid produced during hydrogenation; raises LDLs.

**Elastin**   Yellow connective tissue occurring in limited amounts intramuscularly and in somewhat greater concentrations in deposits outside the muscles.

**Electronic nose**   Testing machine that develops diagrams of the flavor components in a headspace sample.

**Emulsifying agent**   Compound containing both polar and nonpolar groups so that it is drawn to the interface between the two phases of an emulsion to coat the surface of the droplets.

**Emulsion**   Colloidal dispersion of a liquid in another liquid with which it is immiscible.

**Endomysium**   Delicate connective tissue found between fibers.

**Endosperm**   Large inner portion of cereal grains composed largely of starch and some protein.

**Endothermic reaction**   Reaction in which heat is absorbed without an increase in temperature of the reactants.

**Endotoxin**   Poison in cell walls of certain bacteria.

**Enolization**   Reversible reaction between an alkene and a ketone.

**Entamoeba histolytica**   Type of protozoa that can cause amebic dysentery in humans.

**Enterobactereaceae**   Bacteria that can go through the stomach and be viable in the intestines, where they can reproduce and cause illness.

**Enzyme**   Protein capable of catalyzing a specific chemical reaction.

**Epidermal cells**   Layer of cells providing a continuous outer covering for fruits and vegetables.

**Epimysium**   Connective tissue surrounding an entire muscle (many bundles of bundles of fibers).

**Escherichia coli**   Type of coliform bacterium with many different serotypes, some of which produce toxins that cause food-borne illnesses.

**Ethylene gas**   Gas produced in vivo that accelerates ripening of fruits. $H_2C=CH_2$.

**Evaporated milk**  Sterilized, canned milk that has been concentrated to about half its original volume by evaporation under a partial vacuum.

**Evaporation**  Escape of liquid molecules into the surrounding atmosphere.

**Evaporative losses**  Loss of weight from meat during cooking as a result of evaporation.

**Exotoxin**  Poison formed as waste when certain bacteria thrive and multiply.

**Extraction**  Removal of bran and shorts during milling of wheat.

**Extraneous variable**  Variable that is not intended to be part of the experiment and needs to be eliminated from or controlled prior to conducting the experiment.

**Farinograph**  Objective testing equipment that measures the resistance of stirring rods moving through a batter or dough and records the results graphically.

**Fatty acid**  Organic acid containing usually between 4 and 24 carbon atoms.

**Fiber**  Bundle of myofibrils and sarcoplasm encased in the sarcolemma.

**Fibrous protein**  Insoluble, elongated protein molecules.

**Ficin**  Proteolytic enzyme in figs.

**Fish**  Broadly defined as aquatic animals, but more narrowly defined to designate those with fins, gills, a backbone, and a skull.

**Flat-sour spoilage**  Increased acidity caused when viable *Bacillus coagulans* digest sugar without producing gas in canned foods.

**Flavonoids**  Group of chemically related pigments usually containing two phenyl groups and an intermediate 5- or 6-membered ring connecting the two phenyl rings.

**Flavor**  The sensory message blending taste and smell perceptions when food is in the mouth.

**Flavor enhancer**  Additive used to improve food flavor without contributing a specific identifiable taste.

**Flavor inhibitor**  Substance that blocks perception of a taste.

**Flavor potentiator**  Compound that enhances the flavor of other compounds without adding its own unique flavor.

**Flavor profile panel**  Thoroughly trained panel that works as a team to describe flavor of a sample specifically in words.

**Flour**  The fine particles of wheat (or other grain) produced by milling.

**Flukes**  Flatworms (parasites) that can invade the liver, small intestine, or lungs.

**Foam**  Colloidal dispersion of a gas dispersed in a liquid.

**Foam cake**  Cake featuring a large quantity of foam (usually egg white), which results in a light, airy batter and a baked cake with a somewhat coarse texture with moderately large cells; includes angel food, sponge, and chiffon cakes.

**Food Additives Amendment of 1958**  Legislation governing food additives that placed the burden of proof of safety on the food manufacturer.

**Food Allergen Consumer Protection Act of 2004 (FALCPA)**  Legislation requiring the listing of specific food allergens and their sources.

**Food Chemicals Codex**  International directory of food additives.

**Food, Drug, and Cosmetic Act of 1938**  Federal legislation establishing the U.S. Food and Drug Administration and defining its responsibilities.

**Free radical**  Unstable compound containing an unpaired electron.

**Free-range eggs**  Eggs laid by hens that are raised in outside enclosures during the day but in a barn at night.

**Freeze-drying**  Preservation of food by first freezing and then dehydrating the product.

**Freezing in air**  Freezing of food in an extremely cold environment with still air or blasts of air.

**Freeze-thaw stability**  Ability of a starch-thickened product to be frozen and thawed without developing a gritty, crystalline texture.

**Frozen custard**  Ice cream–like product that is a frozen egg yolk–thickened custard.

**Functional food**  Food that may provide health-promoting qualities beyond just the nutrients it provides.

**Fungiform papillae**  Mushroom-like protuberances often containing taste buds and located on the sides and tip of the tongue.

**Fungus**  Lower plant that is parasitic, saprophytic, and lacking in chlorophyll (e.g., molds, mildew, and mushrooms).

***Gambierdiscus toxicus***  Algae that sometimes are eaten by red snapper and other finfish and ultimately produce ciguatoxin in the fish.

**Gamma ray**  Radiant energy of very short wavelength and capable of penetrating food, but not lead.

**Gas–liquid chromatograph (GLC)**  Machine that separates individual compounds from a mixture by passing them with a carrier gas along a special column that adsorbs and releases individual compounds at different rates.

**Gel**  Colloidal dispersion of a liquid dispersed in a solid.

**Gelatinization**  Swelling of starch granules and migration of some amylose into the cooking water when starch is heated in water to thicken various food products.

**Gelation**  Process of forming a gel.

**Gemsweet**  Heat-stable peptide sweetener.

**Genetically modified organisms (GMO)**  Plant or animal foods developed by genetic manipulation to alter nutrient levels or other characteristics; also designated as GM or GMO.

**Genetic engineering**  Biotechnology in which a genetic modification is achieved by removing, adding, or modifying genes.

**Germ (embryo)**  Small portion of cereal grain containing fat and a small amount of protein, as well as thiamine, riboflavin, and other B vitamins.

**Germinal disk**  Blastoderm of the yolk, which is located at the edge of the yolk and is connected to the white yolk.

***Giardia lamblia***  Type of protozoa that, if ingested when they are alive, causes giardiasis in humans.

**Giardiasis**  Parasitic illness caused by *Giardia lamblia*, a type of protozoa, and characterized by severe diarrhea.

**Glassy lactose**  Amorphous (noncrystalline) milk sugar.

**Glassy state**  Solid, inflexible physical state formed at an extremely cold temperature and with limited moisture in an amorphous solid; capable of changing to a rubbery or somewhat elastic physical state.

**Glass transition**  Change of state of a material from a solid glass to a supercooled rubbery or viscous liquid.

**Glass transition temperature ($T_g$)**  Temperature at which an amorphous solid in the glassy state begins to transform to a less rigid state.

**Gliadin**  Protein fraction in wheat gluten that is soluble in alcohol, compact and elliptical in shape, and sticky and fluid.

**Globular proteins**  Native proteins with a tertiary structure that is rather spherical.

**Glucosinolates**  Sulfur-containing irritants in mustard and horseradish.

**Glutamine**  Amino acid prominent in gliadin protein molecules and important in intermolecular hydrogen bonding.

**Glutathione**  Peptide in yeast cells that causes stickiness in dough if the dried yeast is hydrated below 40°C (105°F).

**Gluten**  Complex of gliadin and glutenin that develops in wheat flour mixtures when water is added and the batter or dough is manipulated by stirring, beating, or kneading.

**Gluten flour**  Specialty wheat flour made by adding vital wheat gluten to increase the protein level to about 41 percent.

**Glutenin**  Alcohol-insoluble protein fraction in wheat gluten that is characterized by its fibrous, elongated shape and elastic quality.

**Glycerol**  Polyhydric alcohol containing three carbon atoms, each of which is joined to a hydroxyl group.

**Glycogen**  Complex carbohydrate that serves as the storage form of carbohydrate in animals.

**Glycolipid**  Molecule with a sugar moiety and a lipid portion.

**Gonyaulax catanella**  Algae that produce toxin to cause paralytic shellfish poisoning when infected shellfish are eaten.

**GRAS list**  A list of additives "generally recognized as safe" for use in foods because of long use with no evidence of carcinogenicity.

**Ground substance**  Undifferentiated matrix of plasma proteins and glycoproteins in which fibrous molecules of collagen and/or elastin are bound.

**Ground system**  Bulk of edible portion of plant foods.

**Gums**  Complex carbohydrates of plant origin, usually containing galactose and at least one other sugar or sugar derivative, but excluding glucose.

**Gustatory cells**  Elongated cells in taste buds from which a cilia-like hair extends into the pore of the taste bud.

**HACCP**  Acronym for Hazard Analysis and Critical Control Point System, food safety program.

**Hard meringue**  Egg white foam containing about 50 grams (4 tablespoons) of sugar per egg white; baked to a dry, brittle cookie or dessert shell.

**Hard water**  Water containing salts of calcium and magnesium.

**Haugh unit**  Units used to denote the quality of albumen; correlates thick albumen height with egg weight.

**HAV**  Hepatitis A virus, the cause of a food-borne illness that attacks liver cells.

**Hazard Analysis and Critical Control Point System (HACCP)**  System for analyzing, monitoring, and controlling food safety during production.

**H band**  Region in the center of a sarcomere where only thick myofilaments of myosin occur.

**Heat of fusion**  Heat released when a liquid is transformed into a solid (80 calories per gram of water); also called *heat of solidification*.

**Heat of vaporization**  Heat energy absorbed in the conversion of water into steam (540 calories per gram of water).

**Hectares**  Area equivalent to 10,000 square meters or 2.471 acres.

**Hedonic ratings**  Measures of the degree of pleasure provided by specific characteristics of various food samples.

**Hedonic scale**  Pleasure scale for rating food characteristics.

**Heme**  Compound composed of four adjoining pyrrole rings linked to an atom of iron.

**Hemicelluloses**  Carbohydrate polymers composed of various sugars and uronic acids; structural feature of plant cell walls.

**Hemoglobin**  Large, iron-containing compound consisting of four heme–polypeptide polymers linked together; contributes to meat color.

**Hepatitis A**  Food-borne illness characterized by inflammation of the liver; caused by hepatitis A virus.

**Hexose**  Saccharide with six carbon atoms, the most common size unit.

**High-fructose corn syrup (HFCS)**  Especially sweet corn syrup made using isomerase to convert some glucose to fructose.

**High-pressure liquid chromatograph (HPLC)**  Machine that under pressure separates a sample dissolved in liquid into its individual components as they are adsorbed along a special column and finally eluted individually from the column.

**High-pressure processing (HPP)**  Process using high pressure and water to make food safe and give it a somewhat extended shelf life during refrigerator storage.

**Hilum**  Innermost layer or the nucleus of a starch granule.

**Hold method**  Pasteurization in which milk is heated to 63°C (145°F) and held there for 30 minutes before it is cooled to 7°C (45°F).

**Homogenization**  Mechanical process in which milk is forced through tiny apertures under a pressure of 2,000–2,500 psi, breaking up the fat globules (3–10 microns in diameter) into smaller units (less then 1 micron in diameter) that do not separate from the milk.

**Hot pressing**  Using steam or hot water to heat plant seeds to about 70°C (150°F) to facilitate extraction of lipids from the seeds, a process that also extracts some gums, off flavors, and free fatty acids.

**HTST method**  High-temperature short-time pasteurization in which milk is heated to 72°C (161°F) and held there at least 15 seconds before it is cooled to 10°C (50°F).

**Humectant**  Substance that helps to retain moisture.

**Hydrogen swells**  Spoilage of food and bulging of cans caused by anaerobic, thermophilic microorganisms that produce hydrogen during storage.

**Hydrogenation**  Addition of hydrogen to an unsaturated fatty acid in the presence of a catalyst to reduce the unsaturation of the molecule and raise the melting point.

**Hydrolysis**  Splitting of a molecule by the uptake of a molecule of water.

**Hydrolytic rancidity**  Lipolysis (hydrolysis) of lipids to free fatty acids and glycerol, often catalyzed by lipases.

**Hydroperoxide**  Compound containing a –O-O-H group.

**Hydrophilic**  Attracted to water.

**Hydrophilic/lipophilic balance (HLB)** Twenty-point scale indicating the affinity of an emulsifying agent for oil versus water.

**Hydrophobic** Repelled from water.

**Hygroscopicity** Ability to attract and hold water, which is a characteristic of sugars to varying degrees.

**Hypodermal layer** Layer of cells beneath the epidermal cells.

**Hypothesis** Tentative assumption to test logical or empirical consequences of applying a variable in a research project.

**I band** Light region on either side of the Z line in a sarcomere, consisting of non-overlapping myofilaments of actin.

**Ice cream (also called plain ice cream)** Frozen dessert containing at least 10 percent milk fat and 20 percent total milk solids and no more than 0.5 percent edible stabilizer; flavoring particles must not show.

**Implosion** Violent compression.

**Independent variable** Manipulated variable defined by the researcher.

**Index to volume** Indirect means of comparing volume by measuring the circumference of a cross section of the product.

**Indirect-contact freezing** Freezing accomplished by placing packages of food in contact with cold shelves or by passing liquids through a chilled tube.

**Inferential statistics** Another term for descriptive statistics.

**Infrared spectroscopy** Identification technique in which a pure compound is subjected to infrared (wavelengths from 2,500 to 16,000 nm) energy to vibrate the molecule and create a spectrum of peaks indicating its structural features (e.g., an aldehyde group or benzene ring).

**In-house testing** Evaluations conducted within a food company prior to field testing and test marketing.

**Interesterification** Treatment of a fat, usually lard, with sodium methoxide or another agent to split fatty acids from glycerol and then to reorganize them on glycerol to form different fat molecules with less tendency to form coarse crystals.

**Interfacial tension** The tendency for molecules at the surface of a liquid to remain with the liquid rather than intersperse with molecules of a second adjacent liquid.

**Intermediate crystals** Slightly coarse fat crystals that form when crystals melt and recrystallize.

**International Organization of Standardization (ISO)** International federation of standards boards from 91 participating countries.

**Intraesterification** Catalyzed reaction in which the fatty acids split from glycerol and rejoin in a different configuration, but with the same fatty acids being retained in the molecule.

**Inulin** Complex carbohydrate that is a polymer of fructose.

**Inversion** Formation of invert sugar by either boiling a sugar solution (especially with acid added) or adding an enzyme (invertase) to the cool candy.

**Invert sugar** Sugar formed by hydrolysis of sucrose; a mixture of equal amounts of fructose and glucose.

**Invertase** Enzyme that catalyzes the breakdown of sucrose to invert sugar (fructose and glucose).

**Irradiation** Preservation of food by exposure to beta and gamma rays.

**ISO 9000** Overall document of standards for food quality, established by the ISO.

**Isoelectric point** The pH at which a protein molecule has lost its electrical charge and is most susceptible to denaturation and precipitation.

**Isomalt** Low-calorie sweetener produced from sucrose by enzyme action.

**Jasmine rice** Long-grain, aromatic rice; resists retrogradation in storage.

**Jelmeter** Pipette-like viscometer designed to measure the adequacy of the pectin content of fruit juices used to make jams and jellies.

**Joint FAO/WHO Expert Committee on Food Additives (JECFA)** International committee charged with overseeing issues on food additives in international trade.

**Kefir** Fermented milk that contains about 3 percent alcohol because of fermentation by *Lactobacillus kefir*, which also adds $CO_2$.

**Ketose** Hexose with two carbon atoms external to the 5-membered ring.

**Lactase** Enzyme that catalyzes the breakdown of lactose to equal amounts of glucose and galactose.

***Lactobacillus sanfrancisco*** Bacterium used to produce lactic acid in some bread doughs.

**Lake** Water-insoluble chemical coloring agent certified for use in coloring foods.

**Latebra** Tube connecting the white yolk to the germinal disk in egg yolk.

**Latent heat of crystallization** Heat released during the transition from the liquid state (higher energy) to the frozen state (lower energy).

**Leaking** Draining liquid from a soft meringue to the surface of the filling of a cream pie.

**Labneh** Soft cheese made by separating the curd from yogurt.

**Lecithin** Phospholipid in egg yolk that is an effective emulsifying agent.

**Leucoanthocyanins** Flavonoid pigments that are a subgroup of the flavanols and are often termed *procyanidins*.

**Leucoplast** Plastid in the cytoplasm of plant cells; site of starch storage as granules.

**Level of confidence** Percentage expression of certainty that the results caused by a variable are statistically significant.

**Level of significance** The decimal value below which the results of research will be considered significant, and the null hypothesis can be rejected.

**Levulose** Synonym for fructose, so named because polarized light bend to the left in a fructose solution.

**Lignin** Structural component of some plant foods that is removed to avoid a woody quality in the prepared food.

**Lipids** Nonpolar, water-insoluble compounds composed of carbon, hydrogen, and a small amount of oxygen.

**Linamarin** Cyanogenic glycoside in cassava that is toxic to humans; can be removed by careful processing.

**Line-spread test** Measurement of flow of a viscous liquid or semisolid food by determining the spread of a measured amount of sample in a specified time at 90° intervals on a template of concentric rings.

**Linoleic acid** Essential fatty acid (18 carbons) containing two double bonds.

**Linolenic acid** Fatty acid (18 carbons) containing three double bonds.

**Lipase**  Enzyme that catalyzes the hydrolysis of fat to release free fatty acids from glycerol.

**Lipolysis**  Reaction of a molecule of water with a fat molecule to release a free fatty acid in the presence of lipase or heat.

**Lipovitellin**  High-density lipoprotein in granules in egg yolk, the most abundant granular protein.

***Listeria monocytogenes***  Type of bacteria sometimes found in meats, meat products, unpasteurized milk, and other foods; toxin causes listeriosis.

**Listeriosis**  Food-borne illness caused by *Listeria monocytogenes.*

**Livetin**  Yolk plasma protein found in three forms ($\alpha$, $\beta$, and $\gamma$).

**Logarithmic order of death**  Percentage of bacteria and bacterial spores killed per minute at a constant temperature.

**Long-patent flour**  Wheat flour made from 95 to 100 percent of the flour streams, yielding flour of rather high protein content (e.g., bread flour).

**Low-fat ice cream**  Frozen dairy product containing 2 to 7 percent milk fat and 11 percent total milk solids.

**Low-methoxyl pectinic acids (low-methoxyl pectins)**  Galacturonic acid polymers in which only between an eighth and a fourth of the acid radicals have been esterified with methanol; pectic substance found in fruit that is just beginning to ripen.

**Lycopene**  Acyclic carotene responsible for red color in tomatoes and watermelon and overtones in apricots and other yellow-orange fruits and vegetables; antioxidant that may help prevent some cancers and coronary heart disease.

**Lysozyme**  Albumen protein with an isoelectric point of pH 10.7; notable for its ability to hydrolyze a polysaccharide in the cell wall of some bacteria, thus protecting against contamination by these bacteria.

**Mad cow disease**  Common name for bovine spongiform encephalopathy (BSE).

**Magnetron tube**  Tube in a microwave oven that generates microwaves at a frequency of 915 or 2450 megahertz.

**Maillard reaction**  Nonenzymatic browning that occurs when protein and a sugar are heated or stored together for some time.

**Maltese cross**  A cross consisting of arms of equal length that terminate in a V-shape.

**Marbling**  Small fatty deposits within muscles of meat.

**Mass spectrometry**  Identification technique in which a pure compound is bombarded by a high-energy electron beam to split into various ions that then are sorted and finally recorded as a mass spectrum that reveals the actual compound.

**Mass transfer**  Movement of a food component into or out of a food that is heating.

**Masticometer**  Machine that measures comparative tenderness of meat and other foods by simulating chewing action.

**Mealy**  Fine, granular.

**Mean**  The arithmetic average of scores.

**Meat**  Red meats, including beef, veal, pork, and lamb.

**Measures of central tendency**  Mode, median, and mean.

**Measures of dispersion**  Range, mean deviation, standard deviation, variance, and standard error of the mean or difference between means.

**Mechanical energy**  Energy transferred to food through physical movements, such as beating.

**Median**  The score at the midpoint of data arranged in a sequential array.

**Medium-patent flour**  Wheat flour using about 90 percent of the flour streams, resulting in a somewhat higher protein content and relatively less starch (e.g., all-purpose flour).

**Megahertz**  Measure of frequency defined as 1 million cycles per second.

**Mellorine**  Imitation ice cream in which the milk fat has been removed and replaced by a different fat.

**Melting point**  The temperature at which crystals of a solid fat melt.

**Meniscus**  Curved upper surface of a liquid column that is concave when the containing walls are wetted by the liquid and convex when they are not.

**Meringue**  Egg white foam containing sugar.

**Mesophilic**  Microorganisms with optimal reproduction and survival between 15 and 45°C (59 and 113°F).

**Metmyoglobin**  Brownish-red form of myoglobin formed when the ferrous iron is oxidized to the ferric form and water is complexed to the oxidized iron.

**Metric system**  System of measurements of length, area, volume, and weight using the decimal system (the system of tens).

**Micelle**  Casein aggregate that is comparatively stable and remains colloidally dispersed unless a change, such as a shift toward the isoelectric point or the use of rennin, destabilizes and precipitates casein.

**Microwave**  Comparatively short (1–100 centimeters) electromagnetic wave.

**Middle lamella**  Region between adjacent cells that cements the cells together; composed mostly of pectic substances.

**Millimicron**  Billionth of a meter.

**Milling**  Grinding and refining of cereal grains.

**Miso**  Paste made by fermenting a mold culture with soybeans, salt, and often rice for up to three years.

**Mitochondria**  Organelles in cells involved in respiration and other biochemical processes.

**Mochigome**  Waxy (high amylopectin), sticky rice.

**Mode**  Score or group receiving the most responses or data points.

**Moisture content**  (Initial – dried weight/initial weight) $\times$ 100 = % moisture.

**Molasses**  Sweetener produced as a by-product of the refining of sucrose from sugarcane.

**Mollusks**  Shellfish with a protective shell.

**Mono- and polyunsaturated fatty acids**  Fatty acids with one (mono) or two or more double bonds (polyunsaturated).

**Monosaccharide**  Carbohydrate containing only one saccharide unit.

**Monosodium glutamate**  Flavor potentiator; sodium salt of glutamic acid.

***Morganella morganii***  Bacteria producing scombrotoxin from histidine on the surface of tuna and related fish.

**Mouthfeel**  Textural qualities of a food perceived in the mouth.

**Muffin** Small, rounded quick bread baked from a batter containing a 2:1 flour-to-liquid ratio, plus egg, fat, sugar, baking powder, and salt.

**Muffin method** Mixing method in which the dry ingredients are sifted together in one bowl, the liquid ingredients (including fat) are mixed in another bowl and then poured into the dry ingredients, and the mixture is stirred briefly and baked; the result is a rather coarse, slightly crumbly product that stales readily.

**Munsell system** System of identifying colors on the basis of hue, value, and chroma, using a numerical scale.

**Muscle** Aggregation of bundles of bundles of fibers surrounded by the epimysium.

**Mycomatta** Sheet-like connective tissue in fish.

**Mycotoxin** Poisonous substance produced by some molds that can be lethal when consumed.

**Myofibril** Linear bundle of several myofilaments that contains a number of sarcomeres.

**Myofilament** Simplest level of organization in muscle; classified as thick or thin.

**Myoglobin** Purplish-red pigment consisting of heme-containing ferrous iron and a polypeptide polymer (globin).

**Myosin** Principal myofibrillar protein.

**Myotomes** Fibers in fish; these are thick and about 3 centimeters long.

**Naive panel** Sensory evaluation panel that has not been trained specifically regarding the product evaluation being undertaken in the study.

**Nanotechnology** Development, creation, and application of atomic, molecular, or macromolecular particles between 1 and 100 nanometers in size.

**National Organic Program** Federal legislation passed in 2002 to implement the Organic Food Production Act of 1990.

**Natural** A food product made without chemical or artificial additives.

**Natural cheese** Any cheese made by clotting milk to form a curd and then concentrating the curd by draining the whey; variations are produced by varying the curd concentration and by ripening with or without the addition of selected microorganisms or other ingredients.

**Natto** Fermented cooked soybean product useful as a spread or in soups.

**Natural-processed cocoa and chocolate** Cocoa and chocolate produced from cacao without the addition of alkali.

**Nematode** Roundworms that are parasitic.

**Neotame** Low-calorie dipeptide (aspartic acid and phenylalanine) sweetener.

**Neutralization** Removal of free fatty acids from fats and oils; a step in their refinement.

**Newtonian** Classification of materials having a flow rate that is not affected by shear rate—for example, water and sugar syrups.

**Non-Newtonian** Classification of materials having a flow rate that is influenced by shear rate—for example, chocolate and emulsions.

**Norwalk-like virus (NLV)** Virus causing sudden onset of acute gastrointestinal problems in 12 to 48 hours after eating food contaminated with feces.

**Null hypothesis** Statement that applying a research variable will not make a significant difference in a research project.

**Number of chews** Subjective test in which a judge chews similar bites of food to the same endpoint and records the actual number of chews required to reach that point for each sample.

**Nutraceutical** Sometimes used to describe not only functional foods but also supplements and medicinal herbs.

**Objective evaluation** Measurement of physical properties of a food by the use of mechanical devices.

**Oil-in-water emulsion (o/w)** Colloidal dispersion in which droplets of oil are dispersed in water, for example, mayonnaise.

**Oleic acid** Monounsaturated 18-carbon fatty acid.

**Olfactory epithelium** Yellow, mucus-coated area in the nose containing basal cells, supporting cells, and perikarya.

**Olfactory receptors** Nasal organs capable of detecting aromas.

**Oligosaccharide** Carbohydrate formed by the union of three to ten monosaccharides with the elimination of water.

**Omega-3 eggs** Eggs with increased omega-3 fatty acids (produced by feeding hens a diet high in flaxseed).

**Omega-3 fatty acid** Polyunsaturated fatty acid with the first double bond occurring on the third carbon from the methyl end of the molecule.

**Organic** Legally defined as plant or animal food produced without using growth hormones, antibiotics, or petroleum-based or sewage sludge-based fertilizers.

**Organic Food Production Act of 1990** Federal legislation that regulates production and marketing of organic foods.

**Oryzanols** Class of sterols in rice bran oil of significance for antioxidant properties.

**Ovalbumin** By far the most abundant protein in egg albumin; readily denatured by heat.

**Oven spring** An increase in the volume (usually about 80 percent) of yeast breads during the early part of baking resulting from the expansion of carbon dioxide and the increased production of carbon dioxide stimulated by the oven heat.

**Overrun** Increase in volume (expressed as a percentage) that occurs when ice cream is frozen with agitation.

**Ovomucin** Rather fibrous protein abundant in thick egg white; contributes to the viscous texture of the thick white.

**Oxidative rancidity** Development of off flavors and odors in fats as a result of the uptake of oxygen and the formation of peroxides, hydroperoxides, and numerous other compounds.

**Oxidized starches** Thin-boiling starches produced by alkaline (sodium hypochlorite) treatment, but forming only soft gels.

**Oxymyoglobin** Cherry red form of myoglobin formed by the addition of two oxygen atoms.

**Paired comparison** Difference test in which a specific characteristic is to be evaluated in two samples, and the sample with the greater level of that characteristic is to be identified.

**Papain** Proteolytic enzyme from papaya.

**Papillae** Rough bulges or protuberances in the surface of the tongue, some of which contain taste buds.

**Paralytic shellfish poisoning** Often fatal, paralytic condition caused by ingesting shellfish containing high levels of toxin in their flesh, the result of red-tide feeding.

**Parasite** An organism living in or on a host organism.

**Parenchyma cells**  Predominant type of cell in the fleshy part of fruits and vegetables.

**Parevine**  Imitation ice cream in which both the milk fat and the milk solids have been replaced by nondairy ingredients.

**Pasteurization**  Heat treatment of milk adequate to kill microorganisms that can cause illness in people.

**Pasting**  Changes in a gelatinized starch, including considerable loss of amylase and implosion of the granule.

**Pastry-blend method**  Method in which the flour and fat are creamed (first step); sugar, baking powder, salt, and half the liquid are added (second step); and the last half of the liquid and the egg are combined (third step).

**Pastry flour**  Soft wheat, short- and medium-patent flour with a protein level of about 9.7 percent.

**Pearl tapioca**  Large pellets of partially gelatinized tapioca that are dried, resulting in a product that needs a long soaking period before use but yields a translucent, nonstringy paste.

**Pectic acid**  The smallest of the pectic substances and one lacking methyl esters; occurs in overripe fruits and vegetables; incapable of gelling.

**Pectic substances**  Group of complex carbohydrates in fruits; polymers of galacturonic acid linked by $1,4$-$\alpha$-glycosidic linkages with varying degrees of methylation.

**Pectinates**  Compound resulting from the combination of pectinic acids or pectins with calcium or other ions to form salts that usually enhance gel-forming ability.

**Pectinesterases**  Enzymes that de-esterify protopectin and pectin, a change that reduces gel-forming ability.

**Pectinic acid or pectin**  Methylated polymers of galacturonic acid formed from protopectin as fruit becomes barely ripe; capable of forming a gel.

**Penetrometer**  Machine that measures tenderness by determining the distance a cone or other device penetrates a food during a defined period and using only gravitational force.

**Pentosans**  Polymers of the pentoses xylose and arabinose.

**Pentose**  Saccharide with five carbon atoms.

**Peptization**  Acid hydrolysis of some of the peptide linkages in a protein to yield peptides.

**Percent**  Portion of a hundred.

**Percentile rank**  Relative position of a score within the total array of scores expressed on the basis of hundreths.

**Percent sag**  (Depth in container – depth on plate)/depth in container × 100 = %.

**Periderm**  Layer of cork-like cells protecting vegetable tissues underground.

**Perikarya**  Bodies of olfactory cells in the olfactory epithelium from which dendrites extend to the olfactory vesicles.

**Perimysium**  Connective tissue surrounding a bundle of several fibers.

**Permanent emulsion**  Emulsion containing an amount of emulsifying agent sufficient to enable the emulsion to remain intact during ordinary handling and use.

**Peroxide**  Compound with oxygen attached to oxygen.

**Pheophorbide**  Chlorophyll derivative in which the magnesium and phytyl group have been removed; an olive-drab pigment.

**Pheophytins a and b**  Compounds formed from chlorophyll a and b in which the magnesium ion is replaced by hydrogen, altering the color to greenish-gray for pheophytin a and olive green for pheophytin b.

**Phlobaphene**  Derivative of a polyphenol in cacao that is formed in the presence of oxygen and is responsible for the reddish color sometimes noted in cocoa and chocolate.

**Phloem**  Portion of the vascular system that transports aqueous solutions of substances such as nutrients.

**Phospholipid**  Complex phosphoric ester of a lipid.

**Phosphorylase**  Enzyme in potatoes that promotes sugar formation during cold storage.

**Phosvitin**  Small protein in yolk granules that binds iron in yolk.

**Phytochemicals**  Chemical compounds in plants that are important to promote healthful reactions in the body but are not classified as nutrients required for life and growth.

**Phytyl**  Alcohol component of chlorophyll responsible for its hydrophobic nature.

**Planimeter**  Engineering tool designed to measure distance as its pointer is traced around a pattern.

**Plasmalemma**  Thin membrane between the cell wall and the interior of the cell.

**Plastids**  Organelles in the cytoplasm that contain pigments or starch.

**Polygalacturonase**  Pectic enzyme promoting degradation of pectic substances in avocados, pears, tomatoes, and pineapple.

**Polyhydric alcohols (polyols)**  Alcohols with several hydroxyl groups, enabling them to be used as sweeteners—for example, xylitol and sorbitol.

**Polymerization**  Formation of a variety of polymers, including simple dimers and trimers, when free fatty acids are subjected to intense heat for a long period of time during frying.

**Polyphenoloxidases**  Group of enzymes capable of oxidizing flavonoid compounds to cause browning or other discoloration after harvest.

**Polysaccharide**  Carbohydrate formed by the union of many saccharide units with the elimination of a molecule of water at each point of linkage.

**Poultry**  Fowl, notably turkey, chicken, and duck.

**Prebiotic**  Healthful bacterial culture added during manufacturing to enhance and/or modify a dairy product that does not survive in the digestive tract.

**Preference testing**  Sensory testing designed to provide information on selected characteristics and to indicate preference or acceptability of products.

**Pre-gelatinized starch**  Starch that has been gelatinized and then dehydrated; addition of water produces a thickened product.

**Press fluids**  Juices forced from meat or other food under pressure.

**Previously frozen**  Statement required on meat, poultry, fish, and shellfish if freezing occurred, but the item is thawed to at least $-3.3°C$ ($26°F$) before it is purchased.

**Primary structure**  Covalently bonded backbone chain of a protein: —C-C-N-C-C-N-C-C-N-.

**Probiotic**  Bacterial culture added to a dairy product because of its health-promoting capability and viability in the intestines.

**Prokaryotic**  Cellular organism without a distinct nucleus.

**Process cheese**  Cheese product made by heating natural cheeses with an emulsifier and then cooling in a brick form.

**Process cheese food**   Process cheese product with a moisture content of about 45 percent, which causes the cheese to be comparatively soft, yet firm.

**Process cheese spread**   Spreadable process cheese product with a moisture content of about 50 percent.

**Product Designation of Origin**   Legal designation that the European Union can grant to food products from specific locations.

**Profiling**   Very detailed word description (usually of flavor) developed by a highly trained panel against which subsequent production is evaluated to maintain quality of production.

**Proline**   Imino acid prominent in gliadin and collagen with a cyclic ring structure that restricts protein shape.

**Propenylsulfenic acid**   Compound in onions that causes eye irritation and tears.

**Propyl gallate (PG)**   Antioxidant somewhat effective in vegetable oils, often used in combination with BHA and BHT.

**Protozoa**   Parasitic, single-celled complex microorganism or simple animal; some cause illnesses in humans when fecal contamination of water or food occurs and living organisms are ingested.

**Protopectin**   Non-methylated polymers of galacturonic acid incapable of forming a gel; first pectic substance to be formed in a fruit.

**Protopectinases**   Enzymes in fruits and vegetables capable of catalyzing the hydrolytic cleavage of protopectins to shorter chains of pectins.

**PSE pork**   Pale, soft, and exudative pork from carcasses with a low pH, usually ranging between pH 5.1 and 5.4.

**Pulsed electric field processing (PEF)**   Pasteurization of juice by pulsing it 1,000 times per second in an electrical field of about 35,000 V/cm.

**Pyrrolidine ring**   Organic ring structure containing one atom of nitrogen; linkage to another amino acid through this nitrogen favors formation of a linear, fibrous protein molecule.

**Pyrophosphatase**   Group of enzymes in muscle tissue influencing the water-holding capacity of meat.

**Quantitative descriptive analysis (QDA)**   Development of a thorough description of characteristics of a product and a quantification of their intensity.

**Quick breads**   Breads leavened with a leavening agent other than yeast, ordinarily either with steam or carbon dioxide generated from reaction of an acid and an alkali.

**Quick-rise active dry yeast**   Special strain of *Saccharomyces cerevisiae* available as the active dry yeast and capable of producing carbon dioxide so rapidly in the dough that fermentation time is cut approximately in half.

**Quorn™**   Trade name for products based on hyphae harvested from *Fusarium venenatum* mixed with a binder and then heated and frozen.

**Rad**   Ionizing energy equal to $10^{-5}$ joule per gram of absorbing material.

**Radiation**   Direct transfer of heat energy from its source to the surface of a food.

**Radiant energy**   Energy traveling as electromagnetic waves.

**Rancidity**   Chemical deterioration of a fat caused by the uptake of oxygen (oxidative) or water (hydrolysis).

**Randomized interesterification**   Interesterification accomplished using melted fat.

**Range eggs**   Eggs laid by hens in outside enclosures during the day, but in a barn at night.

**Rank-order test**   Preference or difference test in which all samples are ranked in order of intensity of a specific characteristic.

**Recombinant bovine somatotropin (rbST)**   Genetically engineered hormone that stimulates milk production in cattle.

**Red tide**   Visible evidence of unusually large populations of dinoflagellates (microalgae) capable of causing paralytic shellfish poisoning.

**Reducing sugar**   Sugar having a free carbonyl that can combine with an amine, leading to non-enzymatic browning.

**Regression**   Statistical methods applicable to correlational research.

**Rendering**   Removing fat from animal tissues by either dry or moist heat.

**Rennet**   Unpurified extract from the fourth stomach of unweaned calves, which contains rennin, an enzyme causing milk curds to form.

**Rennin**   Enzyme from the stomach lining of calves that eliminates the protective function of κ-casein in micelles and results in the formation of a curd.

**Resistant starches**   Starches that are not digested until entering the large intestine.

**Restructured meats**   Meats made from meat cuts that are somewhat less expensive; made by creating small particles, adding fat and other ingredients, and shaping into uniform portions.

**Reticulin**   A type of connective tissue protein associated with a fatty acid (myristic acid).

**Retrogradation**   Gradual increase of crystalline aggregates in starch gels during storage that results from breaking of hydrogen bonds between amylose molecules and slow rearrangement into a more orderly configuration and establishment of new hydrogen bonds.

**Reversion**   Development of an off flavor (beany or fishy) in soybean, rapeseed, or various fish oils as a result of a reaction involving only very minor amounts of oxygen.

**Rheology**   Study of deformation and flow qualities of matter. Characteristics of flow and deformation of fats and other substances with flow properties.

**Ripening**   Changes that occur in crystalline candies when they are stored.

**Rods**   Elongated dendrites of photoreceptor neurons that transmit visual images in dim light, revealing movement and varying intensities of black and white.

**Rotating dull knife tenderometer**   Objective testing device that measures relative tenderness of meat by determining the depth of penetration effected by a rotating dull knife.

**Saccharin**   Non-nutritive sweetener.

**Saccharomyces cerevisiae**   Yeast (single-celled plant) used in baking.

**Saccharomyces exigus and Saccharomyces inusitatus**   Yeasts used to produce carbon dioxide in acidic bread doughs.

**Salmonella**   Genus name for several species of gram-negative bacteria that can cause gastrointestinal illnesses, including typhoid fever.

**Salmonella enteritidis**   Form of bacteria that can be incorporated into the yolk of an egg as the hen forms the egg;

potential source of food-borne illness if egg is not heated sufficiently during preparation.

**Salmonella typbi**   Type of *Salmonella* causing typhoid fever.

**Salmonellosis**   General name for illness caused by *Salmonella*, regardless of the specific species.

**Sarcolemma**   Thin, transparent membrane surrounding the bundle of myofibrils that constitute a fiber.

**Sarcomere**   Portion of a myofibril consisting of the area between two Z lines.

**Sarcoplasm**   Jelly-like protein surrounding the myofibrils in muscle fibers.

**SAS-phosphate baking powder**   Leavening containing sodium aluminum sulfate and monocalcium phosphate; double-acting baking powder.

**Saturated fatty acids**   Fatty acids containing all of the hydrogen atoms they can possibly hold.

**Saturated solution**   True solution containing as much solute in solution as is possible to dissolve at that temperature.

**Sclereid**   Type of sclerenchyma cell that gives the somewhat gritty texture to pears and certain other fruits.

**Sclerenchyma cells**   Unique supportive cells with a chewy, fibrous character.

**Scombroid (histamine) poisoning**   Food-borne illness caused by eating spoiled fish with high levels of histamine.

**Scombrotoxin poisoning**   Allergic-type response to ingestion of high levels of histamine, saurine, and other metabolites produced by the action of *Morganella morganii* on tuna and related fish.

**Secondary bonds**   Attractive forces between atoms and functional groups that are less strong than the bonding that occurs when electrons are shared; examples are van der Waal's forces and hydrogen bonding.

**Secondary structure**   Typically the α-helical configuration of the backbone chain of many proteins; held by secondary bonding forces, notably hydrogen bonds; also may be in other forms (e.g., β-pleated sheet).

**Self-rising flour**   Flour (usually soft wheat) to which baking powder and salt have been added during production.

**Semi-permanent emulsion**   Emulsion with rather good stability because of the viscous nature of the liquid constituting the continuous phase.

**Sensory evaluation**   A synonym for subjective evaluation; measurements determined by using the senses of sight, smell, taste, and sometimes touch.

**Serotypes**   Closely related organisms, such as *E. coli*, having a common set of antigens.

**Sharp freezing**   Freezing in still air at a temperature between −23 and −30°C (−9 and −22°F); an outmoded method because of the large ice crystals that form.

**Shear press**   Objective testing machine that measures compressibility, extrusion, and shear of food samples.

**Shellfish**   Subclassification of fish; includes mollusks and crustaceans.

**Sherbet**   Frozen dessert containing acid, from 2 to 5 percent milk solids and no milk fat.

**Shigella boydii**   Type of *Shigella* most likely to cause food-borne illness designated as bacillary dysentery.

**Shortening power**   Ability of a fat to cover a large surface area to minimize the contact between water and gluten during the mixing of batters and doughs.

**Shortometer**   Device designed to measure the tenderness of fairly tender, crisp foods.

**Short-patent flour**   Wheat flour comparatively high in starch and low in protein, the result of using the very fine particles of flour from the center of the endosperm (e.g., cake flour).

**Simple sugars**   Monosaccharides and disaccharides.

**Single sample**   Presenting one sample early in an experimental project to determine acceptability and to aid in the decision on future development of the product.

**Single-stage method**   Mixing method in which all of the ingredients except the egg and half the liquid are added and beaten before the egg and last half of the liquid are beaten in.

**Sinigrin**   Potassium myronate, an isothiocyanate glucoside in cabbage that is broken down to highly pungent allyl isothiocyanate.

**Smiley scale**   Sequential pictures of very happy and continuing through to very unhappy faces used in evaluating food products when respondents are unable to use the language easily.

**Smoke point**   Temperature at which a fat or oil begins to emit some traces of smoke.

**Sodium stearoyl-2-lactylate**   Dough conditioner of particular merit in improving baking quality of triticale and soy flours.

**Soft or tub margarines**   Spreads with melting points lower than those of stick margarines because of a higher content of polyunsaturated fatty acids.

**Soft meringue**   Egg white foam containing about 25 grams (two tablespoons) of sugar per egg white; topping on cream pies.

**Soft water**   Water treated with lime or ion exchange resins (complex sodium salts) to remove the metallic cations.

**Sol**   Colloidal dispersion in which a solid is the dispersed phase and a liquid is the continuous phase.

**Soluble fiber**   Plant gums and pectic substances that undergo some digestion and absorption in the large intestine.

**Solute**   Substance dissolved in a liquid to form a true solution.

**Solvent**   Liquid in which the solute is dissolved to form a true solution.

**Sorghum syrup**   Syrup sweetener produced by boiling the juice of grain sorghum.

**Soy flour**   Finely ground soy flakes.

**Soy grits**   Coarsely ground soy flakes.

**Soymilk**   Beverage made from whole, finely ground defatted soybeans; a beverage designed to compete with milk.

**Soy protein concentrate**   Defatted soy product usually containing about 70 percent protein.

**Soy protein isolate**   Defatted, highly concentrated (up to 95 percent) soy protein; used to make many textured soy products.

**Specific gravity**   Ratio of the density of a food (or other substance) to the density of water.

**Spherulite**   Spherical crystalline body of radiating crystal fibers.

**Sponge cake**   Foam cake containing an egg yolk foam, an egg white foam, sugar, and cake flour.

**Sponge method**   Method of mixing yeast breads in which the yeast, liquid, and part of the flour are mixed and fermented to make a sponge before the rest of the ingredients are added; used commercially with strong flours.

**Spore**   Specialized structure of bacteria capable of retaining viability under extremely adverse conditions.

**Standard deviation**   Square root of the variance.

**Standing time**   Length of time that a food is allowed to stand in a microwave oven without operation of the magnetron tube so the residual heat in the food can be transferred by conduction.

***Staphylococcus aureus***   Bacteria capable of producing an enterotoxin as it grows, which can lead to food-borne illness.

**Starch**   Complex carbohydrate consisting of two fractions—amylose and amylopectin—both of which are polymers of glucose joined together by the elimination of water.

**Starch granule**   Concentric layers of amylose and amylopectin molecules formed in the leucoplasts and held together by hydrogen bonding.

**Starch phosphates**   Starch derivative made by reaction with sodium tripolyphosphate or other phosphate to achieve a thickener with excellent stability and clarity.

**Stearic acid**   Saturated 18-carbon fatty acid.

**Stevioside**   Sweetener extracted from a South American plant; suitable for tabletop sweetener.

**Stick margarines**   Spreads made by hydrogenating plant oils and adding water, milk solids, flavoring, and coloring to achieve a product similar to butter.

**Stilton cheese**   Blue-veined cheese made by an approved process in any of six creameries in Leicestershire, Nottinghamshire, and Derbyshire, England.

**Stirred custard**   Sweetened milk and egg mixture that is heated to form a sol; agitation during heating prevents formation of sufficient intermolecular linkages to form a gel.

**Straight-dough method**   Method of mixing yeast breads in which all of the ingredients are added, mixed, and kneaded prior to fermentation and proofing.

**Strategic Partnership Program—Agroterrorism (SPPA)**   Partnership of federal agencies (USDA, FDA, DHS, and FBI) with states and private industry charged with safeguarding food supplies and commodities.

**Student's *t*-test**   Statistical test to determine the significance of the mean of the experimental group versus the mean of the control group.

**Subjective evaluation**   Evaluation by a panel of individuals using a scoring system based on various characteristics that can be judged by using the senses.

**Sublimation**   Transition from the frozen state directly to the gaseous state without liquefaction.

**Substrate**   A general term for the compound that an enzyme alters.

**Subthreshold level**   Concentration of a taste compound at a level that is not detectable but can influence other taste perceptions.

**Sucralose**   Sweetener made from sucrose and containing three chlorine atoms.

**Sucroglycerides**   Sucrose esters (sucrose and glycerides or free fatty acids) used as a dough conditioner with soy flour.

**Sulfide spoilage**   Canned food contaminated with hydrogen sulfide produced by viable *Clostridium nigrificans*.

**Sulfuring**   Exposure of cut fruits to the smoke fumes created by burning sulfur flowers to retain a bright color during drying.

**Supercooling**   Reduction of the temperature of water below freezing until crystallization begins, after which the temperature rises to 0°C (32°F) because of the latent heat of crystallization.

**Supersaturated solution**   True solution containing more solute than theoretically can be dissolved at that temperature, a situation created by cooling a heated saturated solution carefully.

**Surface tension**   Attraction between molecules at the surface of a liquid.

**Surimi**   Purified and frozen minced fish containing a preservative; intermediate seafood product used in making structured seafood products.

**Suspensoid**   Colloidal dispersion of a gas dispersed in a solid.

**Sweet acidophilus milk**   Unfermented milk to which *L. acidophilus* has been added.

**Sweetened condensed milk**   Canned milk to which sugar is added (contains more than 54 percent carbohydrate because of milk sugar and added sugar); evaporation of about half the water and heat treatment to kill harmful microorganisms precede the canning process.

**Sweet glutinous rice**   General name for waxy, sticky rice varieties; mochigome and calmochi are examples.

**SWOSTHEE (Single-pan Wood Stoves of High Efficiency)**   Acronym for wood-burning stove designed for families to cook food efficiently using wood as the fuel.

**Syneresis**   Weeping or drainage of liquid from a gel.

**Tagatose**   Sweetener derived from dairy products and approved as GRAS.

**Tannins**   Term sometimes used to designate plant phenolic compounds.

**Tapioca**   Root starch derived from cassava, a tropical plant.

**Taste buds**   Tight clusters of gustatory and supportive cells encircling a pore, usually in the upper surface of the tongue; organs capable of detecting sweet, sour, salt, and/or bitter.

**Tempeh**   Fermented cooked soybean product resembling cake.

**Tempering**   Removing heat resulting from crystallization of fats and maintaining a selected temperature to promote the formation of stable, desirable crystals.

**Template**   Pattern guide to ensure accurate cutting of samples from a large sample, such as a cake.

**Temporary emulsion**   Emulsion that has little emulsifying agent and is too fluid to restrict movement of droplets; such instability requires that the ingredients be shaken to form a temporary emulsion immediately before use.

**Tertiary-butylhydroxyquinone (TBHQ)**   Antioxidant often added to animal fats used in baking and frying.

**Tertiary structure**   Distorted convolutions of the helical configuration of a protein; the form in which many proteins occur in nature and which is held by secondary bonding forces.

**Tetrapyrrole**   Complex compound with four unsaturated, 5-membered rings (containing one nitrogen and four carbon atoms) linked by methyne bridges, resulting in a very large molecule with a high degree of resonance because of the excessive number of alternating double bonds.

**Tetrose**   Saccharide with four carbon atoms.

**Textured soy protein (TSP) or textured vegetable protein (TVP)**   Product of a series of steps producing fibers from soybeans.

**Texturometer** Simulation device that measures physical textural properties such as hardness, cohesiveness, and crushability of foods.

**Thaumatin** Natural sweetener from a West African plant—licorice flavor.

**Thearubigens** Dark orange-yellow compounds formed when polyphenoloxidases oxidize epigallocatechin gallate and epigallocatechin to theaflavin gallate and theaflavin for ultimate oxidation to thearubigens in oolong and black teas.

**Thermal death time curve** Comparison of the rate of death of pathogenic microorganisms over a range of processing temperatures.

**Thermophilic** Microorganisms with optimal reproduction and survival between 45 and 95°C (113 and 203°F), but capable of withstanding high processing temperatures.

**Thick albumen** Viscous white forming the middle layer of albumen in an egg.

**Thick myofilament** Thicker, longer type of myofilament; composed of myosin molecules joined together to form a screw-like, thick, and elongated filament.

**Thin albumen** Rather fluid egg white adjacent to the yolk and also to the inner membrane.

**Thin myofilament** Thin filament formed by the helical twisting of two strands of polymerized actin.

**Thin-boiling starch** An acid-hydrolyzed starch containing many debranched amylopectin molecules; useful for making gum drops and other products in which the hot mixture must flow and then cool to form a firm gel.

**Thixotropic** Ability of a gel to become more fluid with increasing shear and then to regain previous viscosity after shearing stops.

**Threshold level** Concentration of a taste compound at a barely detectable level.

**Tocotrienols** Class of sterols related to vitamin E valued for antioxidant properties; found in rice bran and palm oils.

**Tofu** Soybean curd.

**Tofutti** Frozen dessert of sweetened, flavored tofu, the soybean counterpart of ice cream.

**Tonoplast** Membrane separating the protoplasm from the vacuole in a parenchyma cell.

**Top-loading electronic balance** Very accurate, electrically operated balance.

**Torsion balance** Very sensitive (within 0.02 gram) laboratory balance particularly useful for weighing very small quantities or quantities greater than 2 kilograms.

**Trained panel** Sensory evaluation panel that has been thoroughly trained regarding the use of the scorecard and the evaluation of various characteristics.

**Trans configuration** An arrangement in which the hydrogen is attached to the carbon atoms on either end of the double bond from opposite directions, causing a higher melting point than its *cis* counterpart.

**Trehalose** Moderately sweet ingredient used in some foods to improve flavor and/or texture.

**Triangle test** Difference test in which three samples (two of which are the same) are presented, and the odd sample is to be identified.

**Trichinella spiralis** Parasite sometimes found in pork or wild game.

**Trichinosis** Illness caused by eating meat containing viable *Trichinella spiralis*.

**Trigeminal cavity** Olfactory receptors, taste buds, and oral cavity, the three parts of the body required for perceiving flavor.

**Triglyceride (triacylglyceride)** Simple fat containing three fatty acids esterified to glycerol; the most common form of simple fat.

**Triose** Saccharide with three carbon atoms.

**Trip balance** Balance with two pans; the one on the left is used to hold the food being weighed, and the one on the right is used to hold the weights needed to counterbalance the left pan. Riders are also available for counterbalancing.

**Triticale** Hybrid grain produced by crossing wheat and rye.

**Tropocollagen** Fibrous protein consisting of three strands twisted together and containing large amounts of glycine, proline, and hydroxyproline.

**Tropomyosin** Least abundant of the three principal myofibrillar proteins.

**True solution** Dispersion in which ions or molecules no larger than one millimicron are dissolved in a liquid (usually water).

**TVP (or TSP)** Textured vegetable (soy) protein made of fibers of soy protein.

**UHT pasteurization** Extreme pasteurization [138°C (280°F) for at least 2 seconds] that kills all microorganisms and makes possible the storage of milk in a closed, sterile container at room temperature.

**Umami** Taste sensation that enhances savory qualities of flavor but does not have a distinctive flavor itself.

**Universal testing machine** Multipurpose, complex machine capable of measuring various textural properties of food samples.

**Unsaturated solution** True solution capable of dissolving additional solute at the temperature of the solution.

**Unsaturation** Lack of hydrogen relative to the amount that can be held, a situation characterized by a double bond between two carbon atoms in a fatty acid chain.

**Untrained panel** Sensory evaluation panel that has not been trained regarding the use of the scorecard and the evaluation of the various product characteristics.

**Vaccenic acid** *Trans* isomer (t11-18:1) of oleic acid occurring naturally in butterfat; does not raise LDLs.

**Vacuole** Portion of the cell containing most of the water, flavoring components, nutrients, and flavonoid pigments.

**Vapor pressure** Pressure exerted as molecules of a liquid attempt to be in the gaseous rather than the liquid state.

**Variable** Quantity or symbol that has no fixed value.

**Variance** Measure of the dispersion of data; the sum of the squares of the deviation of each value from the mean.

**Vascular system** System in plants that transports water and other essential compounds; composed of xylem and phloem.

**Vibrio cholerae** Bacteria carried by fecal contamination of food and causing cholera when ingested.

**Virus** Chemical macromolecule capable of being engulfed by a body cell and eventually altering or killing the host cell.

**Viscometer** Objective testing device for measuring viscosity of liquids that flow on the basis of rotational resistance or capillary action.

**Vital wheat gluten**   Dried crude gluten.

**Vitelline membrane**   Sac enclosing the yolk.

**Volumeter**   Device for measuring volume of baked products; consists of a reservoir for storing the seeds, a transparent column for measuring volume, and a lower compartment in which the sample is placed.

**Warner–Bratzler shear**   Objective testing device for measuring the force required to shear a sample of meat or other food with measurable tensile strength.

**Water activity ($a_w$)**   Ratio of the vapor pressure of a food sample to the vapor pressure of pure water.

**Water bath canning**   Heat processing of food in containers immersed in water at atmospheric pressure.

**Water-binding capacity**   Amount of water held by muscle protein as bound water; heating reduces capacity.

**Water-in-oil emulsion (w/o)**   Colloidal dispersion in which droplets of water are dispersed in oil, for example, butter.

**Waxy starch**   Starch containing only amylopectin, the result of genetic research and breeding for this composition.

**Wettability**   Ability of a cake or other food to absorb moisture during a controlled period of time; high moisture retention means a cake is sufficiently moist.

**Whey**   Liquid that drains from the curd of clotted milk; contains lactose, proteins, water-soluble vitamins, and some minerals.

**Whipped margarines**   Stick margarines that have been whipped mechanically into a fat foam; increased volume results in fewer calories per given volume.

**White yolk**   Small sphere of light-colored yolk at the center of the yolk.

**Whole wheat flour**   Wheat flour containing most of the bran and shorts.

**Winterizing**   Refining step in which oils are chilled carefully to precipitate and remove fractions with high melting points that would interfere with the flow properties of salad dressings or other products containing the oils.

**Xanthophylls**   Group of carotenoids containing some oxygen, as well as hydrogen and carbon, in a polymer of isoprene.

**Xylan**   A hemicellulose composed of xylose and some glucuronic acid that contributes structure to plant cell walls.

**Xylem**   The water transport system in plants; the tubular cells that move water.

**Yeasts**   Single-celled fungi that reproduce by budding, are active at room temperature, and die at moisture levels below 20 percent.

***Yersinia enterocolitica***   Bacteria sometimes found in raw and undercooked pork and raw milk, which cause yersiniosis.

**Yersiniosis**   Food-borne illness focusing on the intestines caused by consuming infected raw milk or undercooked pork.

**Yogurt**   Clabbered milk product resulting from controlled fermentation by *Streptococcus thermophilus, Placamo bacterium yoghourti,* and *Lactobacillus bulgaricus* or other lactose-fermenting microorganisms to reach a pH ~5.5.

**Yolk index**   Measure of egg quality based on the ratio of yolk height to yolk width.

**Z lines**   Region in a myofibril where the thin myofilaments of actin adjoin, creating a dark line that defines the end of a sarcomere.

# Index